Superconductors in the Power Grid

Related titles

High Temperature Superconductors (HTS) for Energy Applications
(ISBN 978-0-85709-012-6)

High-temperature Superconductors
(ISBN 978-1-84569-578-1)

Woodhead Publishing Series in Energy:
Number 65

Superconductors in the Power Grid

Materials and Applications

Edited by

Christopher Rey

ELSEVIER

AMSTERDAM • BOSTON • CAMBRIDGE • HEIDELBERG
LONDON • NEW YORK • OXFORD • PARIS • SAN DIEGO
SAN FRANCISCO • SINGAPORE • SYDNEY • TOKYO
Woodhead Publishing in an imprint of Elsevier

WP WOODHEAD PUBLISHING

Woodhead Publishing is an imprint of Elsevier
80 High Street, Sawston, Cambridge, CB22 3HJ, UK
225 Wyman Street, Waltham, MA 02451, USA
Langford Lane, Kidlington, OX5 1GB, UK

British Library Cataloguing-in-Publication Data
A catalogue record for this book is available from the British Library

Library of Congress Control Number: 2015932578

ISBN 978-1-78242-029-3 (print)
ISBN 978-1-78242-037-8 (online)

For information on all Woodhead Publishing publications
visit our website at http://store.elsevier.com/

Contents

List of contributors

A. Allais Nexans France, Lyon Cedex, France

K. Allweins Nexans Deutschland GmbH, Hannover, Germany

J.W. Bray GE Global Research, Niskayuna, NY, USA

C.E. Bruzek Nexans France, Clichy Cedex, France

R.G. Buckley Robinson Research Institute, Victoria University of Wellington, Wellington, New Zealand

T.A. Coombs Cambridge University Engineering Department, Cambridge, UK

J.A. Demko LeTourneau University, Longview, TX, USA

D. Dickson Nexans France, Lens Cedex, France

N. Glasson Callaghan Innovation, Christchurch, New Zealand

L. Graber The Florida State University, Tallahassee, FL, USA

Z. Jiang Robinson Research Institute, Victoria University of Wellington, Wellington, New Zealand

C.H. Kim The Florida State University, Tallahassee, FL, USA

N. Lallouet Nexans France, Calais Cedex, France

L.Z. Lin Institute of Electrical Engineering (IEE), Chinese Academy of Sciences (CAS), Beijing, China

A.P. Malozemoff AMSC, Devens, MA, USA

E. Marzahn Nexans Deutschland GmbH, Hannover, Germany

V. Meerovich Ben-Gurion University of the Negev, Be'er Sheva, Israel

S. Pamidi The Florida State University, Tallahassee, FL, USA

M. Pannu Wilson Transformer Company, Melbourne, VIC, Australia

C.M. Rey Energy to Power Solutions (E2P), Knoxville, TN, USA

M.W. Rupich AMSC, Devens, MA, USA

K. Sato Sumitomo Electric Industries, Ltd, Osaka, Japan

V. Sokolovsky Ben-Gurion University of the Negev, Be'er Sheva, Israel

M.P. Staines Robinson Research Institute, Victoria University of Wellington, Wellington, New Zealand

Y.J. Tang Huazhong University of Science and Technology, Wuhan, China

L.Y. Xiao Institute of Electrical Engineering (IEE), Chinese Academy of Sciences (CAS), Beijing, China

Y. Xin Innopower Superconducting Power Cable Company, Yizhuang, Beijing, China

J. Yuan AMSC, Devens, MA, USA

X.H. Zhang Department of Electrical Engineering, Tsinghua University, Beijing, China

Woodhead Publishing Series in Energy

1 **Generating power at high efficiency: Combined cycle technology for sustainable energy production**
Eric Jeffs

2 **Advanced separation techniques for nuclear fuel reprocessing and radioactive waste treatment**
Edited by Kenneth L. Nash and Gregg J. Lumetta

3 **Bioalcohol production: Biochemical conversion of lignocellulosic biomass**
Edited by Keith W. Waldron

4 **Understanding and mitigating ageing in nuclear power plants: Materials and operational aspects of plant life management (PLiM)**
Edited by Philip G. Tipping

5 **Advanced power plant materials, design and technology**
Edited by Dermot Roddy

6 **Stand-alone and hybrid wind energy systems: Technology, energy storage and applications**
Edited by John K. Kaldellis

7 **Biodiesel science and technology: From soil to oil**
Jan C. J. Bart, Natale Palmeri and Stefano Cavallaro

8 **Developments and innovation in carbon dioxide (CO_2) capture and storage technology**
Volume 1: Carbon dioxide (CO_2) capture, transport and industrial applications
Edited by M. Mercedes Maroto-Valer

9 **Geological repository systems for safe disposal of spent nuclear fuels and radioactive waste**
Edited by Joonhong Ahn and Michael J. Apted

10 **Wind energy systems: Optimising design and construction for safe and reliable operation**
Edited by John D. Sørensen and Jens N. Sørensen

11 **Solid oxide fuel cell technology: Principles, performance and operations**
Kevin Huang and John Bannister Goodenough

12 **Handbook of advanced radioactive waste conditioning technologies**
Edited by Michael I. Ojovan

13 **Membranes for clean and renewable power applications**
Edited by Annarosa Gugliuzza and Angelo Basile

14 **Materials for energy efficiency and thermal comfort in buildings**
Edited by Matthew R. Hall

15 **Handbook of biofuels production: Processes and technologies**
Edited by Rafael Luque, Juan Campelo and James Clark

16 **Developments and innovation in carbon dioxide (CO_2) capture and storage technology**
Volume 2: Carbon dioxide (CO_2) storage and utilisation
Edited by M. Mercedes Maroto-Valer
17 **Oxy-fuel combustion for power generation and carbon dioxide (CO_2) capture**
Edited by Ligang Zheng
18 **Small and micro combined heat and power (CHP) systems: Advanced design, performance, materials and applications**
Edited by Robert Beith
19 **Advances in clean hydrocarbon fuel processing: Science and technology**
Edited by M. Rashid Khan
20 **Modern gas turbine systems: High efficiency, low emission, fuel flexible power generation**
Edited by Peter Jansohn
21 **Concentrating solar power technology: Principles, developments and applications**
Edited by Keith Lovegrove and Wes Stein
22 **Nuclear corrosion science and engineering**
Edited by Damien Féron
23 **Power plant life management and performance improvement**
Edited by John E. Oakey
24 **Electrical drives for direct drive renewable energy systems**
Edited by Markus Mueller and Henk Polinder
25 **Advanced membrane science and technology for sustainable energy and environmental applications**
Edited by Angelo Basile and Suzana Pereira Nunes
26 **Irradiation embrittlement of reactor pressure vessels (RPVs) in nuclear power plants**
Edited by Naoki Soneda
27 **High temperature superconductors (HTS) for energy applications**
Edited by Ziad Melhem
28 **Infrastructure and methodologies for the justification of nuclear power programmes**
Edited by Agustín Alonso
29 **Waste to energy conversion technology**
Edited by Naomi B. Klinghoffer and Marco J. Castaldi
30 **Polymer electrolyte membrane and direct methanol fuel cell technology**
Volume 1: Fundamentals and performance of low temperature fuel cells
Edited by Christoph Hartnig and Christina Roth
31 **Polymer electrolyte membrane and direct methanol fuel cell technology**
Volume 2: *In situ* characterization techniques for low temperature fuel cells
Edited by Christoph Hartnig and Christina Roth
32 **Combined cycle systems for near-zero emission power generation**
Edited by Ashok D. Rao
33 **Modern earth buildings: Materials, engineering, construction and applications**
Edited by Matthew R. Hall, Rick Lindsay and Meror Krayenhoff
34 **Metropolitan sustainability: Understanding and improving the urban environment**
Edited by Frank Zeman
35 **Functional materials for sustainable energy applications**
Edited by John A. Kilner, Stephen J. Skinner, Stuart J. C. Irvine and Peter P. Edwards
36 **Nuclear decommissioning: Planning, execution and international experience**
Edited by Michele Laraia
37 **Nuclear fuel cycle science and engineering**
Edited by Ian Crossland

38 **Electricity transmission, distribution and storage systems**
Edited by Ziad Melhem
39 **Advances in biodiesel production: Processes and technologies**
Edited by Rafael Luque and Juan A. Melero
40 **Biomass combustion science, technology and engineering**
Edited by Lasse Rosendahl
41 **Ultra-supercritical coal power plants: Materials, technologies and optimisation**
Edited by Dongke Zhang
42 **Radionuclide behaviour in the natural environment: Science, implications and lessons for the nuclear industry**
Edited by Christophe Poinssot and Horst Geckeis
43 **Calcium and chemical looping technology for power generation and carbon dioxide (CO_2) capture: Solid oxygen- and CO_2-carriers**
Paul Fennell and E. J. Anthony
44 **Materials' ageing and degradation in light water reactors: Mechanisms, and management**
Edited by K. L. Murty
45 **Structural alloys for power plants: Operational challenges and high-temperature materials**
Edited by Amir Shirzadi and Susan Jackson
46 **Biolubricants: Science and technology**
Jan C. J. Bart, Emanuele Gucciardi and Stefano Cavallaro
47 **Advances in wind turbine blade design and materials**
Edited by Povl Brøndsted and Rogier P. L. Nijssen
48 **Radioactive waste management and contaminated site clean-up: Processes, technologies and international experience**
Edited by William E. Lee, Michael I. Ojovan, Carol M. Jantzen
49 **Probabilistic safety assessment for optimum nuclear power plant life management (PLiM): Theory and application of reliability analysis methods for major power plant components**
Gennadij V. Arkadov, Alexander F. Getman and Andrei N. Rodionov
50 **The coal handbook: Towards cleaner production**
Volume 1: Coal production
Edited by Dave Osborne
51 **The coal handbook: Towards cleaner production**
Volume 2: Coal utilisation
Edited by Dave Osborne
52 **The biogas handbook: Science, production and applications**
Edited by Arthur Wellinger, Jerry Murphy and David Baxter
53 **Advances in biorefineries: Biomass and waste supply chain exploitation**
Edited by Keith Waldron
54 **Geological storage of carbon dioxide (CO_2): Geoscience, technologies, environmental aspects and legal frameworks**
Edited by Jon Gluyas and Simon Mathias
55 **Handbook of membrane reactors**
Volume 1: Fundamental materials science, design and optimisation
Edited by Angelo Basile
56 **Handbook of membrane reactors**
Volume 2: Reactor types and industrial applications
Edited by Angelo Basile

78 **Safe and secure transport and storage of radioactive materials**
Edited by Ken Sorenson

79 **Reprocessing and recycling of spent nuclear fuel**
Edited by Robin Taylor

80 **Advances in battery technologies for electric vehicles**
Edited by Bruno Scrosati, Juergen Garche and Werner Tillmetz

81 **Rechargeable lithium batteries: From fundamentals to applications**
Edited by Alejandro Franco

82 **Calcium and chemical looping technology for power generation and carbon dioxide (CO_2) capture**
Edited by Paul Fennell and Ben Anthony

83 **Compendium of hydrogen energy Volume 1: Hydrogen production and purificiation**
Edited by Velu Subramani, Angelo Basile and T. Nejat Veziroglu

84 **Compendium of hydrogen energy Volume 2: Hydrogen storage, transmission, transportation and infrastructure**
Edited by Ram Gupta, Angelo Basile and T. Nejat Veziroglu

85 **Compendium of hydrogen energy Volume 3: Hydrogen energy conversion**
Edited by Frano Barbir, Angelo Basile and T. Nejat Veziroglu

86 **Compendium of hydrogen energy Volume 4: Hydrogen use, safety and the hydrogen economy**
Edited by Michael Ball, Angelo Basile and T. Nejat Veziroglu

87 **Advanced district heating and cooling (DHC) systems**
Edited by Robin Wiltshire

This book is dedicated to my two daughters Samantha and Lilliana,
for providing me a purpose in life,
and
to my mother and father, for love and encouragement, past and present.

Preface

The purpose of this book is to provide an overview of the use and impact of high temperature superconductors (HTS) in the electric power grid. With chapters written by a number of distinguished technical experts from around the world, the book updates a number of earlier books and reviews and introduces some important new material in this rapidly developing field. The book is written at a level geared toward those with a basic science and engineering background, professional R&D managers in energy technology companies, and academic/government researchers, with a basic familiarity with Maxwell equations and electromagnetism. The book starts off with a broad and introductory overview of the electric power grid, of superconductivity basics, and of the use and impact of HTS in the grid starting from the early days of projects sponsored by the US Department of Energy to the present-day sophisticated installations around the world. This chapter concludes with a look at future prospects for superconductor-based devices in the electric power sector.

The book next gives an overview of superconductivity itself, including both low temperature superconductors and HTS, and how their properties, particularly their magnetic behavior and current carrying properties, are both similar and different. It is suggested that the reader unfamiliar with the phenomenon of superconductivity start off by thoroughly reading Chapter 2, before proceeding to the other chapters.

The next two chapters then lead the reader through the fabrication processes of the two types of HTS wires most widely used in electric power devices, namely the so-called (1) first generation (1G) bi-oxide power-in-tube tapes and round wires and (2) second generation (2G) RE–Ba–Cu–O coated conductors.

The book then examines one of the most commercially promising HTS applications for the power grid: power cables, including their various demonstration projects and tests results. A special feature is an in-depth discussion of fault-current-limiting (FCL) cables, which have not been treated in detail in the technical literature heretofore. Four separate chapters are dedicated to HTS cables including (1) HTS three-phase AC cables, including FCL cables, (2) HTS DC cables, (3) HTS gaseous He cooled cables, and (4) cryogenic cooling of HTS cables.

The book next examines a variety of other HTS-enabled devices which bring novel functionality to the electric power sector, including: HTS fault-current limiters, HTS motors and generators, HTS SMES, and HTS transformers. Finally, the book concludes with a brief overview of HTS-based projects in China.

Acknowledgements

As with any large collection of technical and scientific work, there are many contributors and collaborators required to bring the effort to fruition. I would first like to acknowledge and thank all of the authors for contributing their many wonderful chapters and hope that the reader enjoys learning from them, as much as I enjoyed reviewing them. I would also like to thank the publisher for agreeing to edit and publish a book on high temperature superconductivity in the electric power grid. This important and fascinating topic is one that has captured my interest since the early days of its exploration and investigation. I would also like to thank the editors for their tireless and endless chasing down of the status of chapter manuscripts and for the countless e-mails that were exchanged in the editing of this book.

There is one individual that I would like to call out 'a very special thanks and acknowledgement' in the editing of this book on HTS in the electric power grid – that person is Alex Malozemoff. Without the help (and more specifically intervention) of Alex, this book would not have been completed. Towards the end, Alex was the true driving force behind its completion and on multiple occasions had to inspire me to actually finish it. When the effort seemed hopeless and far from its completion, Alex helped review additional chapters and came to the rescue in coauthoring two additional chapters including his excellent work in the books introductory chapter. I take great enjoyment when I hear him tell others that his additional effort in seeing this book to completion was only because he did not want to see his first chapter's effort go to waste! Although he is unaware, I first met Alex at a Gordon Research Conference when I was in graduate school studying the magnetic behaviour of the newly discovered HTS materials. The knowledge and insight that he showed during the early days of this discovery into this phenomenon was impressive. I would have never guessed back then that I would get to coauthor a chapter with him regarding high temperature superconductivity. It was an honour to coauthor Chapter 2 on the fundamentals of superconductivity and you can see his expertise in magnetic phenomena of these materials scattered throughout this chapter.

Finally, I would like to express my appreciation to my two daughters Samantha and Lilliana who bring me joy and happiness which inspires each and every day.

Part One

Fundamentals and materials

The power grid and the impact of high-temperature superconductor technology: an overview

1

A.P. Malozemoff
AMSC, Devens, MA, USA

1.1 Introduction

This book gives a comprehensive description of the fascinating and important ways in which superconductors can impact—and are impacting—the electric power grid. This chapter introduces the basics, including the following: What is the electric power grid? What are superconductors? Also, why is the opportunity for superconductors in the power grid so important?

This chapter gives an overview of the electric power grid, the major challenges it faces in the future, and its major components: overhead lines, underground cables, transformers, electric generators, reactive compensators, voltage regulators, circuit breakers, and fault current limiters. It introduces superconductors and their major properties relevant to applications: the transition temperature, zero resistance, flux exclusion, critical current, and critical fields. Finally, it describes briefly the status and prospects for superconductor power equipment: alternating and direct current cables, fault current limiters, transformers, rotating machinery, and superconductor magnetic energy storage. Overall, although challenges related to cost, reliability, standardization, and utility industry acceptance remain, prospects for broad commercialization of superconductor power grid technology are increasingly promising.

1.2 Overview of the electric power grid

Let us start with the electrical power grid: Electric power is fundamental to modern civilization; it provides our lighting, runs our factories, and powers all our motors, data centers, and telecommunications networks. The electrical grid, or simply "the grid," is the interconnected network that supplies the vast majority of electric power to the world, cleanly and conveniently—"at the flip of a switch" (Schewe, 2007; Schneider, 2010).

The grid consists of all the electrical links between generation and use. Generators convert mechanical energy, supplied from turbines fueled by energy sources such as fossil fuels, nuclear power, and renewable energy, into electric power, which is then

Superconductors in the Power Grid. http://dx.doi.org/10.1016/B978-1-78242-029-3.00001-7

transmitted and distributed to electrical loads through overhead lines and underground cables, supported by transformers, switchgear, and voltage regulators. This vast array of sophisticated equipment is controlled by a complex communications and control network involving supervisory control and data acquisition (SCADA) systems, local and regional control centers, and, increasingly, computerized systems and components for monitoring, controlling, operating, and protecting generation, substation, transmission, and distribution assets. Overlaying this physical system is an economic and business network, as well as a governmental and regulatory network that in many cases extends beyond customer rate transactions to financial and information exchanges across multiple grid entities and their supply and support chains (Schneider, 2010).

From an electrical engineering point of view, the grid is a giant network of impedances—resistors, inductors, and capacitors—interrupted as necessary by circuit breakers and switches. Power lines, cables, transformers, and rotating machines can all be modeled with an equivalent circuit consisting of these basic elements. Electric currents flow through this network in a manner dictated by Kirchoff's laws (Boast, Hale, & Jones, 1993). Increasingly, however, there is a desire to create a smart grid: using modern sensor and communication technology to introduce two-way flows of electricity and information, tracking the grid's operation down to the customer's site, influencing demand, and introducing new technologies that can controllably adjust and steer the current flow in the grid. New devices, often based on power electronics (e.g., flexible AC transmission systems, or FACTS), are being introduced to provide such control. As we shall see, superconductors can potentially contribute here as well.

The electric power grid boasts many superlatives: it is perhaps the greatest engineering achievement of the twentieth century. Spanning entire continents, the grid is the world's most extensive network, arguably eclipsed only recently by the Internet. It is the single greatest purveyor of energy, supplying almost four billion megawatt-hours per year in the United States alone. It is one of the largest and most capital-intensive sectors of the economy, with U.S. assets of almost $1 trillion, yearly U.S. electric bills of more than a quarter of a trillion dollars, and the largest array of customers, including almost every industry, business, and household in the United States (Annual Energy Outlook, 2012; Schneider, 2010).

It is useful to review some terminology right up front (Schewe, 2007). Most of the grid is designed to carry alternating current (AC) power, with voltage and current that oscillate sinusoidally at 60 Hz in the United States and 50 Hz in Europe and which are conventionally described in terms of the root-mean-square (rms) voltage and current. The power P is the product of rms voltage V times rms current I. The power is typically delivered in three phases, each separated by one-third of a cycle. Most AC electric power in the United States is generated at rms voltages up to about 22 kV by electric generators, is then "stepped up" to rms voltages of 138 kV, 345 kV, or higher for transmission across long distances, then "stepped down" to lower voltages for local distribution until finally it can be safely used in the factory, or in the home at rms 120 or 208 V. Although definitions differ worldwide, here we call rms voltages below about 1 kV *low voltage* (LV), rms voltages between 1 and 35 kV are *medium voltage*, rms voltages from 35 to 66 kV are *subtransmission voltage* (sometimes the transition to high voltage is taken to be 72.5 kV), rms voltages from 66 to 345 kV are *high voltage*, rms voltages from 345 to 1000 kV are *extra high voltage*, and rms voltages above 1000 kV (1 MV) are *ultra high voltage*. For simplicity, here we will refer to circuits operating at rms voltages below 66 kV as *distribution* and circuits operating above 66 kV as *transmission*. Transmission grids carry power over long distances, across entire continents if necessary, and feed the more local distribution grids.

The night-time view from the National Aeronautics and Space Administration (NASA) in Figure 1.1 gives a dramatic vision of the entire U.S. grid, both transmission and distribution, through its lighting loads. The U.S. and Canadian grids are principally divided into three main three-phase AC grids or "interconnections," each combining transmission and distribution, and operating out-of-phase (asynchronously) with each other: the Eastern Interconnection, the Western Interconnection, and the Electric Reliability Council of Texas (ERCOT).

The overall concept of the electric power grid and its essential components—overhead lines, cables, transformers, generators, and switchgear—were developed by the end of the nineteenth century, using traditional (conventional) copper and aluminum wires coupled with magnetic iron alloys in transformers and generators. Although its capability has advanced enormously over time, this tried-and-true

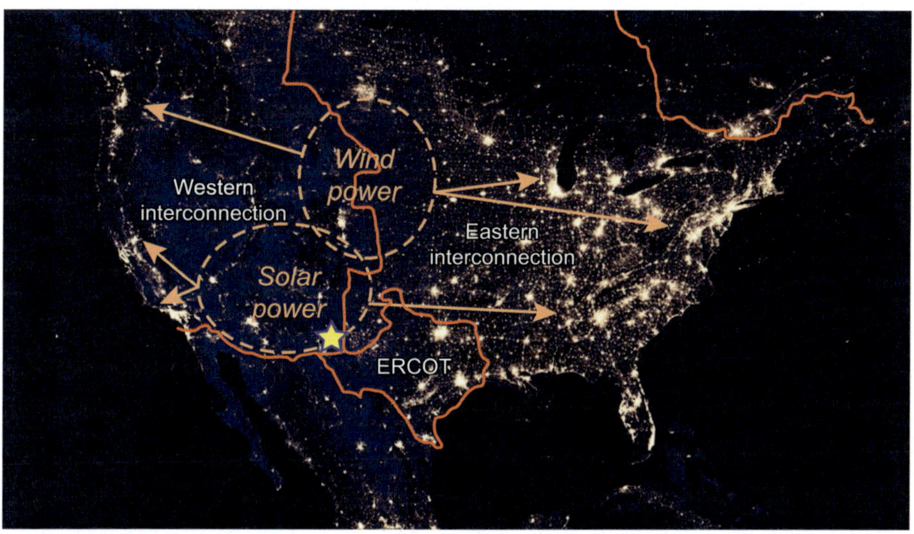

Figure 1.1 Satellite photo of North America at night indicates the scope of the electric power grid and main electric power load centers. Red lines delineate the three main grid interconnections in the United States and Canada – the Eastern and Western Interconnections, and ERCOT covering most of Texas. The three interconnections, asynchronous with each other, meet at the yellow star, near Clovis, New Mexico, location of the proposed Tres Amigas™ SuperStation, a renewable energy hub linking the three interconnections (www.tresamigasllc.com). Areas of major wind and solar power are indicated in orange circles, with arrows suggesting needed transmission to load centers. (Composite diagram by author, based on NASA night-time view (geology.com/articles/satellite-photo-earth-at-night.shtml).)

technology is in some sense over a century old and has provided power reliably and economically throughout this period, even as capacity and total load have grown enormously.

Utilities face many ongoing challenges in their business: maintaining reliable and economic service while matching generating capacity to ever-changing loads, dealing with inevitable power system contingencies and disruptions, adapting to a continually changing mix of energy sources, coordinating independent power producers, navigating the complex regulatory issues that determine the economics of grid operation, and—last but not least—assuring profitability and access to capital for system maintenance and upgrades. Nevertheless, the future brings some major new challenges that will require more than just incremental growth in the grid. These challenges include many of the drivers for introduction of superconductor power equipment, as we shall see below (Malozemoff, 2012a):

1. *Load growth and siting barriers*: The first challenge stems from ongoing demand growth worldwide, especially in the developing world. Even in the United States, where the rate of increase is slowing, demand for electric power has been estimated in a base case to grow from 3.9 billion kilowatt-hours in 2010 to 4.7 billion in 2035, an increase of more than 20% (Annual Energy Outlook, 2012), and possible massive electric vehicle use could drive these numbers much higher. A key issue is that much of this increase is expected to occur in major urban areas, which are increasingly the engines of future economic growth. As we shall see, addressing this critically needed urban power growth is becoming a severe siting challenge for conventional copper and aluminum technology—a challenge sometimes called the "urban power bottleneck," a challenge that can be met by superconductor technology. The load growth and siting challenge is compounded by the fact that much of the transmission and distribution infrastructure in developed grids around the world is nearing the end of its useful specified design life and will need to be replaced.
2. *Environmental and safety issues*: The second major challenge arises from the increasing importance of environmental issues, such as pollution and climate change. These issues put a premium on improved efficiency to minimize pollution and carbon-emitting wasted energy, creating an opportunity for superconductor power equipment if it could be engineered to be sufficiently efficient. Another problem is that the widespread use of highly flammable oil as coolant and dielectric in conventional power equipment risks oil-leak contamination, fire, and explosion—the latter of which is regrettably all too frequent in transformers. Superconductor equipment, cooled by nonflammable liquid nitrogen, eliminates this problem. Furthermore, major renewable energy sources, such as solar in the U.S. Southwest and wind in the Upper Plains, tend to occur far from the major load centers evident in Figure 1.1. As these sources grow in importance to replace fossil fuels and reduce carbon emission, more efficient long-distance transmission of power will become important; here, "lossless" direct current (DC) superconductor cables could be a long-term answer.

   One special renewable energy challenge arises in offshore wind power. Offshore wind is typically stronger and steadier than on land, providing a potentially huge and reliable new source for clean renewable energy. Yet today, the fraction of wind power captured from offshore is less than 2% of the total installed base because of increased installation and maintenance costs of conventional offshore wind turbines. Higher power ratings, in the range of 10 MW or more at blade rotation rates of 10–15 rpm, could solve this problem by providing more output per turbine as compared to a maximum rating of about 5 MW today, and construction costs per megawatt could be reduced as well. But until dramatically smaller and lighter 10-MW generators are developed, this solution lies out of reach; compact superconductor generators may offer the answer.

3. *Reliability and power quality*: Increased reliability and improved power quality is another major requirement for the future, driven in part by the "siliconization" of industry—that is, by the increasing dependence on microprocessor control, which is sensitive to power glitches of even millisecond duration. Another aspect of this problem is that as urban grids grow, fault currents, which arise from short circuits in the grid, also increase (for reasons to be discussed below). In some grids, these are already reaching such high levels that they risk wearing out and even burning out grid components. We shall see how superconductor fault current limiters (FCLs) can address this problem. Brownouts or blackouts can be triggered by excessive voltage drop in an overloaded grid; the reduced impedance of superconductor grid components can mitigate this issue.

Economics is a dominant driver in grid operation. Today's grid has supplied electric power cost-effectively as well as reasonably reliably for over a century. For superconductor equipment to succeed in the grid, it must do likewise. In fact, it must do better. To the extent that superconductor-based equipment has the *same* functionality as existing conventional equipment, it must offer lower cost; at least at present, this is a major challenge. However, the key to the initial penetration of superconductor equipment in the grid is the *new* functionality and higher performance it offers, in many cases enabling solutions inaccessible to conventional technology.

Next, we review the principal components of the grid, along with a closer look at the specific challenges they face in the future.

1.3 Elements of the electric power grid

1.3.1 Long-distance links: overhead lines and underground cables

Perhaps the most fundamental elements of the electric power grid are its long-distance links, which can be either overhead lines or underground cables (note that we specifically use *lines* for overhead and *cables* for underground links; Berthold et al., 1993). These lines or cables consist of long conductors or wires arranged in an appropriate configuration to handle the relevant voltage, current, and mechanical stress. The first overhead electric power transmission was achieved in Germany in 1882, and ever higher voltages were steadily introduced through the twentieth century, recently reaching 1150 kV in a location in Russia. Overhead lines, with typically aluminum wires suspended from wooden or metal towers, are common throughout the world. Their cost per megawatt-mile is low and will be tough for DC superconductor links to beat unless overhead rights-of-way are sufficiently costly and can be avoided by a compact and high-power underground superconductor cable. Also factoring into this trade-off is the growing public resistance to overhead lines due to aesthetic issues and concerns for property values.

Conventional underground cables consist of a central or "phase" conductor at their core, typically wound from multiple copper wires, surrounded by a high-voltage dielectric such as cross-linked polyethylene (XLPE) and a second cylindrical coaxial conductor wound helically from multiple individual wires. The assembly is protected with an outer jacket. Underground cables are particularly favored over overhead lines in urban and suburban environments because of their lower right-of-way requirements and because they avoid weather-related damage and aesthetic concerns.

These conventional overhead line and underground cable technologies have been very successful. But with the increasing power demands described earlier, further improvements are becoming more critical, as detailed here:

1. *Higher efficiency*: Today, approximately 8% of the U.S. grid's almost half-terawatt of annual average power is estimated to be lost in wasteful resistive heating between generator and load. This wasted power is both an economic and environmental issue: Every 1% gain in electrical efficiency would translate to an annual reduction in carbon emissions of 30 million tons (assuming the carbon emission rate of a coal-fired power station; carbonify.com/carbon-calculator). Conventional copper and aluminum technologies are fully optimized and have nowhere to go in lowering their resistivity, whereas superconductors offer a path to lower loss and a corresponding mitigated environmental impact.
2. *Voltage reduction*: It would also be desirable to decrease the operating voltage V of a power link, because safety issues and permitting challenges increase dramatically with increasing voltage. However, at a given power level $P = VI$, decreasing V means carrying more current I, which is not possible in conventional resistive wires without increasing the loss, but it is possible with superconductors.
3. *Addressing the urban power bottleneck*: As mentioned already, there is yet another, perhaps more serious, challenge that is beginning to rear its head in larger urban grids. Safety, aesthetic considerations, and concern for adjacent property values are increasingly preventing the use of overhead lines, especially with high voltage. Under city streets, the existing dense underground infrastructure, including power, water, gas, and telecommunication lines, is making it increasingly difficult to add new power cables, because these cables generate heat and stray electromagnetic fields (emf) and so cannot be located too close to other underground infrastructure. Altogether, these problems create the "urban power bottleneck," making it difficult to bring additional power to city centers and even to some suburban zones. Yet, power demands in large cities, which are increasingly the hubs of economic activity, are growing inexorably. Therefore, circumventing this urban power bottleneck is a central issue for future world economic growth.

As we shall see in Chapters 5–7, superconductivity, with its dramatically reduced resistance, large current-carrying capacity (i.e., high power density), and minimal environmental impact, has the potential to address all of these challenges.

1.3.2 Power transformers

Another key element of the grid is the electric power transformer, enabling a highly efficient increase or decrease of the AC voltage (Pandza, Girgis, & Shaw, 1993). It operates based on the principle of electromagnetic induction (Faraday's law). In a typical core-type transformer, two solenoidal coils, called the primary and secondary, separated by a dielectric insulator, are wound coaxially around a yoke consisting of a soft (easily magnetized) magnetic material, such as iron. An AC rms current $I_{1,\text{rms}}$ and voltage $V_{1,\text{rms}}$ in the primary are then transformed to an AC current $I_{2,\text{rms}}$ at voltage $V_{2,\text{rms}}$ in the secondary, with the voltages in proportion to the ratio of turns in the primary and secondary coils. Total power $P = I_{1,\text{rms}}V_{1,\text{rms}} = I_{2,\text{rms}}V_{2,\text{rms}}$ is conserved except for small resistive and magnetic losses; thus, current is reduced as voltage is increased and vice versa.

Transformers are a familiar sight—small pole-mounted transformers or metal boxes on the ground, to feed homes and small commercial and industrial establishments. The larger power transformers that convert between transmission and distribution voltages are located in carefully protected and enclosed substations, along with the related switchgear, buswork, and reactive compensation, which control grid operation. Substations are the key nodes in the electric power grid.

George Westinghouse's development of a commercial AC transformer was the decisive stroke in making the grid what it is today: an almost entirely AC grid. During the late 1880s, Westinghouse waged the "War of the Currents" with Thomas Edison, who promoted a DC grid (Nichols, 2006). Because current in a link with resistance R generates Joule heating loss I^2R and R increases in proportion to distance, long-distance transport of power with high current can be highly inefficient. Edison lacked an effective way to transform his DC current to high voltage so as to transmit high power with low current and keep resistive loss down. By using transformers to convert AC power to high voltage and low current for long-distance transport and then to transform the power back to lower voltage and higher current for local use, Westinghouse achieved more efficient power transport and so won the War of the Currents. Today's grid uses the same principle. DC is practical mainly for short distances and is widely used in electronics, vehicle systems, and "off-grid" power installations. In the power grid, DC is used today only in undersea links or special ultra-long runs where expensive power-electronic voltage transformation, initially from AC to DC and then ultimately back to AC, is justifiable (see Chapters 5 (Section 5.7.2), 6, and 7).

As we shall see, superconductors change these considerations in a fundamental way. By bringing wire resistance down to near zero, much higher currents can be tolerated in long-distance low-voltage DC superconductor links without significant I^2R losses, and this could give impetus to wider use of DC in the future, particularly for long links such as those bringing renewable energy from distant sources to the main load centers.

Nevertheless, given the existing investment, it is safe to say that the AC grid is here to stay, and so are AC transformers. What are future challenges for transformers? First, although conventional technology has been highly optimized and losses are typically below 2% of the power rating, further loss reduction would be valuable, particularly because as many as 5−10 transformers lie in the path from generator to load (Pandza et al., 1993). Another issue is that the transformer's inductive impedance is one of the main contributions to voltage drop in the grid; reducing that impedance would reduce voltage drop and help to stabilize the grid. Ironically, precisely because it is a dominant contribution to the grid impedance, the transformer's impedance is needed in today's grid for limiting fault currents; therefore, any reduction in impedance must come with some means of fault current limiting, as will be discussed further below. Another persistent problem with conventional transformers is fires and explosions, which feed off the dielectric and cooling oil. As we shall see in Chapter 12, superconductors can help address all these issues.

1.3.3 Electric generators

Yet another key element of the power grid is the electric generator, also developed in the nineteenth century by Faraday, Tesla, and others (Appleyard, Rosenberg, & Kirtley, 1993). It produces electrical energy from mechanical energy, typically from rotation of a rotor driven by any of a number of energy sources, such as a steam turbine, internal combustion of fossil fuels, hydropower, or wind power. In a typical configuration, either permanent magnets or coils on the rotor generate rotating magnetic fields, which induce a voltage—emf or electromotive force—in stator (or armature) coils connected to the grid.

What are future needs for generators? First, like transformers, large electric generators can be highly efficient. However, with the immense amount of electric power they generate, even a half percent improvement in efficiency can translate to significant savings. In many situations, significantly lighter and more compact machines could have a major impact. We have seen one of these in the earlier discussion of offshore wind turbine generators. In the field of transportation including ships, trains, aircraft, and electric vehicles, electrification is a growing theme driven by increased efficiency and functionality, and a lighter and more compact generator could lead to significant advantages in powering the extensive electric grid of ships or aircraft. Here too, superconductors can have a major impact (see Chapter 10).

1.3.4 Reactive compensation and voltage regulation

From an electrical engineering perspective, an ideal AC grid would be purely resistive, having current and voltage perfectly in phase throughout so as to match most efficiently to the resistive loads that consume the power. Close to 40% of those loads are motors, both large and small (Annual Energy Outlook, 2012). Another major load segment, close to 14%, comes from lighting. Heaters, dryers, stoves, electronic devices and computers, televisions, and countless other devices constitute the rest. However, the impedance of an AC grid is not purely resistive; it is in fact dominated by reactances—that is, either inductances or capacitances over which voltage and current are 90° out of phase, with voltage leading current over inductance and lagging over capacitance. Inductance arises particularly where coils are present (in rotating machinery and transformers), but an inductance per unit length characterizes all overhead lines and cables as well. Capacitance arises particularly in cables between the central conductor and its coaxial sheath.

The net grid impedance—a combination of all of these resistances, inductances, and capacitances—causes a voltage drop from the generator to the load. Grids are complex and can have many different configurations, such as radial, looped, and meshed, with substations containing transformers and switchgear forming the nodes. However, at its very simplest, as shown in Figure 1.2, the grid consists of a net AC voltage source V (the emf of the combined electrical generators), a predominantly inductive net grid impedance jX, and a net resistive load R. jX and R form a voltage divider: The larger jX is, the lower is the voltage over R, causing the voltage drop referred to above.

If the voltage over the load drops too far, the grid becomes unstable. Of course, loads rely on a certain voltage level to operate properly. So, voltage drop must be limited, leading to the need for a variety of voltage regulation devices, including step-type voltage regulators and tap-changing transformers (Harlow & Shaw, 1993), as well as the reactive compensation devices discussed later.

Phase shifts arising from reactances reduce grid efficiency because grid currents become larger than necessary to transfer a given power to the load, and this increases I^2R loss. This inefficiency is often characterized by a "power factor" less than 1 (Boast et al., 1993). Because inductance and capacitance have complementary effects in shifting the phase, grid efficiency can be improved by "reactive compensation"—that is, by introducing additional equipment such as capacitor banks or large coil inductors—to compensate the grid's reactive impedance and bring the power factor closer to unity on a steady-state basis (Harlow & Shaw, 1993; Kleeb, Grove, & Shaw, 1993).

Changes in power demand, increasing use of intermittent renewable energy sources such as wind and solar, and the unpredictable power inputs from distributed residential or commercial sources can shift the power factor during the day, and the grid must handle these changes efficiently. Rapid disturbances from short circuits or inrush currents of starting motors can even endanger grid stability. An important relatively recent development is the introduction of FACTS or flexible AC transmission

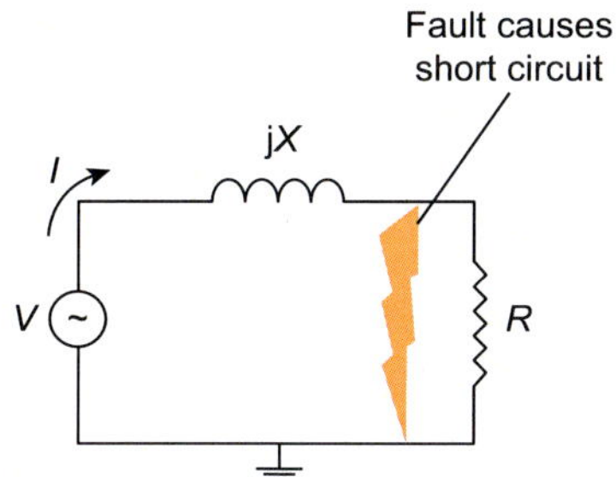

Figure 1.2 A power grid is a network of impedances (resistive, inductive, capacitive), including the loads and impedances of its components— generators, transformers, transmission and distribution lines and cables. According to Kirchoff's laws, total current $I = V/Z = V/(\mathrm{j}X + R)$ is driven by the generator voltage V and determined by the total impedance Z, which includes both load R and the mostly inductive grid impedance jX. As is clear from the figure, jX must be kept low to prevent significant voltage drop over the load R. However, if a fault causes a short circuit of the load, jX is all that is left to control the current; so huge currents $I = V/\mathrm{j}X$, called fault currents, will flow. Since circuit breakers take some time to act, a way must be found to limit excessively large currents and prevent destruction of grid equipment.

systems; many are based on power electronics, enabling active and rapid control of phase and power transfer (Malozemoff, Kehrli, Diaz de Leon, & Kalsi, 2004; Wood et al., 1993). Static VAR compensators (SVCs) controllably switch in or out components of an inductor or capacitor bank, and dynamic synchronous condensers based on rotating machines have a controllable output reactance (Malozemoff et al., 2004). More recent FACTS systems include unified power flow controllers (UPFCs), static synchronous compensators (STATCOMs), and dynamic volt–ampere reactive systems (D-VARs).

These active devices are one aspect of what has become known as the *smart grid*, which involves the increasing use of local sensors and communications to improve grid efficiency and customer interaction. Many of the issues described above could be mitigated with a new generation of power equipment with much reduced reactive impedance, namely a generation based on compact superconductor systems. Because transformers make the single largest contribution to the grid's impedance, the drive for reduced impedance should start with them. Superconductors can also be the basis of a new generation of dynamic synchronous condensers (Malozemoff et al., 2004). Superconductor magnetic energy storage could provide rapid injections of power in situations where real power is critical to grid stability (see Chapter 11).

1.3.5 Circuit breakers and FCLs

Under normal operating conditions, the current I is determined by the generalized Ohm's law or Kirchoff's equation, $I = V/(R + \mathrm{j}X)$. For efficient operation as discussed previously, the grid impedance jX is kept small compared to R. However, if the load impedance is short-circuited as indicated in Figure 1.2, the grid current $I = V/\mathrm{j}X$ increases drastically. This is known as a "fault current." Where grid currents in a transmission or distribution link may be on the order of 1000 A, fault currents can reach tens of thousands of amperes in a distribution grid and as high as 60 or 80 kA in a transmission grid. If they are too large and allowed to continue too long, they burn out grid equipment, which could further cascade to widespread power outages. Such a disaster can and must be prevented.

Circuit breakers, and the switchgear in which they are packaged, are an essential element of the grid; they are necessary to stop fault currents, as well as to isolate segments for repair or service and to introduce new installations (Johnson & Shaw, 1993). For these purposes, various circuit breakers have been developed, from simple fuses to an array of resetting circuit breakers with mechanical contacts opening in a gas or vacuum environment. Because of the huge currents and electric arcs they generate through gas or vacuum, circuit breaker design is sophisticated, with some using special dielectrics, such as sulfur hexafluoride for arc suppression. Of course, circuit breakers themselves have rated limits for currents under normal and grid fault conditions.

A major problem is that as a grid grows and is meshed further, its net impedance generally decreases, and so the effective fault current increases. A simple way to see this is to recognize that electrical generators have output impedances, contributing to the total grid's impedance. As power demands increase, more electrical generators are added to the grid in parallel, and as is well known in electrical engineering, the net impedance $Z = 1/(Z_1^{-1} + Z_2^{-1})$ of parallel impedances Z_1 and Z_2 is always reduced. As a result, in many cities around the world, fault currents are nearing rated limits of the circuit breakers that protect the grid. This inhibits further growth, which is a basic challenge for growing urban grids. An FCL is needed to introduce a large impedance quasi-instantaneously whenever the current becomes too high. A fuse is such a device, but at high powers, the irreversible opening of such a fuse can be an explosive event, and it needs to be replaced after every fault event to enable ongoing grid operation. Significant efforts are being made around the world to develop automatically resetting FCLs using a variety of technologies, such as power-electronic, magnetic, and, as we shall see in the next section, superconducting (see Chapters 5 (Section 5.6), and 9).

1.3.6 Electric energy storage

It is not easy to store electrical energy directly; therefore, the grid has been designed to adjust the output of its generators rapidly to meet time-dependent demand. Yet, with the increasing introduction of intermittent renewable energy sources such as solar and wind power, the desirability of electrical energy storage is increasing. Up to now, the main energy storage techniques have converted electrical energy into potential energy (pumped hydro storage), high pressure (compressed gas storage), or chemical energy (batteries). Because of relatively high inefficiencies, these techniques are still not widely used, although battery technology is advancing rapidly. A more direct means of electrical energy storage is in capacitor banks. Now, superconductors offer a new means of electrical energy storage, in the loss-free circulation of electrical current in a coil, generating magnetic energy; this is the so-called *superconductor magnetic energy storage* (SMES—see Chapter 11). The key challenge in the wider use of these more direct storage means is cost.

1.4 Superconductivity

1.4.1 Superconductor materials

As will be described in more detail in Chapter 2 of this book, superconductors are materials with remarkable properties, such as zero resistance and magnetic flux exclusion (i.e., diamagnetism) below a certain transition temperature T_c. Thousands of materials exhibit superconductivity (Chu et al., 2012), including elements such as mercury (the first one discovered, $T_c = 4.2$ K), lead (7 K), aluminum (1.2 K), and niobium (9 K); the widely used alloys NbTi (10 K) and Nb_3Sn (18 K); the T_c record holder of the metallic so-called low-temperature superconductors (LTS) Nb_3Ge (23 K); organic materials with an ambient-pressure T_c record holder of this group, $RbCs_2C_{60}$ (33 K); and compounds such as MgB_2 (40 K), FeSe (8 K), and doped LaO–FeAs oxypnictides (reaching 52 K). So far, the only family of compounds to breach the 77 K level at which liquid nitrogen boils are the artificially synthesized perovskite-like cuprates, such as $YBa_2Cu_3O_7$ ("YBCO," 93 K), $Bi_2Sr_2Ca_2Cu_3O_{10}$

Table 1.1 Critical temperature T_c and upper critical field H_{c2} at 4 K for several superconductors commercially available in wire form

Superconductor	Max T_c (K)	$H_{c2,4\ K}$ (T)	References
NbTi	10.3	15	T.P. Sheahen (1994) *Introduction to High-Temperature Superconductivity*. Springer
Nb_3Sn	18.3	29.5	A. Godeke et al., the upper critical field of filamentary Nb_3Sn conductors, arXiv:cond-mat/0410463
$Y_{1-x}RE_xBa_2Cu_3O_7$	93	168	T.P. Sheahen (1994) *Introduction to High-Temperature Superconductivity*. Springer
$Bi_{2-x}Pb_xSr_2Ca_2Cu_3O_{10}$	110	200	T.P. Sheahen (1994) *Introduction to High-Temperature Superconductivity*. Springer
MgB_2	40	29	A. Gurevich et al., very high upper critical fields in MgB_2 produced by selective tuning of impurity scattering, arXiv:cond-mat/0305474

Note: The upper critical fields are measured on polycrystalline samples.

("BSCCO-2223," 110 K), and the ambient-pressure T_c record holder $HgBa_2Ca_2Cu_3O_8$ (133 K); these are referred to as *high-temperature superconductors* (HTS), although the iron pnictides and MgB_2 are often included in this category. Properties of several of these materials that are commercially available in wire form are summarized in Table 1.1.

Like the electric power grid, superconductivity boasts many superlatives. Superconductors bring the miracles of the quantum world to a macroscopic level. Their zero resistance is unequaled in nature. DC currents can circulate in loops of superconductor indefinitely. The superconductor order parameter exhibits a phase and exists in imaginary space, phenomena that are completely unintuitive in our macroscopic world. Sensors based on superconductors are the most sensitive known to man (Braginski, 2012). Voltages induced by the Josephson effect are so precise that they are used for voltage standards (Harris & Niemeyer, 2012). Key properties of superconductors that are relevant to applications in the power grid are described in the following sections.

1.4.2 Transition temperature T_c

The main practical advantage of superconductors with high superconductor transition temperature over their low-temperature counterparts is to improve refrigeration efficiency, thereby reducing cost. Refrigeration efficiency drops rapidly with decreasing temperature. The theoretical maximum efficiency is given by the Carnot coefficient of performance $COP_{cooling} = Q_C/W_{in} = T_C/(T_H - T_C)$; here, Q_C is the heat energy removed from a cold environment at temperature T_C to keep it cold and W_{in} is the work energy applied for the refrigeration, exhausting the waste heat at temperature T_H (Duband & Ravex, 1998). For example, a refrigerator cooling to the liquid helium boiling point of 4.2 K and exhausting the heat at room temperature (300 K) has a Carnot COP of only 1.4%, while if it is cooling to the liquid nitrogen boiling point of 77 K, the Carnot COP jumps to 34%. Practical refrigerators are much less efficient than the ideal Carnot level, operating at a third to a tenth of these values, whether at helium or nitrogen temperatures. Nevertheless, the higher Carnot COP of liquid nitrogen, coupled with its abundant supply, drives production costs to a bare minimum: a liter of liquid nitrogen can be purchased at a lower cost than many bottled waters. In comparison, increasing scarcity of liquid helium supply coupled with its far lower COP drives production costs to a premium.

The huge difference between the efficiency and cost of liquid helium and liquid nitrogen temperature refrigeration explains much of the enthusiasm greeting the 1987 discovery of HTS cuprates with transition temperatures above 77 K. Working in the liquid nitrogen rather than helium range is particularly critical for AC power grid applications such as AC cables, FCLs, and transformers because AC currents generate AC losses that significantly increase the low-temperature input heat Q_C and hence the cost of refrigeration. Some applications, such as the rotors of synchronous rotating machinery or DC cables, operate DC (or DC with a small AC ripple), and such HTS systems are sometimes operated at intermediate temperatures from 20 to 40 K to benefit from improved critical current (see below). But even for these applications, a high T_c is of great value for the following reason: The low-temperature specific heat of materials grows with temperature as T^3; it is almost 580 times as large at 35 K as compared to 4.2 K, and more than 6000 times large at 77 K. Higher specific heat enhances thermal stability because the heat disturbance required to quench superconductivity will be correspondingly larger.

Of course, the ideal solution would seem to be a room-temperature superconductor, but countless unsubstantiated claims notwithstanding, no such superconductor has yet been discovered. In fact, even if such a superconductor were discovered, its properties would most likely not be viable for power applications; a phenomenon called thermally activated flux creep is likely to suppress the critical current to unusable levels (Malozemoff, 2012b). Fortunately, presently known HTS cuprates work very well at 77 K, as we shall see below; at this temperature, refrigeration is practical and not so costly.

1.4.3 Zero resistance

This property is of paramount importance, opening up a breakthrough opportunity for power grid efficiency by eliminating I^2R loss, the bane of conventional power equipment. One must hasten to add that superconductors do generate losses in the presence of AC currents, though with proper design these AC losses can be minimized. Furthermore, the flux creep phenomenon mentioned above and discussed further below generates a small voltage in HTS, even in DC operation, but at least in the liquid nitrogen temperature range, this effect is so small that the resulting low resistance is still of tremendous value.

There is a second related and major practical impact of low resistance that is sometimes overlooked. Low resistance enables huge HTS current density (many MA/cm^2), orders of magnitude larger than tolerable in copper, which would melt from I^2R heating at this level. High current density opens up the ability to design more compact systems than corresponding conventional ones, enabling higher capacity cables, lower inductance transformers, and more power-dense rotating equipment. Wherever size and weight are critical, superconductor equipment has a significant advantage.

1.4.4 Flux exclusion: the Meissner effect

Magnetic flux ϕ, measured in tesla-m (Schneider, 2010) or webers, is the magnetic induction field B integrated over the surface area A through which it passes: $\phi = \int B\mathrm{d}A$. In a so-called type I superconductor, the Meissner effect excludes magnetic flux from the bulk of the superconductor except in a very thin surface layer of a magnetic (London) penetration depth (Tinkham, 1996). Actually, most useful superconductors are so-called type II superconductors, which do allow the penetration of flux through the bulk of the material, but in a remarkable way: The flux is quantized into linear structures (hence the terminology "flux lines"), each carrying a flux quantum $\phi = 2.07 \times 10^{-15}$ Wb, with currents circulating around each flux line (hence the alternate name "vortices"). Multiple flux lines can exist within a superconductor, forming an array, the so-called Abrikosov lattice (Tinkham, 1996), in which they are spaced apart by the magnetic forces between them. In equilibrium, this array is usually hexagonal in the plane perpendicular to the flux lines axes. The array's spatial density is proportional to the locally averaged B (i.e., averaged over multiple flux lines). In fact, there are almost always obstacles to the penetration of flux into superconductors, usually arising from defects that "pin" the flux lines to the site of the defect. The result is that type II superconductors can act as effective screens of magnetic fields if those fields are not too large.

1.4.5 Critical current

Any net current in a superconductor must obey Ampere's law, which in its simplest one-dimensional form is given by $\mu_0 J = \mathrm{d}B/\mathrm{d}x$, where μ_0, the permeability of free space, is 1.257×10^{-6} H/m, J is the locally averaged current density, B is the locally averaged magnetic induction field, and x is the measure of position (Tinkham, 1996). This means that corresponding to any net current, there must exist a gradient in the locally averaged magnetic induction field, hence a gradient in the density of the flux lines. Forces proportional to the average current density drive the flux lines to even out ($\mathrm{d}B/\mathrm{d}x = 0$) and bring the current to zero, but this can be prevented by pinning of flux lines at defects.

As long as flux lines remain trapped in the pinning potential wells located at the sites of these defects, the voltage remains zero and the superconductor remains in its superconducting state. However, in the presence of a flux gradient corresponding to an overall local current density vector $\boldsymbol{J}$, a net force $\boldsymbol{J} \times \boldsymbol{B}$, called the Lorentz force, acts on the flux lines; when the Lorentz force exceeds the maximum pinning force, it will drive the flux lines out of their pinning wells, initiating flux motion and hence voltage (emf), according to Faraday's law emf $= d\phi/dt$. The critical point at which flux lines escape out of their pinning wells determines the critical current density J_c of the superconductor. From a practical perspective, J_c is one of the most important parameters, defining how much current the superconductor can carry while remaining in its superconducting state.

It should be noted that because of the thermally activated hopping of flux lines known as flux creep (Malozemoff, 1991), the electric field E in HTS is not precisely zero at finite temperature in the presence of current density J. Typically $E \sim J^n$, where the power n is called the index value, often in the range of 20–30, and the critical current density J_c and corresponding wire critical current I_c are defined by an electric field criterion of 1 μV/cm or less, which is sufficiently small to ensure minimal power loss at and below the critical current in practical applications. Thus defined, the critical current density of HTS materials is found to increase roughly linearly with decreasing temperature, in contrast to LTS, which show more curvature; the difference comes from the larger impact of flux creep in HTS. The index value can also be even lower when the superconductor is highly inhomogeneous.

A tremendous amount of research has been devoted to optimizing defects in practical superconductors to maximize J_c. While progress is ongoing, the 77 K engineering current density (J_e) of HTS tape-shaped wires based on materials such as YBCO and BSCCO-2223 (i.e., the current density including any substrate and stabilizer) is already more than two orders of magnitude higher than the typical room-temperature operating current densities of copper, and high enough to enable practical devices. The development of these HTS wires, with all the required electrical and mechanical properties, is one of the great achievements of the field, and they are now commercially available in considerable quantity from a variety of manufacturers around the world. They and their future development prospects will be described in detail in Chapters 3 and 4. As mentioned above, the high current and power density of these wires enable power equipment with dramatically lower size and weight compared to that based on conventional copper wire. This is a major advantage in many applications.

Because J_c increases with decreasing temperature, as indicated in Figure 1.3, a trade-off with refrigeration cost arises in optimizing the operation temperature. Particularly for some DC applications like rotating machinery, the optimum is often well below 77 K (e.g., in the 30–40 K range). J_c also depends strongly on magnetic field, as indicated in Figure 1.3, and on its direction because the HTS atomic structure and resulting properties are strongly anisotropic. The highly nonlinear resistance that rapidly develops when a superconductor transitions to its normal state as current is increased above the critical current is a property of great value in designing FCLs, because this transition occurs very rapidly and in a fail-safe manner without any external control.

1.4.6 Critical fields

A sufficiently high magnetic field fully quenches the superconductor order parameter (Tinkham, 1996). Together with the temperature and critical current, this so-called upper critical field H_{c2} forms a critical surface in the space of temperature, current,

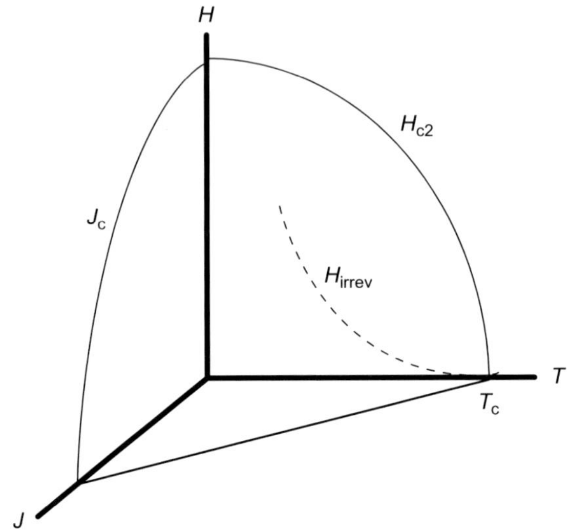

Figure 1.3 Schematic critical surface of a high-temperature superconductor in the space of temperature T, applied field H, and critical current J. The material remains superconducting in the sense of maintaining a local order parameter within the surface defined by the upper critical field H_{c2} and critical current J_c. However above the line $H_{irrev}(T)$ in the $H-T$ plane, long-range order is lost and flux lines can move freely, generating flux flow resistance.

and field, as shown schematically in Figure 1.3. Outside this surface, superconductivity is fully quenched. H_{c2} is extraordinarily high in HTS cuprates, reaching well above 100 T at low temperature, as summarized in Table 1.1. This is of great value in building ultra-high-field magnets for use in research and high-resolution nuclear magnetic resonance. In practice, however, another field, called the *irreversibility* or *melting* field H_{irrev}, which lies far below H_{c2} (see Figure 1.3), marks the point at which HTS materials begin to show higher resistance. It is the field at which pinning ceases to suppress motion of flux lines; above it and below the upper critical field, flux lines persist but are in constant liquid-like motion, enabling flux flow, which gives rise to a "flux flow resistivity." Useful superconducting properties exist only below the irreversibility field. Fortunately, for a superconductor like YBCO, even at 77 K, this field is around 8–10 T, which is easily high enough for most grid applications, such as rotor field coils for rotating machinery.

1.5 Status and prospects of superconductor power equipment

This book explores in detail the design, development, demonstration, and future prospects for all presently investigated superconductor equipment for the electric power grid. Here is a brief synopsis of the types and status of such equipment.

1.5.1 Superconductor AC cables

As will be described in greater detail in Chapter 5 of this book, AC cables are the most advanced of the HTS grid applications, with many successful demonstrations around the world and the first truly commercial (not government supported) installations expected in coming years (Kalsi, 2011). Similar to conventional underground cables, typical HTS cables are assemblies of multiple HTS wires surrounded by a layer of high-voltage dielectric, a screening layer of HTS wires, a flexible cryostat, and an outer jacket. Like a giant snake, they are from 6 to 10 inches in diameter and can be up to a kilometer long without joints. Cooled by liquid nitrogen pumped from a closed cycle refrigeration system at one or both of the cable ends, they have sophisticated high-voltage terminals where the transition is made from superconductor wires to copper leads, and also from liquid nitrogen to room temperature. These cables are designed for use underground, connecting substations in urban or suburban environments. The undergrounding, whether for conventional or superconductor cables, provides important advantages: less exposure to storms or terrorist attacks, as well as reduced right-of-way disruption and improved aesthetics. In addition, the superconductor cables have many advantages over conventional cables, including the following:

1. *Power capacity*: Thanks to the high current-carrying capacity of superconductor wires, such cables can carry much larger currents, typically 2–4 kA_{rms}, in the same cross-section as conventional cables carrying 1 kA_{rms}. This is their major advantage over conventional cables, enabling them to overcome the urban power bottleneck described above.
2. *Ease of installation*: Their screening layer of HTS tapes exploits the screening or flux exclusion property of superconductors described above and eliminates emf, which could otherwise disturb nearby communications cables. In contrast to conventional cables, HTS cables emit no heat into their immediate surroundings and weigh far less per unit length than their conventional counterparts. These properties allow HTS cables to be installed much more easily and cost-effectively than conventional cables. They can be placed close to other cables without interference, or if necessary they can be dug deep underground. By contrast, to ensure adequate dissipation of heat, conventional high-power cables must be located near the surface, often surrounded by special thermally conducting sand and a meter or more away from neighboring underground infrastructure. Compact HTS cables can also be retrofitted through existing conduits, significantly enhancing power capacity.
3. *Lower voltage operation*: The high current capacity of HTS enables operation at lower voltages than equivalently rated copper cables. In urban or suburban environments, this is a major safety as well as permitting advantage because delays in permitting high-voltage cable can be very long and costly.
4. *Fault current limiting*: Another important option in HTS cable design is to introduce fault current limiting functionality directly into the cable, as will be described further below. This is particularly valuable, enabling substation-to-substation links at distribution voltage in urban environments where fault currents are already nearing their acceptable limit. Without the limiting feature, such links would reduce the grid's residual impedance and dangerously increase the magnitude of fault currents. Fault current limiting links offer multiple advantages: They provide alternative sources of power if there is a failure in the substation's high-voltage supply. Because in many urban grids entire neighborhoods depend on a given substation, such substation-to-substation links increase reliability of the electric power

for that neighborhood. Protecting in this way critical assets, such as New York City's financial center, is the motivation behind the Hydra Project undertaken by American Superconductor and Con Edison, with the Ultera cable company and support of the U.S. Department of Homeland Security (Maguire et al., 2009). The alternative power source also allows utilities to reduce the number of extra transformers stationed at each substation in case of a transformer failure; New York requires two extra transformers for redundancy, a considerable investment in cost and space.

5. *Low impedance*: Low inductive impedance of HTS cables originates from their compactness and tight containment of magnetic field in the narrow shell between the inner conductor and the screening layer. When such cables lie in parallel with conventional cables in a grid, they will take on more load and unload an otherwise overloaded conventional circuit. Furthermore, when such cables are coupled with a commercially available device called a phase angle regulator (PAR), power transmitted through the cable can be directly and conveniently controlled (Malozemoff et al., 2004). As described earlier, grids are huge networks of impedances, through which current flows are dictated by Kirchoff's laws. Of course, current flows can be abruptly changed by turning links on and off with circuit breakers. However, the HTS-cable/PAR combination allows the current flows to be dialed in with much finer control. This would be an important step toward a smart grid in which current flows can be controlled by operators rather than simply by Kirchoff's laws. While such current flow control has been possible with power-electronic systems, such as the UPFC, the HTS-cable/PAR system promises to be more economical and reliable. Low cable impedance is critical in this regard because it enables control with much smaller phase angle adjustments and hence much less costly PARs. This novel concept of power flow control has been proposed and modeled but not yet tested in the grid.
6. *Increased capacitive charging length*: A limitation of conventional underground cables arises from the capacitance between conductor and shield and the so-called line charging current, which must flow during every current cycle to charge up this capacitance. Because cable capacitance is proportional to length, this effect sets a limit, called the capacitive charging length, on the practical length over which the cable can operate, which is in the range of 20 miles for 230 or 345 kV circuits and typically shorter for lower voltages (puc.nh.gov/2008IceStorms/, 2008). The more compact dimensions of HTS cables mean lower capacitance and their higher current capacity handles higher line charging currents, all of which leads to significantly longer capacitive charging lengths and hence the option of significantly longer underground cable runs between substations.

Although HTS AC cable systems are admittedly more complex and the cables themselves more costly per kilometer than the latest generation of conventional cables based on XLPE dielectric, their many unique advantages enumerated above can predominate in addressing certain critical utility needs.

1.5.2 Superconductor FCLs

Like HTS AC cables, HTS FCLs have been widely demonstrated as described in Chapter 9. (Bock et al., 2011; Kalsi, 2011; Neumueller et al., 2009; Noe & Steurer, 2007). Among many possible principles of HTS FCL operation, the most common one is to exploit the highly nonlinear resistance that develops when current surges above the superconductor's critical current, causing a superconductor-to-resistive transition. This transition, instantaneous on the AC time scale, turns an HTS wire into a resistor, immediately limiting the fault current. A typical configuration of a stand-alone HTS FCL consists of a stack of bifilar-wound pancake superconductor coils (Neumueller et al., 2009). The bifilar winding (with current direction alternating from layer to layer) ensures a low inductance and hence a negligible grid impedance during normal operation. The HTS coils can be connected in parallel and series in such a way as to meet current and voltage requirements. The module is submerged in liquid nitrogen, which may be pressurized for high-voltage operation. The most challenging design issue is to avoid overheating and damaging the coils in their resistive state before a circuit breaker cuts off the remaining current. These challenges have been met and Nexans has already pioneered a number of demonstrations and initial commercial installations in Europe (Bock et al., 2011).

Of course, there are many alternative means to limit current. Circuit breakers are widely used but take a finite time to act and themselves need to be protected when fault currents are too high. Fuses are simplest and lowest cost, but most require manual intervention because they are destroyed in each event, and handling high voltage is nontrivial. Relative to fuses, HTS FCLs have the advantage of resetting relatively quickly, once the HTS coils cool back down. Power-electronic current limiters are also being developed, but HTS FCLs have the advantage of being self-triggering and fail-safe, responding quasi-instantaneously without relying on power-electronic activation. Adding inductors into the grid can reduce fault currents, but this increases voltage drop and worsens system stability and transient recovery, while HTS FCLs can be designed with very low inductance and very little voltage drop. Further discussion of the many advantages and challenges for HTS FCLs is given in a later chapter.

One of the most important developments in fault current limitation is making use of the HTS wires already incorporated in cables and transformers to design current-limiting cables (as described above) and transformers. The advantage of this approach is that it avoids the need for an additional piece of equipment—the stand-alone FCL—and provides unique functionality (see Chapter 5, Section 5.6).

1.5.3 Superconductor power transformers

Despite longstanding efforts to demonstrate HTS power transformers, they still lag behind AC cables and FCLs in commercial readiness (Kalsi, 2011). Yet, they have many potential advantages, such as the following:

1. *Reduced AC loss*: A key goal for HTS transformers is to achieve lower losses than conventional transformers. Unfortunately, this is not so easy because conventional transformers are highly optimized and quite efficient already; also, the AC currents in HTS coils generate non-negligible AC loss, which is further amplified by the unfavorable coefficient of refrigeration performance (COP) discussed previously. In cables, the dominant AC magnetic fields generated by the currents are parallel to the surface of the helically wound tapes, which helps keep losses to a minimum, while the magnetic fields near the ends of transformer coils are directed perpendicular to the tape surface, generating large losses that rise in proportion to the width of the tape.

A worldwide effort has addressed mechanisms of AC loss, and some progress has been made in developing low AC-loss wire by means of filamentization, twisting, and transposition (Amemiya et al., 2004). A transposed Roebel cable made of tape-shaped wires has also been successfully developed (Lakshmi et al., 2011), although combining filamentization with the Roebel cable geometry is still an unsolved challenge in practice. Careful balancing of the two ends of a layer-wound solenoid (i.e., a helically wound coil with multiple layers) can obviate the need for transposition (Iwakuma et al., 2011), which is otherwise difficult to do in a tape-shaped wire. These developments raise hope for HTS transformers with lower AC loss than comparably rated conventional ones. However, until this is achieved, HTS transformers will not be commercially successful.

2. *Compactness and low inductance*: Another potential benefit of HTS transformers is their compactness or, equivalently, their high power density, thanks to the high wire current density. This is not only a space-saving feature; it also leads to lower inductance because inductance is determined by the total energy $\int(B^2/2\mu_0)\mathrm{d}V$ stored in the magnetic field B during each AC cycle and hence by the volume V over which the magnetic fields are integrated. As pointed out previously, the low inductance reduces voltage drop, reducing the need for tap changers and contributing to grid stability. However, this improvement must be coupled with fault current limitation because lower transformer inductance increases the fault current.
3. *Fault current limiting*: To protect a grid with low-inductance HTS transformers, a standalone FCL could be added. However, introducing current limiting functionality directly within the transformer by making use of the transformer's HTS wires would be a major cost advantage, just as in the case of AC cables described above. This concept is being pursued in most of the present transformer development projects (Berger, Noe, & Kudymov, 2011; Hayakawa et al., 2011). This is a design challenge because the variation in magnetic field from place to place within the transformer coil results in a variation in the critical current, as is evident from Figure 1.3. Then, when the fault current reaches the lowest of the local critical currents, only that portion of the coil switches to a resistive state, concentrating the resistive losses in one section of the coil and risking burnout. However, progress is steadily being made.
4. *Safety, reliability, and environmental footprint*: Fire risk is reduced in HTS transformers because they are cooled with liquid nitrogen rather than oil. Also, the low HTS operating temperature largely eliminates serious thermal degradation issues of conventional technology.

Further discussion is given in Chapter 12.

1.5.4 Superconductor rotating machinery

HTS motors and generators constitute a significant opportunity for commercial applications (Kalsi, 2011). Large generators based on rotating machinery supply the dominant portion of electric power to the grid. Motors greater than 1000 horsepower, the range most advantageous for HTS application, consume around 30% of total U.S. electric energy. Dynamic synchronous condensers are another opportunity: these are generators without a power source but which can supply reactive power for mitigation of voltage dips and sags in the grid. Significant demonstrations of large rotating HTS machinery have already been achieved, and the technology looks ready for commercial application.

The typical design of HTS synchronous rotating machinery involves a rotor with HTS coils operating on a quasi-DC basis and a stator with copper coils connected to the grid. All-superconductor machines with HTS in both rotor and stator coils have been proposed; these would have further advantages if sufficiently low AC loss wire were available. Unfortunately, stator coils experience large AC currents and magnetic fields in the Tesla range, which would generate even larger AC loss than in power transformers, where fields are in the 0.1 T range; therefore, the problem of achieving low AC loss in rotating machine stators looks daunting. Nevertheless the "half-superconductor" design with HTS only in the rotor has plentiful advantages, including the following:

1. *Enhanced efficiency*: Because the stator field coils operate essentially DC, with only a small ripple in most designs, their AC loss is low and easily handled with standard small cryogenic systems like Gifford–McMahon refrigerators, coupled with a gaseous helium loop or thermo-siphon system. By eliminating the resistive and eddy-current losses of conventional copper rotor coils, HTS machines cut total loss by close to half compared to conventional equivalently rated machines. This advantage is further enhanced in partial load operation.
2. *Reduced size and weight*: As in the cables and transformers discussed above, high HTS wire current density enables dramatically more compact and lightweight designs than conventional machines. For example, the 36.5-MW ship propulsion motor demonstrated for the U.S. Navy weighed one quarter of an equivalently rated conventional machine, including all supporting components, such as refrigeration (Gamble, Snitchler, & MacDonald, 2011). In shipboard and other transportation environments, such size and weight advantages are obviously important, but they are also valuable in spatially constricted generation stations. Lightweight HTS generators are also key to larger ratings of offshore wind turbines, as discussed previously (Snitchler, Gamble, King, & Winn, 2011).
3. *Reduced vibration and noise*: This HTS benefit comes both from the smaller size and from the elimination of cogging torques arising from magnetic switching of conventional iron stator teeth. This is particularly important for sensitive navy applications.
4. *Enhanced coil lifetime*: Thermal load cycling of copper coils, a major failure mode in conventional rotating machinery, is eliminated in the cooled coils of HTS machines.
5. *Large dynamic stability range*: The various benefits of HTS rotating machines have also been applied to dynamic synchronous condensers, which can inject reactive power into the grid to counter various grid disturbances (Malozemoff et al., 2004). The low synchronous reactance of these machines gives them a large dynamic stability range (Kalsi, 2011). While one 8-MVAR HTS system was exhaustively and successfully tested in the grid, compensating the voltage disturbances from an arc furnace in a steel plant (Kalsi et al., 2007), economic competitiveness will require much larger systems that have not yet been built.

Further details on HTS rotating machines are given in Chapter 10.

1.5.5 Superconductor magnetic energy storage

Until the recent 2012–2014 first commercial HTS fault current limiter projects by Nexans, the only truly commercial superconductor product for the power grid had been a SMES system by Superconductivity Inc. and American Superconductor, which addressed power quality problems by compensating voltage dips and sags (Schöttler & Coney, 1998). The system used a 2.6-MJ magnet based on LTS (NbTi wire), and more than 10 units were installed in various applications around the United States and internationally. The value of this application has been quite controversial, however, and

American Superconductor finally gave up on this product in favor of an all-power-electronic system called D-VAR, a type of STATCOM which uses power transistors without any energy storage for reactive compensation and which has become a leading system worldwide for compensating voltage dips and sags (Kehrli & Ross, 2003).

Nevertheless, SMES development continues around the world, mainly focused on applications requiring injections of real power, as reviewed in more detail in a later chapter. Initial interest had focused on diurnal electrical energy storage, which would require gigajoules of energy storage, but the enormity of such a project and its cost make this an unlikely prospect. An intermediate need is for energy storage to help smooth power output from intermittent sources, such as wind and solar, and one project is under way in the United States to develop technology for such a SMES (Arpa-e.energy.gov/?q=arpa). Controlling frequency oscillations is another grid application. There are also interesting military applications such as supplying bursts of energy for rail guns. Because energy storage is determined by $\int(B^2/2\mu_0)\mathrm{d}V$, high fields are desirable, and here the high critical and irreversibility fields of HTS materials offer a distinct advantage.

Another application of superconductivity to energy storage is in flywheels suspended through the flux-trapping properties of superconductors. This application is reviewed in another chapter. Overall, commercial HTS SMES is an interesting but still distant prospect (see Chapter 11).

1.5.6 Superconductor DC cables

In addition to AC cable applications, DC cables offer important advantages in several applications (see also Chapters 6 and 7). One is the transport of multiple gigawatts of power over long distances from renewable energy sources to load centers. In its grandest version, this concept has been called a Supergrid, providing a redundancy in the overall electric infrastructure, allowing power from one region to serve another if the local primary source is disrupted. The elimination of resistive loss in HTS DC cables is the major advantage here, but it has to be combined with excellent thermal insulation to keep thermal losses very low. Another major advantage, as with other underground cables, is the narrow right-of-way achieved by burying these links underground; this reduces land-acquisition and installation costs, avoids the aesthetic issues and perceived health risks, and increases safety from storms and other disturbances. Of course, long-distance DC cable must compete with long-distance overhead AC and DC lines, which feature well-optimized low-cost technology. Nevertheless public opposition to overhead lines is creating ever greater challenges in permitting new corridors and avoiding lawsuits, heightening the advantage that underground HTS DC cables can bring. Unfortunately, the truly enormous amount of wire required for, say, a 1000-km DC HTS link means that such an application must await considerable expansion of the HTS wire manufacturing capacity from where it is today.

Another interesting proposed application of HTS DC cables is as an ultra-high-power link between asynchronous grids. The Tres Amigas SuperStation (TASS) at Clovis, New Mexico, is planned to link the three main U.S. grid interconnects at their vertex (see Figure 1.1) with AC-to-DC-to-AC links so as to transport renewable energy from one region to the other in response to imbalances of supply and demand (www.tresamigasllc.com). This will require many gigawatts of capacity, something safely achievable in HTS DC cable, while the main competitor, SF_6-based cable, suffers from the fact that SF_6 has been identified as the most potent of all greenhouse gases. Installation of HTS DC cables at TASS will wait, however, until an initial lower-power installation using conventional technology is completed.

Although HTS DC cables still appear to be a long way off, initial development work has started with a successful 200-m demonstration in Japan and planned extensions to 500 m and eventually 2 km (Yamaguchi et al., 2011). More details on DC cable applications are given in Chapters 6 and 7 of this book.

1.6 Conclusion and future trends

HTS power grid technology is in active development around the world, and as described above, several applications are moving close to commercialization with government subsidy and support no longer needed. Of course, it has taken over a quarter of a century to get to this point, since the initial discovery of HTS in 1986. And there are still significant challenges which must be overcome to truly revolutionize the electric power grid with HTS power technology:

1. *Cost:* This is the most obvious challenge. A measure of the cost-effectiveness of HTS wires is the cost-to-performance metric \$/kA m, that is, dollars (or euro or yen...) per kiloamperemeter, where the kA is usually interpreted as the critical current at 77 K and self-field. Alternatively, especially for rotating machine or magnet applications, the critical current under actual operational conditions of field (several Tesla) and temperature (30−40 K) is used, which can be quite different. In either case, the basic concept behind the kiloamps in this metric is that with higher critical current, correspondingly less wire is required. This drives ongoing R&D efforts to achieve ever higher critical current density without increasing the wire cost, and in parallel finding ways to reduce wire production cost per meter.

   Scale is an essential aspect of driving down the manufacturing cost-curve; so as the industry grows, significant cost reduction is expected. Presently, HTS costs (prices) are in the range of several 100 \$/kA m, while competing copper costs are typically only several tens of \$/kA m, an almost order-of-magnitude difference. At these levels, HTS wire becomes a major component of overall systems cost, making HTS power equipment significantly more expensive than equivalently rated conventional equipment. Of course this comparison ignores the key functional advantages of HTS power equipment, advantages which can justify a higher price—though only up to a point. HTS cables, for example, may be several times the price of conventional cables per kilometer, and their other advantages—higher power density, lower heat and emf emissions, small right-of-way and fault current limiting characteristics—must justify the price premium. This trade-off applies to all HTS power equipment at the present time. And because HTS technology is so new, accurately assessing the justifiable price premium is often difficult.

   Refrigeration is another significant cost component, even with liquid nitrogen systems, and while conventional power systems also require cooling, the HTS refrigeration cost is higher. Although cryogenic technology is steadily advancing to higher efficiency at lower cost, the HTS power industry is still too small for such advances to achieve enough manufacturing scale for significant cost impact. The relative refrigeration costs are minimized in equipment carrying higher power.

2. *Safety:* Although HTS technology offers significant safety and environmental advantages, it also entails danger in working with ultra-cold temperatures and the risk of oxygen deficiency in enclosed environments if significant amounts of nitrogen gas are accidentally released.
3. *Reliability and standardization:* Another major challenge for HTS technology arises from the fact that implementation of the technology is still in its infancy. The very rapidity with which the technology is advancing, both in the materials and wire technology and in the design and demonstration of power equipment, creates a problem for the reliability-driven utility industry. It is difficult for customers to meaningfully assess reliability when, by the time one system is tested, a new generation of improved and modified devices may be offered, requiring a new round of testing. Standards for the wire and equipment are needed: CIGRE, IEEE, NEMA, VAMAS, etc. And in most cases, a commercial business will require established second sources.

   Superconductor technology brings some important reliability benefits: The consistently low temperature and vacuum jacket surrounding the superconductor, and the negligibly low resistive losses, eliminate the most problematic thermal failure modes of conventional technology, arising from thermal stresses, diffusion of contaminants, and corrosion. Against this must be weighed the critical dependency on complex refrigeration and vacuum technology. When the superconductor systems must be warmed to room temperature for repair or maintenance, they are subject to substantial stresses arising from thermal contraction and expansion, and the cycle times for warming and recooling are relatively long. But significant progress has been made in using redundancy and effective system choices to engineer competitive failure rates.
4. *Novelty:* A related challenge is the utility industry's natural caution in introducing any new technology into the grid. Conventional technology works and has a century of reliability experience behind it! Therefore, convincing utility decision makers to take a risk on a radically new and initially unfamiliar technology is extremely difficult. However, a decade of successful HTS demonstrations is gradually familiarizing utility planners with the capabilities and benefits of HTS power equipment. New demands on the power grid, described above, are also driving them to accept new technologies to solve problems not easily solved by conventional means. Indeed, the first commercial installations will most likely occur in situations where there is little or no option for a conventional solution, such as the installation of HTS AC cable for additional power in urban streets already so chock-a-block with underground infrastructure that conventional installation costs are prohibitive.

All in all, impressive progress toward commercialization of HTS power technology is being made steadily, year after year. Its advantages are commanding. Costs are declining while credibility is growing. And the need for new solutions is growing. So the superconductor revolution in the power grid is beginning to look inevitable. This book aims to lay out all this progress clearly and so hopefully to help accelerate bringing the many benefits of superconductor technology to the electric power grid.

Acknowledgments

The author thanks Chris Rey, Patrick Duggan, Syed Ahmed, Narend Reddy, Rob Rouse, and Kerry Farrell for their careful reading and many suggestions.

References

Amemiya, N., Kasai, S., Yoda, K., Jiang, Z., Levin, G. A., Barnes, P. N. et al. (2004). AC loss reduction of YBCO coated conductors by multifilamentary structure. *Superconductor Science and Technology, 17*, 1464–1471.

Annual energy outlook. (June 25, 2012). *U.S. Energy Information Administration, report #DOE/EIA-0383.* www.eia.gov/forecasts/aeo/MT_electric.cfm.

Appleyard, R. E., Rosenberg, L. T., & Kirtley, J. L. (1993). Alternating current generators. Section 7. In D. G. Fink & H. W. Beaty (Eds.), *Standard handbook for electrical engineers* (13th ed.). New York: McGraw-Hill.

Berger, A., Noe, M., & Kudymov, A. (2011). Test results of 60 kVA current limiting transformer with full recovery under load. *IEEE Transactions on Applied Superconductivity, 21*(3), 1384–1387.

Berthold, L. O., DiGiolia, A. M., Jr., Douglass, D. A., Grant, I. S., & Moran, J. A., Jr. (1993). Transmission systems. Section 14. In D. G. Fink & H. W. Beaty (Eds.), *Standard handbook for electrical engineers* (13th ed.). New York: McGraw-Hill.

Boast, W. B., Hale, H. W., & Jones, E. C., Jr. (1993). Electric and magnetic circuits. Section 2. In D. G. Fink & H. W. Beaty (Eds.), *Standard handbook for electrical engineers* (13th ed.). New York: McGraw-Hill.

Bock, J., Bludau, M., Dommerque, R., Hobl, A., Kraemer, S., Rikel, M. O., et al. (2011). HTS fault current limiters – first commercial devices for distribution level grids in Europe. *IEEE Transactions on Applied Superconductivity, 21*(3), 1202–1205.

Braginski, A. I. (2012). SQUIDS and detectors. In H. Rogalla & P. H. Kes (Eds.), *100 Years of superconductivity* (pp. 311–373). Boca Raton, FL: CRC Press.

Chu, C. W. (2012). Materials. In H. Rogalla & P. H. Kes (Eds.), *100 Years of superconductivity* (pp. 233–307). Boca Raton, FL: CRC Press.

Duband, L., & Ravex, A. (1998). Small cryocoolers. In J. G. Weisend (Ed.), *Handbook of cryogenic engineering* (pp. 287–320). Philadelphia: Taylor & Francis.

Gamble, B., Snitchler, G., & MacDonald, T. (2011). Full power test of as 36.5 MW HTS ship propulsion motor. *IEEE Transactions on Applied Superconductivity, 21*(3), 1083–1088.

Harlow, J. H., & Shaw, K. A. (1993). Voltage regulators. Section 10. In D. G. Fink & H. W. Beaty (Eds.), *Standard handbook for electrical engineers* (13th ed.). New York: McGraw-Hill.

Harris, R. E., & Niemeyer, J. (2012). Quantum metrology. In H. Rogalla & P. H. Kes (Eds.), *100 Years of superconductivity* (pp. 515–557). Boca Raton, FL: CRC Press.

Hayakawa, M., Kojima, H., Hanai, M., & Okubo, H. (2011). Progress in development of superconducting fault current limiting transformer. *IEEE Transactions on Applied Superconductivity, 21*(3), 1397–1400.

Iwakuma, M., et al. (2011). Development of a rebco superconducting transformer with current limiting function. *IEEE Transactions on Applied Superconductivity, 21*(3), 1405–1408.

Johnson, D. S., & Shaw, K. A. (1993). Circuit breakers. In D. G. Fink & H. W. Beaty (Eds.), *Standard handbook for electrical engineers* (13th ed.). New York: McGraw-Hill. Section 10.

Kalsi, S. S. (2011). *Applications of high temperature superconductors to electric power equipment.* Hoboken, NJ: John Wiley & Sons.

Kalsi, S., Madura, D., Snitchler, G., Ross, M., Voccio, J., & Ingram, M. (2007). Discussion of test results of a superconductor synchronous condenser on a utility grid. *IEEE Transactions on Applied Superconductivity, 17*(2), 2026–2029.

Kehrli, A., & Ross, M. (2003). Understanding integration issues at wind farms and solutions using voltage source converter FACTS technology. *Proceedings of IEEE Power Engineering Society General Meeting, 3*, 1828.

Kleeb, R. E., Grove, L. C., & Shaw, K. A. (1993). Power capacitors. In D. G. Fink & H. W. Beaty (Eds.), *Standard handbook for electrical engineers* (13th ed.). New York: McGraw-Hill.

Lakshmi, L. S., Long, N. J., Badcock, R. A., Staines, M. P., Jiang, Z., Thakur, K. P., et al. (2011). Magnetic and transport AC losses in HTS Roebel cable. *IEEE Transactions on Applied Superconductivity, 21*(3), 3311–3315.

Maguire, J., Folts, D., Yuan, J., Lindsay, D., Knoll, D., & Bratt, S. (2009). Development and demonstration of a fault current limiting HTS cable to be installed in the Con Edison grid. *IEEE Transactions on Applied Superconductivity, 19*, 1740–1743.

Malozemoff, A. P. (1991). Flux creep in high tempature superconductors. *Physica C, 185-189*, 264–269.

Malozemoff, A. P. (2012a). Second-generation high-temperature superconductor wires for the electric power grid. In D. R. Clarke, M. Rühle, & D. N. Seidman (Eds.), *Annual review of materials research* (Vol. 42, pp. 373–398). Palo Alto, CA, Annual Review.

Malozemoff, A. P. (2012b). New materials requirements for superconductor grid technology. *Physics Procedia, 36*, 1429–1433.

Malozemoff, A. P., Kehrli, A., Diaz de Leon, J., & Kalsi, S. (2004). Superconductor technologies for a controllable and reliable high capacity grid. In *Proceedings of PSCE04 New York, October 10–13, 2004*. IEEE Catalog No. 04EX943C.

Neumueller, H. W., Schmidt, W., Kraemer, H.-P., Otto, A., Maguire, J., & Yuan, J. (2009). Development of resistive fault current limiters based on YBCO coated conductors. *IEEE Transactions on Applied Superconductivity, 19*, 1950–1955.

Nichols, T. (2006). *AC/DC: The savage tale of the first standards war*. New York: John Wiley and Sons.

Noe, M., & Steurer, M. (2007). High-temperature superconductor fault current limjiters: concepts, applications, and development status. *Superconductor Science and Technology, 20*, R15–R29.

Pandza, C. M., Girgis, R. S., & Shaw, K. A. (1993). Transformers. In D. G. Fink & H. W. Beaty (Eds.), *Standard handbook for electrical engineers* (13th ed.). New York: McGraw-Hill. Section 10.

Schewe, P. F. (2007). *The grid: A journey through the heart of our electrified world*. Washington DC: Joseph Henry Press.

Schneider, T. (2010). Electric power grid technologies. In G. Crabtree & A. P. Malozemoff (Eds.), *Science for Energy technologies*. Washington DC, U. S. Department of Energy. http://www.sc.doe.gov/bes/reports/list.html.

Schöttler, R., & Coney, R. G. (1998). Betriebsverfahrung mit mikro-SMES im kommerziellen Industrieeinsatz. In *EESAT '98 Proceedings* (pp. 175–180).

Snitchler, G., Gamble, B., King, C., & Winn, P. (2011). 10 MW class superconductor wind turbine generators. *IEEE Transactions on Applied Superconductivity, 21*(3), 1089–1092.

Tinkham, M. (1996). *Introduction to superconductivity*. New York: McGraw-Hill.

Wood, P., Gyugyi, L., Brewster, J. B., Heinrich, T. M., Oates, R. M., Pelley, B. R., et al. (1993). Electronics. Section 22. In D. G. Fink & H. W. Beaty (Eds.), *Standard handbook for electrical enkgineers* (13th ed.). New York: McGraw-Hill.

Yamaguchi, S., Kawahara, T., Hamabe, M., Watanabe, H., Ivanov, Y., Sun, J., et al. (2011). Experiment of 200-meter superconducting DC cable system in Chubu University. *Physica C, 471*, 1300–1303.

Fundamentals of superconductivity

2

C.M. Rey[1], *A.P. Malozemoff*[2]
[1]Energy to Power Solutions (E2P), Knoxville, TN, USA; [2]AMSC, Devens, MA, USA

2.1 History

In this chapter, the fundamental concepts of superconductor phenomena are introduced to provide a foundation for the remaining chapters on high-temperature superconductor power equipment. The discussion focuses on type I and type II superconductors, their corresponding magnetic behaviors, low-temperature superconductors, and high-temperature superconductors. The chapter is based on publications by A.C. Rose-Innes, E.H. Roderick, and C. Kittel, supplemented by significant newer material arising from the novel properties of high-temperature superconductors.

In the early 1900s, soon after Kamerlingh Onnes had discovered how to liquefy helium, he began an investigation concerning the electrical resistance of very pure metals at low temperatures. At that time, it was unknown how the electrical resistance would behave at very low temperatures. Predictions ranged from the electrical resistance continuing to linearly decrease with temperature toward zero, leveling out at some residual resistance value, or starting to increase at some point due to other electron scattering mechanisms. One of the purest metals available at that time was mercury. In 1911, Kamerlingh Onnes was measuring the electrical resistance of pure mercury as a function of temperature when he discovered that the mercury's resistance suddenly dropped to zero below 4 K (De Bruyn Ouboter, Van Delft, & Kes, 2012; Kamerlingh Onnes, 1911). Figure 2.1 shows a reproduction of the electrical resistance versus temperature plot for pure Hg as measured by Kamerlingh Onnes. He realized that below 4 K the mercury entered a new state, which he called "supraconductivity" (De Bruyn Ouboter et al., 2012). No one had predicted this remarkable and fascinating phenomenon, which still intrigues many today.

Although many materials were subsequently discovered to show the phenomenon of superconductivity at sufficiently low temperatures, it was assumed for many years that only one type of superconductor existed. Only much later was it realized that two quite distinct types of superconductors exist. The two types of superconductors have many properties in common, but their most distinguishing differences show up in their behavior in applied magnetic fields. The type I superconductors (formerly referred to as "soft" superconductors), usually elements, lose their superconducting properties in relatively weak magnetic fields, whereas the type II superconductors (formerly referred to as "hard" superconductors), usually alloys, withstand very strong magnetic fields before losing their superconducting properties. A few pure elements are notable exceptions, including niobium, vanadium, and technetium; these are type II with Ginzburg–Landau parameters (see Section 2.8) of $\kappa \sim 0.78$, 0.82, and

Superconductors in the Power Grid. http://dx.doi.org/10.1016/B978-1-78242-029-3.00002-9

0.92, respectively (the crossover from type I to type II occurs at $\kappa = 1/\sqrt{2}$). Each type of superconductor is sufficiently different and must be treated separately. Discussion of the more complex type II superconductivity relevant to power applications is given in Sections 2.8–2.11 of this chapter.

2.2 Meissner effect

In 1933, Meissner and Ochensenfield (1933) began investigating the magnetic properties of type I superconductors. They discovered that if you cool a superconductor in an applied steady-state magnetic field H, then at the superconducting transition temperature T_c, the magnetic field lines are expelled and the superconductor behaves like a perfect diamagnet with magnetization $M = -H/4\pi$ or $M = -H$ (cgs/mks). This type of magnetization measurement is sometimes referred to as a field-cooled (FC) experiment and is schematically shown in Figure 2.2 (left). This behavior is far different than a zero-field-cooled (ZFC) experiment and cannot be explained by simply assuming that a superconductor is a perfectly conducting (infinite mean free path) medium. Instead, the Meissner effect implies that the flux density B inside the material is identically zero ($B = 0$) for

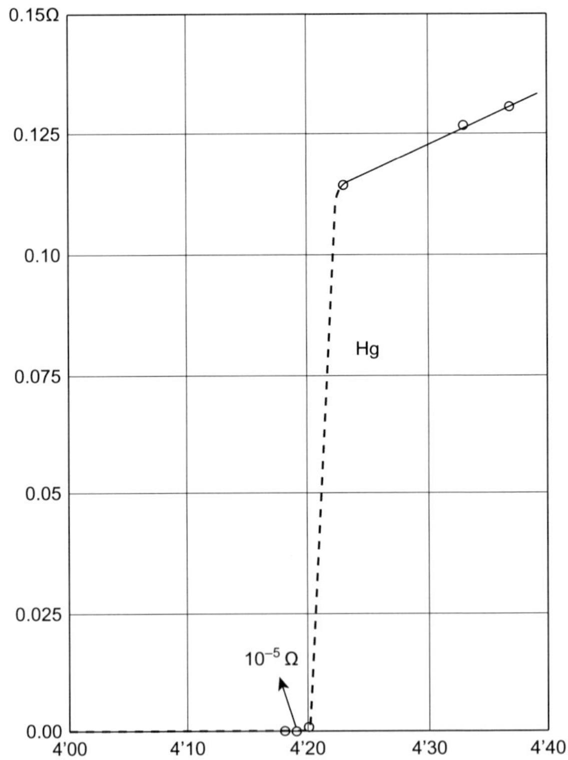

Figure 2.1 Reproduction of the original plot of the electrical resistance of Hg versus temperature by Kamerlingh Onnes (1911).

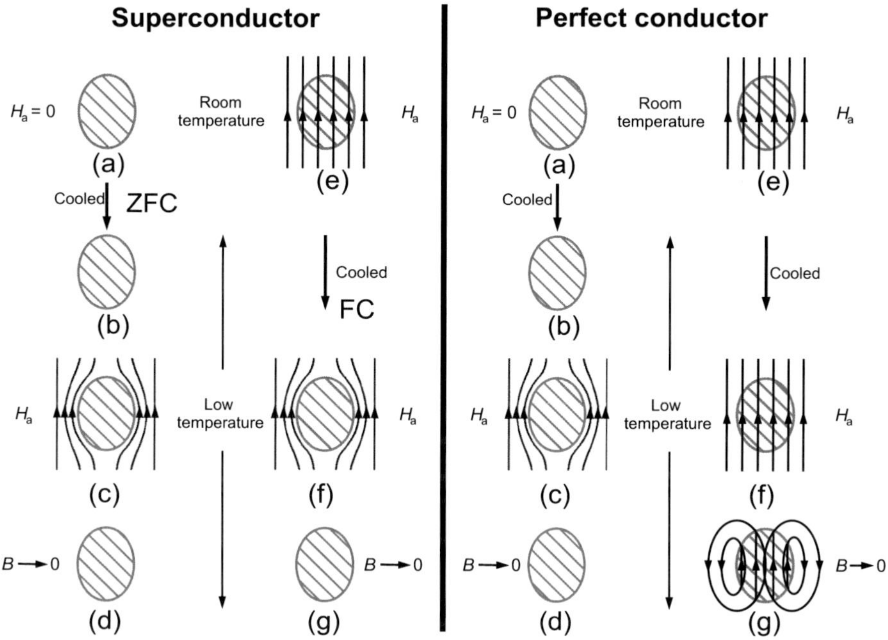

Figure 2.2 Schematic behavior of applied field H_a and magnetic flux density B in field-cooled (FC) and zero-field-cooled (ZFC) experiments on a type I superconductor and non-superconducting material with perfect conductivity. While the ZFC experiment (a → b), followed by application of a field H_a (c) and then removal of the field (d), does not distinguish between the two types of materials, the true signature of conventional superconductivity is perfect diamagnetism, as demonstrated by the FC experiment (path e → f → g, cooling of the sample in a field and then removal of the field) known as the Meissner effect. Magnetic flux is excluded from the interior of the specimen ($B = 0$) by supercurrents flowing within the penetration depth at the surface in the presence of H_a.
Adapted from Rose-Innes and Rhoderick (1978, pp. 18 and 20).

temperatures below T_c. Hence, $B = H + 4\pi M$ or $B = \mu_0(H + M) = 0$ (cgs/mks), where $\mu_0 = 4\pi \times 10^{-7}$ Henries/meter is the permeability of free space. If the superconductor were simply a perfect conductor with infinite conductivity ($\sigma = \infty$) and were cooled below T_c in the presence of a steady-state magnetic field H, there would be no magnetic flux expulsion ($B = 0$) at T_c because there was no time varying magnetic field dH/dt. The perfect conductor cooled in a background steady-state magnetic field would simply pass through T_c as if nothing happened. If, on the other hand, a perfect conductor were cooled in zero magnetic field and subsequently a magnetic field were applied (i.e., dH/d$t > 0$), the perfect conductor would repel flux. This type of magnetization experiment is referred to as ZFC and is schematically shown in Figure 2.2 (right). Thus, the perfect diamagnetism observed from the Meissner effect with the FC magnetization experiment is a true signature of superconductivity. Of course, all of these descriptions should be qualified by the fact that a thin layer (characterized by a magnetic or London penetration depth) at the surface of the superconductor does admit magnetic flux, as will be explained below.

2.3 London equations and magnetic penetration depth

In 1935, the London brothers (London & London, 1935) were able to mathematically describe the Meissner effect by assuming that the current density J in the superconducting state is directly proportional to the vector potential A of the local magnetic field B, where $\boldsymbol{B} = \nabla \times \boldsymbol{A}$ (in the London gauge, $\nabla \cdot A = 0$). The London equation is given by $J = -1/(4\pi\lambda_L^2)A$ or $J = -1/(\mu_0\lambda_L^2)A$ (cgs/mks), where λ_L has the dimensions of length.

Under static conditions, one of Maxwell's equations (Ampere's law) reduces to $\nabla \times \boldsymbol{B} = 4\pi\boldsymbol{J}$ or $\nabla \times \boldsymbol{B} = \mu_0\boldsymbol{J}$ (cgs/mks). By taking the curl of both sides, this equation reduces to $-\nabla^2 B = 4\pi\nabla \times \boldsymbol{J}$ or $-\nabla^2 B = \mu_0\nabla \times \boldsymbol{J}$ (cgs/mks). By combining this equation with the curl of the London equation, we obtain $\nabla^2 B = B/\lambda_L^2$. The solution of this equation is a flux density B, which decays exponentially with distance from the external superconductor surface. In a one-dimensional, semiinfinite superconducting medium occupying the positive side of the x-axis, the solution to this equation for the magnetic flux density inside this medium would be $B(x) = B_0 \exp(-x/\lambda_L)$, where $B_0 = \mu_0 H$ is the parallel field at the plane boundary (e.g., see Kittel, 1986). In this example, λ_L measures the distance at which $B(x)$ has fallen to $1/e$ of its initial value, which is known as the magnetic or London penetration depth.

2.4 Critical currents in type I superconductors

A type I superconductor, by definition, is a material that exhibits perfect flux expulsion in its interior. Physically, the Meissner effect arises because resistanceless currents flow on the surface of the superconductor to exactly cancel B throughout the volume of the specimen. Thus, if the applied field H is increased, the surface shielding current will also increase to keep $B = 0$ inside the specimen. There is, however, an upper limit to the magnitude of surface (within a distance of λ_L) shielding currents that a type I superconductor may sustain. The limiting magnetic field is known as the critical magnetic field H_c and the corresponding current density is known as the critical current density J_c. The critical current density may be expressed in terms of the critical field by using the curl of the London equation—namely, $\nabla \times \boldsymbol{J} = (-1/4\pi\lambda_L^2)\boldsymbol{B}$ or $\nabla \times \boldsymbol{J} = \left(-\frac{1}{\mu_0\lambda_L^2}\right)\boldsymbol{B}$ (cgs/mks). This equation relates the supercurrent density J to the magnetic flux density B at any point in the superconductor.

An inverse to this effect exists. In the absence of an external field, the self-field generated by the transport current flowing at the surface must not exceed a critical magnetic field. If the self-field exceeds this critical field, the superconductor will reversibly pass from its superconducting state to its normal state. This is known as the Silsbee criterion (Silsbee, 1916). For a type I superconductor with a reversible magnetization (see Section 2.5), the Silsbee criterion represents the maximum current density J_c that a type I superconductor can carry before returning to the normal (nonsuperconducting) state. This criterion for J_c holds true for many of the (pure) elemental superconductors and is quite low, making type I materials not suitable for electrical power applications requiring high transport currents.

2.5 Magnetization in type I superconductors

A type I superconductor (formerly known as a "soft" superconductor) will exhibit a Meissner effect (i.e., perfect diamagnetism, $B = 0$) when subject to an applied magnetic field, independent of whether the material was cooled in zero field (ZFC) or in field (FC), making the material magnetic history independent. This "perfect diamagnetic" response is a reversible thermodynamical process, and the principles of thermodynamic phase transitions must apply.

The energy difference between the normal and superconducting state may be determined as follows. Consider a long, thin rod with a long axis in the direction of H. This will allow demagnetizing effects caused by the fringing fields at the ends of the specimen to be neglected. When any magnetic material is placed in a magnetic field, its Gibbs free energy per unit volume G changes by an amount $\Delta G(H) = -\int_0^H M\mathrm{d}H$ or $\Delta G(H) = -\mu_0 \int_0^H M\mathrm{d}H$ (cgs/mks). For a type I superconductor in an applied field, $M = -H/4\pi$ or $M = -H$ (cgs/mks), and the Gibbs free energy increases by $H^2/8\pi$ or $\mu_0 H^2/2$ (cgs/mks). The normal state of a superconductor is commonly only slightly magnetic and acquires no appreciable change in its free energy. Thus, if H is increased sufficiently, the superconductor will reversibly change from its superconducting state to its normal state. This change will occur when $G_s(T,H) > G_n(T,0)$ or $H_c^2/8\pi > [G_n(T,0) - G_s(T,0)]$ or $\mu_0 H_c^2/2 > [G_n(T,0) - G_s(T,0)]$ (cgs/mks). For a type I superconductor, the maximum magnetic field strength that can be applied to a superconductor, if it is to remain in its superconducting state, is the *thermodynamic critical field* H_c, where $H_c(T) = \{8\pi [G_n(T,0) - G_s(T,0)]\}^{1/2}$ or $H_c(T) = \{2\mu_0[G_n(T,0) - G_s(T,0)]\}^{1/2}$ (cgs/mks). A typical magnetization curve for a type I superconductor is shown in Figure 2.3.

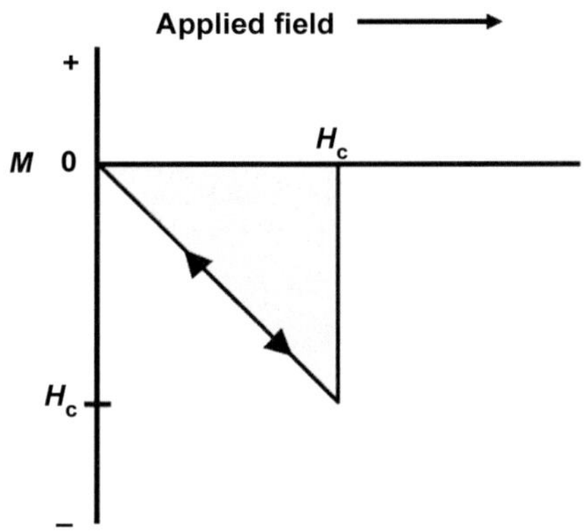

Figure 2.3 Typical M versus H curve for a type I superconductor. Note the complete magnetic flux exclusion ($B = 0$ or $M = -H$) and the abrupt reversible transition at $H = H_c$.
Adapted from Kittel (1986, p. 324).

The critical magnetic field of a superconductor is found to vary with the temperature. As the temperature is lowered, this value increases. Likewise, as the temperature is increased, the value of $H_c(T)$ decreases to zero at T_c. The temperature dependence of $H_c(T)$ is given by the general expression $H_c(T) = H_{c0}\ [1 - (T/T_c)^{\alpha}]$, where H_{c0} is the measured critical magnetic field at zero temperature ($T \rightarrow 0$) and α has a theoretical value of 2. In practice, α is an experimentally fitted parameter typically close to 2 for most superconductors. T_c and H_c are intrinsic properties of a superconductor, and using the given temperature-dependent properties, H_c can be calculated for any value of temperature below T_c. Similarly, the temperature dependence for the critical current density is given by $J_c(T) = J_{c0}\ [1 - (T/T_c)^{\beta}]$, where J_{c0} is the critical current density at zero temperature and β is experimentally fitted with values typically close to unity.

2.6 Intermediate state

2.6.1 Demagnetizing coefficient

The demagnetizing magnetic field is the magnetic field H_d generated by the magnetization M within the magnet. The total magnetic field in the magnet is a vector sum of the demagnetizing field and the magnetic field generated by any free or displacement currents. The term demagnetizing field reflects the tendency of this field to reduce the total magnetic moment of the specimen. The demagnetizing field can be very difficult to calculate for arbitrarily shaped objects, even in the case of a uniform magnetizing field. For the special case of ellipsoids, which includes objects such as spheres, long thin rods, and flat plates, H_d is linearly related to M by a geometry-dependent constant called the demagnetizing factor n. For a long thin rod placed in a uniform magnetic field along its long axis, the demagnetizing factor approaches zero. For a sphere placed in a uniform magnetic field, the demagnetizing factor is 1/3; for a flat plate with magnetization perpendicular to the plane, it approaches unity. The result is that in certain sample shapes, the demagnetizing field can concentrate the magnetic field lines in certain localized regions of the specimen relative to the bulk sample. These shape or demagnetizing effects must be considered and accounted for in superconducting applications operating in the presence of magnetic fields.

Consider now the magnetic properties of a type I superconductor with a nonzero demagnetizing factor n. To keep the mathematics as simple as possible, a spherical geometry will be assumed. A demagnetizing factor n assigned to any superconducting body is defined by $n = (H_i - H)/M$, where H is the applied field and H_i is the field strength inside the specimen. The value of n ranges from $0 \rightarrow 4\pi$ or $0 \rightarrow 1$ (cgs/mks). The effective value of B inside such a body is the total flux passing through the specimen divided by its maximal cross-sectional area. However, we must now consider that to satisfy the electromagnetic criteria when the applied field is sufficiently high, the sample will need to break up into domains of normal and superconducting regions, each having the respective areas X_n and X_s. This is called the *intermediate* state, which must be distinguished from the mixed state (discussed later for type II superconductors). In a type I superconductor, $B = 0$ inside any superconducting region (except in the magnetic penetration depth at the surface). For the sample, B is zB_n, where B_n is the local flux density in the normal regions, and $z = X_n/(X_n + X_s)$ is the fractional cross-section which is in the normal state. The local magnetic field in all regions of the specimen should not exceed the critical magnetic field H_c or the sample can enter this intermediate state with alternating regions of normal and superconducting material.

2.6.2 Surface energy

To qualitatively understand many of the magnetic properties of superconductors, we must introduce the concept of a surface energy that can exist between the normal and superconducting (n/s) phases. The Gibbs energy will now be either increased or decreased by the additional contribution coming from the surface energy. The magnitude of the surface energy per unit area depends on the material and its corresponding relevant length scales of the magnetic penetration depth λ_L and its coherence length ξ, which will be discussed in the next section. If the surface energy is positive, the Gibbs energy is minimized by decreasing the area at the n/s boundary. The converse is true for a material with a negative surface energy, where energy is released on forming an n/s boundary. If the Gibbs energy is lowered by forming an n/s boundary (negative surface energy), then the superconductor will want to split into a large number of n/s interfaces to increase the area at the n/s boundary. This will occur even in the absence of a demagnetizing field at fields $<H_c$.

2.7 Coherence length

2.7.1 History

Pippard (1953) introduced the concept of a coherence length ξ while studying the nonlocal generalization of London's equation. This concept of a coherence length was later used to explain the origin of the surface energy at the normal/superconducting boundary. Pippard reasoned that the density n_s of the highly ordered superelectrons cannot change rapidly with position. Instead, n_s can only change appreciably within a distance he called the coherence length. Loosely speaking, the coherence length can be viewed as the spatial distance over which the superconducting state (i.e., electron–electron pairing length scale) can exist.

2.7.2 Magnitude

Pippard demonstrated how the concept of a coherence length may be derived from thermodynamic principles; however, its magnitude can be estimated by using an uncertainty principle argument. The electrons that can play a role in the superconductivity have an energy within $\sim kT_c$ of the Fermi energy, where $k = 1.38 \times 10^{-23}$ J/K is Boltzmann's constant. The momentum range of these electrons is $\Delta p \sim kT_c/v_F$, where

v_F is the Fermi velocity. Using the uncertainty relation, one determines the uncertainty in the real space position, which has a characteristic length $\xi_0 = ahv_F/2\pi kT_c$ where a is a numerical constant of order 1, to be determined for each superconducting material, and $h = 6.63 \times 10^{-34}$ J-s is Planck's constant. For pure elemental type I superconductors, the coherence length is quite large: $\xi_0 \sim 1000$ nm. For high-temperature superconductor (HTS) materials, the coherence length is quite short: $\xi_0 \sim 1-4$ nm in their *ab*-plane and less than 1 nm along the *c*-axis (see below).

2.7.3 Dependence on purity

An intrinsic characteristic of the coherence length and the penetration depth is their dependence on the purity of the material or essentially the electron mean free path ℓ. The coherence length in a perfectly pure material (denoted as ξ_0) is larger than in an impure specimen (denoted as ξ). We define both a "clean" and "dirty" limit of a superconductor, which depends on the magnitude of ℓ and ξ_0. In the clean limit, $\ell >> \xi_0$, but in the dirty limit, $\ell << \xi_0$. When a superconductor is very impure or in the dirty limit, then $\xi \sim (\xi_0 \ell)^{1/2}$ and $\lambda \sim \lambda_L (\xi_0/\ell)^{1/2}$ so that $\lambda/\xi \sim \lambda_L/\ell$. Because ξ_0 is so small in HTS materials, they tend to be in the clean limit.

2.8 Type II superconductors

2.8.1 History

For many years, anomalous magnetic behavior—that is, behavior inconsistent with the understanding of type I superconductors—was observed in superconducting alloys and impure metals. These anomalies were ascribed to impurity effects. In 1957, Abrikosov (1957) published a groundbreaking paper explaining these anomalies as inherent features of another type of superconductor with a radically different set of properties. This other class of superconductivity became known as type II superconductivity (or, formerly, "hard" superconductors).

2.8.2 Ginzburg–Landau parameter

The criterion as to whether to classify a material as either type I or type II can be explained in terms of the surface energy—that is, whether the surface energy is positive or negative. Both ξ and λ provide separate contributions to the surface energy by raising or lowering the free energy of the superconductor relative to the normal (i.e., nonsuperconducting) region. Due to the presence of the ordered super-electrons, the free energy density of the superconducting state is lowered by an amount $G_n - G_s$. However, because the superconducting region has acquired a magnetization which cancels B inside, there is a positive magnetic contribution to G_s equal to $H_c^2/8\pi$ or $\mu_0 H_c^2/2$ (cgs/mks). At the boundary of the normal/superconducting phases, the degree of electron pair order rises over a distance ξ; however, counteracting this is the increase in G_s stemming from the magnetic contribution, which rises over a distance of λ. Thus, the order introduced by entering the superconducting state decreases the free energy G_s; however, because the material transitions from essentially a nonmagnetic state to a perfect diamagnet with relative permeability $\mu_r = M/H = -1$, the presence of a magnetic field increases the free energy G_s.

Abrikosov developed the theory of type II superconductors (Abrikosov, 1957) based on the famous phenomenological theory of Ginzburg and Landau (1950). In this theory, the ratio of λ/ξ, defined as κ, determines the value of the surface energy. If κ is less than $1/\sqrt{2}$, then the surface energy is positive; if κ is greater, the surface energy is negative.

2.9 The mixed state: H_{c1} and H_{c2}

2.9.1 Quantized flux lines or vortices

If the surface energy between the normal and superconducting phases is negative, then it is energetically favorable for the superconducting body to split into a large number of normal regions whose boundaries lie parallel to the applied magnetic field. This is called the mixed state. In the normal regions, the material is no longer superconducting (i.e., perfect diamagnetism) and the external applied magnetic field can penetrate the superconductor in this region. The extent of the magnetic field decays over a distance of λ_L (see Figure 2.4, left). The normal regions form in the shape of cylinders (often referred to as normal cores), threading the bulk of the superconductor. The cylindrical shape is preferred because of its high ratio of surface area to volume. The normal cores have a very small radius ($\sim\xi$) because of their high ratio of surface area to volume. The magnetic flux in the normal core, having the same direction as the external applied magnetic field, is generated by a vortex of persistent supercurrent that circulates around the normal core with a sense of rotation that is opposite to the bulk diamagnetic response of the superconductor (Figure 2.4, middle). The radius of the vortex is of order λ, and because λ is much larger than ξ in HTS materials, the current vortex

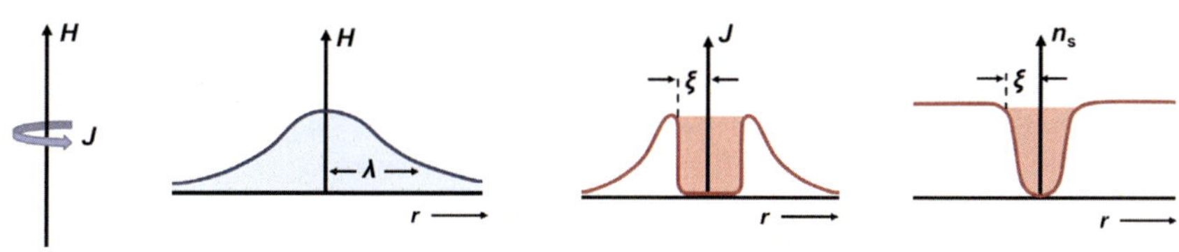

Figure 2.4 Structure of a flux line or vortex in a Type II superconductor, with contributions from positive surface energy with corresponding length scale λ and negative surface energy with its corresponding length scale ξ. In the far left, the external applied magnetic field penetrates the bulk of the superconductor as a flux line carrying a flux quantum, which is maintained by a circulating persistent current $J(r)$. In the middle left, the magnetic field created by the persistent current J extends out a radius λ. In the middle right, the normal core extends a radius ξ with the hole in the number n_s of superconducting electrons also extending a radius ξ (far right).

spreads far outside the tiny normal core. The value of magnetic flux threading the normal core is quantized in discrete units given by $\Phi_0 \sim 2.07 \times 10^{-7}$ G cm (Kittel, 1986).

In this chapter, we use the terms *flux lines* and *vortices*, and occasionally *fluxons* or *fluxoids*, interchangeably to denote these remarkable quantized linear structures that can thread type II superconductors in their mixed state.

The picture of the cylindrical normal core is only approximate; however, because it does not have sharp and fixed boundaries. As discussed in Section 2.7, the number n_s of superconducting electrons falls over a distance $\sim\xi$ (Figure 2.4, right). The concept of the normal core with the applied magnetic field penetrating its center is extremely important in the design and fabrication of practical superconductors, for if the normal cores move or are set in motion, this can result in a time-changing magnetic flux. From Maxwell's equations, a time-changing magnetic flux in turn results in an electric field and hence unwanted dissipation in the superconductor. Thus, a major task in processing practical superconductors is to find ways to pin the flux lines in place.

2.9.2 The Abrikosov lattice

The vortex persistent current circulating around the normal core interacts with the magnetic field produced by its neighboring vortices; as a result, the vortices have a mutual repulsion force between them. Because of this mutual interaction, the vortices threading an ideal (undefected) type II superconductor arrange themselves into a periodic array at high enough fields, and they have been directly observed by a variety of decoration or electron microscopy techniques (see Figure 2.5). This periodic array is known as the fluxon, vortex, or Abrikosov lattice. The energetically most favorable configuration for the vortex lattice in an isotropic and perfect Type II superconductor is a triangular close-packed lattice (Abrikosov, 1957). The distance between vortices, known as the vortex lattice parameter a, is often less than 10^{-5} cm and is equal to $(2\Phi_0/\sqrt{3}B)^{1/2}$. Note that in a few materials the vortex lattice is known to be square, presumably because of intrinsic anisotropy in the material.

The mixed state is bounded by two critical field strengths H_{c1} and H_{c2}. Below the lower critical field strength H_{c1}, the material is fully superconducting ($B = 0$), except for the magnetic penetration depth at the surface, and quantized vortices are excluded from the bulk. Above the upper critical field strength H_{c2}, the material is fully normal.

An approximate value for H_{c1} can be obtained by considering the value of the Ginzburg–Landau parameter κ. The mixed state will be energetically favorable if $H = H_c\xi/\lambda$, where H_c is the thermodynamic critical field. This gives a value for $H_{c1} \sim H_c/\kappa$. As κ increases, the relative ratio of H_{c1} to H_c decreases, and with the high values of κ in HTS materials, H_{c1} is very small.

An estimate of H_{c2} can be made by examining the properties of ξ. Recall that ξ is the length scale over which superconductivity may rise or fall from the normal state. At H_{c2}, the normal cores are packed together as densely as ξ will allow. Each core admits one quantized unit of flux, $\Phi_0 \sim 2.07 \times 10^{-7}$ G cm (Kittel, 1986, see Section 2.13); thus $H_{c2} \sim \Phi_0/2\pi\xi$ (Kittel, 1986), where $\pi\xi^2$ (Kittel, 1986) is the cross-sectional area of the normal core.

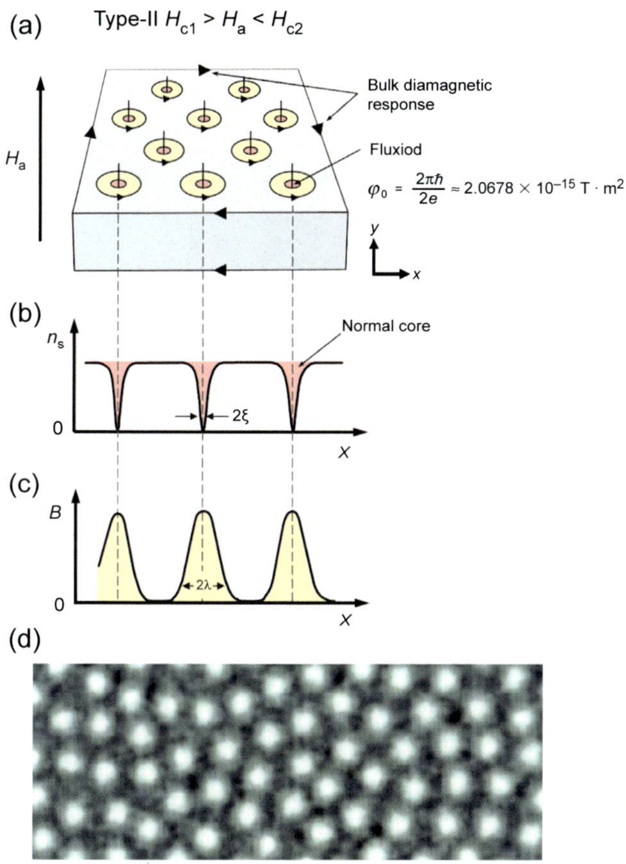

Figure 2.5 (a) The mixed state of a type II superconductor in an applied field $H_a > H_{c1}$, with normal cores threading the bulk of the material. The surface current flowing around the periphery maintains the bulk diamagnetism. (b) Variation with position of the concentration of superconducting electrons. (c) Variation of the magnetic flux density. (d) Plan view of a flux line or vortex lattice in an HTS material, revealed by electron microscopy. On average, each flux line is surrounded by six other flux lines, as predicted by the Abrikosov theory. However, many defects in the lattice are also evident, with some flux lines having only five nearest neighbors. These defects are caused by random pinning of the flux lines on structural defects within the material.
Figures 2.5a–c adapted from Rose-Innes and Rhoderick (1978, p. 189).

The mixed state of a type II material should not be confused with the intermediate state, which can exist in either type of superconductor. The appearance of the intermediate state depends solely on the shape of the superconducting body and is caused by the demagnetizing field changing the relative magnetic field strength throughout the specimen. The mixed state, however, is an intrinsic property of a type II superconductor and would appear even if a body has no demagnetizing factor (e.g., a long thin rod). The mixed state appears in type II superconductors because of their negative surface energy. In addition, the structure of the intermediate state has very coarse features and dimensions that could be visible with the naked eye. The mixed state's dimensions have a much finer scale ($\sim 10^{-5}$ cm).

2.10 Reversible magnetization in type II superconductors

The magnetic properties of an ideal, nondefected type II superconductor are described by three regions of applied field. Below H_{c1}, a type II superconductor behaves identically as a type I superconductor; that is, it has a perfect Meissner effect with $M = -H/4\pi$, $M = -H$ (cgs/mks). Above H_{c1}, quantized flux lines or vortices penetrate into the material, so that B in the material is no longer zero and M decreases. For $H_{c1} < H < H_{c2}$, the number of vortices and their distribution in the sample are determined by the mutual repulsion among the vortices and the magnitude of the applied field, and an Abrikosov lattice is formed. An electron micrograph of a flux line or vortex lattice is shown in Figure 2.5d.

As H increases toward H_{c2}, the normal cores pack closer together, increasing the average value of B in the superconductor. Near H_{c2}, B and M change linearly with H. At the value H_{c2}, in conventional theory (see Section 2.11 for modifications in HTS materials), there is a discontinuous change in the slope of the magnetization curve. Above H_{c2}, the material is in the normal state, which is virtually nonmagnetic. A typical magnetization curve for an ideal, undefected type II superconductor is shown in Figure 2.6.

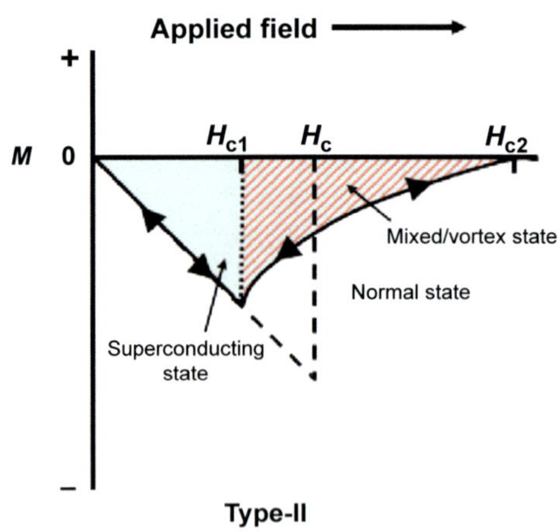

Figure 2.6 Typical magnetization curve for an undefected type II superconductor. For $H < H_{c1}$, the sample is in the superconducting state exhibiting perfect diamagnetism; for $H_{c1} < H < H_{c2}$, the sample is in the mixed/vortex state; and for $H > H_{c2}$, the sample returns to the normal state. Adapted from Kittel (1986, p. 324).

However, this picture is dramatically changed in HTS materials, where thermal activation destroys the long-range order of the Abrikosov lattice above a certain temperature or magnetic field, creating a huge region of the phase diagram in the $H-T$ plane in which the flux lines or vortices exist in a "liquid-like" state. This novel phase diagram, illustrated schematically in Figure 2.13, is discussed in greater detail in Section 2.11.

2.11 Critical currents and irreversible magnetic properties of type II superconductors

Owl explained about flux pinning *and* creep. *He had explained this to Pooh and Christopher Robin once before, and had been waiting ever since for a chance to do it again, because it is a thing you can easily explain twice before anybody knows what you are talking about…*
(After A.A. Milne, Winnie-the-Pooh, *in Blatter, Feigel'man, Geshkenbein, Larkin, & Vinokur, 1994)*

Critical currents and irreversible magnetic properties in type II superconductors involve multiple complicated mechanisms. Unlike T_c, H_{c1}, and H_{c2} which are intrinsic properties of superconductors, J_c tends to be an extrinsic property controlled by defects that are greatly affected by the fabrication process. Furthermore, especially in the new HTS materials, J_c is affected by thermal activation leading to flux creep and a power-law $I-V$ curve, so that J_c must now be interpreted as a quantity determined by an arbitrary electric field criterion. The irreversible magnetic properties of type II superconductors are in turn controlled by the critical current density and are explained, at least to a first approximation, by the so-called Bean critical state model (Bean, 1962, 1964). This section introduces these complex phenomena in greater, although still minimal, detail; the full picture can be found in the wonderful reviews by Campbell and Evetts (1972) and Blatter et al. (1994). In particular, thermal activation effects and flux creep have revolutionized our understanding of the classical theory of superconductors and have a major impact on properties relevant to applications.

2.11.1 Critical currents and pinning in defected type II superconductors: the T = 0 limit

2.11.1.1 Flux line gradient

Except for the most perfect specimens, the Silsbee criterion (Silsbee, 1916), discussed previously, is largely irrelevant when determining the critical current density J_c of type II superconductors because J_c is overwhelmingly determined by defects. This is fortunate because the defect-controlled J_c can be far higher, and current can flow through the entire bulk of the material, not just on the surface as in type I materials. The higher J_c values enable practical HTS superconductors for electric grid applications.

One can see how defects in a superconductor come into play from Ampere's law, which in vector notation and mks units is $\nabla \times \boldsymbol{B} = \mu_0 \boldsymbol{J}$, and in a one-dimensional form is $dB/dx = \mu_0 J$. Thus, any bulk current flow or current density J must correspond to a gradient in B. In a type I superconductor, the resistanceless current only flows on the surface of the material within a penetration depth λ_L. However, in a type II superconductor in the mixed state, resistanceless current can flow throughout the bulk if two conditions are met: (1) flux and hence B can be introduced into the volume of the superconducting body so that it is filled with flux lines and (2) there is a gradient in the flux line density.

However, because the equilibrium state of the flux-line lattice in a perfect undefected superconductor has uniform density (the Abrikosov lattice), the only way to sustain a gradient is through defects that pin the flux lines in potential wells, with defects of a size comparable to the coherence length and the normal core of the flux line being most effective. The pinning is theoretically understood from the reduced free energy at the defect.

2.11.1.2 Maximum pinning force and critical current density

As current density through a type II superconductor is increased, the flux-line gradient grows steeper according to Ampere's law, creating ever-stronger forces to push the flux lines out of their defect potential wells. When that force exceeds the maximum gradient in the potential energy of the well (i.e., the maximum pinning force), the flux lines are pushed out of the well. A crude analogy is a washboard holding marbles; as the washboard tips and the point is reached where no barrier remains, the marbles roll out. At $T = 0$, this is the point that determines the critical current density J_c. The maximum pinning force can be written as $\boldsymbol{F}_p = \boldsymbol{J}_c \times \boldsymbol{B}$, where B is proportional to the local flux line areal density. Thus, the magnitude of the pinning force can be determined from a measurement of J_c in the presence of a given average flux level B. An example of data on an HTS second-generation yttrium barium copper oxide (YBCO) wire is shown in Figure 2.7 with two different concentrations of the impurity Zr, which affects the defect concentration in the material through the spontaneous formation of $BaZrO_3$ defect columns during film growth.

2.11.1.3 V–I curve and pinning optimization

If the applied current density exceeds J_c, the flux lines move; thus, by Lenz's law, they generate voltage. The superconductor V–I curve looks like a hockey stick: zero voltage up to a critical current, and then an abrupt increase in voltage. Such a critical current is easy to measure. We note that an inhomogeneous distribution of J_c in a material can smear out this sharp V–I curve, and the right kind of distribution can cause a power law dependence $V \sim I^n$, which is often observed in superconductors (Warnes & Larbalestier, 1986). Of course, in this case, the power or index value n is expected to depend strongly on the particular inhomogeneity distribution. Here, we discuss another mechanism for the power-law V–I curve, which arises from flux creep and is particularly relevant to HTS materials.

Given the importance of high J_c in practical applications, it is no surprise that many studies address both the theoretical determination of J_c for different types of defects, as

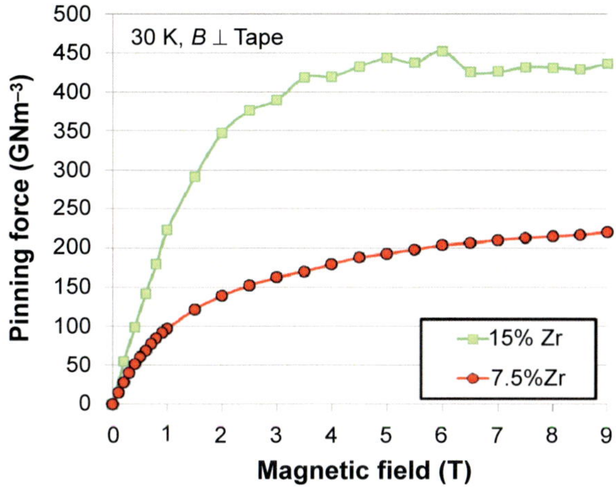

Figure 2.7 Pinning force density (GN/m^3) versus applied magnetic field at $T = 30$ K with field perpendicular to the plane of YBCO-coated conductor tape. In these chemical vapor-deposited films, extra Zr forms columns of $BaZrO_3$, which are strong pinning centers.
Courtesy of V. Selvamanickam, University of Houston.

well as the experimental optimization of defects to pin the flux lines most strongly and increase J_c (Blatter et al., 1994; Campbell & Evetts, 1972). It is a complex problem because interactions between flux lines come into play, causing J_c to be determined by collective phenomena. Therefore, it is not necessary to pin every flux-line core at an impurity site for all flux lines to remain stationary in the face of a flux-line gradient.

Tremendous progress has been made, both theoretically and experimentally; now practical materials, including HTS materials like YBCO, carry huge current densities (Malozemoff et al., 2012). Much of the pinning force in these materials arises from naturally occurring dislocations, crystal twin boundaries, stacking faults, chemical precipitates, voids, and the like. A significant research and development (R&D) effort also goes into developing artificial pinning centers, such as patterning defects into substrates, which are then replicated in deposited HTS films. A large literature also addresses irradiation by protons, neutrons, and heavy ions—damage from which can significantly increase both the magnitude of the pinning force F_p and the number of these pinning centers (Civale, 1997; Weber, 2011). Disorder in the location of pinning centers is a critical factor in determining F_p, leading to the breaking up of the Abrikosov lattice into the famous Larkin-Ovchinnikov (Larkin & Yu., 1964) and Fulde-Ferrell domains (Fulde & Ferrell, 1964). This creates a veritable "zoology" of different types of glassy behavior, usually referred to as a *vortex glass* (Blatter et al., 1994).

2.11.1.4 Lorentz force

Another way to understand this phenomenon is in terms of the so-called Lorentz force per length $\boldsymbol{F}_L$, which acts on each vortex in the presence of bulk current density J. Note that the locally averaged magnetic induction field $\boldsymbol{B}$ is proportional to the areal flux line density. Thus, $\boldsymbol{B} = n\boldsymbol{\Phi}_0$, where n is the number of flux lines per unit area and $\boldsymbol{\Phi}_0$ is taken as a vector pointing in the direction of flux lines and hence of $\boldsymbol{B}$. Then, from Ampere's law, one can derive the elegant vector formula:

$$\boldsymbol{F}_\mathrm{L} = \boldsymbol{J}x\boldsymbol{\Phi}_0 = J\Phi_0 \sin\theta \tag{2.1}$$

The vector relationship follows the right-hand rule, so that the force is perpendicular to the plane defined by $\boldsymbol{J}$ and $\boldsymbol{\Phi}_0$. In the scalar form of the formula, θ is the angle between $\boldsymbol{J}$ and $\boldsymbol{\Phi}_0$; therefore, for applied fields perpendicular to the axis of a wire, this reduces to $\boldsymbol{F}_\mathrm{L} = J\Phi_0$. If the pinning force per length F_p is the maximum slope of a defect's free energy well, the criterion for pinning in the presence of a current density J passing through the sample is $F_\mathrm{p} > F_\mathrm{L}$.

2.11.2 Bean critical state model

2.11.2.1 Magnetic moment and magnetization

The penetration of flux lines into defected type II superconductors gives rise to fascinating and important irreversible magnetic behavior.

It is useful to review the basic definitions of magnetic moment and magnetization. The magnetic moment m is the product of a loop of current I and the area A it surrounds ($m = IA$). In a ferromagnet, for example, each magnetic atom carries a spin, which is effectively a loop current. However, because each atom carries the same spin, all the local currents, when averaged over many spins inside the material, cancel out and only a net surface current flows, as shown schematically in Figure 2.8 (left). That surface

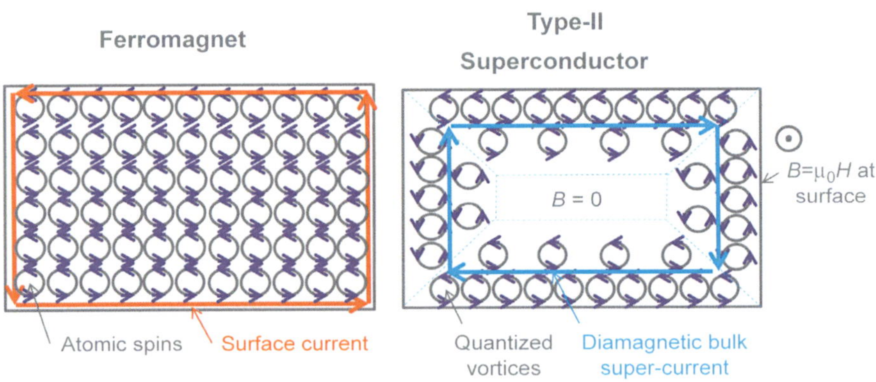

Figure 2.8 Schematic current flows in a ferromagnet and type II superconductor. In the ferromagnet, atomic spin currents uniformly distributed cancel out in the interior but create a net surface current that, when multiplied by the area, gives the magnetic moment. In a type II superconductor, quantized flux lines with their vortex current loops penetrate into the bulk, but with a density gradient that corresponds by Ampere's law to a diamagnetic current at the critical current density. If the external field H is not too high, a region in the center remains screened ($B = 0$).

current multiplied by the area of the sample determines m, and the magnetization M is just the magnetic moment m per unit volume.

2.11.2.2 Basic principles of the Bean model: the sand-pile analogy

Consider now the cross-section of a long rectangular superconductor rod with an applied field H applied parallel to its axis. For the case of HTS materials, H_{c1} is small and we ignore it in the present discussion. By continuity of the surface field, the flux line density B at the rod surface must equal $\mu_0 H$. Therefore, flux lines parallel to the rod axis enter the superconductor; when viewed end-on, in the cross-section of Figure 2.8 (right), each flux line is surrounded by a vortex of current, analogous to the spin current of ferromagnetic atoms. Because this penetration of flux lines into the material creates a flux gradient, a corresponding current must, by Ampere's law, flow perpendicular to the flux lines. Indeed, as can be seen by examining neighboring local loops in Figure 2.8 (right), the local currents do not cancel out as in the ferromagnet, but because of the flux line density gradient, a net local current flows. This net of the local vortex currents is in fact the physical source of the bulk current flow in type II superconductors. Furthermore, because flux lines penetrate from all sides, this current flow forms a loop, as indicated schematically by the blue arrows in the figure. The current flow direction is such as to create a moment opposing the applied field; in other words, this is a diamagnetic current that effectively shields the sample interior, although not as completely as the Meissner current of a type I superconductor.

As long as the average slope dB/dx of the flux front exceeds $\mu_0 J_c$, the Lorentz force of the current pushes flux lines out of the material's pinning wells, causing the flux front to advance and the flux gradient to decline. This process only stops when the flux gradient dB/dx equals $\mu_0 J_c$, at which point the defect potential wells just balance out the Lorentz force. If the applied field is increased, once again more flux lines must enter the material, and the flux front advances until it stabilizes a bit further in, but again with the same slope $\mu_0 J_c$.

This physical process, with the simple and fundamental insight that the flux gradient always stabilizes at $\mu_0 J_c$, and with any modifications to the flux configuration initiated only by applied field changes at the surface, is the essence of the famous Bean model (Bean, 1964; Campbell & Evetts, 1972). This model provides a simple and intuitive way to calculate magnetic properties of defected type II superconductors. Another well-known analogy to this irreversible flux penetration process is the so-called sand-pile model. Think of a child's sandbox with sand added at the edges. Loose and dry sand has a characteristic slope, like a talus slope. As sand is added at the edge (analogous to increasing the external applied field), the sand advances and once again stabilizes at the characteristic slope. If sand is removed from the high side of the slope (analogous to decreasing the external applied field), a new slope forms with the opposite inclination and gradually eats in to the original slope as more and more sand is removed. Once this physical picture is understood, one can begin to apply the model to a tremendous diversity of important situations.

2.11.2.3 Magnetic hysteresis loop of a type II superconductor tape

In the case of Figure 2.8 (right), with the "sand" coming in from all four sides, it is immediately evident that a kind of "picture frame" configuration develops, as suggested by the dashed lines in the figure, surrounding a flux-free region in the center. Let us now consider the case where the width w is much larger than thickness d, so that the rod looks like a tape. The magnetic properties are then dominated by the tape width; although the ends still play an important role, as this is where the current "turns around," they can be ignored to a good approximation.

The field and current directions in the tape are shown schematically in Figure 2.9(a), with a set of flux penetration profiles with increasing applied field shown in Figure 2.9(b) as a function of position through the thickness d. From Ampere's law, it is immediately evident that the distance of flux front penetration from each surface is just $x = H/J_c$. When

$$H_{\text{fullpenetration}} = J_c d/2 \tag{2.2}$$

the flux fronts meet at the center, fully penetrating the sample. Because the current now flows in the volume of the superconductor, to evaluate the total moment, it is necessary to integrate over all the current loops as a function of their positions from the edge of the sample. Equivalently, one can use the basic relationship $M = (B_{ave}/\mu_0) - H$ to derive the magnetization M by averaging over the flux penetration profile $B(x)$.

Calculating M now becomes a simple evaluation of the area between $\mu_0 H$ and the flux contour B in Figure 2.9(b). The result for $M(H)$, starting from flux-free material, is as follows:

$$M = \left[\left(H^2/J_c d\right) - H\right], \quad H < J_c d/2 \tag{2.3}$$

At the full penetration field and above, it is immediately evident from the shaded area in Figure 2.9(b) that M remains constant at

$$M = -\mu_0 J_c d/4, \quad H > J_c d/2 \tag{2.4}$$

Equation (2.4) shows that M is proportional to J_c, providing a widely used magnetic method of determining J_c. Of course, this relation only holds in specific experimental configurations and under adequately high applied fields; therefore, one must make sure proper conditions apply in a given experiment.

When the field is now reduced, the slope of B at the surface reverses as shown in Figure 2.9(c), and accordingly a region of reversed current flow develops near the surface. Once the field has been reduced by $J_c d$, the magnetization $M = +\mu_0 J_c d/4$ is now positive and constant for a further reduction of the applied field, as evident from the shaded area in the figure.

The full hysteresis loop is shown schematically in Figure 2.9(d). However, in comparing to experiment, many additional effects need to be taken into account. A key one is the field-dependence of $J_c(B)$, which declines strongly with $|B|$ in most

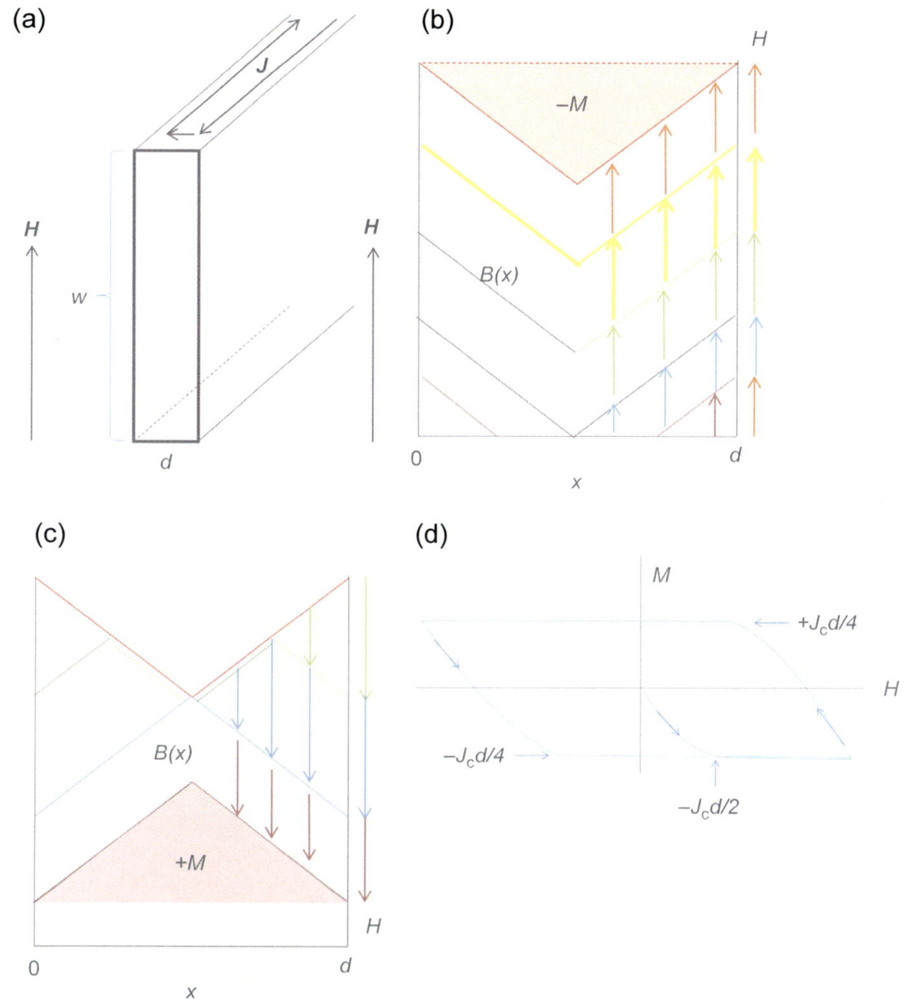

Figure 2.9 (a) Schematic orientations of applied field ***H*** and current flow ***J*** in a finite length of tape of width *w* and thickness *d*. (b) Schematic flux penetration profiles into the thickness of the tape as *H* is increased, according to the Bean model. (c) Schematic flux profiles as *H* is decreased. (d) Schematic hysteretic magnetization loop for the process illustrated in (b) and (c).

HTS materials. This creates a peak in $|M(H)|$ near $H = 0$. HTS materials are also highly anisotropic; therefore, the hysteresis loop depends strongly on the direction of *H*. Demagnetization effects must also be taken into account, such as when the field is applied perpendicular to the HTS tape plane.

2.11.2.4 Alternating current loss in the Bean model

The alternating current (AC) hysteretic power loss *P* per volume can be derived from the hysteresis loop area as a function of the peak-to-peak applied field H_{p-p}. For the simple case in Figure 2.9, at sufficiently high applied AC fields and at a frequency f, the loss is simply as follows:

$$P = \mu_0 J_c d f H_{p-p}/2, \quad H_{p-p} \gg J_c d \tag{2.5}$$

At lower fields, one finds (Bean, 1964; Campbell & Evetts, 1972) that

$$P = \mu_0 H_{p-p}^3 f/12 J_c d, \quad H_{p-p} < J_c d \tag{2.6}$$

These classic Bean model formulas have been widely used in interpreting AC experiments (see Chapter 5, Section 5.3).

The Bean model has also been applied by Norris (1970) for the case of an AC current in an elliptical or tape-shaped wire. For example, the formula for the AC loss per length of a tape-shaped wire is

$$P = (\mu_0/6\pi) I_c^2 f F^4, \quad F < 1 \tag{2.7}$$

where $F = I_{peak}/I_c$, I_c is the critical current of the wire and I_{peak} is the magnitude of the peak current. The Norris formulas are also widely used in studies of wire AC loss (see Chapter 5, Section 5.3).

2.11.3 *Critical current in defected type II superconductors at $T > 0$: flux creep*

2.11.3.1 *Anderson–Kim flux creep theory*

The phenomenon of flux creep plays a vastly more important role in HTS materials than in the conventional low-temperature metallic superconductors (LTS), primarily because of their small coherence lengths ξ and high anisotropy. This is true even at $T = 0$ because the HTS coherence length, and consequently also the barrier to flux line escape from its potential well, have atomic dimensions, opening the possibility for quantum mechanical tunneling through the barrier. Such tunneling leads to flux creep at $T = 0$ and hence the appearance of a slight voltage, even below J_c. This phenomenon has in fact been observed in YBCO (Fruchter et al., 1991). Nevertheless, it is small compared to the much larger flux creep caused by thermal activation.

At finite temperatures, thermal energy can cause flux lines to hop out of their pinning potential wells following the familiar Arrhenius activation law. Because the barrier is reduced by the Lorentz force of Eqn (2.1), the J-dependent barrier is, in the simplest model (Anderson & Kim, 1964; Malozemoff, 1991; Yeshurun, Malozemoff, & Shaulov, 1996), as follows:

$$U_J = U_0[1 - J/J_{c0}] \tag{2.8}$$

Then, the rate of decay of a current of density J circulating within a superconductor is

$$dJ/dt \sim t_0^{-1}\exp(-U_J/kT) \quad (2.9)$$

where t_0 is a hopping attempt time, T is temperature, and k is Boltzmann's constant. This leads to the famous Anderson–Kim formula for flux creep (Anderson & Kim, 1964):

$$J(T,t) = J_{c0}\left[1 - (kT/U_0)\ln(t/t_{\text{eff}})\right] \quad (2.10)$$

where t_{eff} is proportional to t_0 along with some geometrical factors. The current decay is commonly measured through the magnetization decay, using the proportionality of J and M in appropriate experimental configurations, as described above. Equation (2.10) predicts the logarithmic time dependence commonly observed in experiment.

2.11.3.2 Vortex glass theory of flux creep and current density relaxation

More sophisticated vortex glass theory takes into account collective flux-line pinning arising from the magnetic forces between flux lines, coupled with disorder in defect locations, leading to (Feigel'man, Geshkenbein, Larkin, & Vinokur, 1989; Fisher, 1989):

$$J(T,t) = J_{c0}(T)\Big/[1 + (\mu kT/U_0)\ln(t/t_{\text{eff}})]^{1/\mu} \quad (2.11)$$

where μ is a vortex glass exponent of order 1. This formula shows that the critical current density at a given timescale is reduced from the value J_{c0} without flux creep by the reduction factor in the bracket. Because of the low values of U_0 and the high temperatures, the reduction in most regions of practical interest for HTS materials is substantial—factors of 2–4.

Thus, flux creep plays a huge role in determining the critical currents of HTS materials, although sufficiently strong pinning has been developed in practical materials to still provide high enough current density values for practical applications. Flux-creep reduction of J_c is much smaller in LTS.

One may wonder what remains of the Bean model when flux creep has such a drastic effect on the apparent critical current density. In fact, if one uses the critical current density corresponding to the time scale of the measurement, the Bean model continues to be a useful approximation for the irreversible magnetic properties of type II superconductors, including the HTS materials.

It is also useful to define a normalized, unitless flux-creep rate S, where

$$S = -(1/J)dJ/d\ln t = -d\ln J/d\ln t = kT/[U_0 + \mu kT\ln(t/t_{\text{eff}})] \quad (2.12)$$

At low temperatures, S reduces to the simple Anderson–Kim result kT/U_0, showing the intuitive linear increase of normalized relaxation rate with temperature. But at higher temperatures, Eqn (2.12) reduces to.

$$S = 1/[\mu \ln(t/t_{\text{eff}})] \tag{2.13}$$

This remarkable result predicts a plateau as a function of temperature. Taking an atomic hopping time of $t_{\text{eff}} \sim 10^{-10}$ s, a measurement time $t \sim 1000$ s (as in typical magnetic relaxation measurements), and $\mu = 1$, one finds $S \sim 0.03$ (Malozemoff, 1991).

The behavior predicted by Eqn (2.12) is seen to a first approximation in HTS materials, as shown in Figure 2.10 for an YBCO single crystal (Civale et al., 1990; Thompson et al., 1993). The normalized relaxation rate S climbs linearly at low temperature and rolls over to a plateau at higher temperatures, before beginning to diverge as T_c is approached. The plateau S value is $\sim$0.022–0.026, not far from the predicted 0.03. Most remarkably, studies of many different YBCO materials, including single crystals and films with different defect distributions, all show an almost "universal" value for $S \sim 0.022$–0.026 (Malozemoff & Fisher, 1990). Although many detailed questions remain to be answered, these results provide good support for the vortex glass theory of flux creep.

An additional interesting feature in the temperature dependence of flux creep is a small maximum, which appears in the plateau region of $S(T)$ in HTS materials that have been either irradiated to create columnar defects or have intrinsic columnar defect structures, such as the $BaZrO_3$ columns in the sample of Figure 2.7 (Maiorov et al., 2009). Columnar defects are particularly strong pinning centers because they match the columnar geometry of flux lines. Nevertheless, a new flux creep mechanism

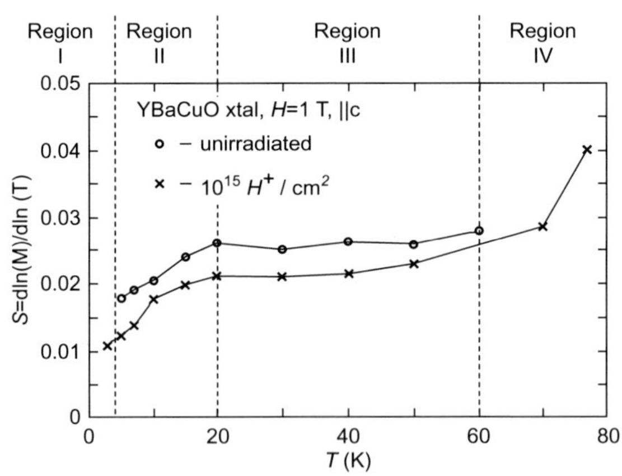

Figure 2.10 Normalized magnetic relaxation rate S as a function of temperature T for a $YBa_2Cu_3O_7$ crystal, unirradiated, and proton-irradiated. Four regions are evident. At the lowest temperatures, quantum flux creep gives a finite S. In region II, S climbs roughly linearly with T, as predicted by the Anderson–Kim thermally activated flux creep theory. In region III, a plateau in $S(T)$ is explained by vortex glass theory. In region IV, relaxation accelerates as the irreversibility line is approached.

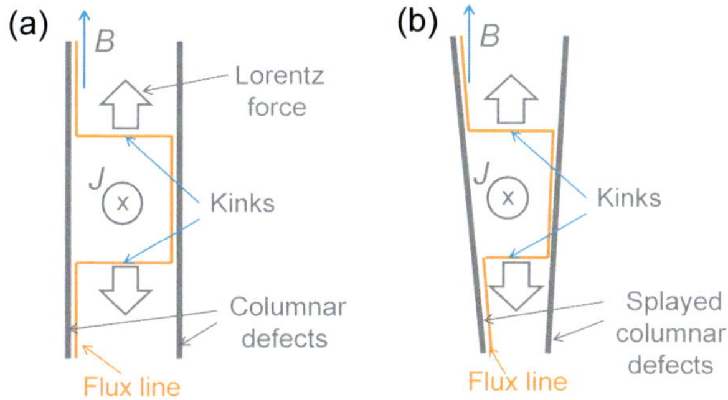

Figure 2.11 Schematic kinks in flux lines between two columnar defects. (a) Columns are parallel and there is no barrier to kink motion. (b) Columns are splayed, and an energy barrier arises from the increasing length as the upper kink moves up. The lower kink eventually experiences a barrier as well once it reaches the point where the distance between the columns (assumed to lie in different planes) begins to increase again.

appears when these defect columns are all parallel. This is because if a flux line segment is driven by the Lorentz force to jump to a neighboring defect column, kinks are formed between the two columns, as shown in Figure 2.11(a); as the Lorentz force drives these kinks to slide along the column axes, no energy barrier prevents their motion. A means to prevent this mechanism of accelerated flux creep is to splay the columns—that is, to introduce columns having a distribution of angles. This can be done in a controlled manner in irradiation experiments. It is evident from inspection of Figure 2.11(b) that the increasing length of the kink as it moves up between splayed columnar defects creates an energy barrier to its motion. Experiments with splayed irradiation confirm suppression of the $S(T)$ peak (Civale et al., 1994).

2.11.3.3 Transport properties and the electric field criterion for critical current

Another remarkable prediction of vortex glass theory is that in the collective pinning regime, $V \sim J^n$. Now by Lenz's law, $V \sim \mathrm{d}\phi/\mathrm{d}t$. Because the flux $\phi \sim M \sim J$, one can show the following (Malozemoff, 1991; Fisher, 1989):

$$S = 1/(n-1) \tag{2.14}$$

Thus, magnetic relaxation properties are intimately related to transport properties. In support of Eqn (2.14), the measured index value in many transport measurements on YBCO materials cluster around 30, which implies $S = 0.03$, remarkably close to the magnetic relaxation values. While the conventional explanation of a power law V–I curve implicates sample inhomogeneity (Warnes & Larbalestier, 1986), this novel flux-creep mechanism explains why the power law exponent n lies so often in the same range. Interestingly, studies of strained first-generation wires show the index falling with strain, implying that large enough inhomogeneity can overwhelm the flux creep mechanism and dominate the index value (Malozemoff et al., 1992). Thus, both mechanisms can contribute in a given sample.

The $V-I$ curve, with its continuous and smooth power law behavior, undermines the notion of a "critical" current density J_c. Nevertheless, because the power law is quite steep, it remains useful to identify some value of voltage per length, or electric field, to determine J_c or I_c. A useful criterion should correspond to voltage levels and time scales of a given application or measurement, and the one most widely used is 1 μV/cm. However, for magnet applications, where a much lower current is often needed to minimize heating, a criterion that is one or two orders of magnitude lower has been used. By convention, the community continues to use the terms *critical current* and *critical current density* for these admittedly criterion-dependent values.

2.11.4 Irreversibility line and the vortex liquid

2.11.4.1 Vortex glass melting

The impact of thermal activation and thermal fluctuations on type II superconductors is often evaluated from the so-called Ginzburg number $G = (kT_c/H_c^2\varepsilon\xi^3)^2/2$, which is the squared ratio of the thermal energy to the superconductor condensation energy in the volume of a Cooper pair (Blatter et al., 1994; Gurevich, 2011). Here, ε is the inverse electron mass anisotropy. For typical low-temperature superconductors, $G \sim 10^{-7}$; for HTS materials, with their low ξ and ε, $G \sim 10^{-2}$. This enormous increase in G portends many new phenomena never seen in low-temperature superconductors.

Perhaps the most significant effect arising from thermal activation is that above a certain temperature as flux creep rises to extreme levels, the disordered flux line lattice or vortex glass "melts," causing a transition into a liquid-like state often called the *vortex liquid*. In the liquid state, magnetic irreversibility is no longer possible and critical current density disappears. The melting temperature is found to depend strongly on flux density, so that an irreversibility line arises in the $H-T$ plane (examples of which are shown in Figure 2.12(b), right; Larbalestier et al., 2014). The irreversibility line $H_{irrev}(T)$ is highly anisotropic in oriented HTS materials; it occurs at ever lower temperatures, the more anisotropic the material. The line depends on the time scale or frequency with which it is measured (Malozemoff, Worthington, Yeshurun, Holtzberg, & Kes, 1988). For all HTS materials, as shown schematically in Figure 2.13, $H_{irrev}(T)$ lies far below the upper critical field $H_{c2}(T)$, where superconductivity in the form of vortices around quantized flux lines completely disappears. The irreversibility line and vortex liquid have also led to a major reinterpretation of the nature of H_{c2}: it is no longer a "critical" phase transition but rather a crossover between the vortex liquid and the fully nonsuperconducting state (Hao et al., 1991; Malozemoff et al., 1988).

Indeed, these novel phenomena have led to a major reinterpretation of superconductivity itself: Is the vortex liquid region below $H_{c2}(T)$ a superconducting state, where a "superconducting" gap with Cooper pairs and lossless vortex currents exist, but where the Meissner effect and zero resistance do not? Even below $H_{irrev}(T)$, is it really superconductivity when there is a finite voltage due to flux creep, and the flux

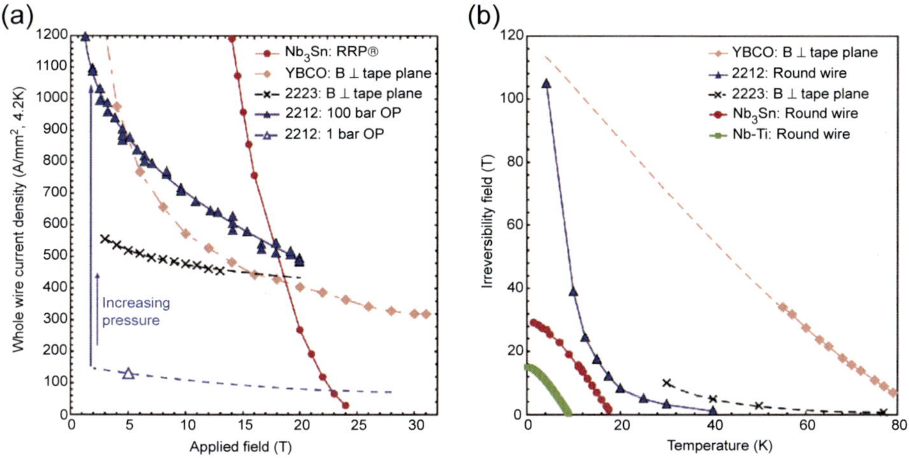

Figure 2.12 Plots of the whole-wire current density versus applied magnetic field (a) and temperature-dependent irreversibility field H_{irrev} (b) for low-temperature superconductor Nb_3Sn and NbTi round wire, HTS Bi-2212 round wire (processed under an overpressure [OP]), and Bi-2223 and YBCO flat tape with field perpendicular to the tape plane.
Courtesy of D. Larbalestier and P.J. Lee, Applied Superconductivity Center.

exclusion or diamagnetism is time-dependent, gradually (although very slowly) relaxing away? Thus, there are fundamental questions about the nature of the superconducting state and under what conditions there is truly a phase transition. In practice, the technical community continues to refer to all this behavior below $H_{c2}(T)$ as "superconducting." Whatever it is, the materials showing superconducting properties are being successfully applied in electric power applications.

The irreversibility line has major consequences for power and magnet applications at a given operating temperature, because it limits the applied field at which the material can sustain without losing its supercurrent (Gurevich, 2011; Malozemoff et al., 1988). For example, the field that HTS YBCO-based magnets can achieve at 77 K is below 10 T, whereas H_{c2} can be many times as large.

Most theories of the irreversibility line focus on a melting model (Blatter et al., 1994; Gurevich, 2011; Tinkham, 1996). In one version (Tinkham, 1996), melting occurs when the thermal energy kT equals the sum of the elastic displacement energy and the energy required to stretch a flux line against its line tension. This leads to a prediction that the flux level B at which melting occurs increases as $(T_c-T)^2$ —in other words, with an upward curving behavior as temperature is lowered below T_c. The theory predicts that the melting line lies far below $H_{c2}(T)$ in HTS materials because of their high anisotropy as well as the high temperatures (Tinkham, 1996). Other theories interpret the irreversibility line as a glassy phase transition between a vortex-glass phase below the line and a vortex liquid above. Evidence for this transition comes from scaling of the $V-I$ curves (Koch et al., 1989).

All these features are found in experiments on HTS materials: the upward curving behavior in the $H-T$ plane as shown in Figure 2.12 (although the power-law dependence on $T_c - T$ is usually found to have an exponent of ~3/2 or 4/3 rather than 2 as predicted by the melting model), melting far below H_{c2}, the frequency dependence (Malozemoff et al., 1988), and the loss of supercurrent and critical current density with a transition to flux flow resistivity.

2.11.4.2 *Flux flow resistivity*

Whenever current is driven through the vortex liquid, the Lorentz force drives the flux lines in a phenomenon called flux flow. By Lenz's law $V \sim d\phi/dt$; the moving flux generates voltage and hence a flux flow resistivity. The first theory of flux flow resistivity was made by Bardeen & Stephen (1965), who assumed vortex cores of radius $\sim\xi$ and effective area $2\pi\xi^2$ (Tinkham, 1996; where an extra factor of 2 has been added to account for dissipation outside the core) were normal regions having the full normal-state resistivity ρ_n. They determined the flux-flow resistivity ρ_f from the fractional area occupied by these cores. Because the flux line density is just B/Φ_0 and the area per flux line is then Φ_0/B, they obtained the following simple and intuitive result:

$$\rho_f/\rho_n = 2\pi\xi^2/(\Phi_0/B) = B/B_{c2} \tag{2.15}$$

using $B_{c2} = \Phi_0/2\pi\xi^2$. Because B is usually much smaller than B_{c2} in typical experiments or applications, the flux flow resistivity is small compared to ρ_n. Nevertheless, the flux flow voltage is much larger than flux creep voltage below the irreversibility line; therefore, sample heating can become significant. It should be noted that while Eqn (2.15) has been verified in some measurements, more often than not it compares poorly to experiment. Many other factors play a role, such as the energy spacings between the lowest quantum states inside the flux-line core; therefore, theoretical prediction of the flux flow resistivity remains a complex topic.

One can now visualize the full V–I curve of an HTS material, which starts with the power law behavior $V \sim I^n$, rolling over to the linear flux flow resistivity behavior $V \sim \rho_f I$, and eventually rising to the normal behavior when the current becomes so large that the gradient in flux causes some regions of the sample to reach $B = B_{c2}$. This kind of progression is of great importance for the behavior of a fault current limiter (FCL), in which a large current drives the superconductor out of its superconducting state, first passing into the flux flow regime. Because the current is limited in an FCL, heating rather than simply very high currents is what drives the sample into its normal state.

The behavior of vortex liquids is a complex topic, especially in the highly anisotropic HTS materials, including the concepts of Bose liquids, vortex liquid duality, disorder effects, and scaling behavior of the nonlinear resistance near the irreversibility line. We refer the reader to the excellent review by Blatter et al. (1994) for these fascinating topics.

2.11.5 *Upper limit on J_c*

There exists an upper theoretical limit on the maximum current that a superconductor may carry, even in the presence of infinite pinning force strength F_p. This maximum upper limit is known as the "de-pairing" current density J_d. As discussed in Section 2.5, there is a corresponding decrease in entropy and hence free energy when a material transitions from the normal state to the superconducting state. This decrease in free energy is smooth at the transition temperature T_c with no latent heat evolved. Instead, this normal-to-superconducting transition is a second-order phase transition as can be seen in heat capacity measurements as a function of temperature. In contrast to this lowering of the energy by entering the superconducting state, there is the corresponding increase in energy (i.e., kinetic energy) of the paired electrons when they are transporting current. Thus, the de-pairing current is the current at which the kinetic energy from the transport current exceeds that of the reduction in energy from entering the superconducting state and the corresponding pairing of electrons in the Cooper pair. At high-enough transport currents, it is therefore energetically favorable for the superconductor to transition back to the normal state. The change in energy during scattering is maximized when the momentum change is maximized. This transition occurs when a carrier is scattered from one point on the Fermi surface to a diametrically opposite one, in total reversal of direction. J_d can be shown to be equal to $H_c(T)/\lambda_L$. This theoretical limit of J_d can only be reached for very high F_p, with values of J_d approaching 10^{12-13} A/m^2 in HTS samples.

2.11.6 The critical surface of a type II superconductor

Now that the concepts of critical temperature T_c, upper critical magnetic field H_{c2}, irreversibility line $H_{irrev}(T)$, and critical current density J_{c0} have been discussed, all properties can be summarized with a single concept of a critical surface of a superconductor (see Figure 2.13). Shown in the figure are three axes representing temperature (y-axis), applied magnetic field (x-axis), and current density (z-axis). According to conventional theory for low-temperature superconductors, for values of T, H, or J that lie below the critical surface of $H_{c2}(T)$ and $J_{c0}(T)$, the material remains superconducting. Above the critical surface, the material transitions from the superconducting state back to its normal resistive nonsuperconducting state.

However, we have seen that thermal activation and pinning center disorder cause a drastic modification in the critical surface for HTS materials. They give rise to an irreversibility line in the $H-T$ plane which lies far below $H_{c2}(T)$, as shown schematically in Figure 2.13. Significant supercurrent can only exist below the irreversibility line, but even here a tiny voltage can persist due to flux creep. Above the line, in the vortex-liquid region, superconductivity exists in the sense that lossless vortex currents flow around flux lines and an energy gap persists throughout much of the sample. However, the Meissner effect and the characteristic near-zero resistivity to macroscopic current flow are lost. The upper critical field $H_{c2}(T)$ is no longer considered "critical" in the sense of a phase transition; it is a crossover. The most widely accepted picture is that the irreversibility line is actually a glassy phase transition between a vortex-glass phase below the line and a vortex liquid above. In the vortex glass region, the critical current is strongly suppressed by flux creep, although not enough to prevent useful applications.

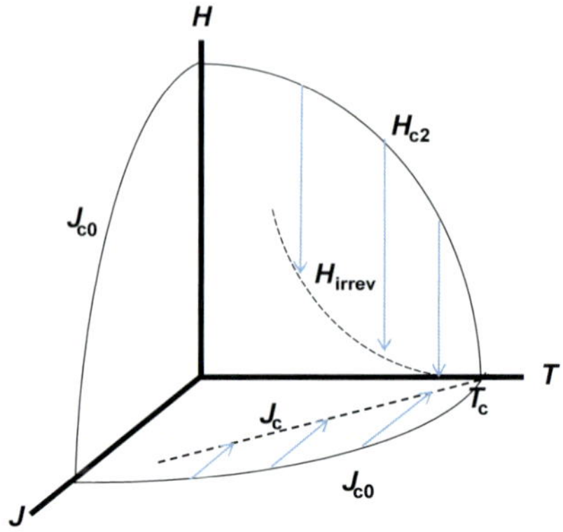

Figure 2.13 Critical surface of a superconductor in temperature T, applied magnetic field H, and current density J, below which supercurrent can persist. In low-temperature superconductors, the critical surface is very close to the upper critical field H_{c2} and the flux-creep-free critical current density J_{c0}. In HTS, thermal activation and flux creep cause a reduction in the critical surface as indicated by the blue arrows, with the surface now defined by an irreversibility line $H_{irrev}(T)$ and a flux-creep-reduced critical current density $J_c(T)$. However, a superconducting gap in the density of electronic states and vortex currents around quantized vortex lines can persist in the entire region under H_{c2}. Between H_{c2} and H_{irrev}, the flux lines are in a liquid-like state called a vortex liquid; below H_{irrev}, in the presence of disorder in pinning center locations, the flux lines are in a disordered state called a vortex glass.

Thus, for any device operating on the electric utility grid, the device must be designed with enough safety margin that for a given load line, the superconducting device remains well below the irreversibility line to retain the characteristic superconductor properties of near-zero resistance and diamagnetic screening. There are only a few HTS electrical devices that operate with exceptions to this general rule, where the sharp change in electrical resistance experienced by the superconductor when transitioning from the vortex liquid to vortex-liquid state is purposely used to "limit" currents in fixed voltage systems (e.g., the FCL or fault-current-limiting cables and transformers in Chapters 5 (Section 5.6), 9, and 12).

2.12 Entropy and free energy

In Section 2.5, it can be seen that the Gibbs free energy density in the normal state is independent of the strength of the applied field H_a. This is true because the material is essentially nonmagnetic in the normal state and acquires no magnetization in an applied magnetic field. When a type I material enters its superconducting state (i.e., perfect diamagnetism) below T_c, the application of an applied magnetic field H_a raises the free energy by $\frac{1}{2}\,\mu_0 H_a^2$. Thus, the difference in the Gibbs free energy between the normal (nonmagnetic) state and the superconducting (diamagnetic) state is given by $G_n - G_s(H) = \frac{1}{2}\,\mu_0(H_c^2 - H_a^2)$. Using the definition for the Gibbs free

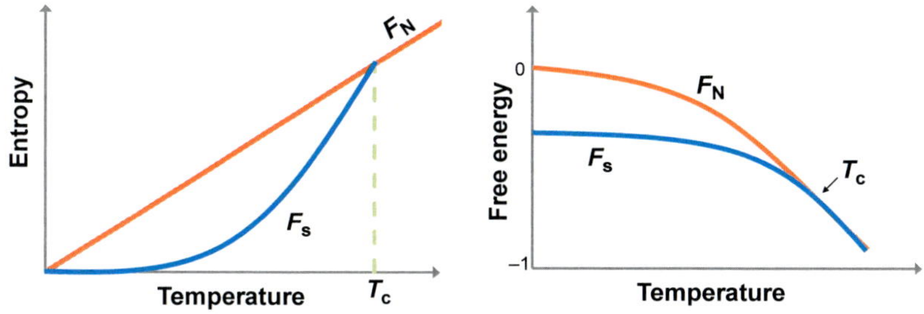

Figure 2.14 Left: A qualitative plot of the difference in entropy between the (more ordered) superconducting state and the normal state. Right: Similarly, a plot of the free energy versus temperature of a superconductor and its normal state.
Adapted from (Kittel, 1986, p. 326.)

energy of $G = U - TS + PV - \mu_0 H_a M$, where U is the internal energy, S is the entropy, P is the pressure, V is the volume, and M is the magnetization, and the relation for entropy per unit volume $S = -(\partial G/\partial T)_{p,Ha}$, it can be shown that the difference in entropy between the normal and superconducting states is given by the following:

$$S_n - S_s = -\mu_0 H_c \mathrm{d}H_c/\mathrm{d}T \tag{2.16}$$

Because $H_c(T)$ always decreases with increasing temperature, the quantity on the right-hand side of Eqn (2.16) must always be positive. This implies that there is always an associated decrease in the entropy when going from the normal state to the superconducting state. Because entropy is associated with disorder, this qualitatively demonstrates that the superconducting state is a more ordered state with lower free energy than the normal state (see Figure 2.14).

2.13 Bardeen, Cooper and Schrieffer (BCS) theory

There is no generally accepted microscopic explanation of high-temperature superconductivity. The qualitative features of the very successful microscopic theory of low-T_c metallic superconductors are outlined in this section.

In 1957, Bardeen, Cooper, and Schrieffer (BCS) presented the first fully quantum mechanical theory to completely describe the microscopic mechanism involved in superconductivity of LTS (Bardeen, Cooper, & Schrieffer, 1957). They were later awarded the Nobel Prize in Physics in 1972. As outlined by Kittel (1986), the assumptions and accomplishments of this theory were as follows:

1. The mechanism governing the elemental or intermetallic type of superconductivity is an electron−lattice−electron interaction mediated by the phonons (quantized lattice vibrations) in the material. This indirect interaction occurs when one electron interacts with the crystal lattice and deforms it. The second electron sees this deformation and adjusts itself to take advantage of the redistribution of energy. Thus, the two electrons attract each other via the

lattice deformation or phonons. A common analogy used in this phonon-mediated electron–electron interaction is that of a bowling ball placed in the center of a soft mattress with electrons moving around the outer periphery. Prior to the placement of the bowling ball in the center of the mattress, the electrons at the outer bands of the Fermi surface move around and behave as normal metals within an electrostatically screened Coulomb potential. Once the bowling ball (i.e., the phonon) is placed in the center of the mattress, its presence causes an indentation which deforms the mattress (i.e., crystal lattice). The electrons traveling at the outer bands of the Fermi surface feel this lattice deformation caused by the presence of the phonon and are hence "attracted" to one another via the phonon, forming pairs, which are called Cooper pairs. The Cooper pairs then behave in a "cooperative" fashion.

2. The attractive interaction between electrons leads to a ground state separated from excited states by an energy gap denoted by Δ. The existence of an energy gap Δ is the cause of many unusual electromagnetic and optical properties in these materials. The energy gap in a classical BCS type electron–phonon mediated superconductor is given by $E_g = 2\Delta \sim 3.5k_BT_c$. The attractive interaction assumed is a generic interaction that could be produced by any pairing mechanism. This is significant because the electron pairing mechanism for the HTSs, still highly controversial, likely involves something other than just phonons.
3. Because the BCS ground state involves two electrons, any flux is quantized in units of $2e$ rather than e. The magnetic flux threading the normal cores of a type II superconductor in its mixed state (see Figure 2.5) is quantized with this effective unit of charge. The flux quantum is defined by $\Phi_0 = hc/2e \sim 2.0678 \times 10^{-7}\ \text{G cm}^2$ or $2.0678 \times 10^{-15}\ \text{T m}^2$.

2.14 Low-temperature metallic superconductors (LTS): NbTi, Nb_3Sn, and MgB_2

No book on HTS applications in the electric grid would be complete without at least briefly mentioning LTS (Scanlan, Malozemoff, & Larbalestier, 2004). Although there is no strict guideline defining what is the cutoff temperature defining LTS and HTS, the general trend has been to call materials that appear to obey classical BCS phonon-mediated superconductivity as LTS and others, with T_c above about 30 K, as HTS. A notable exception is the well-known binary compound MgB_2, which has been discovered to be a superconductor with $T_c \sim 39$ K in zero field (Nagamatsu, Nakagawa, Muranaka, Zenitani, & Akimitsu, 2001). MgB_2, by all accounts, appears to be the ultimate strongly coupled BCS phonon-mediated superconductor, with a T_c higher than some HTS materials. Depending upon the publication, it has been included in the HTS classification based upon its relatively high T_c, but more appropriately as an LTS material based upon its electron-pairing mechanism.

Despite the tremendous progress of HTS materials since their first discovery in 1986, LTS materials still dominate the commercial industrial market. By far the two most common LTS materials widely used today are NbTi, with a T_c of ~ 10.2 K in zero B-field, and Nb_3Sn, with a $T_c \sim 18.3$ K in zero B-field. They have the simple crystal structures shown in Figure 2.15. Examples of commercial multifilamentary wires made from these materials in copper matrices are shown in Figure 2.16. Even with its relatively low T_c and $H_{c2} \sim 10.2$ T, NbTi still remains the true workhorse of the industry. NbTi wire's high ductility, ease of fabrication, and relatively low cost per

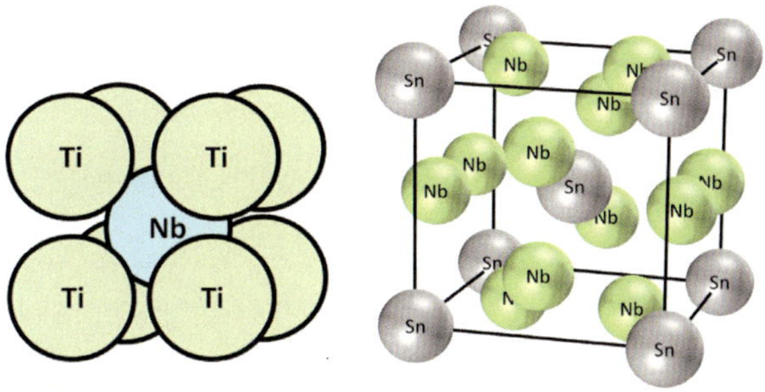

Figure 2.15 Body-centered-cubic (BCC) crystal structure of the LTS material NbTi (left) and the A-15 crystal structure for Nb_3Sn, where A site is the Nb and the B site is the Sn (right).

unit length make it the superconductor of choice in most commercial applications, with the largest applications being in magnetic resonance imaging and nuclear magnetic resonance. NbTi is also widely used in research and high-energy physics magnets. The first and so far only commercial electric power grid application that employed NbTi was superconducting magnetic energy storage for compensation of voltage dips and sags to improve power quality, but this product has been discontinued in favor of reactive power solutions (see Chapter 1, Section 1.5.5; Malozemoff, 2012). Annual production of NbTi exceeds 2000 tons/year.

For applications where either higher T_c or higher magnetic fields ($H_{c2} \sim 22-25$ T) are required, such as in fusion magnets or magnets for high-field research, Nb_3Sn is the next most widely implemented LTS conductor. There are 76 known A-15 compounds, of which 46 are known to be superconducting, with the Nb and V compounds possessing the highest T_c values (Smathers, 1990). The crystal structure A_3B is formed by a body-centered cubic arrangement of B atoms and two A atoms centered at every face (Figure 2.15, right). Nb_3Sn is a far more difficult material to work with than NbTi because of its brittleness and sensitivity to mechanical strain. For this reason, superconducting magnets fabricated with Nb_3Sn conductor are often wound prior to reaction and subsequently reacted (at >650 °C) to form the superconducting Nb_3Sn

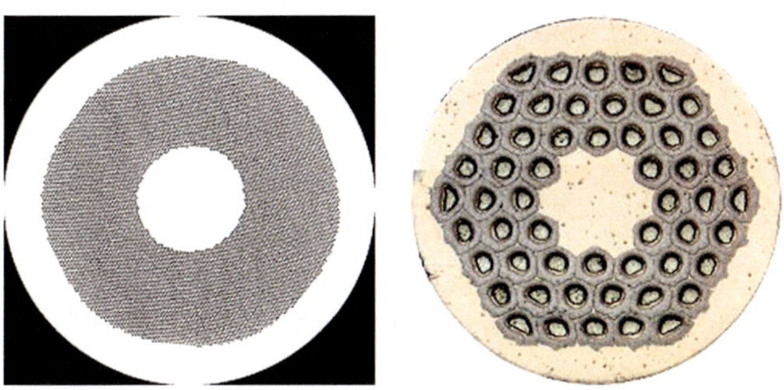

Figure 2.16 Typical ~0.8 mm diameter commercial multifilamentary NbTi wire (left) and Nb_3Sn wire (right).

phase. This type of *wind-and-react* process adds considerable cost and complexity to any superconducting system.

Nb_3Sn has experienced a virtual explosion in both R&D expenditure and production capacity. Production capacities have increased more than an order of magnitude from ~15 tons/year in 2007 to >400 tons/year (http://www.iter.org/newsline/262/1541). These advances in performance and capacity have been driven largely by the International Thermonuclear Experimental Reactor (ITER) fusion effort and high-energy physics accelerator magnet efforts required for luminosity upgrades at the large hadron collider located at the Euoprean Laboratory for Particle Physics (CERN: Conseil Européen pour la Recherche Nucléaire).

Superconductivity was discovered in the binary compound magnesium di-boride or MgB_2 by the Akimitsu group in 2001 (Nagamatsu et al., 2001). Interestingly, MgB_2 was first synthesized back in 1953 (Jones & Marsh, 1954), but its superconducting properties were not discovered until 2001. This "rediscovery" of MgB_2 proves the adage in experimental superconductivity: "If you ever have an old material sitting on the shelf somewhere, test it for superconductivity; you never know what you may find."

As stated previously, despite its unusually high T_c, MgB_2 appears to be the ultimate strongly coupled LTS with a classical BCS type electron–phonon mediated interaction. Many of the unusual electronic properties of MgB_2 appear to be explained by the presence of two different bands of electrons at the Fermi level: strongly superconducting σ-bonding and more weakly superconducting π-bonding, with the possible existence of two energy gaps (Larbalestier et al., 2001).

The commercial interest in MgB_2 stems from its relatively high T_c, its high H_{c2} (when chemically doped; e.g., with ~5% C), and its relative ease and low cost of fabrication. MgB_2 is a binary compound whose magnetic and current-carrying properties can readily be enhanced by chemical doping. It has a simple hexagonal crystal structure (see Figure 2.17) and is easily synthesized (see Figure 2.18). Round multifilamentary wires of MgB_2 can be readily made from the so-called powder-in-tube (PIT) process (see Figure 2.19), which has been successfully used in the fabrication of other round

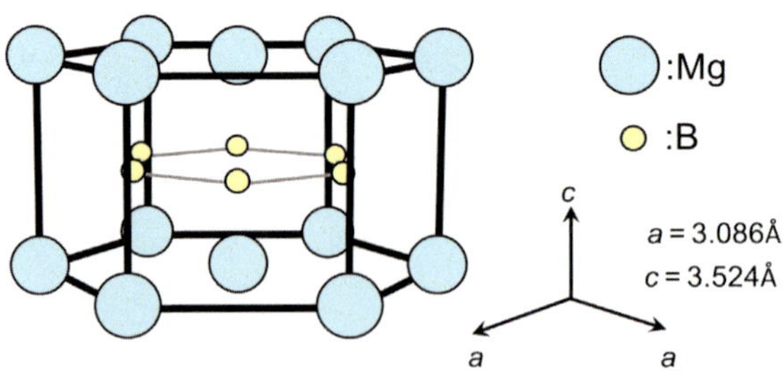

Figure 2.17 The hexagonal crystal lattice structure of the binary compound MgB_2 with lattice constants of $a = 0.3086$ nm and $c = 0.3524$ nm, with a T_c ~39 K in zero applied magnetic field. Adapted from Nagamatsu et al. (2001).

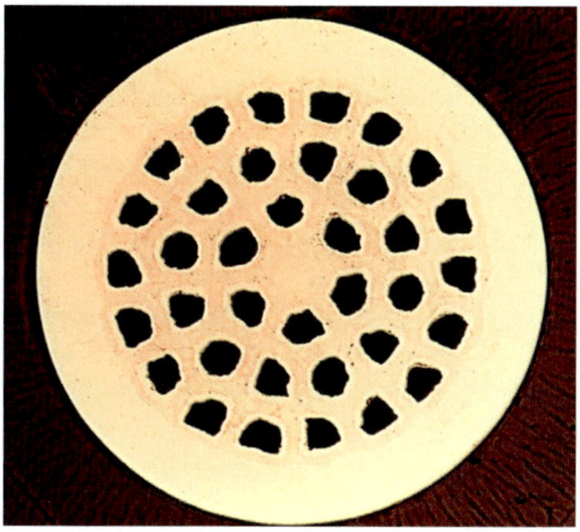

Figure 2.18 A typical commercial 36 multifilament MgB_2 round ~1 mm diameter wire with 60-μm filaments, fabricated using the powder-in-tube method.
Courtesy of M. Tomisc, HyperTech, Columbus, OH.

multifilamentary superconducting wires, such as Nb_3Sn and Bi-2212 (see Figures 2.16 and 2.18; Fluekiger et al., 2012). MgB_2 is a candidate for superconducting electric power equipment, including AC and DC power cables, motors, generators, transformers, and FCLs because of its relatively low cost of materials and fabrication, although its inability to access the LN_2 temperature range and the high refrigeration penalty for handling AC losses at lower temperatures are significant issues for AC applications such as AC cables, transformers, and FCLs.

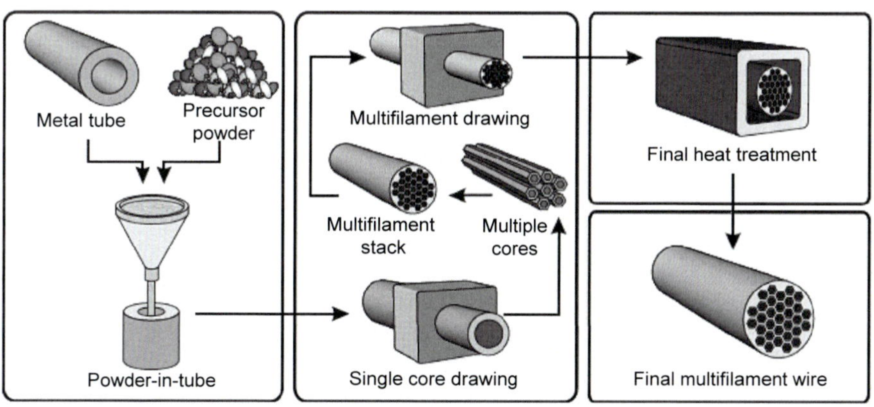

Figure 2.19 Schematic of the basic steps of the powder-in-tube (PIT) process used in the fabrication of round multifilamentary wire processing. Variants of the PIT process are used in fabrication of round MgB_2 wire, round Bi-2212 wire, and some Nb_3Sn. Combined with a rolling process, it is also used for BSCCO-2223 tape.

2.15 High-temperature superconductivity

2.15.1 History

In 1986, J.G. Bednorz and K.A. Mueller discovered high-temperature superconductivity in $LaBaCuO_x$ (Bednorz & Mueller, 1986), for which they were quickly awarded the Nobel Prize in physics. This material had a T_c onset of 35 K and about a 20 K width in the resistance between T_c onset and zero. A year later, the groups of M.K. Wu and C.W. Chu (Wu et al., 1987) discovered superconductivity in the Y−Ba−Cu−O system, in which $YBa_2Cu_3O_7$ ("YBCO-123") has a T_c onset of 93 K and a transition width <5 K (higher quality present-day materials have transition widths as low as 1 K). Almost all the rare earths can be substituted for Y with similar atomic structures and high T_c values, and these materials are often referred to as *REBCO*. YBCO-REBCO were the first materials discovered with T_c values above the liquid nitrogen boiling point of 77 K, and they have continued to be leading materials for electric power applications.

Many more families of compounds have since been discovered, with even higher T_c values. Superconductivity was discovered by Maeda et al, (1988) and Tallon et al. (1988) in the system Bi−Sr−Ca−Cu−O, with $Bi_2Sr_2Ca_2Cu_3O_{10}$ ("BSCCO-2223") having the highest T_c onset of 120 K and a 10 K width (in its typical form, BSCCO-2223 has a T_c of 110 K). The Tl−Ba−Ca−Cu−O system, discovered by Sheng and Hermann (1988), reached an onset temperature of 122−125 K with a 20 K width. To date, the Hg-Ba-Ca-Cu-O system has the highest T_c of all, reaching 135 K at ambient pressure and ~160 K under high pressure (Schilling et al, 1993).

2.15.2 Crystal structure

Because this book deals primarily with HTS electrical power applications, the discussion is limited to the following types of superconductors. The crystal structure for the $YBa_2Cu_3O_{7-x}$ compound (left-hand side) with $T_c \sim 93$ K possesses both planes of CuO_2 and chains of Cu-O (see Figure 2.20, left side). X-ray analysis shows that the $YBa_2Cu_3O_{7-x}$ compound has an orthorhombic unit cell with approximately seven oxygen atoms per formula unit. The lattice parameters of the orthorhombic cell are $a = 3.8856$ Å, $b = 3.815$ Å, and $c = 11.6804$ Å (see Figure 2.20).

The crystal structure for the $Bi_2Sr_2Ca_2Cu_3O_{10}$ compound (right-side) with T_c ~110 K has similar CuO_2 planes. Note the complexity of the HTS crystal structure relative to the LTS NbTi and Nb_3Sn materials shown in Figure 2.15. HTS tapes using the BSCCO-2223 compound in a silver composite have been successfully fabricated using the PIT method with rolling into a tape shape, with especially high critical current densities achieved under high-pressure oxygen anneals (see also Figures 2.19 and 2.21). These wires, produced commercially, are widely used in electric power applications. A more thorough discussion is given in Chapter 3.

Early attempts at using the PIT process for YBCO-123 were quite unsuccessful due to the strong suppression of critical current at high-angle grain boundaries. But now, tapes based on a coated-conductor geometry, with highly aligned YBCO films, have

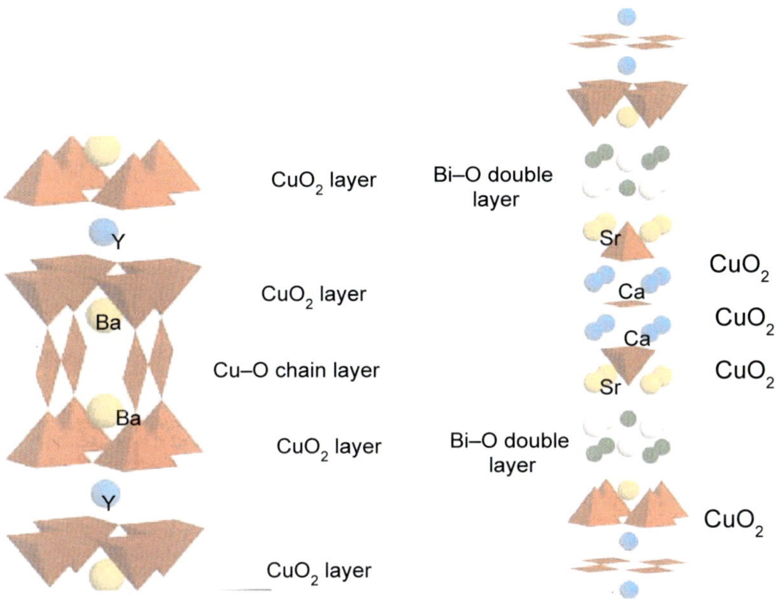

Figure 2.20 Unit cell structure of Y-Ba-Cu-O-123 (left) and Bi-Sr-Ca-Cu-O-2223 (right). In YBCO, oxygen atoms sit at the vertices of the tetrahedra and squares, and Cu sits at the center of the base of the tetrahedral and in the middle of the squares. The CuO_2 planes play a major role in the superconductivity, while the CuO chains in YBCO act as electron donors or acceptors, which are filled or emptied depending on the oxygen concentration of the material.

solved the grain boundary problem and are produced commercially with excellent properties, gaining wide use in electric power applications. A more thorough discussion of these wires is given in Chapter 4.

Another important wire for certain energy applications is round wire based on $Bi_2Sr_2Ca_1Cu_2O_8$ ("BSCCO-2212") in a silver matrix (Rupich et al., 2012). BSCCO-2212 has one less CuO_2 plane than BSCCO-2223 and a slightly lower $T_c \sim 85$ K. The ability to make round wire with this compound is based on a special feature of its phase diagram, allowing formation directly from a melt phase. Although its T_c is comparable to that of YBCO-123, its high anisotropy leads to a much lower-lying irreversibility line, constraining its in-field applications to the 20 K range and below. Nevertheless, very high engineering (full wire) critical current densities have been achieved (Larbalestier et al., 2014), and the advantage of cabling round wires rather than tapes makes BSCCO-2212 a strong candidate for high-field magnet cables. The round wires have diameters around 1 mm.

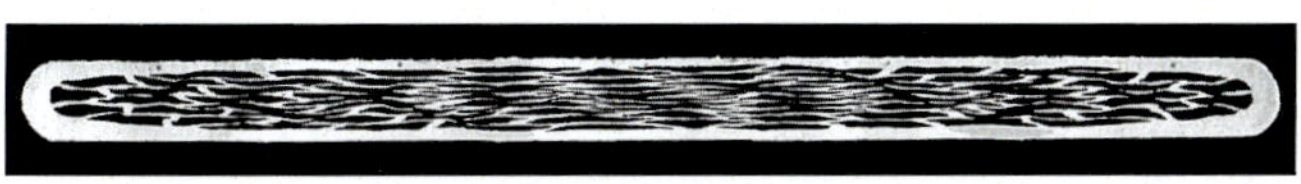

Figure 2.21 Typical multifilamentary BSCCO-2223 PIT tapes (dimension 0.3mm × 4 mm). Courtesy of Sumitomo Electric Industries, Ltd.

Even though Tl and Hg-based HTS materials have even higher T_c values than YBCO or BSCCO, they have not so far found commercial use because of their even lower irreversibility lines (due to even higher anisotropy), difficult processing, poor stability, and the toxic nature of Tl and Hg.

2.15.3 Anisotropy

One of the key characteristics of HTS materials is the large anisotropy in their superconducting properties. This anisotropy shows up in nearly every fundamental parameter, including the penetration depth λ and the coherence length ξ. Substantially different values were measured depending upon whether measurements were taken parallel to the *ab* crystal planes or parallel to the *c*-axis. This anisotropy was naturally also seen in the critical fields H_{c1} and H_{c2}.

Strong anisotropy also appears in the critical current J_c, which can be an issue in certain electric power applications. Critical currents can no longer just be reported as a function of temperature and magnetic field $J_c(T,H)$, but also have to be reported as a function of the applied magnetic field direction relative to the crystallographic axes (i.e., $J_c(T,H,\theta)$). This is very different from the more isotropic LTS materials, such as NbTi and Nb_3Sn, which show virtually no anisotropy in their fundamental parameters. Shown in Figures 2.22 and 2.23 are the critical currents versus applied magnetic field of a REBCO-coated conductor at various temperatures, with the magnetic field applied parallel to the *ab*-plane (Figure 2.22) and parallel to the *c*-axis (Figure 2.22). These figures show the large difference in J_c depending upon applied

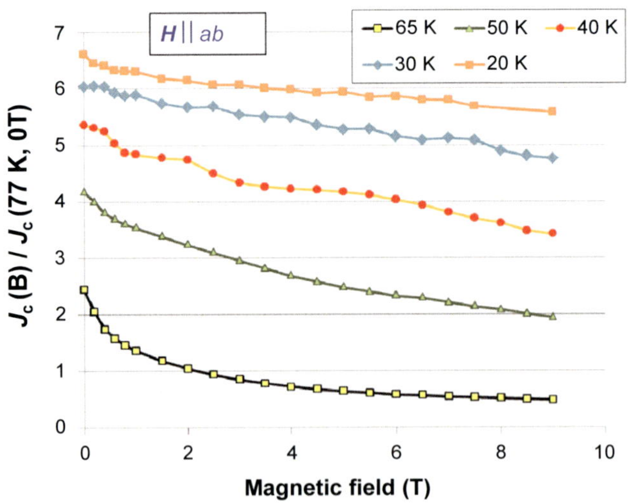

Figure 2.22 Normalized J_c versus H for applied magnetic field parallel to the *ab* plane of a REBCO-coated conductor tape at various temperatures. Results have been normalized to their values at 77 K and self-field.
Courtesy of V. Selvamanickam, University of Houston.

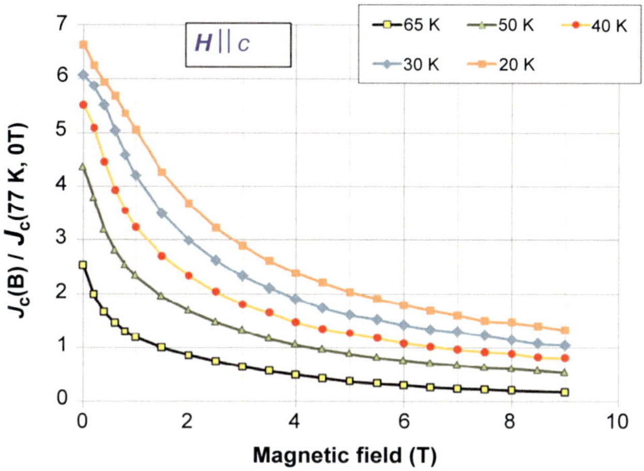

Figure 2.23 Normalized J_c versus H for applied magnetic field parallel to the c-axis of a REBCO-coated conductor tape at various temperatures. Results are normalized to the value at 77 K and self-field.
Courtesy of V. Selvamanickam, University of Houston.

field direction. The behavior at intermediate angles, shown in Figure 2.24, is also complex and has been the object of much study (Holesinger et al., 2008). As is often the practice in HTS electrical device design, $J_c(T,H)$ at a given temperature and applied magnetic field has been normalized to its J_c value at 77 K and in its own self-generated magnetic field ("self-field"). The full data on $J_c(T,H,\theta)$ are important because different

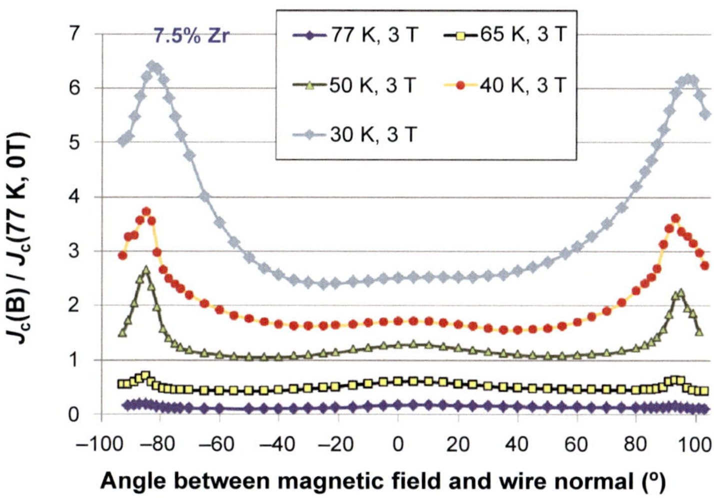

Figure 2.24 Normalized J_c versus angle θ between a 3 T applied magnetic field and a REBCO-coated conductor tape doped with 7.5% Zr at various temperatures.
Courtesy of V. Selvamanickam, University of Houston.

electrical power devices require operation at different magnetic field levels and orientations, and also at different operating temperatures.

For most magnet/coil type applications in the electric power grid (e.g., transformers, energy storage, motors, generators, FCLs), it is the lowest J_c that ultimately determines the maximum current the device can carry because magnetic fields impinge over an entire range of angles to the *ab*-planes of the superconducting tape. This lower J_c caused by the anisotropy negatively affects size (footprint), mass, and cost of the device because the lower J_c requires more turns in a magnet for an equivalent number of ampere-turns. One notable exception is superconducting cable, where the azimuthal magnetic field generated by the cable during operation is both small in magnitude and also in the most favorable direction for maximum current-carrying capacity of the HTS material (see also Chapter 5, Section 5.2.1). For the Y-Ba-Cu-O tapes, this is predominantly in the plane of the tape.

2.16 Comparison of HTS to LTS properties and summary of fundamental parameters

Numerous HTS materials have been discovered since the initial discovery by Bednorz & Mueller (1986). Likewise, there have been many theoretical models attempting to explain the microscopic behavior in HTS materials and the origin of the electron–electron pairing mechanism. A number of experiments have been performed, which indicate that the high T_c superconducting mechanism may be described by a variant of the BCS theory, but just as many theories point to an entirely new electron–electron pairing mechanism. Fortunately, the many well-developed HTS devices fabricated for the electric power grid described in the later chapters of this book have done so without the benefit or understanding of the microscopic theory of HTS. It is the phenomenological theory based on the Ginzburg–Landau equations, supplemented with a new understanding of thermal activation effects, that have been most valuable in understanding and optimizing HTS materials performance for grid applications.

A few key results from experiments pertaining to HTS are listed here:

- The flux quantum has been directly measured and confirmed to equal the theoretical prediction $\Phi_0 = hc/2e$, which indicates electron pairing.
- Measurements of electron tunneling and electromagnetic absorption have confirmed the existence of an energy gap Δ, whose magnitude at $T \ll T_c$ is larger than that predicted by BCS theory. However, if pseudo-gap and superconducting fluctuation effects suppressing T_c are taken into account, the ratio of 2Δ to kT_c is much closer to the BCS prediction (Tallon, Storey, & Mallett, 2012).
- The temperature dependence of the penetration depth has the same qualitative features as those predicted by BCS theory, but the wave-function symmetry is d-wave rather than the more conventional s-wave.
- Specific heat data show a Ginzburg–Landau second-order phase transition at T_c in zero field. At low temperatures, the specific heat data possess a linear term, whereas BCS theory predicts an exponentially small electronic term.

Table 2.1 Summary of measured parameters of LTS and HTS materials*

Quantity	NbTi	Nb_3Sn	$YBa_2Cu_3O_{7-x}$	$Bi_2Sr_2Ca_2Cu_3O_x$
T_c (K)	10	18.3	93	108–110
$H_{c1\parallel c}$(mT, $T \rightarrow 0$)			90–110	
$H_{c1\parallel ab}$(mT, $T \rightarrow 0$)			24	
$H_{c2\parallel c}$(T, $T \rightarrow 0$)	12	22–25	120	
$H_{c2\parallel ab}$ (T, $T \rightarrow 0$)	12	22–25	250	800
H_c (T, $T \rightarrow 0$)		0.52	1.1	
ξ_c (nm)	3–4	3.6–6	0.3–2(0.65)	0.6
ξ_{ab} (nm)	3–4	3.6–6	0.5–3(3.1)	
$\lambda^{\perp}$ (nm)	240	65–80	120–150	
$\lambda_{\parallel}$ (nm)	240	65–80	700–800	
$2\Delta(0)/k_BT_c$		3.6–4.4	2.2–2.8	

* For practical purposes, the LTS materials exhibit isotropic materials properties. As such, there is no meaning to "parallel" and "perpendicular"; values along all principal axes are the same, although some subtle differences have been discussed in the literature (Suenaga, Ghosh, Xu, & Welsh, 1991). All parameters are taken in the limit of zero temperature.
From Ginsberg (1989).

- High-resolution X-ray data show no change in crystal structure at T_c, indicating no first-order transition.

In Table 2.1, we have listed some of the measured fundamental parameters for several industrially used superconductors. Note the large anisotropy in many of the fundamental parameters of HTS materials, whereas most classic LTS materials are isotropic.

2.17 Practical superconductors

2.17.1 Price–performance ratio

One of the most common and effective ways to evaluate and compare the technical and economic viability of a superconducting wire or tape is to calculate its price–performance ratio, which is typically measured in dollars per kiloamp-meter or $/kA m. This metric of comparison is also useful to electric utilities when weighing the technical and economic benefits of superconductors versus conventional Cu- and Al-based technologies. In this simple calculation, the price per meter ($/m) of a superconducting wire or tape is divided by its I_c performance at a specified temperature and magnetic field. Confusion can quickly arise, however, when comparing LTS to HTS materials price—performance ratio because of two factors: (1) the extremely large differences in temperature and magnetic field magnitude at which the comparisons can be made and (2) the large anisotropy that exists in HTS materials, necessitating the magnetic field direction or angle to be specified. It is common, for example, in the high-energy physics accelerator magnet community, to state price—performance ratios for LTS NbTi at the operating parameters 4.2 K, 5 T; for the fusion magnet community, however, a more common reference point for LTS Nb_3Sn is 4.2 K, 12 T. Because these LTS materials are isotropic, no magnetic field direction is stated, which makes direct comparison with HTS difficult since the magnetic field direction for HTS materials must also be specified. The HTS community seems to have mostly settled on the convention of reporting price—performance at 77 K, self-field, and then translating that value to the operating T, B, and θ for a particular application.

In addition, these price—performance values can further vary even when all parameters of T, B, and θ are specified consistently due to variations in superconducting wire or tape architecture, such as amount of normal metal stabilizer present, monofilament versus multifilament, number and size of filaments, insulated or noninsulated, etc. The authors have specifically refrained from specifying precise values ($/kA m) for the various commercial LTS and HTS wires and tapes because of the many variables necessary for a consistent comparison, as well as the rapidly changing costs of HTS, which continue to decline with progress in HTS wire manufacturing and more widespread acceptance of these HTS-based devices. But present HTS values are still an order of magnitude or more than LTS values.

2.17.2 Why so few practical superconductors?

Since the discovery of superconductivity in 1911, thousands of materials have been discovered to exhibit superconductivity. There are more than 27 pure elements on the periodic table that exhibit superconductivity under the right conditions of low-enough temperature and high-enough pressure (Kittel, 1986). Yet, even though there are so many superconductors, there are basically only eight that have any practical use in electric power applications to date. These are listed in Table 2.2.

The question naturally arises: If there are so many superconductors, why are so few used in practical superconducting applications? While the answer is complex, it basically comes down to the fact that in order to fabricate a useful superconducting device, many practical engineering aspects need to be satisfied. For the device to operate appropriately, it must have a superconducting transition temperature that is cryogenically readily accessible and not too low. Many superconducting devices operate in a bath of pool-boiling or subcooled liquid cryogen, such as liquid nitrogen ($\sim 65 \rightarrow 77$ K) or liquid helium ($\sim 1.2 \rightarrow 4.2$ K). Thus, the selected superconductor must have a T_c well above the boiling temperature of the cryogenic fluid to account for any parasitic heat loads and/or unwanted transient temperature excursions. Next, the selected superconductor must have a high-enough $J_c(T,B)$ in the expected background operating magnetic field such that the device is not too large or heavy.

Table 2.2 Superconductors used in practical power devices

Superconductor	Transition temperature (K)
Nb	9.5
NbTi	9.5
Nb_3Sn	18.3
Nb_3Al	18.8
MgB_2	39
$REBa_2Cu_3O_7$	93
$Bi_2Sr_2Ca_1Cu_2O_8$	85
$Bi_2Sr_2Ca_2Cu_3O_{10}$	108–110

Nb has primarily been used in high-Q cavities, while all the others have been used in wire form for a variety of magnets, coils, and cables. The listed transition temperature is in zero-applied field.

In short, the energy density per unit volume or weight of the superconducting device must be significantly greater than what can be obtained by a conventional non-superconducting copper or aluminum wire-based system. The method and cost of fabrication of the superconducting wire must be within realistic expectations for the contemplated device. Mechanical properties also play a significant role in the down-selection of superconducting wires, because the wires are subject to powerful Lorentz forces when operating in large fields, and they can also experience stresses during cool-down because of thermal contraction. Ductility is strongly preferred; however, even extremely brittle materials such as Nb_3Sn, Nb_3Al, and REBCO have been successfully adapted for use in high-field superconducting devices by imbedding them in composite metallic structures and, in the case of Nb_3Sn and Nb_3Al, winding the device before reacting to form the brittle superconducting phase. In summary, for a superconducting material to be of practical use, it must have a combination of properties, including T_c, J_c, H_{c2} (or H_{irrev}), mechanical strain tolerance (ε), chemical stability, means for practical manufacturing, and a price–performance that surpass any conventional non-superconducting alternative.

2.17.3 Iron pnictides and iron chalcogenides

A new class of superconducting materials has been discovered based upon layered Fe–As compounds (Kamihara, Watanabe, Hirano, & Hosono, 2006). These new compounds are the so-called iron pnictides or iron chalcogenides because of their crystal structure. Superconducting transition temperatures as high as ~55 K have been found in a compound of SmFeAsO (Ren et al., 2008). It remains to be seen if any of these new classes of materials is worth exploring for practical wire fabrication and implementation into superconducting devices for the electric power grid. In addition to having adequate T_c, J_c, H_{c2} (and H_{irrev}), anisotropy in J_c and $H_{c2,}$ mechanical properties, and price–performance for a given application, their performance in at least one area must exceed that of the existing superconducting materials listed in Table 2.2 in order to justify the significant time and fiscal expenditure required to turn them into practical useful wire architectures; that advantage has not yet been identified.

Acknowledgment

The authors wish to thank Ben Andrews for his help in the production of figures for this chapter.

References

Abrikosov, A. A. (1957). On the magnetic properties of superconductors of the second group. *Zh Eksperim.i Teor Fiz, 32*, 1442 [Soviet Physics. JETP, 1957, 5, 1174].

Anderson, P. W., & Kim, Y. B. (1964). Hard superconductivity: theory of the motion of Abrikosov flux lines. *Reviews of Modern Physics, 36*, 39–43.

Bardeen, J., Cooper, L. N., & Schrieffer, J. R. (1957). Theory of superconductivity. *Physical Review, 108*, 1175–1204.

Bardeen, J., & Stephen, M. J. (1965). Theory of the motion of vortices in superconductors. *Physical Review, 140*, 1197.

Bean, C. P. (1962). Magnetization of hard superconductors. *Physical Review Letters, 8*, 250–253.

Bean, C. P. (1964). Magnetization of high-field superconductors. *Reviews of Modern Physics, 36*, 31–39.

Bednorz, J. G., & Mueller, K. A. (1986). Possible high Tc superconductivity in the Ba-La-Cu-O system. *Zeitschrift für Physik B, 64*, 189–193.

Blatter, G., Feigel'man, M. V., Geshkenbein, V. B., Larkin, A. I., & Vinokur, V. M. (1994). Vortex matter. *Reviews of Modern Physics, 66*, 1125–1388.

Campbell, A. M., & Evetts, J. E. (1972). Flux vortices and transport currents in type II superconductors. *Advances in Physics, 21*, 199–428.

Civale, L. (1997). Vortex pinning and creep in high temperature superconductors with columnar defects. *Superconductor Science and Technology, 10*, A11.

Civale, L., Krusin-Elbaum, L., Thompson, J., Wheeler, R., Marwick, A. D., Kirk, M. A., et al. (1994). Reducing vortex motion in YBaCuO crystals with splay in columnar defects. *Physical Review B, 50*, 4102.

Civale, L., Marwick, A. D., McElfresh, M. W., Worthington, T. K., Malozemoff, A. P., Holtzberg, F. H., et al. (1990). Defect independence of the irreversibility line in proton-irradiated Y-Ba-Cu-O crystals. *Physical Review Letters, 65*, 1164–1167.

De Bruyn Ouboter, R., Van Delft, D., & Kes, P. H. (2012). Discovery and Early History of Superconductivity. In H. Rogalla, & P. H. Kes (Eds.), *100 years of superconductivity* (pp. 1–29). Boca Raton, FL: CRC Press.

Feigel'man, M. V., Geshkenbein, V. B., Larkin, A. I., & Vinokur, V. M. (1989). Theory of collective flux creep. *Physical Review Letters, 63*, 2303–2306.

Fisher, M. P. A. (1989). Vortex glass superconductivity: a possible new phase in bulk high Tc oxides. *Physical Review Letters, 62*, 1415–1418.

Fluekiger, R., & Kumakura, H. (2012). The MgB2 story. In H. Rogalla, & P. H. Kes (Eds.), *100 years of superconductivity* (pp. 702–712). Boca Raton, FL: CRC Press.

Fulde, P., & Ferrell, R. A. (1964). Superconductivity in a strong spin-exchange field. *Physical Review, 135*, A550.

Fruchter, L., Malozemoff, A. P., Campbell, I. A., Sanchez, J., Konczykowski, M., Griessen, R., et al. (1991). Low temperature magnetic relaxation in YBaCuO – evidence for quantum tunneling of vortices? *Physical Review B, 43*, 8709.

Ginsberg, D. M. (1989). *Physical properties of high temperature superconductors I*. New Jersey: World Scientific, 46–48.

Ginzburg, V. L., & Landau, L. D. (1950). On the theory of superconductivity. *Zh Eksperim Teor Fiz, 20*, 1064.

Gurevich, A. (2011). To use or not to use cool superconductors. *Nature Materials, 10*, 255–259.

Hao, Z., Clem, J. R., McElfresh, M. W., Civale, L., Malozemoff, A. P., & Holtzberg, F. (1991). Model for the reversible magnetization of high-κ type II superconductors: application to high-T_c superconductors. *Physical Review B, 43*, 2844–2852.

Holesinger, T., Civale, L., Maiorov, B., Feldmann, M., Coulter, Y., Miller, D., et al. (2008). Progress in nano-engineered microstructures for tunable high current high-temperature superconducting wires. *Advanced Materials, 20*, 391–407.

Jones, M. E., & Marsh, R. E. (1954). The preparation and structure of magnesium boride, MgB2. *Journal of the American Chemical Society, 76*(5), 1434.

Kamerlingh Onnes, H. (1911–1912). On the sudden change in the rate at which the resistance of mercury disappears. *Communications from the Physical Laboratory of the University of Leiden, 124c (Nov. 1911); English translation in KNAW Proceedings, 14II*. Amsterdam 1912, pp. 818–821.

Kamihara, Y., Watanabe, T., Hirano, M., & Hosono, H. (2006). Iron-based layered superconductor LaOFeP. *Journal of the American Chemical Society, 128*(31), 10012–10013.

Kittel, C. (1986). *Introduction to solid state physics* (6th ed.). New York: John Wiley & Sons.

Koch, R. H., Foglietti, V., Gallagher, W. J., Koren, G., Gupta, A., & Fisher, M. P. A. (1989). Experimental evidence for vortex-glass superconductivity in Y-Ba-Cu-O. *Physical Review Letters, 63*, 1511.

Larbalestier, D. C., Cooley, L. D., Rikel, M. O., Jiang, J., Patnaik, S., Cai, X. Y., et al. (2001). Strongly linked current flow in polycrystalline forms of the superconductor MgB2. *Nature, 410*(6825), 186–189.

Larbalestier, D., Jiang, J., Trociewitz, U., Kametani, F., Scheuerlein, C., Dalban-Canassy, M., Matras, M., Chen, P., et al. (2014). Isotropic round-wire multi-filament cuprate superconductor for generation of magnetic fields above 30 T. *Nature Materials, 13*, 375–381.

Larkin, A. I., & Ovchinnikov, Yu. N. (1964). Nonuniform state of superconductors. *Zh Eksp Teor Fiz, 47*, 1136.

London, F., & London, H. (1935). The electromagnetic equations of the supraconductor. *Proceedings of the Royal Society of London, A149*, 71.

Maeda, H., Tanaka, Y., Fukutomi, M., & Asano, T. (1988). A new high-Tc oxide superconductor without a rare earth element. *Japanese Journal of Applied Physics, 27*, L209.

Maiorov, B., Baily, S. A., Zhou, H., Ugurlu, O., Kennison, J. A., Dowden, P. C., et al. (2009). Synergetic combination of different types of defects to optimize pinning landscape using BaZrO3-doped YBa2Cu3O7. *Nature Materials, 8*, 398.

Malozemoff, A. P. (1991). Flux creep in high temperature superconductors. *Physica C, 185-9*, 264–269.

Malozemoff, A. P. (2012). Second-generation high-temperature superconductor wires for the electric power grid. *Annual Review of Materials Research, 42*, 373–397.

Malozemoff, A. P., Carter, W. L., Gannon, J., Joshi, C. H., Miles, P., Minot, M., et al. (1992). Progress in the development of bismuth-based high-temperature superconducting tapes. *Cryogenics, 32*, 478–484.

Malozemoff, A. P., & Fisher, M. P. A. (1990). Universality in the current decay and flux creep of YBaCuO high temperature superconductors. *Physical Review B, 42*, 6784.

Malozemoff, A. P., Worthington, T. K., Yeshurun, Y., Holtzberg, F., & Kes, P. H. (1988). Frequency dependence of the ac susceptibility in a Y-Ba-CuO crystal: a reinterpretation of H_{c2}. *Physical Review B, 38*, 7203–7206.

Malozemoff, A. P., & Yamada, Y. (2012). Coated conductor: second generation HTS wire. In H. Rogalla, & P. Kes (Eds.), *100 years of superconductivity* (pp. 689–702). Boca Raton, FL: CRC Press, Taylor & Francis Group.

Meissner, W., & Ochensenfield, R. (1933). Ein neuer Effeckt bei Eintritt der Supraleitfaehigkeit. *Naturwissenschaften, 21*, 787.

Nagamatsu, J., Nakagawa, N., Muranaka, T., Zenitani, Y., & Akimitsu, J. (2001). Superconductivity at 39 K in magnesium diboride. *Nature, 410*, 63–6424.

Norris, W. T. (1970). Calculation of hysteresis losses in hard superconductors carrying ac: isolated conductors and edges of thin sheets. *Journal of Physics D, 3*, 489.

Pippard, A. B. (1953). An experimental and theoretical study of the relation between magnetic field and current in a superconductor. *Proceedings of the Royal Society of London, 216*, 1127.

Ren, Z.-A., Che, G.-C., Dong, X.-L., Yang, J., Lu, W., Yi, W., et al. (2008). Superconductivity and phase diagram in iron-based arsenic-oxides ReFeAsO1-δ (Re = rare-earth metal) without fluorine doping. *Europhysics Letters, 83*, 17002.

Rose-Innes, A. C., & Rhoderick, E. H. (1978). *Introduction to superconductivity* (Vol. 6). New York: Pergamon Press.

Rupich, M. W., & Hellstrom, E. E. (2012). Bi-Ca-Sr-Cu-O wire. In H. Rogalla, & P. H. Kes (Eds.), *100 years of superconductivity* (pp. 671–689). Boca Raton, FL: CRC Press.

Scanlan, R. M., Malozemoff, A. P., & Larbalestier, D. C. (2004). Superconducting materials for large scale applications. *Proceedings of the IEEE, 92*, 1639–1654.

Schilling, A., Cantoni, M., Guo, J. D., & Ott, H. R. (1993). Superconductivity above 130 K in the Hg-Ba-Ca-Cu-O system. *Nature, 363*, 56–58.

Sheng, Z. Z., & Hermann, A. M. (1988). Bulk superconductivity at 120 K in the Tl-Ca/Ba-Cu-O system. *Nature, 332*, 138.

See http://www.iter.org/newsline/262/1541.

Silsbee, F. B. (1916). A note on electrical conduction in metals at low temperatures. *Journal of the Washington Academy of Sciences, 6*, 597.

Smathers, D. B. (1990). A15 superconductors. In *American society for metals: vol. 2. Metals Handbook* (10th ed.) (pp. 1060–1076).

Suenaga, M., Ghosh, A. K., Xu, Y., & Welsh, D. O. (1991). Irreversibility temperatures of Nb3Sn and NbTi. *Physical Review Letters, 66*, 1777.

Tallon, J. L., Buckley, R. G., Gilberd, P. W., Presland, M. R., & Brown, I.W.M. (1988). High-T_c superconducting phases in the series $Bi_{2.1}(Ca,Sr)_{n+1}Cu_nO_{2n+4+\delta}$. *Nature, 333*, 153–156.

Tallon, J. L., Storey, J. G., & Mallett, B. (2012). The design of high-T_c superconductors — room temperature superconductivity? *Physica C, 482*, 45–49.

Thompson, J. R., Sun, Y. R., Civale, L., Malozemoff, A. P., McElfresh, M. W., Marwick, A. D., et al. (1993). Effect of flux creep on the temperature dependence of the current density in YBaCuO crystals. *Physical Review B, 47*, 14440–14447.

Tinkham, M. (1996). *Introduction to superconductivity* (2nd ed.). New York: McGraw-Hill, Inc., 336–337.

Warnes, W. H., & Larbalestier, D. C. (1986). Critical current distributions in superconducting composites. *Cryogenics, 26*, 643–653.

Weber, H. W. (2011). Radiation effects on superconducting fusion magnet components. *International Journal of Modern Physics E, 20*, 1325–1378.

Wu, M. K., Ashburn, J. R., Torng, C. J., Hor, P. H., Meng, R. L., Gao, L., et al. (1987). Superconductivity at 93 K in a new mixed phase Y-Ba-Cu-O compound system at ambient pressure. *Physical Review Letters, 58*(9), 908–910.

Yeshurun, Y., Malozemoff, A. P., & Shaulov, A. (1996). Magnetic relaxation in high temperature superconductors. *Reviews of Modern Physics, 68*, 911–950.

Bismuth-based oxide (BSCCO) high-temperature superconducting wires for power grid applications: properties and fabrication

3

K. Sato
Sumitomo Electric Industries, Ltd, Osaka, Japan

3.1 Introduction

In 1988, Maeda, Tanaka, Fukutomi, and Asano (1988) discovered bismuth-based oxide high-temperature superconductors, which are written in the form of Bi—Sr—Ca—Cu—O (BSCCO). Since then, two types of BSCCO superconducting wires have been developed for commercial applications: $(Bi, Pb)_2Sr_2Ca_2Cu_3O_{10+x}$ (Bi-2223) and $Bi_2Sr_2CaCu_2O_{8+x}$ (Bi-2212). Both Bi-2223 and Bi-2212 wires are fabricated by the powder-in-tube (PIT) process, in which bismuth-based oxide powders are put into silver tubes, drawn to round wires, and then rolled to tape-shaped wires using conventional plastic deformation processing techniques. Due to the long length, wire availability, and mass production capability of these fabricating processes for Bi-2223 and Bi-2212 wires, application prototypes have been built since the middle of the 1990s.

It is well known that high-temperature oxide superconductors need to be aligned in the *c*-plane to have a high critical current (I_c) due to the anisotropy of their crystal structure. Due to their plate-like grain structure, Bi-2223 wires can be readily aligned during the rolling process, whereas Bi-2212 is more easily aligned during the partial melting heat treatment at its final fabricating process. Hence, Bi-2223 wires have a flat tape geometry, whereas the shape of Bi-2212 wires can be either round or flat tape. The advantage of the round wire geometry over the flat tape is that the I_c of the round wires tends to be more spatially isotropic than the highly anisotropic flat tapes. This has an enormous advantage in some high-field magnet applications. To the contrary, $YBa_2Cu_3O_{7-x}$ (YBCO), which was discovered in 1987, needs not only alignment in the *c*-plane but also alignment in the *ab*-plane to obtain a high critical current, leading to a far more complex multilayer deposition process for YBCO wire fabrication. YBCO wires still cannot be fabricated consistently and reliably in long lengths of several hundreds of meters and will take several more years to develop.

The superconducting critical temperature (T_c) of Bi-2223 wires is approximately 110 K. Bi-2223 wires are suitable for electrical grid applications such as power cables, motors/generators, transformers, and fault current limiters (FCLs) operated at liquid

Superconductors in the Power Grid. http://dx.doi.org/10.1016/B978-1-78242-029-3.00003-0

nitrogen temperature (65−77 K), high-field magnets, motors operated at 20−30 K, and super high-field magnets producing more than 20 T operated at liquid helium temperature (4.2 K or below). The critical temperature of Bi-2212 wires is around 80−90 K. Its round wire shape, with its more isotropic I_c properties and high critical currents at 4.2 K, are leading researchers to build super high-field magnets producing over 20 T using Bi-2212. Heine, Tenbrink, and Thöner (1989) first introduced Bi-2212 wires produced by the PIT process, which could transport large superconducting currents under high magnetic fields. Sato, Hikata, and Iwasa (1990) reported that Bi-2223 wires produced by the PIT process could transport large superconducting currents under high magnetic fields. Also, Sato et al. (1991) reported that the multifilamentary configuration of Bi-2223 wires could be flexible after final heat treatment and could be used for magnets and power cables using the react-and-wind method.

In this chapter, the electromagnetic, mechanical, and thermal properties of Bi-2223 and Bi-2212 wires, their fabrication processes, typical prototypes, and future trends are described.

3.2 Properties of bismuth-based oxide (BSCCO)

3.2.1 Crystal structure

Figure 3.1 shows the crystal structure of $Bi_2Sr_2CuO_{6+x}$ (Bi-2201), Bi-2212, and Bi-2223 bismuth-based superconductors, which are discussed in this chapter. All three crystal structures have a long layered structure along the *c*-axis; only the CuO_2 layer and Ca is incremented by one layer in the order of Bi-2201, Bi-2212, and Bi-2223. The bismuth-based superconductor is written in the chemical formula from this as Bi_2Sr_2-$Ca_{n-1}Cu_nO_y$ ($n = 1, 2, 3, \ldots$). In this formula, each constituent element ratio is represented by an integer ratio. In practice, the element ratio is not an integer ratio for the wire/tape fabrication process. The Sr site tends to be replaced (i.e., doped) with Ca and Bi, depending on the manufacturing method.

Bi-2212 and Bi-2223 have a CuO_2 plane along their *ab*-planes, and the hole in this plane has become a carrier in creating superconducting currents. The nonstoichiometry of metallic elements and oxygen affects the superconducting properties. The ratio of metallic elements affects their microstructures, which control superconducting current flow, and oxygen affects the electronic charge carrier density, which controls the critical temperature and magnetic field dependence of critical currents.

3.2.2 Electromagnetic properties

Among the electromagnetic properties, critical currents under magnetic fields at specified temperatures are most important for electrical grid applications. Bi-2212 and Bi-2223 wires are fabricated through the PIT process and composed of plate-like polycrystals having grain boundaries, which tend to align well during the rolling process. YBCO wires fabricated through the PIT process are also composed of polycrystals and have grain boundaries. Dimos, Chaudhari, Mannhart, and LeGoues (1988) reported

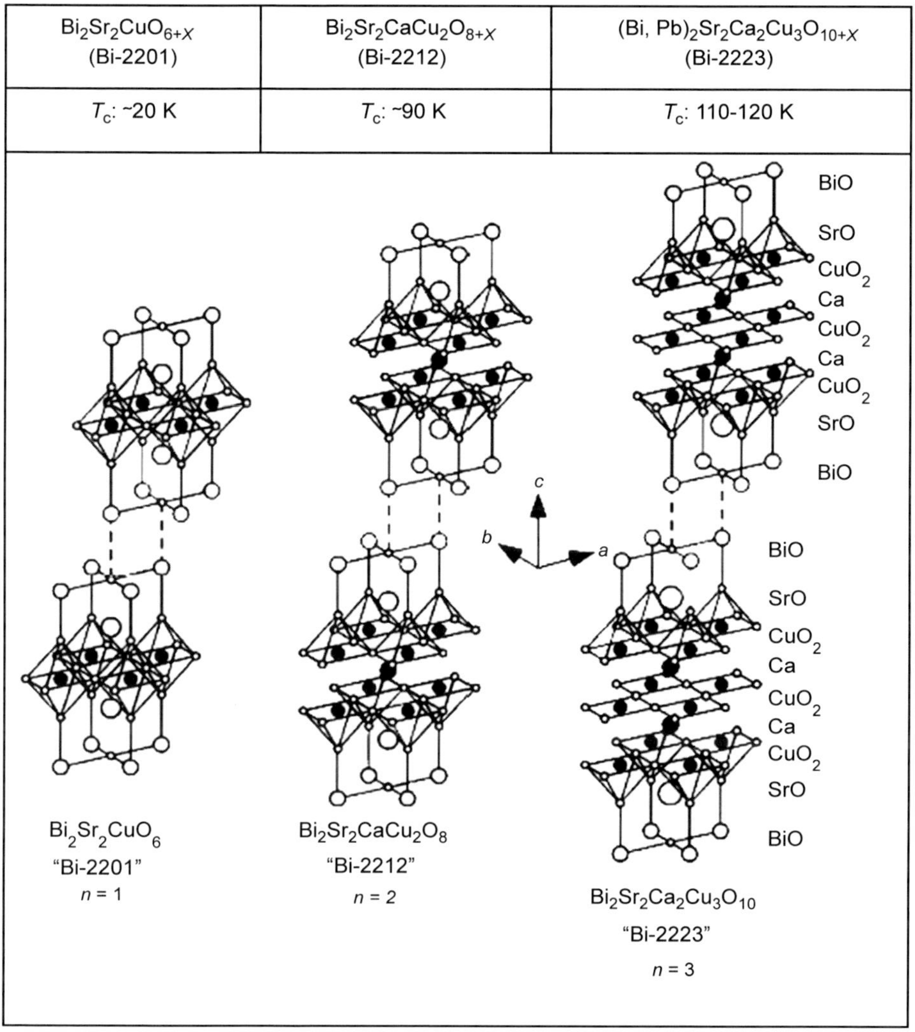

Figure 3.1 Crystal structure of Bi-2201, Bi-2212, and Bi-2223 bismuth-based superconductors.

that the YBCO grain boundaries have a misorientation angle larger than $\sim 10°$, which diminished the critical current by one to two orders of magnitude. On the contrary, in bismuth-based superconductors, Li, Tsay, Suenaga, Gu, and Koshizuka (1997) first reported that in Bi-2212 wires, critical current across the grain boundaries is essentially independent of twist angle. Critical currents across [001] twist grain boundaries are as high as their constituent single crystals (Li et al., 1999). One possible origin is considered to be the expansion and softening of the twist boundary, which is likely to facilitate the insertion there of interstitial oxygen ions, thus resulting in oxygen segregation and overdoping of the boundary (Wu, Zhu, Tafto, Welch, & Suenaga, 2002).

For electric power cable applications, Bi-2223 wires are one of the most promising among high-temperature superconducting wires, especially because of their high critical currents under magnetic fields of around 0.1–0.2 T at 77 K and their high critical temperature. In the power cable configuration, the magnetic field is quite low, often less than 0.2 T, and the magnetic field direction is parallel to the wire surface. Figure 3.2 shows critical currents under magnetic fields at 66, 77, and 90 K, in which the temperature range is typical for power cable operation. For magnetic fields parallel (*B//*) to the wire surface, Bi-2223 wires show nearly the same level of critical currents at up to 0.2 T.

For an intermediate temperature range from 77 K to 20 K, magnet applications can be considered. Figures 3.3 and 3.4 show the magnetic field dependences of the critical currents of Bi-2223 wires. Due to their anisotropic nature, critical currents decrease more rapidly under magnetic fields perpendicular ($B\perp$) to the wire surface in comparison with those of parallel fields. In a typical solenoid magnet geometry with a high aspect ratio (length/diameter), at the center of the magnet, the magnetic field strength parallel to the wire surface is much larger than that of the perpendicular wire surface at the fringe of the magnet. However, in most magnet designs, critical currents under magnetic fields perpendicular to the wire surface determine its rated operating current. In Figure 3.4, the engineering current densities (J_e), which are overall current densities including all constituting materials of the wires, are 4.0×10^4 A/cm^2 at 3 T and more than 2.0×10^4 A/cm^2 at 10 T, both at 20 K, which are promising values for magnets that can be cooled with a 20-K refrigerator.

For magnet applications beyond 20 T, higher I_c performance is required. Bi-2212 and Bi-2223 wires are expected to be used at 4.2 K or below due to their high irreversible magnetic fields (i.e., about 100 T). In particular, Bi-2212 wires can be produced in the shape of round wires, which makes them very unique and can be used for very high magnetic field applications, such as nuclear magnetic resonance (NMR) magnets and accelerators. Figure 3.5 shows the magnetic field dependence of I_c of Bi-2212

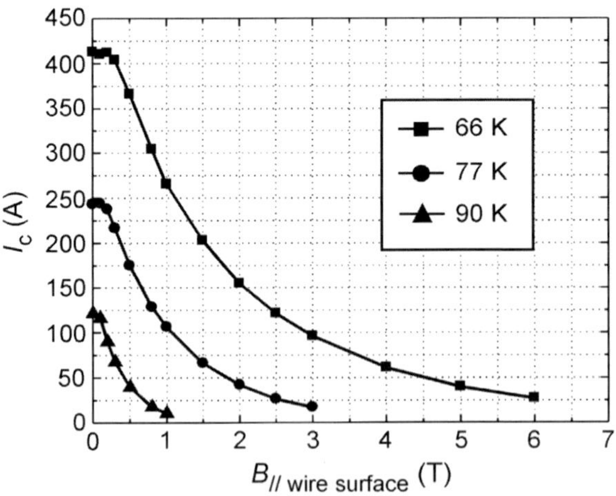

Figure 3.2 Critical currents of Bi-2223 wires under magnetic fields at 66, 77, and 90 K.

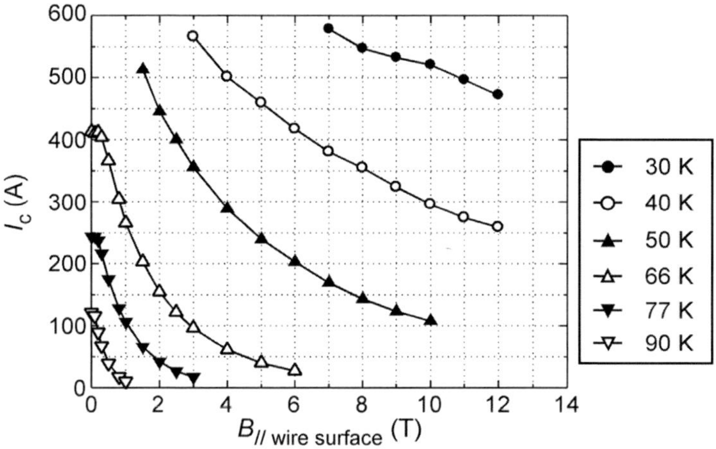

Figure 3.3 Magnetic field dependence of critical currents of Bi-2223 wires (magnetic field parallel to wire surface).

(Yang et al., 2011), Bi-2223, YBCO, and $GdBa_2Cu_3O_{7-x}$ (GdBCO) high-temperature superconducting wires that are available for coil windings, as summarized by Kiyoshi and Miyazoe (2012). Table 3.1 shows the dimensions and properties of the specimens (Kobayashi, Nakashima, Yamazaki, & Sato, 2012). Bi-2212 and Bi-2223 wires showed excellent critical currents at above 5 T, showing them to be good candidates for high-field magnet windings.

Chen et al. (2013) and Larbalestier et al. (2013) showed the example of Bi-2212 wire applications for high-field magnets. Using 30 m of Bi-2212 round wire (diameter: 1.4 mm), a small coil, 36.3 mm in outer diameter and 71.2 mm in height, was built and tested at 1.8 K. In the 31.2-T background magnetic field, this small coil could generate

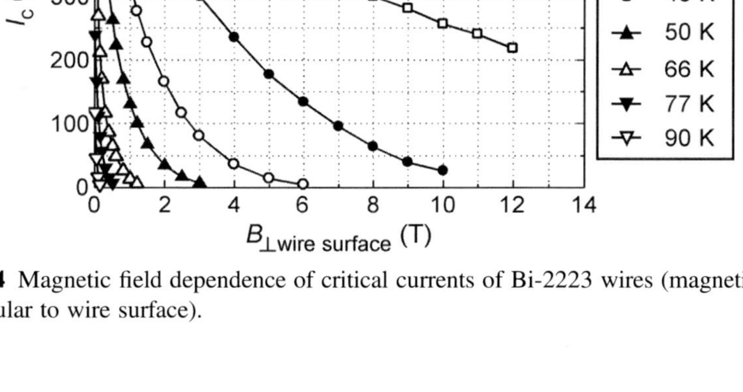

Figure 3.4 Magnetic field dependence of critical currents of Bi-2223 wires (magnetic field perpendicular to wire surface).

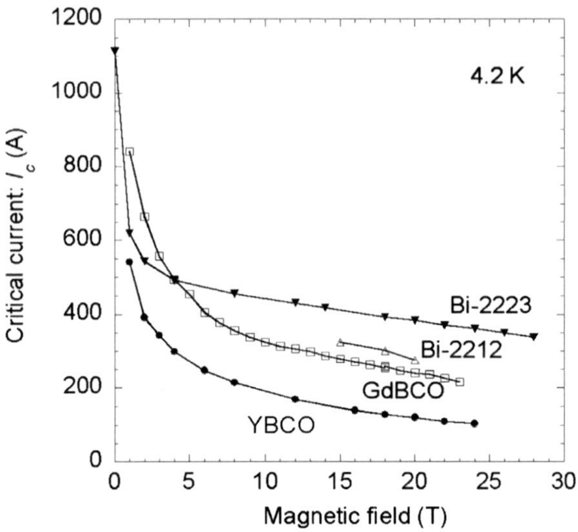

Figure 3.5 Magnetic field dependence of Bi-2212, Bi-2223, YBCO, and GdBCO wires.

Table 3.1 Dimensions and properties of specimens

Wire	Dimension	Property
Bi-2212	Diameter: 1.5 mm	—
Bi-2223	0.29 × 4.43 mm	I_c(77.3 K, sf): 198 A
YBCO	0.1 × 4 mm	I_c(77.3 K, sf): 82 A
GdBCO	0.253 × 5.05 mm	I_c(77.3 K, sf): 252 A

2.6 T with a current of 388 A. This critical current is even higher than that of Figure 3.5, due to the enhanced densification of Bi-2212 filaments processed with a 10-bar overpressure furnace.

Kiyoshi et al. (2011) showed an example of Bi-2223 wire applications for very-high-field magnets. Using 3.2 km of Bi-2223 wire (0.41 mm × 4.5 mm), a coil with an outer diameter of 127 and 840 mm in height was built for the innermost coil of 1.03-GHz NMR magnets (24.2 T). A Bi-2223 innermost coil will generate 3.6 T (operating current: 244.4 A at 1.8 K or below) with a background magnetic field of 20.6 T.

3.2.3 Mechanical properties

Hojo et al. (2010), Osamura et al. (2009), and Yamazaki et al. (2012) extensively studied tensile stress tolerance of Bi-2223 wire and tensile fatigue property. Lamination techniques with high-strength copper-alloy sheets (50-μm thickness) or stainless steel (SS) sheets (20-μm thickness) for both sides of bare Bi-2223 wire gave approximately

300-MPa tensile stress tolerance and 0.4−0.5% tensile strain tolerance, which was defined to retain 95% critical current compared with those of nonloaded critical currents.

Further improvements in tensile stress tolerance could be feasible by using the following: (1) thicker lamination sheets, (2) stronger lamination materials and a bigger difference of the coefficient of thermal expansion between a bare Bi-2223 wire and lamination materials, and (3) bigger pretension, which stretches lamination sheets, when soldering lamination sheets to a bare Bi-2223 wire. Indeed, direct evidence of the high strain tolerance of critical currents by means of the pretensioned lamination technique was confirmed by Osamura et al. (2013). Using 50-μm-thick SS sheets and 161 N pretension, Bi-2223 wire has a tensile stress tolerance of 455 MPa and 0.506% strain before losing 5% of critical current.

Using new lamination material with a bigger Young's modulus than SS and pretension technique, Bi-2223 wire can reach up to 500 MPa (or more) tensile stress tolerance with thinner lamination sheets (less than 50 μm thick). Figure 3.6 shows the tensile tolerance of 540 MPa (strain of 0.58%) using this new lamination material at 50-μm thickness.

Hojo et al. (2010) revealed that when the static irreversible stress as defined by a 1% decrease in the critical current at the unloading step (315 MPa and 0.4% strain) was selected as the maximum stress for fatigue loading, the critical current decreased by only 2% after the stress cycles of 10^6 cycles. Thus, the SS-laminated Bi-2223 wire showed excellent mechanical properties, even under fatigue loading.

Regarding the bending strain characteristics of Bi-2223 wires, a double-bending procedure is commonly applied. The procedure is as follows: one end of the wire is fixed at the mandrel with a certain diameter and bent along the mandrel from the fixed end to the free end. The wire is turned to the opposite surface and bent in the same way. Then, I_c is measured in a nearly straight shape.

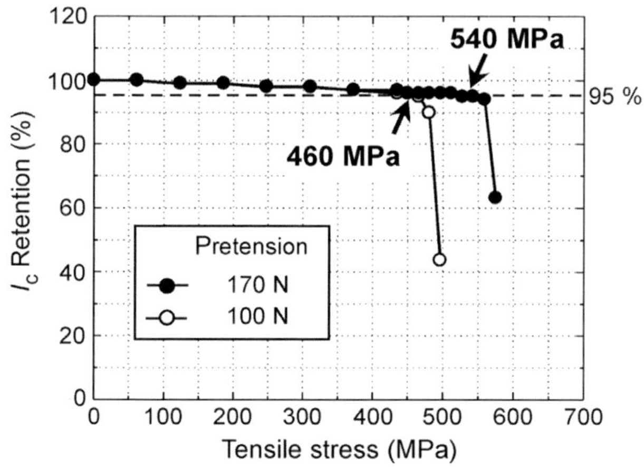

Figure 3.6 Tensile tolerance of Bi-2223 wire reinforced with new lamination material (50 μm thick).

Typical bending diameters for Bi-2223 wires retaining 95% of I_c are 70 mm for a bare wire (0.23-mm thickness) and 60 mm for SS and copper-alloy laminated wire (0.3- and 0.36-mm thickness, respectively).

In the case of Bi-2212 wires, some wires can exhibit a reversible tensile stress strain of around 0.3%, and some wires did not show reversible behavior (Cheggour, Lu, Holesinger, Jiang, & Goodrich, 2012). The origin of this complex behavior may be due to the porosity in the filaments. The elimination of porosity in the filaments can improve not only mechanical properties but also critical currents, which will be noted in a later section.

3.2.4 Thermal properties

From the point of view of current leads application, the overall thermal conductivity of the HTS wire/tape is very important. For HTS current lead applications of the current leads to superconducting magnets, the thermal conductivity of the current leads must be low enough to reduce heat transfer via thermal conduction from current leads to magnets. Therefore, the standard silver (Ag) matrix encasing the Bi-2223 filaments is not suitable because of its high thermal conductivity. In the case of Bi-2223 wires for current leads application, Bi-2223 oxide filaments are typically encased in an Ag−Au alloy matrix to provide a lower thermal conducting matrix. Fujishiro, Ikebe, Noto, Sasaoka, and Nomura (1993) reported that the thermal conductivity of Ag−Au alloy wires decreased markedly at low temperatures with increasing Au content. Naito, Fujishiro, and Yamada (2009) measured the temperature dependence of thermal conductivity of Bi-2223 wires that were sheathed with Ag and Ag−5.4 wt% Au alloy. At approximately 77 K, the thermal conductivities of Ag- and Ag−5.4 wt% Au alloy-sheathed Bi-2223 wires were about 260 and 80 W/(m K), respectively. The thermal conductivity of the Ag−5.4 wt% Au alloy-sheathed wire is very low and monotonically decreases at temperatures from 250 K to 10 K, as shown in Figure 3.7.

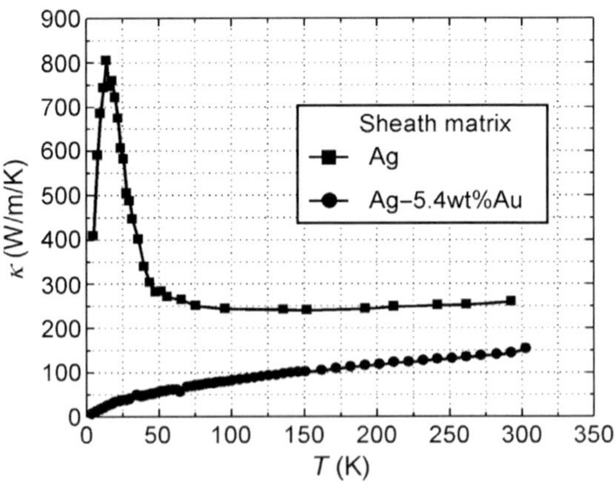

Figure 3.7 Thermal conductivity of Ag-sheathed Bi-2223 wire and Ag−5.4 wt% Au alloy-sheathed Bi-2223 wire.

For large current applications, such as fusion magnets and accelerator magnets, Ando et al. (1998) first observed that conductively cooled Bi-2223 wires with Ag–Au alloy sheath could reduce liquid helium consumption up to 90% compared with that of conventional copper current leads operating at the same current level.

3.2.5 AC loss

When an AC magnetic field or an AC current is applied to superconducting wires, heat generation takes place due to hysteresis, coupling between superconducting elements, and eddy current losses from metallic materials used in superconducting wires. In some important applications, a large heat generation is expected and the superconducting wires used are specifically designed to prevent the breakdown of the superconducting state from the heat generated.

For magnet applications, especially for transformer applications, Funaki et al. (1998) revealed that a transposition of multiple parallel wires for windings could make uniform current distribution among parallel wound wires, resulting in the reduction of AC loss.

When an AC current is supplied to the superconducting layer formed from multiple layers of power cable conductors, normally nonuniform current distribution occurs. In this case, most of the current flows to the outer layers, whose inductance is low, and causes an increase of the AC loss. As a countermeasure against the occurrence of nonuniform current distribution, a spiral winding pitch adjustment technique is applied to the cable conductors. In this technique, the spiral winding pitch of the layers is adjusted to match the impedance of each layer in order to make uniform current distribution to each superconducting wire. The AC loss of a multilayered conductor can be minimized by this technique (Honjo et al., 2003).

A four-layer conductor was developed for the Yokohama HTS cable project to reduce the AC loss of power cable conductors (Yumura et al., 2013). To reduce AC loss, low-loss and copper-alloy laminated Bi-2223 wires (2.7 mm × 0.33 mm, with a smaller cross-sectional area than standard-size wires) are used in the third and fourth conducting layers. This low-loss-type wire contains twisted superconducting filaments, thus reducing AC loss from the parallel magnetic field to one-third of that of standard size and copper-alloy laminated wires (4.5 mm × 0.35 mm), which were used in the first and second conducting layers and shielding layers. The AC loss was 0.8 W/m/phase at AC 2 kA; thereby, it was confirmed that the AC loss characteristics of the cable conductor that was developed met the required specification of less than 1 W/m/phase.

3.3 Fabrication of BSCCO superconducting cables and wires

Although both Bi-2212 and Bi-2223 wires are fabricated by the PIT process, their crystal alignment techniques are quite different. Alignment of Bi-2223 crystals can be achieved during the rolling process and alignment of Bi-2212 crystals can be achieved during the solidification after a partial melting process.

3.3.1 Bi-2223 wires

Figure 3.8 shows a schematic drawing for the Bi-2223 wire fabrication process. A mixture of Bi-2212 phase and other oxides is put into a silver tube and drawing is performed. This composite, which is called monofilament wire, is again put into a silver-alloy tube. The Ag alloy is necessary for sufficiently enhancing the mechanical strength of the wire for the fabrication process. Usually, approximately 100 monofilament wires are then stacked, and these are called multifilamentary wires. Next, a drawing and rolling process is added to create a flat tape-shaped wire. Then, the first sintering process is performed to convert a mixture of Bi-2212 phase and other oxides to Bi-2223 crystal. After the first sintering, an intermediate rolling step is performed to make a dense microstructure, and then the second sintering is performed. At the second sintering, high-pressure gas (a mixture of oxygen and argon, as high as 300 bar with a high temperature of around 850 °C) is simultaneously applied to reach a relative density of nearly 100% of the Bi-2223 filament (Kobayashi et al., 2005). This process is called the controlled overpressure (CT-OP) sintering process. Filaments of 100% relative density enable the Bi-2223 wire to obtain improved critical currents, improved mechanical strength, and excellent uniformity at more than 1500- to 2000-m unit length. Bi-2223 wire fabricated through this process has been given the trade name DI-BSCCO (see, e.g., http://global-sei.com/super).

Figure 3.9 shows the improved microstructure of DI-BSCCO compared to that of conventional Bi-2223 wire. Bi-2223 wire processed without the CT-OP process results in a large amount of voids and a large amount of different phases. To the contrary, the microstructure of DI-BSCCO is very dense and has a lesser amount of

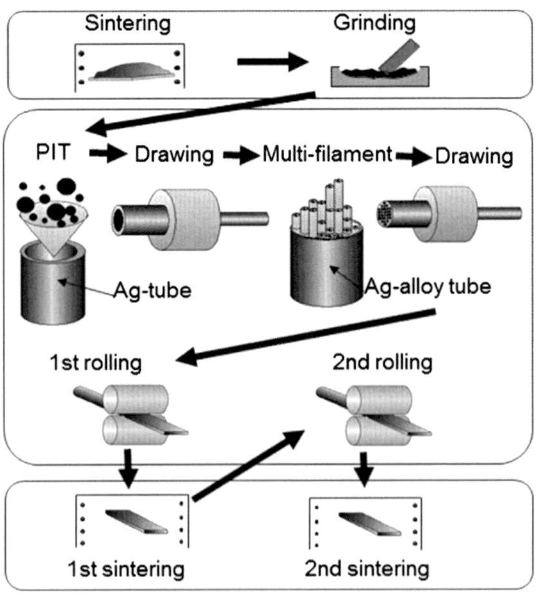

Figure 3.8 Schematic drawing for the Bi-2223 wire process.

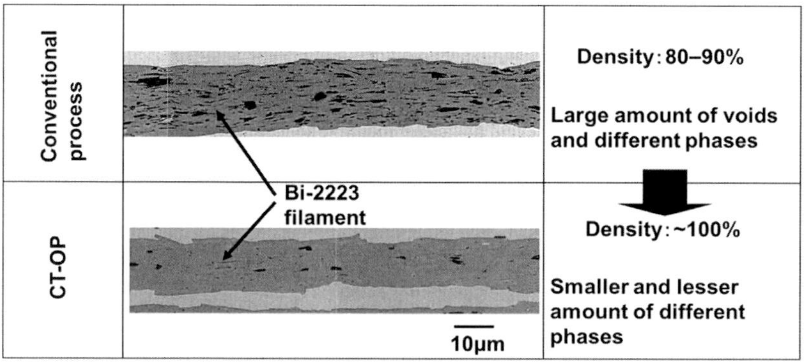

Figure 3.9 Microstructure of DI-BSCCO and conventional Bi-2223 wire.

nonsuperconducting phases. During the second sintering, postannealing is applied to control oxygen content in order to get an adequate doping level of oxygen. Silver is transparent to oxygen, so the oxygen content inside the Bi-2223 filaments can be easily controlled. An optimal doping level is used to achieve maximum critical temperature, and overdoping is used to achieve high critical current under the magnetic field at a lower temperature (below 30 K).

For large current conductors used for electric power cables and current leads, Bi-2223 wires are typically wound spirally as multilayers on the flexible former in the case of power cables, and are solder bonded in stacked Bi-2223 wires in the case of current leads. Figure 3.10 shows spirally wound Bi-2223 wires for power cable conductors. The number of Bi-2223 layers, the number of Bi-2223 wires per layer, and the twist pitch of the Bi-2223 wires per layer for each cable conductor are carefully calculated and subsequently controlled during cable conductor fabrication so as to minimize the loss generated by AC currents and ensure that mechanical strain tolerances are not exceeded.

Figure 3.10 Spirally winding Bi-2223 wires for power cable conductors.

3.3.2 Bi-2212 wires

Figure 3.11 shows a schematic drawing for the Bi-2212 wire process. A major issue for Bi-2212 long-length wire fabrication has been excessively low critical currents—less than 50% of that of short sample values. A series of investigations were carried out for bubble formation (Kametani et al., 2011), cold isostatic pressing (CIP) technique (Jiang et al., 2011), heat treatment control (Shen et al., 2011), filament connectivity (Dalban-Canassy et al., 2012), and wire expansion and filament densification (Malagoli et al., 2013). The results of these investigations are summarized as follows: (1) lower critical currents of long-length Bi-2212 wires are due to porosity agglomerated into bubbles during partial melting, which leads to narrow current paths and poor grain connectivity; (2) internal gas pressure during partial melting increased; (3) the origin of internal gas could be carbon impurity, N_2, and moisture; (4) a longer heat treatment time of more than 2.5 h between partial melting and nucleation (872 °C) of Bi-2212 decreases critical currents; and (5) CIP before partial melting and/or partial melting under a high-pressure environment could densify Bi-2212 filaments and improve critical currents. Short CIP Bi-2212 wires that were 0.78 mm in diameter (8 cm long) exhibited critical currents of 385.4−386.4 A ($Je = 807-809\ \text{A/mm}^2$) at 4.2 K and 5 T.

Bi-2212 wires are round and multifilamentary wires with isotropic superconducting properties and can be made into Rutherford cables for extra-high-field magnets and accelerator magnets (Godeke et al., 2010). Figure 3.12 shows an example of a cross-section of a Rutherford cable.

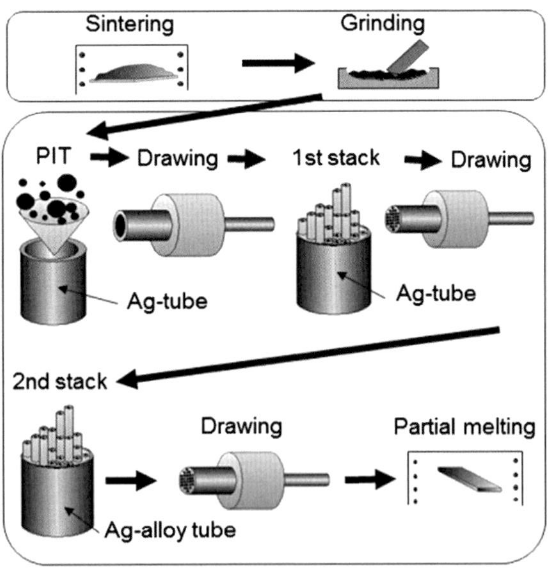

Figure 3.11 Schematic drawing for the Bi-2212 wire process.

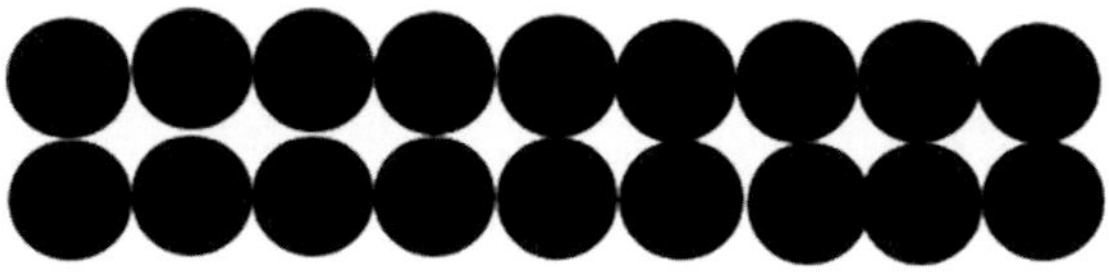

Figure 3.12 Example of a cross-section of a Rutherford cable (strand diameter is ~0.8 mm).

3.4 Applications of BSCCO superconducting cables and wires

There are already daily working apparatus and components made of Bi-2223 wires. The first example is the application for HTS current leads. Bi-2223 with Ag–Au matrices that have been fabricated for HTS current leads for accelerator magnets for the large hadron collider (LHC) at CERN (the European Organization for Nuclear Research) are listed in Table 3.2. The largest rated current is 13,000 A with well over 1000 HTS current leads installed in the LHC machine (Ballarino, 2008). These HTS current leads have been used stably since 2010 and have experienced more than 1000 current cycles. Bi-2223 current leads help to reduce liquid helium consumption as well as the overall room temperature with electrical power from 450 kW compared with 1290 kW, which is what would be needed when using copper current leads.

The second example is superconducting power cable. High-temperature superconducting power cables can transport large currents with low loss and can be compact in size.

The key features of superconducting power cables can be summarized as follows:

1. Net CO_2 reduction due to lower loss compared with conventional copper cables
2. Smaller footprint and more compact cross-section of underground space due to higher power density (reduction of construction cost)
3. Uprating of existing grid
4. Elimination of substations due to lower voltage
5. Nonflammability
6. No electromagnetic interference for coaxial cable or balanced three-phase design
7. 75% mass-per-unit length reduction compared with copper cable of equivalent power rating

The merits of using superconducting power cables can be extended in the case of DC superconducting power cables. Table 3.3 shows some typical examples of AC

Table 3.2 Bi-2223 current leads at CERN

Rated current	Number of leads
13,000 A	64
6000 A	298
600 A	820

Table 3.3 **Typical examples of superconducting power cable projects**

Area	Project	Voltage (kV)	Current (kA)	Length (m)	Site	Note
Japan	TEPCO/CRIEPI/Sumitomo	66	1	100	Lab	Finished
	Chubu University (DC)	20	2	200	Lab	Ongoing
	Yokohama	66	2	240	Grid	Ongoing
	Sumitomo	3.3	0.4	70	Factory	Ongoing
	RTRI (DC)	1.5	5	30	Lab	Ongoing
	METI (DC)	10	5	2000	Grid	Plan
USA	Albany	34.5	0.8	350	Grid	Finished
	Ohio	13.8	3	200	Grid	Finished
	LIPA	138	2.4	600	Grid	Ongoing
EU	Denmark	30	0.2	30	Grid	Finished
	VNIIKP	20	1.4	200	Grid	Plan
	Essen	10	2.3	1000	Grid	Plan
	St. Petersburg (DC)	20	2.5	2500	Grid	Plan
China	Yunnan	35	2	33.5	Grid	Finished
	Lánzhōu	10.5	1.5	75	Factory	Ongoing
	IEE/CAS (DC)	1.3	10	380	Factory	Ongoing
Korea	KEPCO	22.9	1.25	100	Lab	Ongoing
	DAPAS1	22.9	1.25	100	Lab	Finished
	DAPAS2	154	3.75	30	Lab	Finished

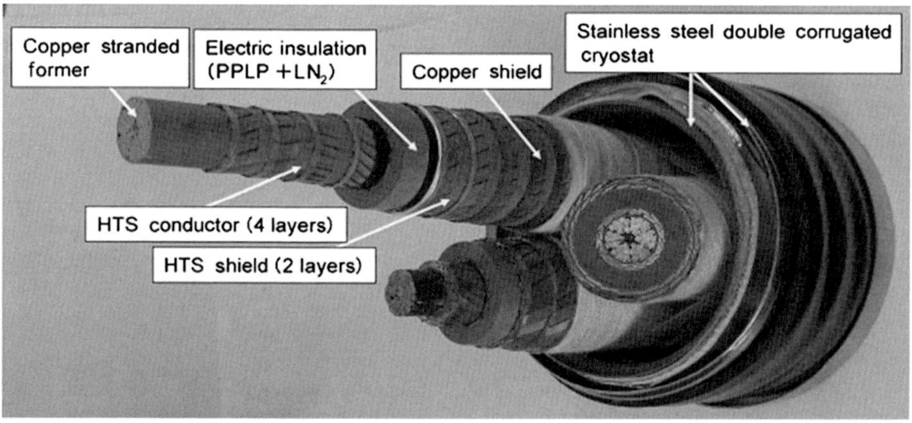

Figure 3.13 AC superconducting power cable.

and DC superconducting power cable projects using Bi-2223 wires in Japan, the United States, the European Union, China, and Korea. Figure 3.13 shows a photograph of an AC superconducting power cable.

The HTS cable shown in Figure 3.13 is composed of a Bi-2223 conductor, Bi-2223 shielding conductor, copper former, copper shielding conductor, electrical insulation, and double-corrugated SS pipes. The space between the inner and outer pipe is evacuated to prevent heat invasion from the outside affecting the liquid nitrogen. High-pressure liquid nitrogen flows inside the inner SS pipe to cool the Bi-2223 conductors. The high-pressure liquid nitrogen helps to reduce the possibility of bubble formation. Figure 3.14 shows the view when laying the Bi-2223 cable into the duct on the Yokohama HTS cable project. The HTS cable was pulled into the duct using the same method as conventional copper cable.

Figure 3.14 Laying Bi-2223 cable into the duct on the Yokohama project.

The third example of the use of Bi-oxide wire in electrical grid applications is for motor and generator applications. Some key advantages of superconducting motors and generators over conventional machines are high torque, compact size, reduced weight, and energy reduction through increased efficiency. Table 3.4 shows some typical examples of motor and generator projects around the world. Many of these projects aim to develop ship propulsion motors and generators in Germany, the United States, Japan, Korea, and China.

One project is to develop a lightweight, high-efficiency superconducting motor for use in future electric cars in Japan. Figure 3.15 shows a photograph of a superconducting motor incorporated with a cryocooler. Figure 3.16 shows a photograph of a superconducting electric car. The motor in this machine uses a series-wound DC motor. The field coil (i.e., stator coil) is made of Bi-2223 winding, which is cooled with liquid nitrogen, with an iron core consisting of four claw poles, where the armature rotor is conventional copper winding. The maximum torque is 136 N m at 1540 rpm. The output of a superconducting motor is 30 kW at 2200 rpm. The maximum speed of this car is 80 km/h.

The fourth example of the use of Bi-oxide wire is for high-field magnets. This constitutes a very wide range of applications, such as FCLs, magnetic resonance imaging (MRI), NMR, and a variety of magnets. Table 3.5 shows some typical examples of Bi-2223 magnet application projects. Some magnets, which were

Table 3.4 Typical example of Bi-2223 motor and generator projects

Project		Output	rpm	Torque	Cooling
Tokyo University of Marine Science and Technology	Motor	10 kW	850	0.137 kN m	30 K; Ne
Kyoto University	Motor	20 kW	–	–	77 K; N_2
Sumitomo	Motor	30 kW	2200	0.136 kN m	77 K; N_2
Siemens	Motor	400 kW	1500	2.5 kN m	27 K; Ne
IHI	Motor	400 kW	250	–	70 K; N_2
Kawasaki Heavy	Motor	1 MW	190	50 kN m	30 K; He
KERI/Doosan	Motor	1 MW	3600	2.65 kN m @1800 rpm	30 K; Ne
Wuhan Institute of Marine Electric Propulsion	Motor	1 MW	500	–	30 K; Ne
Siemens	Motor	4 MW	120	320 kN m	27 K; Ne
AMSC	Motor	5 MW	230	–	30 K; He
AMSC	Motor	36.5 MW	120	2900 kN m	30 K; He
Siemens	Generator	4 MW	3600	10 kN m	27 K; Ne

Figure 3.15 Superconducting motor incorporated with cryocooler.

incorporated in magnetic billet heaters, are commercial products in Germany. FCLs have been developed and have been evaluated in power grids in Italy and China.

For Bi-2212 applications, a series of small coils have been built and tested for high-field magnets. Due to the low bending strains, Bi-2212 coils are typically fabricated using the so-called wind-and-react method. Using this coil fabrication technique, coils are first wound using the unreacted (i.e., "green") wire, which is highly flexible, and then subsequently reacted by high-temperature sintering to form the more brittle superconducting Bi-2212 phase. The high-temperature reaction requires an electrical insulation system capable of surviving these temperatures. To meet this challenge, a new electrical insulation material has been developed to make the overall coil current density high (Chen et al., 2013). Using this new insulation material consisting of a titanium-oxide polymer, researchers succeeded in achieving 33.8 T with a backup magnetic field of 31.2 T at the National High Magnetic Field Laboratory in Florida, USA.

Figure 3.16 Superconducting car.

Table 3.5 **Typical Bi-2223 magnet application projects**

Apparatus	Project	Feature	Top	Type	Note
FCL	CESI	9.4 kV	77 K	Resistive	Grid
	InnoPower	35/220 kV	77 K	Saturated iron	Grid
MRI	GE	Limb size	27 K	Layer wound	–
	Kyoto University	3 T (brain)	20 K	LHe free	2008–2012
NMR	NIMS	24.2 T/1.03 GHz	1.8 K	Layer wound	2007–2013
High-field magnet	Toshiba	1.4 T	20 K	Helmholtz	Single-crystal pulling
	Zenergy	1.8 m long	30 K	Billet heater	Commercial
	Sumitomo	8 T	20 K	200-mm bore	2006
	Sumitomo	5 T	20 K	100-mm bore	*B–H* curve tracer

3.5 Future trends

The present status on critical currents under magnetic fields of various superconducting wires is summarized in a comparative graph, plotted by P.J. Lee (available at http://magnet.fsu.edu/~lee/plot/plot.htm). Bi-2212 and Bi-2223 wires show higher critical current density above 20 T, which is higher than those of Nb-based superconductors. Further improvements in the critical currents of Bi-2212 and Bi-2223 should facilitate the introduction of these wires in more commercial applications. Figure 3.17 shows a history of the improvement of critical currents in Bi-2223 wires. The improvement of critical currents is indispensable for the enlargement of applications and economical aspects, expressed as \$/Am—the price of superconducting wires (\$) divided by critical currents (A) and wire length (m).

For applications, DC superconducting power cables are expected to be the next focus for the future of the grid. DC cables will be the best choice for transporting electricity generated by renewable energy, such as solar cells and wind generators (Sato, Kobayashi, & Nakashima, 2012). This is due to the use of DC electricity in nature for solar cells and for long distances from electricity consumption areas, in the case of wind generators. In fact, DC submarine and inland cables are the main electricity transportation media from offshore wind generators in the North Sea area to the substations in inland Germany. The length of the DC cables is approximately 200 km. The low loss, large current, and compact-sized features of superconducting DC cables could be applied in the electricity transmission and distribution from renewable energy in the near future.

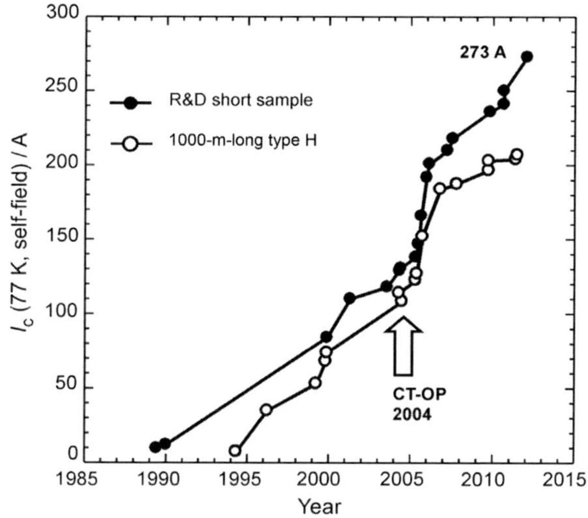

Figure 3.17 History of the improvement of critical currents in Bi-2223 wires.

Acknowledgments

The author thanks Kitaguchi, H., Shimoyama, J., Kiyoshi, T., Miyazoe, A., Nakashima, T., Yamazaki, K., Masuda, T., Shinzato, T., and Oyama, H. for their evaluation on superconducting wires and for supplying materials.

References

Ando, T., Isono, T., Hamada, K., Tsuji, H., Yasukawa, Y., Yamada, H., et al. (1998). Design and testing of 10kA current leads for fusion magnets using high temperature superconductors. In *15th international conference on magnet technology* (p. 847). Beijing: Science Press.

Ballarino, A. (2008). Large-capacity current leads. *Physica C, 468*, 2143. http://dx.doi.org/10.1016/j.physc.2008.05.217.

Cheggour, N., Lu, X. F., Holesinger, T. C., Jiang, J., & Goodrich, L. F. (2012). Reversible effect of strain on transport critical current in $Bi_2Sr_2CaCu_2O_{8+x}$ superconducting wires: a modified descriptive strain model. *Superconductor Science and Technology, 25*, 015001. http://dx.doi.org/10.1088/0953-2048/25/1/015001.

Chen, P., Trociewitz, U. P., Dalban-Canassy, M., Jiang, J., Hellstrom, E. E., & Larbalestier, D. C. (2013). Performance of titanium oxide–polymer insulation in superconducting coils made of Bi-2212/Ag-alloy round wire. *Superconductor Science and Technology, 26*, 075009. http://dx.doi.org/10.1088/0953-2048/26/7/075009.

Dalban-Canassy, M., Myers, D. A., Trociewitz, U. P., Jiang, J., Hellstrom, E. E., Viouchkov, Y., et al. (2012). A study of the local variation of the critical current in Ag-alloy clad, round wire $Bi_2Sr_2CaCu_2O_8C_x$ multi-layer solenoids. *Superconductor Science and Technology, 25*, 115015. http://dx.doi.org/10.1088/0953-2048/25/11/115015.

Dimos, D., Chaudhari, P., Mannhart, J., & LeGoues, F. K. (1988). Orientation dependence of grain-boundary critical currents in $YBa_2Cu_3O_{7-\delta}$ bicrystals. *Physical Review Letters, 61*, 219. http://dx.doi.org/10.1103/PhysRevLett.61.219.

Fujishiro, H., Ikebe, M., Noto, K., Sasaoka, T., & Nomura, K. (1993). Thermal and electrical properties of Ag-Au and Ag-Cu alloy tapes for the metal stabilizer of oxide superconductors. *Cryogenics, 33*, 1086. http://dx.doi.org/10.1016/0011-2275(93)90215-A.

Funaki, K., Iwakuma, M., Kajikawa, K., Takeo, M., Suehiro, J., Hara, M., et al. (1998). Development of a 500 kVA-class oxide-superconducting power transformer operated at liquid-nitrogen temperature. *Cryogenics, 38*, 211. http://dx.doi.org/10.1016/S0011-2275(97)00134-3.

Godeke, A., Acosta, P., Cheng, D., Dietderich, D. R., Mentink, M. G. T., Prestemon, S. O., et al. (2010). Wind-and-react Bi-2212 coil development for accelerator magnets. *Superconductor Science and Technology, 23*, 034022. http://dx.doi.org/10.1088/0953-2048/23/3/034022.

Heine, K., Tenbrink, J., & Thöner, M. (1989). High-field critical current densities in $Bi_2Sr_2Ca_1Cu_2O_{8+x}$/Ag wires. *Applied Physics Letters, 55*, 2441. http://dx.doi.org/10.1063/1.102295.

Hojo, M., Osawa, K., Adachi, T., Inoue, Y., Osamura, K., Ochiai, S., et al. (2010). Effect of fatigue loading on critical current in stainless steel—laminated DI-BSCCO superconducting composite tape. *Physica C, 470*, 1373. http://dx.doi.org/10.1016/j.physc.2010.05.116.

Honjo, S., Shimodate, M., Takahashi, Y., Masuda, T., Yumura, H., Suzawa, C., et al. (2003). Electric properties of a 66 kV 3-core superconducting power cable system. *IEEE Transactions on Applied Superconductivity, 13*, 1952. http://dx.doi.org/10.1109/TASC.2003.812975.

Jiang, J., Starch, W. L., Hannion, M., Kametani, F., Trociewitz, U. P., Hellstrom, E. E., et al. (2011). Doubled critical current density in Bi-2212 round wires by reduction of the residual bubble density. *Superconductor Science and Technology, 24*, 082001. http://dx.doi.org/10.1088/0953-2048/24/8/082001.

Kametani, F., Shen, T., Jiang, J., Scheuerlein, C., Malagoli, A., Michiel, M. D., et al. (2011). Bubble formation within filaments of melt-processed Bi2212 wires and its strongly negative effect on the critical current density. *Superconductor Science and Technology, 24*, 075009. http://dx.doi.org/10.1088/0953-2048/24/7/075009.

Kiyoshi, T., Choi, S., Matsumoto, S., Zaitsu, K., Hase, T., Miyazaki, T., et al. (2011). Bi-2223 innermost coil for 1.3 GHz NMR magnet. *IEEE Transactions on Applied Superconductivity, 21*, 2110. http://dx.doi.org/10.1109/TASC.2010.2082475.

Kiyoshi, T., & Miyazoe, A. (2012). *Private communication.*

Kobayashi, S., Kato, T., Yamazaki, K., Ohkura, K., Fujino, K., Fujikami, J., et al. (2005). Controlled over pressure processing of Bi2223 long length wires. *IEEE Transactions on Applied Superconductivity, 15*, 2534. http://dx.doi.org/10.1109/TASC.2005.847520.

Kobayashi, S., Nakashima, T., Yamazaki, K., & Sato, K. (2012). High-temperature superconducting materials III—Bismuth-based oxide high-temperature superconducting wires. *Teion Kogaku (Journal of Cryogenics and Superconductivity Society of Japan), 47*, 422.

Larbalestier, D.C., Jiang, J., Trociewitz, U.P., Kametani, F., Scheuerlein, C., Dalban-Canassy, M., et al. (2013). A transformative superconducting magnet technology for fields well above 30 T using macroscopically untextured, isotropic round wire multifilament $Bi_2Sr_2CaCu_2O_{8-x}$ conductor. arXiv:1305.1269v1. http://arxiv.org/abs/1305.1269v1.

Li, Q., Tsay, Y. N., Suenaga, M., Gu, G. D., & Koshizuka, N. (1997). Superconducting coupling along the c-axis of [001] twist grain-boundaries in $Bi_2Sr_2CaCuO_{8+\delta}$ bicrystals. *Physica C, 282—287*, 1495. http://dx.doi.org/10.1016/S0921-4534(97)00854-X.

Li, Q., Tsay, Y. N., Zhu, Y., Suenaga, M., Gu, G. D., & Koshizuka, N. (1999). Supercurrent transport across [001] twist grain boundaries in $Bi_2Sr_2CaCu_2O_{8+\delta}$ bicrystals. *Superconductor Science and Technology, 12*, 1046. http://dx.doi.org/10.1088/0953-2048/12/12/302.

Maeda, H., Tanaka, Y., Fukutomi, M., & Asano, T. (1988). A new high-Tc oxide superconductor without a rare earth element. *Japanese Journal of Applied Physics, 27*, L209. http://dx.doi.org/10.1143/JJAP.27.L209.

Malagoli, A., Lee, P. J., Ghosh, A. K., Scheuerlein, C., Michiel, M. D., Jiang, J., et al. (2013). Evidence for length-dependent wire expansion, filament dedensification and consequent degradation of critical current density in Ag-alloy sheathed Bi-2212 wires. *Superconductor Science and Technology, 26*, 055018. http://dx.doi.org/10.1088/0953-2048/26/5/055018.

Naito, T., Fujishiro, H., & Yamada, Y. (2009). Thermal conductivity of single and multi-stacked DI-BSCCO tapes. *Cryogenics, 49*, 429. http://dx.doi.org/10.1016/j.cryogenics.2009.06.005.

Osamura, K., Machiya, S., Ochiai, S., Osabe, G., Yamazaki, K., & Fujikami, J. (2013). Direct evidence of the high strain tolerance of the critical current of DI-BSCCO tapes fabricated by means of the pretensioned lamination technique. *Superconductor Science and Technology, 26*, 045012. http://dx.doi.org/10.1088/0953-2048/26/4/045012.

Osamura, K., Machiya, S., Suzuki, H., Ochiai, S., Adachi, H., Hayashi, K., et al. (2009). Improvement of reversible strain limit for critical current of DI-BSCCO due to lamination technique. *IEEE Transactions on Applied Superconductivity, 19*, 3026. http://dx.doi.org/10.1109/TASC.2009.2019020.

Sato, K., Hikata, T., & Iwasa, Y. (1990). Critical currents of superconducting BiPbSrCaCuO tapes in the magnetic flux density range 0-19.75 T at 4.2, 15 and 20K. *Applied Physics Letters, 57*, 1928. http://dx.doi.org/10.1063/1.104146.

Sato, K., Hikata, T., Mukai, H., Ueyama, M., Shibuta, N., Kato, T., et al. (1991). High-Jc silver-sheathed Bi-based superconducting wires. *IEEE Transactions on Magnetics, 27*, 1231. http://dx.doi.org/10.1109/20.133408.

Sato, K., Kobayashi, S., & Nakashima, T. (2012). Present status and future perspective of bismuth-based high-temperature superconducting wires realizing application systems. *Japanese Journal of Applied Physics, 51*, 010006. http://dx.doi.org/10.1143/JJAP.51.010006.

Shen, T., Jiang, J., Kametani, F., Trociewitz, U. P., Larbalestier, D. C., & Hellstrom, E. E. (2011). Heat treatment control of $Ag-Bi_2Sr_2CaCu_2O_x$ multifilamentary round wire: investigation of time in the melt. *Superconductor Science and Technology, 24*, 115009. http://dx.doi.org/10.1088/0953-2048/24/11/115009.

Wu, L., Zhu, Y., Tafto, J., Welch, D. O., & Suenaga, M. (2002). Quantitave analysis of twist boundaries and stacking faults in Bi-based superconductors by parallel recording of dark-field images with a coherent electron source. *Physical Review B, 66*, 104517. http://dx.doi.org/10.1103/PhysRevB.66.104517.

Yamazaki, K., Kagiyama, T., Kikuchi, M., Yamade, S., Nakashima, T., Kobayashi, S., et al. (2012). Recent progress on improvement to mechanical properties of DI-BSCCO wire. *Superconductor Science and Technology, 25*, 054015. http://dx.doi.org/10.1088/0953-2048/25/5/054015.

Yang, Y., Young, E. A., Falorio, I., Bailey, W. O. S., Friend, C. M., Twin, A., et al. (2011). Quench characteristics of Bi2212 solenoid insert coils in background field up to 20 T. *IEEE Transactions on Applied Superconductivity, 21*, 2432. http://dx.doi.org/10.1109/TASC.2010.2092742.

Yumura, H., Ashibe, Y., Ohya, M., Itoh, H., Watanabe, M., Masuda, T., et al. (2013). Update of YOKOHAMA HTS Cable project. *IEEE Transactions on Applied Superconductivity, 23*, 5402306. http://dx.doi.org/10.1109/TASC.2013.2245931.

Second-generation (2G) coated high-temperature superconducting cables and wires for power grid applications

4

M.W. Rupich
AMSC, Devens, MA, USA

4.1 Introduction

Twenty-five years after the discovery of the high-temperature cuprate-based superconductors, practical wires based on the rare earth (RE) $REBa_2Cu_3O_{7-x}$ family of materials, referred to as second-generation (2G) wire, can now be produced in lengths exceeding 500 m with critical currents (I_c) over 400 A/cm-width (77 K, self-field) by multiple manufacturers (Fleshler et al., in press; Fujikura; Hazelton, 2013a; Lee, 2013; Moon & Hazelton, 2013; Shiohara, Taneda, & Yoshizumi, 2012). The commercial development of the 2G manufacturing technology was a remarkable technical accomplishment, requiring the deposition of multilayer epitaxial oxide films with a near-perfect crystallographic orientation over 100 m in length. Amazingly, multiple organizations successfully developed vastly different texturing and deposition technologies into high-rate roll-to-roll manufacturing processes that achieve this ultra-high degree of crystalline perfection on a routine basis. Currently, three manufacturers (AMSC, SuperPower, and SuNam) are producing 2G wire on a routine basis, while numerous other organizations are striving to develop comparable manufacturing capabilities. The number of organizations still focused on 2G high-temperature superconducting (HTS) wire development after 25 years attests to the strong belief that the 2G HTS wire technology represents an important field that promises to have a major and growing impact on a range of energy applications.

After many years of research and development, 2G wires are now being tested or introduced in applications including transmission and distribution cables, fault current limiters (FCL), motors and generators, transformers, superconducting magnetic energy storage, magnetic resonance imaging, and numerous defense related applications. Although the technical viability of HTS wire technology for electric power applications is established, beyond doubt, a major obstacle remaining for the widespread acceptance of the 2G HTS wire technology is cost. The usual metric used to define commercial viability of the 2G HTS wire is the cost-to-performance ratio, which is typically measured in dollars/kiloampere-meter ($/kA m). For HTS cable applications, the performance metric, measured in kA, generally refers to the critical current (I_c) at 77 K, self-field; for HTS coil-based applications (rotating machines or magnets), the

Superconductors in the Power Grid. http://dx.doi.org/10.1016/B978-1-78242-029-3.00004-2

performance metric refers to the critical current at the operating temperature and magnetic field. In general, it is projected that the 2G HTS wire cost needs to reach the tens of \$/kA m level to compete broadly with conventional copper or permanent magnet-based technologies.

During the early years of 2G HTS wire development, potential manufacturers realized that a roll-to-roll technology was the only viable path for the development of low-cost commercial 2G HTS wire. However, there was a considerable divergence of opinion about the long-term cost and technical feasibility of potential texturing and deposition technologies, with different manufacturers favoring different approaches. Although a variety of different process technologies have been demonstrated to have comparable performance, it is still uncertain if any particular route will have a significant cost advantage in large-scale manufacturing.

Because 2G HTS wire is produced in multiple widths with different formulations and thicknesses of the superconducting layer, the common industry metric for describing the wire performance is the critical current (measured at 77 K and self-field) divided by the wire width (amperes/cm-width). This performance metric accounts for both the thickness and J_c of the superconducting layer, allowing a direct comparison of the performance of wires with different HTS thicknesses, formulations, or wire widths.

This chapter reviews fundamental YBCO properties and how these lead to the current 2G HTS wire architectures and manufacturing technologies. The chapter also describes the different paths chosen for the key texturing and deposition technologies and briefly reviews the major 2G HTS wire-based applications under consideration.

4.2 Second-generation (2G) materials and wire design

4.2.1 YBCO properties

The discovery of the cuprate-based $YBa_2Cu_3O_{7-x}$ HTS material with superconducting transition temperatures (T_c) > 90 K in 1987 led to considerable excitement and major efforts to form these materials into wires usable in a multitude of applications. However, it was soon realized that the cuprates are complex materials that would require a formidable scientific and engineering effort to understand their fundamental properties and incorporate them into useful wire architectures.

High-temperature superconductors, based on the $REBa_2Cu_3O_{7-x}$ (REBCO) family of materials (where RE is a rare earth element), are complex oxide compounds based on a perovskite-type crystal structure containing an RE atom located at the center of the lattice, sandwiched between a square planar arrangement of CuO_2 units sharing four vertices, as shown in Figure 4.1 (where RE = Y). Cu–O chains, sharing two vertices, are arranged perpendicular to the planes along the b-axis and can have variable oxygen content. The oxygen content, expressed by $(7-x)$ in the chemical formula of $REBa_2Cu_3O_{7-x}$, is determined by the amount of oxygen on the CuO chains and plays an important role in the superconducting properties of the material. At low oxygen content (large x), the crystal structure is tetragonal and the material is electrically insulating. Higher oxygen contents ($x < 0.65$) result in the formation of Cu–O chains along the b-axis, converting the material into the superconducting

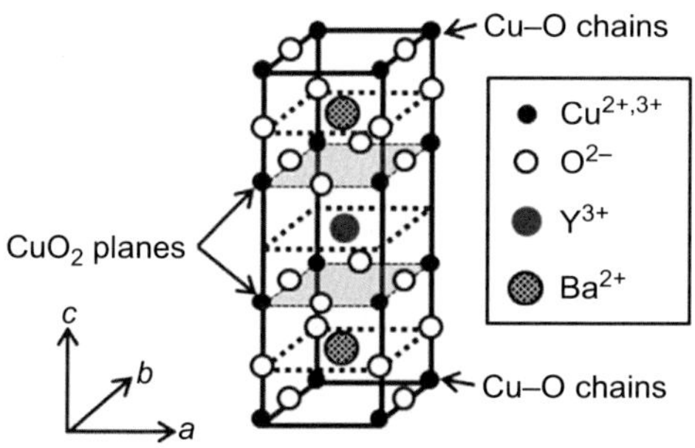

Figure 4.1 The $YBa_2Cu_3O_{7-x}$ cuprate superconductors crystalize in a defect perovskite structure in which the layers of the RE atoms are sandwiched between the two adjoining pyramid type CuO_2 layers. There are also one-dimensional chains of Cu—O—Cu oriented along the *b*-axis of the crystal.

orthorhombic structure, with a maximum superconducting transition temperature (T_c) occurring for $x \sim 0.07$. Superconductivity in the orthorhombic material is confined to the CuO_2 layers, while the CuO chains act as charge reservoirs providing carriers to the CuO_2 planes. This planar structure results in a strong anisotropy in the electromagnetic properties of the material, with supercurrents in the $a-b$ plane being several orders of magnitude greater than supercurrents along the *c*-axis (Cooper and Gray, 1994; Jia et al., 2011).

The first HTS wire produced using the REBCO material was fabricated by Jin et al. using an oxide-powder-in-tube approach (Jin, Sherwood, van Dover, Tiefel, & Johnson, 1987). However, the wire, which consisted of a polycrystalline YBCO core, carried a critical current of only 175 A/cm^2 (77 K, self-field), which is several orders of magnitude lower than required for practical conductors. Soon after, studies of the REBCO materials found that the cause of the low critical currents originated in the fundamental physics of the cuprate materials; specifically, the short coherence length and the correlated character of their electron systems render them sensitive to grain boundary misorientations.

The coherence length, defined in the BCS-type model as $\xi = a\hbar v_F/kT_c$ (where a is a constant of order unity, $\hbar$ is Planck's constant divided by 2π, and v_F is the Fermi velocity, and T_c is the superconductor transition temperature) is a fundamental property of the HTS material and can be loosely viewed as the spatial distance over which the paired superconducting electrons interact. In REBCO materials, ξ is in the range of Angstroms (Tinkham, 1996); thus, any disruption in the lattice periodicity can affect the superconductor gap, obstructing current flow. High-angle grain boundaries, such as those present in polycrystalline samples, are essentially planar defects that drastically restrict the current flow from grain to grain. Consequently, the major obstacle to the manufacture of commercial lengths of YBCO wire is that grain boundaries, formed by the misalignment of neighboring YBCO grains, form weak links that obstruct the current flow (Mueller & Deutscher, 1987).

Studies by Dimos, Chaudhari, and Mannhart (1990), using YBCO films deposited epitaxially on bi-crystals with a specific grain boundary misorientation, demonstrated that when the misalignment angle between superconducting YBCO grains was less than 4°, the critical current density (J_c) in polycrystalline samples approached that of YBCO films grown on single crystals (Dimos et al., 1990). These experiments also showed that for misalignment angles above 4°, the J_c across the grain boundary decreases exponentially with the grain boundary misorientation angle (Hilgenkamp & Mannhart, 2002). The group also found the J_c across grain boundaries was independent of the type (twist, tilt) of grain misorientation (Dimos et al., 1990). These results established that, to achieve high critical currents, the YBCO material needs to be biaxially textured, requiring nearly perfect alignment of all three crystallographic axes.

For many years, it was thought that the manufacture of kilometer-length wire using the REBCO materials would require the production of a kilometer-length single crystal—a challenge beyond the realm of imagination. Thus, the development of HTS-based wires focused on the BiSrCaCuO-based materials (i.e., $Bi(Pb)_2Sr_2Ca_2Cu_3O_x$), typically referred to as the first-generation or 1G wire, which could be uniaxially textured (*c*-axis grain alignment only) through thermomechanical processing using powder-in-tube methods (Hikata, Sato, & Hitotsuyanagi, 1989). Although there is no alignment of the *ab*-planes, the critical current in the wire is enhanced by the brick-wall type stacking of the platelet-shaped grains, providing an enormous effective grain boundary area that compensates for the weak-linked grain boundaries (Bulaevskii, Clem, Glazman, & Malozemoff, 1992; Bulaevskii, Daeman, Maley, & Coulter, 1993; Hensel et al., 1993; Riley, Malozemoff, Li, Fleshler, & Holesinger, 1997).

Efforts to achieve an epitaxial texture in long lengths of the HTS materials met with little success until 1991 when Iijima, at Fujikura, demonstrated that it was possible to grow biaxially textured materials (in biaxially textured materials, all three axis are oriented), such as yttria-stabilized zirconia (YSZ), on untextured substrates using an argon-ion beam aimed at an angle to a substrate during the vapor phase deposition process (Iijima, Tanabe, Ikeno, & Kohno, 1991; Iijima, Tanabe, Kohno & Ikeno, 1992; Iijima, Kakimoto, Kimura, Takeda, & Saitoh, 2001). These textured YSZ films, which appeared like a YSZ single crystal, served as templates for depositing epitaxial YBCO films with low grain boundary misorientations and thus high J_c's. They also demonstrated that this texturing process can be adapted to a roll-to-roll configuration using flexible metal substrates, establishing the first practical route to manufacturing long lengths of highly textured YBCO films.

This fundamental concept of using a textured template with an epitaxial HTS layer gave birth to 2G HTS wire. It was quickly recognized by the international scientific community that materials on long-length flexible substrates could be textured in multiple ways; in fact, not only buffer layers but also the substrates themselves could be textured. Numerous processes were also developed to deposit the epitaxial HTS layer on the textured templates. This led to a tremendous acceleration of efforts around the world to develop and scale up 2G HTS wire manufacturing processes.

4.2.2 Wire architecture

All 2G wire is now produced with a composite tape architecture comprised of a flexible textured template onto which an epitaxial film of $REBa_2Cu_3O_{7-x}$ is deposited, as

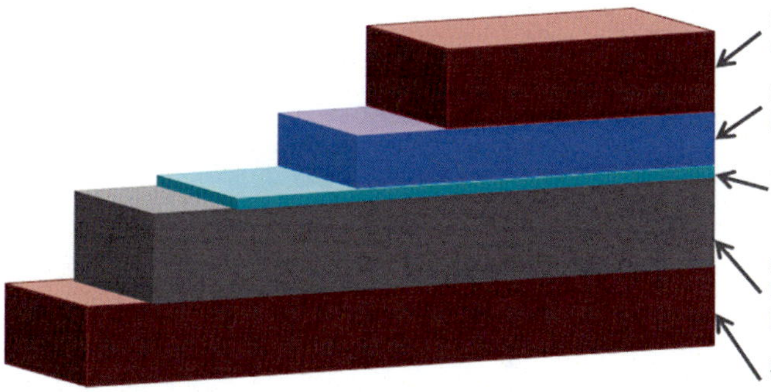

Figure 4.2 All 2G HTS wires are based on a composite architecture comprised of a metal substrate, biaxially textured oxide buffer layers, REBCO layer, and a metallic stabilizer layer (not to scale). Specific materials, formulations, and thicknesses of the individual layers vary from manufacturer to manufacturer.

shown in Figure 4.2. The composite tape is surrounded with a normal metal that serves both as an electrical and thermal stabilizer and provides a low resistance electrical contact to the REBCO. The template has a multilayer architecture consisting of a flexible metallic substrate onto which are deposited a series of oxide layers that isolate the YBCO layer from the metal substrate and provide a surface for the epitaxial nucleation and growth of the YBCO. Although all 2G HTS wires are based on this composite structure, different methods have been developed to produce the biaxially textured substrates, deposit the HTS layer, and stabilize the wire. As a result, the specific materials and architectures of the substrate, oxide layers, and stabilizer layers vary with the actual manufacturing processes and manufacturer. There have been numerous efforts to develop round wire architectures compatible with the biaxial texture requirements of the REBCO; however, these have not evolved past small laboratory samples (Dechoux et al., 2012).

Today, 2G HTS wire is manufactured using roll-to-roll processing in widths of 10 to 50 mm. These tapes are then slit to multiple, narrower strips, which are fabricated into the actual wires. Typically, wires for cable applications are fabricated in a width of ~4 mm, whereas wires for coil applications are manufactured in a width of 10–12 mm. Figure 4.3 illustrates the generic process showing the major manufacturing steps (template, superconductor, and wire) employed by all manufacturers for the preparation of 2G HTS wire.

4.3 2G wire fabrication approaches

Although the basic architecture of all 2G HTS wire produced today consists of a composite tape structure, different manufacturers have chosen different technologies for producing both the multilayer template and the REBCO layer. The three approaches developed for producing a template consisting of a flexible metal

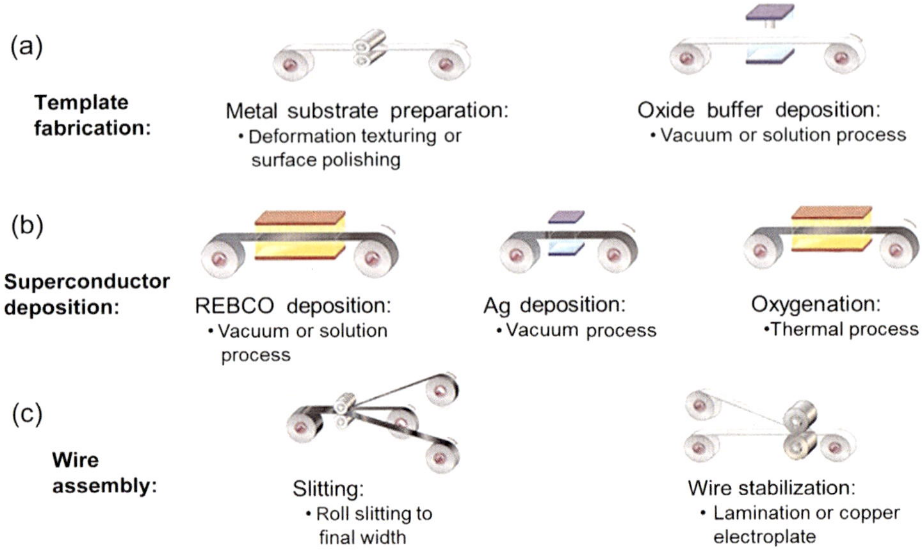

Figure 4.3 Schematic diagram illustrating basic process steps for (a) template fabrication, (b) superconductor deposition, and (c) final wire assembly used in the manufacturing of all commercial 2G HTS wire.

substrate and supporting a biaxial oxide film include the following: (1) ion-beam-assisted deposition (IBAD), (2) rolling-assisted biaxially textured substrates (RABiTS™), and (3) inclined-substrate deposition (ISD). Although all three processes have been demonstrated in long-length manufacturing, they rely on significantly different technologies. The greatest challenge with each technology has been controlling the texturing process so that the individual grains of the template are crystallographically aligned over its entire length and width. Remarkably, all three processes have been successfully scaled for roll-to-roll manufacturing, and the RABiTS and IBAD templates are now routinely manufactured in lengths of more than 500 m.

The four methods developed for depositing the heteroepitaxial REBCO (where *heteroepitaxial* refers to oriented growth of a material on a substrate of a different material) layer include the following: (1) pulsed laser deposition (PLD), (2) metal organic deposition (MOD), (3) metal organic chemical vapor deposition (MOCVD), and (4) reactive co-evaporation (RCE). In general, the template and REBCO processes can be mixed and matched. However, depending on the specific combination, details of the processes may need to be adjusted. Each of these processes is described in the following sections.

4.3.1 Processing approaches

Unlike 1G HTS wire that is manufactured using conventional powder-in-tube wire drawing, rolling, and batch heat treatments, 2G HTS wire, which consists of thin films on a flexible substrate, is amenable to roll-to-roll (R2R) processing. R2R processing, also referred to as reel-to-reel or web processing, is a rapidly emerging manufacturing technique for flexible electronics, displays, photovoltaics, and other materials. Its major advantages include high rates and throughput and lower materials cost compared to conventional batch processes. The 2G HTS wire manufacturers were among the first to adopt the R2R processing technology for large-scale manufacturing. Because there was no existing infrastructure for R2R manufacturing of flexible electrical materials, particularly ones with the stringent biaxial texture requirements of 2G wire, the 2G scientific and engineering community had to focus on designing custom tools as well as developing the basic HTS technology.

Once the basic template and REBCO deposition processes were established, companies focused on selecting specific processes that were amenable to high-rate R2R processing. In the 1990s, it was uncertain whether either the RABiTS or IBAD process could be successfully scaled to produce kilometer lengths of templates with near single crystal quality over their length and width. Although different organizations ultimately chose different combinations of technologies for both the template and REBCO processes, the major manufacturers have demonstrated the ability to produce 2G HTS wire with comparable performance. One of the primary focuses of the development has been the cost of the manufacturing process. From the beginning, it was recognized that for cost reasons, it was necessary to maximize the process (or deposition) zone area in order to maximize throughput (m^2/h or, in the industry jargon, meters of wire 4-mm-wide wire/h). Again, different paths were taken to address this issue.

AMSC chose to focus on starting with a wide substrate and passing it through the entire R2R process. At the end of the process, the wide-web is slit into multiple strips, which are then converted into individual wires, as illustrated in Figure 4.4. AMSC's current process is designed around a 46-mm-wide starting strip, which is slit into multiple narrower wires. AMSC has also designed its manufacturing equipment to process strips up to 100 mm wide. A key requirement of this approach is maintaining uniform properties across the substrate width. The wide-web process is particularly advantageous with solution-based deposition approaches, which use low-cost ovens with process zones up to 10 m in length, thus enabling exceptionally high throughput (m^2/h) compared to the plasma vapor deposition (PVD) approaches in which depositions zones are generally restricted to under a meter in length.

In contrast, SuperPower chose to start with a narrower 12-mm-wide strip that is slit into three individual wires of 4 mm width at the end of the process. To increase the deposition area, SuperPower developed a helix system in which the narrow tape passes through the deposition zone multiple times, as illustrated in Figure 4.4. Because the strip passes through the deposition zone multiple times, the line speed can be increased by the number of passes. In practice, a seven-pass helix system with a 12-mm tape achieves the same throughput (m^2/h) as is achieved with a 100-mm wide-web processed in a deposition zone with the same length (assuming equivalent deposition rates). An advantage of the helix system is that any inhomogeneities across the width of the deposition zone are averaged out; however, it adds complexity to the transport system and requires a higher strength substrate to avoid mechanical damage. Additionally, the helix system

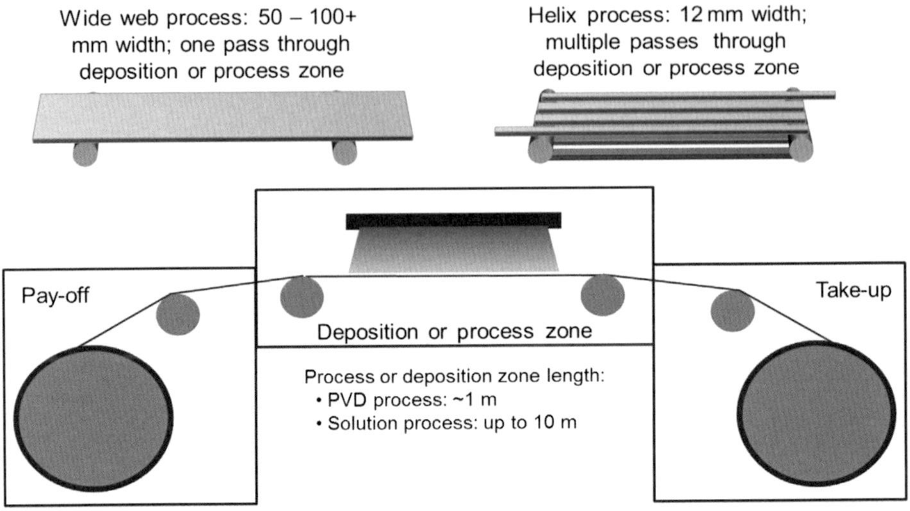

Figure 4.4 In the wide-web process, developed by AMSC, all processing is done on a wide (up to 100 mm) substrate, which at the end of the process is slit into multiple wires of 4–10 mm width. In the helix system, developed by SuperPower, the narrow (~12 mm) substrate passes through the deposition zone multiple times, effectively increasing the length of the zone by the number of passes. In principle, for deposition (or process) zones of the same physical length and width, both processes yield a comparable throughput (m^2/h).

is not readily adaptable to solution-based deposition processes, which favor ovens with long process zones.

The wide-web process has been the choice of manufacturers using the RABiTS process, while those using the IBAD and ISD processes have favored the helix approach.

4.3.2 Template technologies

The template controls the nucleation and biaxial grain alignment of the REBCO and is the key component that allows roll-to-roll manufacturing of today's 2G HTS wire. In addition to enabling the growth of the biaxial REBCO, the template must also be chemically compatible with the REBCO, stable at the REBCO growth conditions, amenable to formation into tapes, and provide sufficient mechanical stability to the fragile REBCO films. Early attempts to grow REBCO on the surface of metal foils such as Ni or Ag were unsuccessful because of the difficulty of controlling the hetero-epitaxy of the oxide films and interfacial reactions between the REBCO and substrate. Buffering the metal substrates with oxide films such as YSZ (*c*-axis texture only) minimized the interfacial reactions, but the resulting *c*-axis-oriented (fiber-textured) YBCO films had J_c values that were orders of magnitude lower than those obtained on single crystal substrates (Russo, Reade, McMillian, & Olsen, 1990; Witanachchi, Patel, Zhu, Kwok, & Shaw, 1990). The major breakthrough in the template technology, as described earlier, came in 1991 when Iijima produced a biaxially textured YSZ film on polished polycrystalline Hastelloy C-276, which supported the subsequent growth of biaxially textured REBCO films with high J_c values (Iijima et al., 1991).

4.3.2.1 IBAD templates

The IBAD process is based on the use of an off-normal ion beam bombardment of a film during its vapor phase deposition, as illustrated in Figure 4.5. The method, first used for improving the in-plane texture of Nb films on fused silica (Yu, Harper, Cuomo, & Smith, 1985), was extended to the deposition of YSZ on a polished polycrystalline Hastelloy C-276 substrate by Iijima in 1991. The texture in the IBAD process arises as the off-normal ions sputter away grain orientations that are not favorable, allowing the formation of a highly oriented, biaxially textured layer on a completely untextured substrate.

Once this texturing process was demonstrated, it was rapidly developed by Fujikura, Los Alamos National Laboratory, Bruker High Temperature Superconductor, SuperPower, and others using YSZ and gadolinium zirconate, $Gd_2Zr_2O_7$ (GZO) films (Foltyn et al., 1999; Iijima, Kakimoto, Takeda, & Saitoh, 2001; Iijima et al., 2004; Ma, Li, Fisher, Koritala, & Balachandran, 2002; Selvamanickam et al., 2003; Usoskin et al., 2003; Wu et al., 1995). However, these materials require a thickness on the order of a micron to develop a sufficient degree of texture; also, because the IBAD texturing process with these materials was slow, long times were required to achieve the texture necessary to eliminate the current impeding weak links between the YBCO grains. The combination of the film thickness and long texturing times presented a formidable challenge to developing an economically viable roll-to-roll manufacturing process.

However, in 1999, Hammond (Stanford University) discovered that MgO developed an excellent biaxial texture almost immediately after the onset of nucleation, reaching a constant texture within the first 10 nm (Wang, Do, Beasley, Geballe, & Hammond, 1997). This development provided the path to a rapid texturing process essential for high-rate roll-to-roll manufacturing. This MgO texturing process was adapted to a R2R process by the group at Los Alamos National Laboratory and

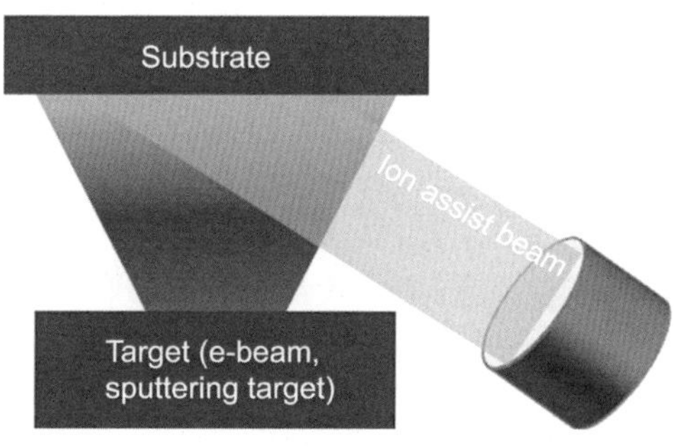

Figure 4.5 Schematic illustration of the IBAD process, which utilizes an off-axis ion beam to texture the growing oxide film on an untextured substrate.

subsequently transferred to SuperPower (Arendt & Foltyn, 2004). Today, nearly all IBAD templates are based on the MgO material.

A prerequisite for achieving the necessary texture in the IBAD process, particularly with the thin MgO layer, is that the surface roughness (R_a) of the metal substrate should be ~1 nm or less. Typically, an electropolishing process is used to smooth the substrate surface. After the electropolishing, a thin Al_2O_3 layer is deposited that functions as a barrier to block diffusion of potentially detrimental constituents of the metal substrate into the HTS layer during subsequent high-temperature processes. Finally, an amorphous Y_2O_3 layer is deposited onto the Al_2O_3 layer, providing a smooth, noncrystalline surface that enhances the rapid texturing of the IBAD-deposited MgO layer.

After deposition of the 10-nm biaxial-textured IBAD-MgO layer, a homoepitaxial (where *homoepitaxial* refers to oriented growth of a material on a substrate of the same material) MgO layer is deposited to stabilize the epitaxy, followed by a final oxide layer, such as $LaMnO_3$, CeO_2, or $SrTiO_3$, which provides a good lattice match and chemical compatibility for the subsequent growth of the REBCO film (Paranthaman et al., 2003). The basic architecture of the IBAD template developed by SuperPower for long-length manufacturing is shown in Figure 4.6. The SuperPower process uses a combination of magnetron sputtering and IBAD to deposit the multilayer buffer stack on the metal substrate. As seen in Figure 4.7, the entire multilayer buffer stack is ~200 nm thick, allowing for rapid roll-to-roll deposition over long lengths, resulting in production rates of 300–760 m/h of 12-mm-wide tape and lengths over 1 km (Xiong et al., 2009). SuperPower has reported in-plane textures of 6–7° on substrates over 1 km in length. Other groups including SuNam, working with Seoul University and KERI in Korea (Ko et al., 2007), ISTEC-SRL Nagoya (Yamada et al., 2009), and Fujikura in Japan (Hanyu et al., 2010; Kutami et al., 2009) have also produced IBAD-MgO templates with comparable lengths and rates; however, the composition or thickness of the metal substrate and layers are often changed to optimize the properties for the specific HTS deposition process.

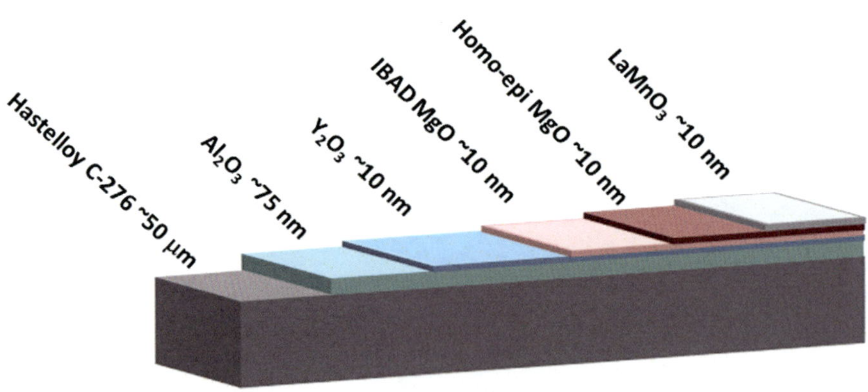

Figure 4.6 Schematic illustration of the of the IBAD-based template architecture developed by SuperPower (not to scale).

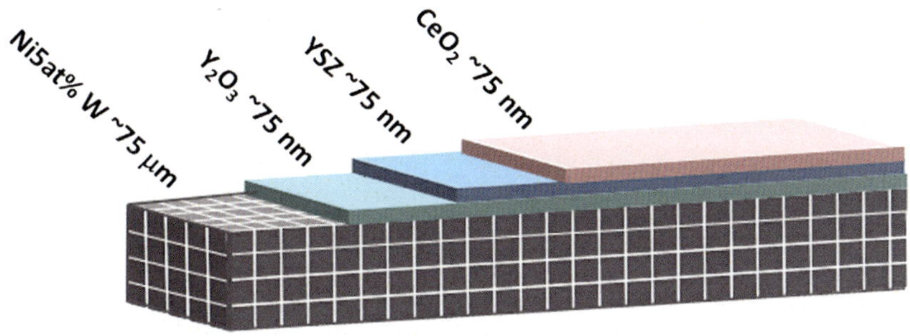

Figure 4.7 Schematic illustration of the of the RABiTS-based template architecture developed by AMSC (not to scale).

The slowest step in the IBAD process is the electropolishing of the metal substrate. Alternative planarization technologies have been explored, including solution routes using a hydrolysable metal compound to form a glassy oxide layer on an unpolished alloy substrate surface through dip coating and heat treatment (Qiao et al., 2011). Although this process has been shown to produce the surface smoothness required for the IBAD process, it has not currently been incorporated into any long-length manufacturing processes.

A primary advantage of the IBAD technology is that, because virtually any substrate can be used, it is possible to select low-cost off-the-shelf materials that are nonmagnetic, high-strength, resistive or conductive, and easy to handle. Additionally, because the IBAD templates have a small grain size, potential percolation issues around large area, weak link grain boundaries are minimized, allowing the fabrication of very narrow wires or patterning of filaments.

4.3.2.2 Rolling-assisted biaxially textured substrates template

The second method to achieve a highly textured template is the RABiTS™ method, introduced by Goyal et al. (1996) of Oak Ridge National Laboratory and extensively developed by AMSC, Evico and IFW-Dresden, Zenergy, Nexans, Furukawa, and Sumitomo (Eickemeyer et al., 2008). In contrast to the IBAD process, the RABiTS technology (Izumi et al., 2009) is based on generating a cube texture in a metal substrate by conventional rolling deformation and recrystallization techniques as it is converted into a thin tape (Goyal et al., 1996; Norton et al., 1996; Specht et al., 1998). The origin of this process dates back to the early 1920s when cold working was found to produce a preferred crystallographic orientations in metals (Barrett & Massalski, 1996). Numerous cubic metals or alloys can be textured by this technique; however, face-centered cubic materials such as Cu and Ni are preferred because the thermomechanical deformation process produces a very sharp cube texture with the cube plane parallel to the plane of the tape and a cube edge parallel to the rolling direction. The resulting structure resembles a single crystal with a mosaic of slightly misoriented subgrains.

In addition, the substrate material must also have sufficient oxidation resistance to withstand subsequent higher temperature buffer and REBCO deposition steps. This makes Ni or Ni-based alloys the material of choice. The original RABiTS process was demonstrated with pure Ni tapes (Goyal et al., 2001). However, its low tensile strength made handling of the thin Ni tapes a challenge during the high-temperature R2R processing. Additionally, the ferromagnetism of the Ni substrate produces hysteretic losses in 2G wire when used in alternating current (AC) applications (Goyal, Paranthaman, & Schoop, 2004). Thus, the choice of substrate moved to Ni alloys, particularly Ni−5 at% W, which has a sharp cube texture, a tensile strength of 250 MPa, reduced ferromagnetism, and sufficient oxidation resistance (Eickemeyer et al., 2010; Verebelyi et al., 2003). The Ni−5 at% W alloy was the initial choice of AMSC, which now produces the textured substrate in kilometer lengths and 100-mm widths. The Ni−5 at% W substrates are about 98% cube textured with in- and out-of-plane textures of 5° and 6° respectively. Although higher weight-percent W alloys are nonmagnetic with no ferromagnetic loss and higher tensile strength, their reduced stacking fault energy makes it difficult to achieve a single cube texture using conventional thermomechanical processing.

Because the Ni alloys are not chemically compatible with the direct deposition of the REBCO layer during the high-temperature growth processes, an oxide buffer layer is required to isolate the HTS layer from the metal substrate. The buffer layer stack must also provide a chemically compatible surface for the epitaxial nucleation and growth of the REBCO film, as well as maintain the biaxial texture of the metal substrate. The typical buffer layer stack on the RABiTS substrate consists of three layers: a seed layer that replicates the biaxial texture of the substrate, a barrier layer that inhibits diffusion of the substrate constituents to the REBCO layer during the high-temperature processing, and a cap layer that is chemically compatible with the epitaxial REBCO nucleation and growth.

The deposition of the seed layer is the most challenging process. The oxide layer and the deposition conditions must be selected such that formation of a native oxide on the substrate is not thermodynamically stable. Y_2O_3 or CeO_2, deposited by rf or reactive sputtering, e-beam evaporation, or PLD techniques, are the typical choice for the seed layer. Cantoni also showed that the formation of a $c(2 \times 2)$ sulfur super structure on the metal alloy surface ensured the formation of a single in-plane texture with these seed layer materials (Cantoni et al., 2001, 2002, 2003). Under the correct deposition conditions, the out-of-plane texture of the Y_2O_3 can be improved by $\sim 1°$ relative to that of the Ni−5 at% W substrate.

Once the seed layer is deposited, the barrier and cap layers can be deposited by a variety of vacuum-based deposition techniques. The most commonly used barrier layer is YSZ, while CeO_2 is used for the cap layer.

The basic architecture of the RABiTS template developed by AMSC for long-length manufacturing is shown in Figure 4.7. The AMSC process uses conventional rolling and annealing techniques to produce the substrate, and reactive sputtering for deposition of the buffer layer stack. As seen in Figure 4.7, each buffer layer is ~ 75 nm thick, allowing for rapid roll-to-roll deposition over long lengths. AMSC's RABiTS templates are typically produced with in-plane and out-of-plane textures of

4° and 5°, respectively, in a length over 600 m and a width of 46 mm. Sumitomo and Deutsche Nanoschrift are also developing similar RABiTS template technologies; however, the composition and thickness of the buffer layers may be changed to optimize the properties for the specific HTS deposition process.

Significant effort has been focused on developing solution-based routes for depositing the buffer layers (Cordero-Cabrera et al., 2007; Obradors et al., 2004, 2006; Paranthaman et al., 2008). These nonvacuum routes offer the opportunity to significantly reduce the materials and capital costs for the buffer layer deposition. Although numerous organizations have developed processes for the seed, barrier, and cap layers, the resulting buffer layers do not match the quality of the vacuum-based buffer layers, particularly the texture enhancement on the seed layer.

A drawback to the current commercial RABiTS template is the magnetism of the Ni−5 at% W material used for the substrate. Although the Ni−5 at% W alloy is only slightly ferromagnetic, it still results in a hysteretic loss when used in AC applications. In contrast, the typical IBAD substrate (Hastelloy C-276) is completely nonmagnetic and does not contribute to any hysteretic loss. To overcome this issue, AMSC has pioneered the development of a Ni−9 at% W alloy substrate that is nonmagnetic and has properties comparable to the standard Ni−5 at% W substrate (Rupich et al., 2013). An alternative approach, developed by Chubu Electric (Kashima et al., 2009) and Sumitomo Electric Industries (SEI) (Ohya et al., 2010), is based on a composite strip consisting of a stainless-steel core bonded to a highly textured Cu strip, which is coated with a thin Ni layer. 2G HTS wire prepared from both of these nonmagnetic substrates have losses comparable to the IBAD-based materials.

4.3.2.3 ISD templates

The third method of producing a textured template, initially introduced by Hasegawa et al. of SEI (Hasegawa et al., 1996), is referred to as inclined substrate deposition or ISD and involves the deposition of an oxide film (YSZ) at an angle slightly tilted from the plane of the substrate plane. This process was also developed by several organizations using MgO (Balachandran et al., 2003; Hasegawa, Hobara, Nakamura, Izumi, & Shiohara, 2001; Ma et al., 2004; Prusseit et al., 2003). The texture develops due to a shadowing effect that allows preferential growth of the MgO along the [001]-axis, which is slightly tilted from the substrate normal (Dürrschnabel et al., 2012). The texture evolution and tilt of the MgO, determined primarily by the angle, deposition rate, and film thickness, are critical because they have a strong influence on the subsequent superconductor film growth (Bauer, Metzger, Semerad, Berberich, & Kinder, 1999). Typically, growth rates in excess of 500 nm/min and films thickness of over 1 μm are required for highly textured films.

Theva, the only company continuing to pursue the ISD technology, has developed a roll-to-roll process capable of producing the MgO layer on continuous lengths of Hastelloy C276 tapes that supports the subsequent growth of REBCO layers deposited by RCE (Dürrschnabel et al., 2012; Prusseit et al., 2006).

4.3.3 Superconductor layer

Four distinct deposition approaches have been developed for the superconducting layer, including: (1) PLD, (2) MOD, (3) MOCVD, and (4) RCE. Although each technique is based on a distinct approach, all have been demonstrated in roll-to-roll processes with good performance.

A variety of different chemical formulations have also been developed for the REBCO layer in order to enhance deposition rates, growth conditions, or J_c for each specific deposition technique. Most of the manufacturers using the *in situ* based growth processes (PLD, MOCVD, and RCE) use a $GdBa_2Cu_3O_{7-x}$ composition while the *ex situ* solution routes (MOD) use a $YBa_2Cu_3O_{7-x}$ composition. Additionally, a variety of dopants have been added to these basic compositions to modify the flux pinning defect structure and thereby optimize the J_c of the HTS layer at specific temperature and magnetic field regimes. These chemical modifications include the addition of RE_2O_3-based nanoparticles and self-assembled $BaZrO_3$ nanoparticles or columns. AMSC uses a Dy_2O_3 nanoparticle-doped YBCO composition to enhance the flux pinning (Strickland et al., 2008), while SuperPower uses a combination of RE_2O_3 nanoparticle and $BaZrO_3$ nanorod additions (Chen et al., 2009).

4.3.3.1 Pulsed laser deposition

PLD is an *in situ* growth process that uses a high-power excimer laser (KrF or XeCl) to vaporize a REBCO target, which is then deposited from the plasma onto the template as a thin film. A major advantage of the PLD process is that the composition of the target is preserved during the deposition process, and by rastering the laser between targets with different compositions, it is easy to vary the stoichiometry or even deposit different layers of material during film growth. The PLD process was first used for the deposition of YBCO films in 1988 (Dijkkamp et al., 1987) and has become widely used as a research tool that produced some of the highest I_c REBCO films (1400 A/cm w at 75 K, self-field) reported (Foltyn et al., 2005).

In the early days of 2G HTS wire development, PLD was considered to be too expensive for manufacturing, primarily due to the cost of the excimer lasers and the small deposition zone of the research tools. However, as the PLD process was studied in the laboratory and laser costs decreased, a number of organizations, including Fujikura, Gottingen University, ISTEC-SRL Nagoya, Bruker, Sumitomo, and Los Alamos (Abiru, Shingai, Konishi, & Ohmatsu, 2011; Usoskin & Freyhardt, 2004; Usoskin et al., 2007; Watanabe et al., 2005), adapted the process for R2R manufacturing. The most successful approach has been the development of multi-plume and multi-turn deposition process (Ibi et al., 2006). By rastering the laser beam over a large target and splitting the beam through the use of mirrors, the deposition zone is increased, allowing the use of a multi-pass helix system or wide-web. Both Fujikura and Sumitomo Electric (Fujino et al., 2003; Ibi et al., 2006; Ohmatsu, Hahakura, Hasegawa, & Ueyama, 2005; Ueyama, Hahakura, Hasegawa, & Ohmatsu, 2006) have scaled this process for the manufacture of long-length wire. An alternate approach pioneered by Bruker (BEST) in Germany uses a cylindrical target (Watanabe et al., 2005) to increase the deposition area of the tape substrate. A number of organizations, especially in Japan, use a GdBCO-based material in the PLD process because of higher deposition rates and higher J_c compared to the standard YBCO composition (Yamadaet al, 2007).

4.3.3.2 Metal organic deposition

MOD is a low-cost, *ex situ* technique in which a solution-based precursor is coated onto a substrate and subsequently converted into the product film. Although solution-based coating techniques are well-established industrial processes for the low-cost coating of wide, continuous lengths of flexible substrates (examples include photographic paper and film), they had never been used for the deposition of high-quality, epitaxial ceramic films over flexible kilometer-length substrates.

A number of precursors have been used with the MOD process; however, the most widely used, referred to as the TFA precursor, is based on a methanol solution of the RE, Ba, and Cu trifluoroacetate salts developed by Gupta (Gupta, Cooper, Jaganathan, & Giess, 1988; Gupta, Jaganathan, et al., 1988) and refined by Cima (McIntyre & Cima, 1994; McIntyre, Cima, & Ng, 1990; McIntyre, Cima, & Roshko, 1995; McIntyre, Cima, Smith, et al., 1992). The basic MOD process consists of three distinct steps: (1) precursor coating, (2) pyrolysis of the organics, and (3) YBCO nucleation and growth (Matsuda, Nakaoka, Izumi, Yamada, & Shiohara, 2008; Obradors et al., 2012; Solovyov, Wiesmann, & Suenaga, 2001; Yoshizumi et al., 2008).

A variety of industrial coating techniques, including slot die coating, dip coating, and inkjet printing (Bartolomé et al., 2013; Katayama, Nakahata, Yoshizumi, Izumi, & Shiohara, 2010; Rupich et al., 2010), are used for depositing the TFA precursor onto the flexible template. Once the precursor is deposited on the template, the solvent is removed in a low-temperature drying process, and then the film is pyrolyzed in a wet oxygen environment, leaving a film comprised of copper oxide and RE-Ba oxyfluoride (REBaOF) phases. The metal oxyfluoride film is then converted to the REBCO phase by reaction with H_2O and O_2, according to Eqns (4.1) and (4.2):

$$\mathrm{RE}_n\mathrm{Ba}_m\mathrm{O}_x\mathrm{F}_y \xrightarrow{\mathrm{H_2O}} (0.5n)\mathrm{RE_2O_3} + (m)\mathrm{BaF_2} + (y-2m)\mathrm{HF}(g) \tag{4.1}$$

$$0.5\mathrm{RE_2O_3} + 2\mathrm{BaF_2} + 3\mathrm{CuO} \xrightarrow{\mathrm{H_2O/O_2}} \mathrm{REBa_2Cu_3O_{7-x}} + 4\mathrm{HF}(g) \tag{4.2}$$

The pyrolysis is the slowest step in the process and also limits the thickness of the film. The pyrolysis of the TFA precursor films results in >50% reduction in film thickness and is accompanied by an increase of stress within the film. The stress development, which depends on the film thickness and pyrolysis rate, along with rapid gas evolution resulting from decomposition of the trifluoroacetate ligands, can result in severe cracking or delamination of the precursor film. The initial pyrolysis process developed by Cima required in excess of 15 h to produce a crack-free, adherent film that would yield a 0.2-μm-thick YBCO film, but it was not capable of producing thicker films. The major advance that allowed adoption of the MOD process for R2R manufacturing was the development of a modified trifluoroacetate-based precursor by AMSC and others that undergoes a rapid pyrolysis process without cracking or delamination (Fuji et al., 2005; Li et al., 2012; Tada et al., 2008a; Tokunaga et al., 2004; Yoshida et al., 2008). AMSC now routinely processes films over a micron thick in a single-pass R2R pyrolysis step requiring <0.5 h residence time (Rupich et al., 2010, 2013). An alternate approach involves the repetitive coating and rapid pyrolysis of multiple thin films. ISTEC has used a 10×0.2 μm coat/pyrolysis process to prepare films with 2-μm-thick REBCO layers (Izumi et al., 2009).

Unlike the vapor phase deposition processes in which the film grows layer-by-layer from the vapor phase, the MOD process requires nucleation of the REBCO phase at the bottom of the precursor and subsequent epitaxial growth through the film thickness. Any nucleation of the REBCO phase that occurs in the bulk precursor film over top of the growth front results in the formation of randomly oriented REBCO grains that disrupt the epitaxial growth (Rupich, Li, Sathyamurthy, Thieme, & Fleshler, 2011; Tada et al., 2008b). Controlling the REBCO nucleation and growth kinetics, which depends on the equilibrium $K_{eq} = [HF]^4/[H_2O]^2$, requires precise control of the H_2O_g and HF_g concentrations at the tape surface. Adapting this growth process to a wide-web R2R system demands precise control of the gas concentrations along the length and width of the tape. Using computational fluid dynamic modeling, AMSC has designed R2R production-scale furnaces, which achieve uniform growth rates exceeding 10 Å/sec over the entire width of the wide-web RABiTS templates. The ISTEC group has used a similar approach to grow MOD-based films on narrow IBAD templates (Nomoto et al., 2005).

4.3.3.3 Metal organic chemical vapor deposition

MOCVD is a versatile *in situ* deposition process widely used for the epitaxial growth of a variety of semiconductor, optical, and other electronic films. The technique, which originated with the work of Manasevit (1968), is based on the injection of a precursor material into the reactor chamber, where it is vaporized and undergoes a pyrolysis reaction as it is deposited as a thin film on a substrate. The technique was broadly developed in the 1980s.

Shortly after the discovery of YBCO in 1987, work began on the use of MOCVD as a deposition technique for YBCO films. However, at that time, the only materials available for the Y, Ba, and Cu precursors were the solid phase 2,2,6,6-tetramethyl-3,5-heptanedionates, which have high melting points and low vapor pressures. These properties made it extremely difficult to achieve the vapor phase concentration necessary for high rates or to maintain the stoichiometry of the vapor-phase composition. The major advance in the development of the MOCVD process was the design of liquid-phase precursors that can be dissolved in an organic solvent, volatized in a flash vaporizer, and transported to the reactor by an inert carrier gas. These single-source materials, based on the RE, Ba, and Cu tetramethyl heptanedionates (tmhd) (Selvamanickam et al., 2001), have enabled the adaption of the MOCVD technique to the high rate deposition of the REBCO films. The typical R2R MOCVD process consists of a liquid precursor feed, a flash vaporizer, and a showerhead to direct the precursor to the deposition zone.

Unlike the PLD technique, which is a direct-line-of-sight deposition process with a limited deposition zone, the deposition area in the MOCVD technique is determined by the dimensions and gas flow dynamics of the shower head. SuperPower developed a R2R system that relies on a helix template transport system that allows a wide deposition zone. In contrast, Chuba Electric (Mori et al., 2006; Watanabe et al., 2007) developed a single-pass process utilizing a series of sequential deposition chambers. Both systems have successfully produced high-quality REBCO films over long lengths of substrates.

4.3.3.4 Reactive co-evaporation

RCE is based on the classic thermal evaporation technique used for depositing metal films. This technique was extended to the multi-component oxide HTS films by the Technical University of Munchen and Theva using an innovative two-stage deposition approach (Berberich et al., 1994). The technique, first used for the deposition of REBCO films on single crystal substrates, is based on the evaporation of a thin layer of the metal components onto the substrate in a low pressure zone followed by oxidation in a high oxygen pressure zone. Rapid cycling of the substrate between the deposition and oxygen zones, using a rotating stage, allows the layer-by-layer growth of the REBCO film.

Theva extended this technique to a R2R deposition process using a helix transport system and an oxygen shuttle that is rapidly cycled over the tape surface several times per second as the metals are deposited by electron beam evaporation on the moving template. Workers at KAIST (Lee et al., 2004) and KERI (Oh et al., 2008) in Korea later replaced the shuttle system with a large rotating drum, wound with the template, which was rapidly rotated through deposition and oxygen zones.

In 2005, Hammond (MacManus-Driscoll et al., 2007; Ohnishi, Huh, & Hammond, 2004) discovered an alternate approach to the RCE process that has been converted into a R2R process by SuNam (Moon, 2010). This alternate process involves the co-evaporation of the entire metal film in a single step, forming an amorphous, glassy sub oxide film. This amorphous film is then moved into a high temperature zone where it rapidly converts to the superconducting REBCO phase, as illustrated in Figure 4.8.

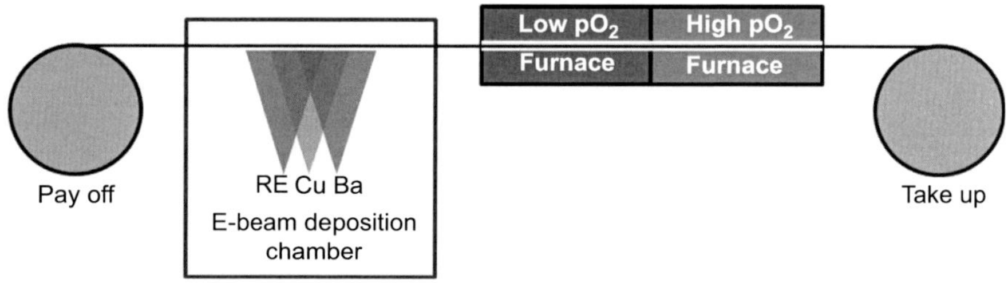

Figure 4.8 Schematic illustration of the RCE process developed by SuNam. In the process, the metals are deposited in one step then passed sequentially into high temperature, low pO_2, and high pO_2 zones where the REBCO forms via a liquid assisted growth process.

The conversion process involves a liquid assisted growth, and similar to the MOD process, the initial nucleation occurs on the template surface followed by epitaxial growth through the film thickness. The growth rate, which can exceed 30 nm/s, makes this the fastest REBCO growth process known; however, the fast rate also presents challenges controlling the epitaxial growth over long lengths.

4.3.4 Wide strip to narrow wire conversion

Once the REBCO has been epitaxially deposited on the template, the next step in the process is depositing a Ag layer on the surface. The Ag layer provides a low resistance electrical contact to the REBCO film and environmental protection. Typically, the Ag layer, with a thickness of 1−3 μm, is deposited by sputtering or e-beam evaporation and annealed at temperatures above 450°C in an oxygen atmosphere. In addition to producing a low resistance contact at the Ag/REBCO interface, the annealing process is needed to oxygenate the REBCO, thereby forming the orthorhombic superconducting phase.

After the Ag deposition, the strips, which range from about 10 to 50 mm in width (depending on the manufacturing approach), are roll-slit to the width of the final wire, depending on the application. AMSC's wide-web process yields up to ten 4-mm-wide wires or four 10 mm-wide-wires (the outer most edges are typically discarded in the slitting process). These individual wires are referred to as insert wires in AMSC's process. The narrower strips, processed with the helix approach, generally yield two to three narrow or one wide insert from each processed strip. As manufacturers using the wide-web process move to wider strips, the slitting process will yield correspondingly more insert wires (Li et al., 2009). The ability to produce a large number of individual insert wires simultaneously from a single processed strip is a major cost advantage of the wide-web process.

The final, and one of the most critical steps, in the process is converting the insert wire into a robust, hermetically sealed wire that (1) is impervious to moisture, liquid cryogens, and other environmental degradation, (2) has the electrical stability to withstand quenches experienced in resistive FCL, and (3) has the mechanical properties to withstand the rigors of the cable and coil fabrication processes, strains during repetitive cooling and heating cycles, and the stresses imparted during years or even decades of use. In cable applications, the wire must withstand high mechanical tensions and bending strains during stranding and, even more critically, the tensions arising from thermal contraction when the cable is cooled to operating temperature. HTS wires designed for coil-based applications must also withstand tensile strains perpendicular to the tape plane (transverse delamination strength) which arise from differential contraction between the wire and its support (e.g., epoxy) as well as Lorentz forces arising from high magnetic fields (Hazelton, 2013b).

Again, different manufacturers have chosen different paths to fabricating the final 2G HTS wire. AMSC uses a lamination process in which a metal foil is bonded to each side of the insert wire and the edges are sealed with solder fillets to provide a hermetic seal and impart exceptional *c*-axis tensile strength. The laminated wire architecture with the solder fillets at the wire edges allows the wire to readily tolerate

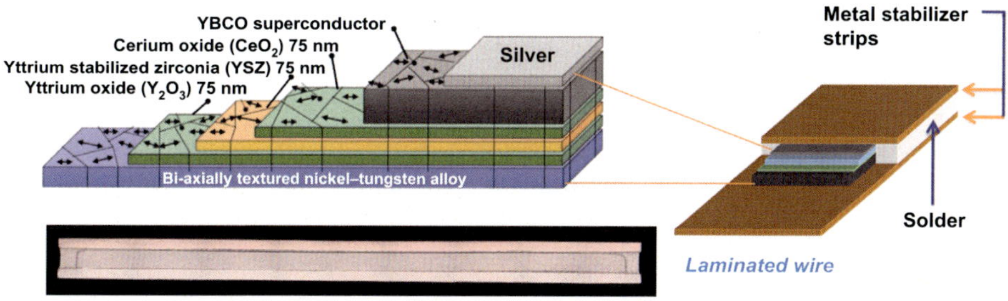

Figure 4.9 Schematic illustration of AMSC's 2G insert and laminated wire architectures. A micrograph of a 2G HTS wire cross-section is shown in the bottom left.
Courtesy AMSC.

large *c*-axis strains during coil fabrication and is compatible with use in an epoxy matrix. The lamination process also provides the flexibility to customize the electrical or mechanical properties of the final wire. Typically, high conductivity copper is used for wire intended for conduction-cooled lower temperature coil applications, while brass is favored in liquid nitrogen applications such as cables, and stainless steel is used for resistive FCL (Schmidt et al., 2007). In AMSC's 2G HTS wire, which uses a Ni−5 at% W substrate (yield stress of 250 MPa), the mechanical properties are determined by the properties of the lamina and solder fillet dimensions. Figure 4.9 shows the architecture and cross-section of AMSC's laminated 2G wire. The lamination process also provides the opportunity to place two insert strips between the lamina, resulting in a wire with double the I_c of the standard process.

In contrast, the approach of SuperPower and most other manufacturers is to electroplate a copper layer around the entire insert strip. Typically, a copper thickness of 25−50 μm is deposited to provide the needed electrical stability and hermeticity for most applications. The tensile strength of the substrate (Hastelloy C-276; yield stress of 600 MPa) dictates the strength of the final wire, while the yield strength and elastic modulus are affected by the thickness of the copper stabilizer. Figure 4.10 shows the

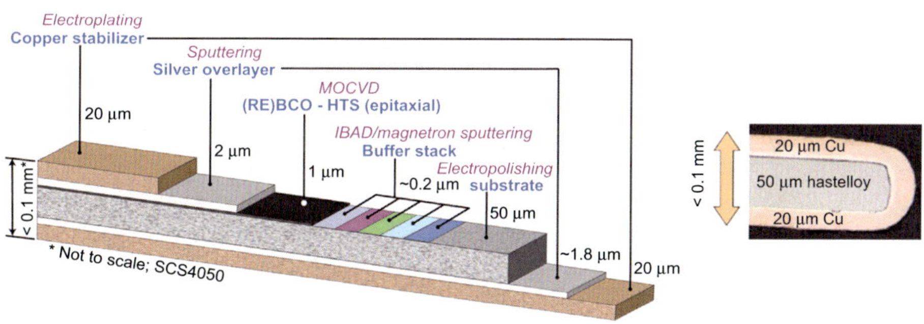

Figure 4.10 Schematic illustration of SuperPower's 2G wire architecture (left). A micrograph of a 2G HTS wire cross-section is shown on the right.
Courtesy SuperPower.

architecture and cross-section of SuperPower's copper plated 2G HTS wire. For select applications, such as resistive FCL, the wire is provided with only a Ag coating and thicker substrate.

The reliability of the 2G wire delivered to users is critical, particularly as 2G wire is being used in an increasing number of applications. Failure of a wire, either electrical or mechanical, in a demonstration project can be catastrophic with the application developer deciding not to pursue further development. Consequently, manufacturers have implemented quality programs and extensive testing to insure the quality of the wire. AMSC characterizes the critical current along the length of every wire at overlapping 1 m intervals at 77 K using a DC current before delivery to customers. In addition, wires are subjected to extensive testing to confirm the dimensions and mechanical properties. For example, to insure that the 4.4-mm-wide brass cable wire can handle the cable winding operations, samples are subjected to a torture test consisting of sequential reverse double bends around 6 inch rollers, a 360° twist and helical winding around a 19-mm diameter former at 10 lb tension. The test is carried out twice, once with the wire oriented so that the substrate faces radially inward to the former and once with the substrate oriented radially outward (Fleshler et al., in press). Copper laminated coil wire is tested to confirm that the *c*-axis strength exceeds a minimum specification, insuring the wire can tolerate the fabrication and cool down in an epoxy impregnated coil without a delamination failure.

Although the commercially available 2G HTS wires are similar in dimensions and structure, the end user must recognize that different wire architectures may respond differently to stresses experienced during coil fabrication, cabling, or use and when necessary, adjust their handling procedures and designs as appropriate for the specific wire.

4.4 2G manufacturers and wire properties

To date, AMSC and SuperPower have led the development of 2G HTS wire manufacturing; however, a number of other manufacturers have or are in the process of developing long length manufacturing capabilities. Table 4.1 lists the major 2G wire manufacturers along with the processes used for the template and HTS layers. AMSC and SuperPower (now part of Furukawa Electric Co., Ltd.) were the first two companies to establish R2R manufacturing lines and have provided most of the 2G HTS wire to the various application projects. SuNam has also established a long-length manufacturing line and is now offering wire to commercial customers. Other organizations, including Fujikura and Theva, have demonstrated long length, high I_c wires, but are yet to provide large quantities for general applications. The remaining organizations on the list are either producing wire for internal use, developing R2R manufacturing capabilities or focusing on short length R&D.

Table 4.1 Companies involved in the manufacture or development of 2G HTS wire

Commercial sales of long length wire		
American Superconductor	RABiTS/MOD	http://www.amsc.com/
Super power (subsidiary of Furukawa Electric Co.	IBAD/MOCVD	http://www.superpower-inc.com/
SuMAN	IBAD/RCE	http://www.i-SuNAM.com/
Limited quantity, short length, or development		
Fujikura	IBAD/PLD	http://www.fujikura.com/solutions/superconductingwire/
Theva	ISD/RCE	http://www.theva.com
Bruker Corporation	IBAD/PLD	http://www.bruker.com/products/superconductors-and-metal-composite-materials.html
Deutche Nanoschrift (part of BASF)	RABiTS/MOD	http://www.d-nano.com/en/
Superconductor Technologies Inc.	IBAD/RCE	http://www.suptech.com/home.php
Sumitomo	RABiTS/PLD	http://global-sei.com/super/hts_e/index.html
MetOx	RABiTS/MOCVD	http://www.metox.biz/
Super Ox	TBD	http://www.superox.ru/en/

AMSC and SuperPower both offer wires specifically designed for cable or coil applications. The cable wire contains an HTS layer specifically optimized for superior performance at around 77 K, the liquid nitrogen temperature regime, in very low magnetic fields. The HTS layer in the coil wire is optimized for applications such as motors and generators operating between 77 and 4 K in higher magnetic fields. SuNam currently offers a single formulation for both cable and coil applications. The formulations and architectures available from the three manufacturers are summarized in Table 4.2.

Although most organizations have reported wire performance approaching 500 A/cm w over 100's of meter length, it must be realized that these generally represent the top distribution from production, or in some cases, one off R&D

Table 4.2 **Formulation and Architecture of Commercial 2G HTS Wire**

	Manufacturer		
Wire property	**AMSC**	**SuperPower**	**SuMAN**
Material/formulation	YBCO/cable and coil formulation	GdBCO/cable and coil formulation	GdBCO/single formulation
Stabilizer layer/width	Brass (lamination)/4 mm Cu (lamination)/4, 10 mm SS (lamination)/4, 10 mm	Cu (electroplate)/3, 4, 6, 12 mm Stabilizer free/3, 4, 6, 12 mm	Cu (electroplate)/4, 12 mm Brass (lamination)/4 mm Stabilizer free/4, 12 mm

demonstrations. Thus the companies generally offer wire with minimum currents of >100 A (4 mm wire). Higher current wires are generally available upon request from all the manufacturers; however, the lengths and quantities may be limited, depending on the precise requirements.

4.5 Applications (brief review of major applications for 2G wire)

The 1G HTS wire based on the multi-filamentary BSCCO composition was used in numerous demonstration projects and validated the capabilities of HTS wire. 2G HTS wire promises to greatly expand the market since its lower cost and superior performance will enable new applications. As the performance of the 2G wire continues to improve and cost decreases, the 2G HTS wire is envisioned to find use in a variety of applications illustrated in Figure 4.11. However, the major applications being addressed today are in the electric power sector.

The near lossless current flow in the 2G wire allows the manufacture of cables with much higher power density than conventional copper, providing a means to increase the capacity of urban centers and to transmit power from distant renewable energy sources.

Energy	Defense	Research	Medical
Generators (wind, utility, hydro) Motors Cables (ac, dc) FCL Transformers SMES	Ship propulsion Degaussing Mine detection Directed energy	High field magnets Fusion Current leads	MRI NMR

Figure 4.11 Projected markets and applications for HTS wires.

2G HTS wire also enables the design of power-dense superconductor generators which can provide a light weight, cost-effective solution for off-shore wind power, a major potential source of sustainable energy. Finally since the superconductors become resistive when the current exceeds the critical current, they can be used in FCL's. A brief summary of the three major power sector applications is presented below. More in-depth reviews of these and other applications are provided in later chapters.

4.5.1 Fault current limiters

As electricity requirements continue to grow around the world, particularly in urban areas, and new power generation sources are brought on line, current surges, or fault currents, are increasing in frequency and severity. These fault currents, which result from short circuits caused by lightning, accidental contact between the lines or the ground, etc., are reaching levels that could exceed the safe operating levels of circuit breakers and other power equipment currently employed to limit the fault currents.

Superconducting Fault Current Limiters (SFCL) made with 2G HTS wire are seen as potential solutions to this problem and multiple designs have been proposed for commercial systems. The most widely investigated design is called a "resistive" FCL (Figure 4.12) in which a current exceeding the critical current of the HTS material causes the HTS material to transition to a very high resistance state. When the HTS layer becomes normal, the fault current is shunted into a higher resistance layer, effectively reducing the amplitude of the fault current for a fixed voltage source. Unlike 1G HTS, which is fabricated in a Ag matrix, 2G HTS wire with its high resistivity substrate is uniquely suited for this application. In addition, the composite architecture which allows replacing the conductive copper stabilizer with a high resistance stainless steel layer provides a means to further increase the wire resistivity and the capacity to absorb heat generated during the fault. Alternate approaches based on inductive and magnetic saturation principles are also being investigated.

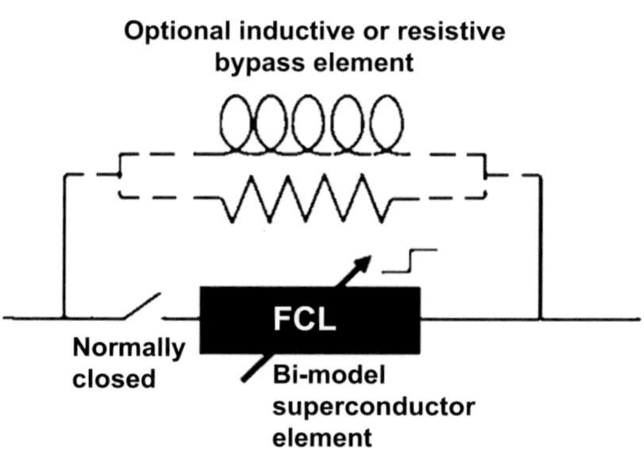

Figure 4.12 Schematic operation of a Stand-Alone Resistive Superconductor Fault Current Limiter utilizing 2G HTS wire. When a fault current occurs, the HTS element functions as a fast switch, shunting the current through the resistive bypass element.
Courtesy AMSC.

FCL's have been one of the most successful applications for 2G wire with a number of successful demonstrations of the technology around the world (Na et al., 2009; Neumueller et al., 2009; Tixador & Nguyen, 2012; Yazawa et al., 2009). Nexans and AMSC are offering a Medium-voltage SFCLs system for in grid use Europe and the US (AMSC).

4.5.2 Cables

HTS cables, which can provide power-dense electrical transmission, can efficiently carry large amounts of electrical power over long distances, and are often envisioned as a potential replacement for conventional high-voltage cables as the backbone of a transmission grid. However, a nearer term application is providing power over short distances in congested urban areas where space is a premium or siting and installation of conventional cables is difficult.

Over last dozen years, a number of AC HTS cable demonstration projects have been installed and tested in the grid around the world. The longest, a 600 m, 138 kV transmission level voltage cable, was installed in Holbrook, NY in 2006 by AMSC and Nexans. The first HTS cable systems used 1G wire produced by AMSC, SEI, or Furukawa. However, over the last few years, cable projects have begun to migrate to 2G wire as it has become available in the quantity and lengths required for the projects. The first short cable demonstration using 2G wire was carried out as part of the Albany cable project in the United States by SEI and SuperPower (Yumura et al., 2009). This was followed by other cable projects in Europe lead by Nexans (Super3C Project, 2009) and in Japan by SEI and Furukawa (Mukoyama et al., 2009; Ohya et al., 2009). A project is currently underway to retrofit one phase of Holbrook cable (1G wire) with a 2G HTS-based cable (Maguire, Folts, et al., 2009; Maguire, Schmidt, Bratt, Welsh, & Yuan, 2009). Recently, LS cable has installed a 410 m long, 22.9 kV cable in the Incheon substation in Korea (Figure 4.13) using 2G wire from AMSC (Lim et al., 2012; Sohn et al., 2010). Numerous other projects are under consideration around the world.

A fault-current limiting cable technology which combines the high current capacity of an HTS cable with inherent FCL functionality has been demonstrated by AMSC, Southwire, Con Edison and Oak Ridge National Laboratory and supported by the Department of Homeland Security as a measure to increase electric grid resiliency to disruptive faults (either natural causes or intentional acts of vandalism) (Maguire et al., 2010; Rey et al., 2010). After initial laboratory tests, a three-phase 300 m long 13.8 kV cable with a phase current of 4000 A_{rms} rating is planned for installation between two substations in New York City.

4.5.3 Rotating machines

HTS wire has been considered and tested for use in rotating machines, such as motors and generators. The design of the HTS machines are generally based on synchronous designs where the HTS rotor field coils operate with dc currents while the stator coils, which experience high ac currents, are made of copper. The HTS rotors can generate a

Figure 4.13 The Icheon substation on Korea is the site of a 410-m-long, 22.9-kV, 1.25-kA, 50 MVA 2G HTS-based cable installed by LS Cable using wire from AMSC. The cable is connected to KEPCO's grid.
Courtesy LS Cable.

very high torque in a compact machine, significantly reducing the size and weight in large machines. In addition, the reduced losses in the HTS rotors increase the machine efficiency. AMSC, in collaboration with Northrop-Grumman designed, built and successfully tested a 36.5 MW HTS ship propulsion motor (based on 1G HTS wire) for the U.S. Navy in 2009. The motor is less than half the size of a conventional machine of comparable size and can reduce the ship weight by almost 200 metric tons.

The increasing importance of renewable energy and the growth of wind power have spurred interest in the development of a compact direct-drive HTS generator for wind turbines. It is now believed that off-shore wind turbines rated at 10 MW and greater may be the first commercial market for HTS-based machines. The economics of off-shore wind farms favor larger and larger power turbines; however, the weight of conventional generators located hundreds of feet above the tower base in the wind turbine nacelle has restricted power ratings to about 5 MW. In contrast, the size and weight advantage of HTS-based generators, illustrated in Figure 4.14, can enable 10 MW turbines that will be cost competitive with smaller conventional systems. Unlike cables and FCL, full-scale prototype HTS generators for wind turbines have not been built. One of the major obstacles has been the cost of the HTS wire for commercial applications. However, recent advances in improving the performance of 2G HTS wire exposed to strong magnetic fields, such as those present in rotating machine applications, promise to significantly reduce the cost and quantity of wire required for these applications. In the meantime, there is a growing effort to begin testing designs and prototypes of subscale components.

4.6 Conclusion and future trends

Over the past two decades, vast strides have been made in transforming the brittle ceramic superconductors into flexible, robust wires that carry an extraordinary amount

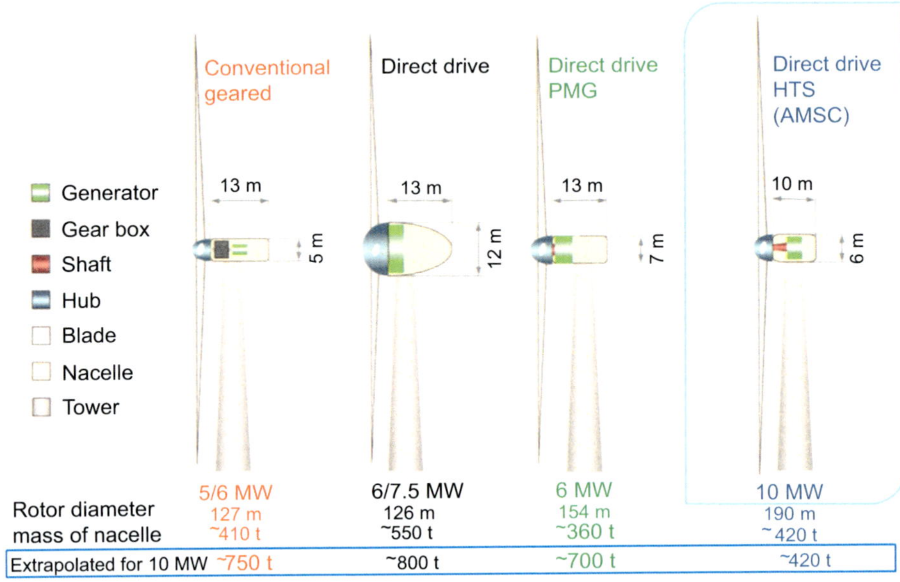

Figure 4.14 A 10 MW, direct drive 2G HTS-based wind turbine generator is projected to be significantly lighter and smaller than conventional geared, direct drive or permanent magnet direct drive (PMG) generators.
Courtesy AMSC.

of current. A number of manufacturers are now producing 2G HTS wire with the length and performance needed for initial large-scale demonstration projects and more potential manufacturers are appearing every year. Although there is much excitement in the field and new demonstration projects are constantly appearing, it must be remembered that revolutionary materials-based technology development entails high risk and uncertainty compared to incremental innovations. Moving beyond the demonstration projects to the broad commercial acceptance of the new technology can take additional years, necessitating a sustained effort involving government-industry commitments to keep the technology moving forward. Prospective customers such as electric utilities require a stable climate for investment in research, development, and demonstration projects.

While application efforts continue, 2G HTS wire manufacturers remain focused on continuing to improve the current carrying capability of the wire, particularly at the temperature and operating magnetic fields required for rotating machines and high field magnet applications. Significant progress has been made in this area in the last few years with the performance of commercial long length wires more than doubling in the 4–50 K range. However, additional improvements will be needed as applications progress. These increases in performance will come from a combination of skillful engineering of the pinning microstructure and increases in the thickness of the HTS layer.

Pinning: Recent improvements in pinning have resulted from the introduction of a mixture of correlated defects, oriented along the *c*-axis, and point defects. These have been accomplished by the *in situ* growth of $BaZrO_3$ nano-columns, introduction of twin boundaries, addition of nanoparticles throughout the HTS layer, and formation of *ab*-plane defects (Choi, Shin, Joo, & Yoo, 2013; Gutierrez et al., 2007; Kang, Leonard, Martin, Li, & Goyal, 2007; MacManus-Driscoll et al., 2004; Miura, Yoshizumi, Izumi, & Shiohara, 2010; Specht et al., 2006; Strickland, Talantsev, Xia, et al., 2009). These engineered pinning microstructures have been particularly effective at low temperatures and high fields, with wire performance approaching 1000 A/cm w in magnetic fields (*H*//*c*-axis) around 2 T at 30 K. A challenge to manufacturers is controlling these nano-scale structures over the kilometer lengths of wire produced on the R2R manufacturing lines.

While these recent advances are transferred to the manufacturing lines, manufactures need to remain focused on other techniques that can provide further pinning improvements. One area that has shown promise is irradiation by ions ranging from protons to gold (Civale, 1997; Jia et al., 2013; Matsui et al., 2012; Strickland, Talantsev, Long, et al., 2009). Although significant enhancements have been reported, these techniques are not amendable to low-cost manufacturing. Therefore, it is important that scientists continue to characterize and understand the types of defects that can provide the needed pinning enhancement and develop scalable methods to apply them to a R2R manufacturing technology.

Film Thickness: The most straight forward route to increasing the current is to increase the thickness of the HTS layer. However, in practice this becomes a serious challenge. The main issue is finding a method of increasing the HTS layer thickness without offsetting the benefit with additional manufacturing or materials cost. Manufacturers are currently producing HTS layers in the 1−1.5 μm thickness range. Research needs to focus on increasing deposition rates, maintaining constant J_c as the thickness increases, and avoiding the need for multi-pass deposition processes. In addition, as thicker HTS layers are introduced into routine manufacturing, it is critical that there is no degradation of the mechanical properties of the composite structure and that the wire architecture (e.g., stabilizer thickness) can accommodate the higher currents.

Cost: Successful, cost-effective improvements in the I_c of the wire (both self-field and at the operating conditions) will directly reduce cost. However, manufacturers need to focus on additional cost reductions. A significant cost reduction will result from incremental improvements in yield, material cost, equipment utilization, etc., as volume increases. However, it is also critical that research remains focused on developing new or improved deposition techniques that reduce equipment cost, or allow a large increase in throughput. Recent efforts to develop a R2R technology based on the RCE process are focused in this direction; however, it still remains to be seen if such approaches are any more cost-effective than the established MOD, MOCVD, and PLD approaches.

One potential area for significant cost reduction is development of a solution-based process for forming the buffer layers on the RABiTS substrate (Obradors et al., 2006). Such a solution-based process has a number of potential advantages for high volume, low-cost R2R manufacturing and is of growing interest for other applications such as ferroelectrics, photovoltaics, etc. (Bäcker, Falter, & Holzapfel, 2013). The advantages of the solution route compared to a PVD based process include easy control of the chemistry, near 100% material utilization, fast deposition rates, the ability to deposit uniform films over wide (10−100 cm) substrates and inexpensive non-vacuum equipment. There are currently a number of efforts focused on this technology; however, to date the performance of the oxide buffer layers has not matched that of the best PVD-deposited layers. Success in this area will require an improvement in the understanding of the epitaxial nucleation and growth in these *ex-situ* processes in addition to the development of R2R process equipment.

The future: The development of 2G HTS wire manufacturing has progressed rapidly over the past few years and multiple manufacturers are now able to supply wire in the quantities and with the performance needed for demonstration projects and initial commercial applications. As 2G HTS wire moves into the hands of application engineers, its credibility is being establish and its advantages are becoming clear. At the same time, active R&D programs around the world are promising significant improvements in performance and manufacturing capabilities. As the wire manufacturing capability is improved, performance increases and capacity grows, 2G HTS wire cost will continue to decline, opening the 2G HTS market to a broad range of new applications.

4.7 Sources of further information and advice

Journals and databases

NIST Superconducting Materials Database: http://www.ceramics.nist.gov/srd/hts/htsquery.htm.
KEIRIN Superconductivity Papers: http://sprcnd.cloudapp.net/top_e.php.
Superconductor Science and Technology: http://iopscience.iop.org/0953-2048.
Superconductor Week: http://www.superconductorweek.com/

Organizations and research laboratories

Superconductors.org: http://superconductors.org/Index.htm.
International Superconductivity Technology Center (ISTEC): http://www.istec.or.jp/index-E.html.
European Society for Applied Superconductivity (ESAS): http://www.esas.org/.
Consortium of European Companies Determined to use Superconductivity (Conectus): http://www.conectus.org/index.html.
Coalition for the Commercial Applications of Superconductors (CCAS): http://www.ccas-web.org/.
Karlsruhe Institute of Technology: http://www.itep.kit.edu/english/66.php.
ICMAB: http://departments.icmab.es/suman/.

EUROTAPES: http://eurotapes.eu/summary/summary.
Texas Center for Superconductivity: http://tcsuh.com/.
Oak Ridge National Laboratory: http://web.ornl.gov/sci/htsc/index.html.
National High Field Magnetic Laboratory: http://www.magnet.fsu.edu/.

References

Abiru, K., Shingai, Y., Konishi, M., & Ohmatsu, K. (2011). *IEEE Transactions on Applied Superconductivity, 21*, 2941.

http://www.amsc.com/library/SAFCL_BRO_0812.pdf.

Arendt, P. N., & Foltyn, S. R. (2004). *MRS Bulletin, 29*, 543–550.

Bäcker, M., Falter, M., & Holzapfel, B. (2013). Superconducting films. In T. Schneller, R. Waser, M. Kosec, & D. Payne (Eds.), *Chemical solution deposition of functional oxide thin films*. Heidelberg: Springer Wien.

Balachandran, U., Ma, B., Li, M., Fisher, B. L., Koritala, R. E., Miller, D. J., et al. (2003). *Physica C, 392*, 806–814.

Barrett, C. S., & Massalski, T. B. (1996). In *Structure of metals* (3rd ed.). New York: McGraw-Hill.

Bartolomé, E., Vlad, V. R., Calleja, A., Aklalouch, M., Guzmán, R., Arbiol, J., et al. (2013). *Superconductor Science and Technology, 26*, 125004.

Bauer, M., Metzger, R., Semerad, R., Berberich, P., & Kinder, H. (1999). *MRS Proceedings, 585*.

Berberich, P., Uts, B., Prusseit, W., & Kinder, H. (1994). *Physica C, 219*, 497.

Bulaevskii, L. N., Clem, J. R., Glazman, L. I., & Malozemoff, A. P. (1992). *Physical Review B, 45*, 2545.

Bulaevskii, L. N., Daemen, L. L., Maley, M. P., & Coulter, J. Y. (1993). *Physical Review B, 48*, 13798.

Cantoni, C., Christen, D. K., Feenstra, R., Norton, D. P., Goyal, A., Ownbyand, G. W., et al. (2001). *Applied Physics Letters, 79*, 3077.

Cantoni, C., Christen, D. K., Goyal, A., Heatherly, L., List, F. A., Ownby, G. W., et al. (2003). *IEEE Transactions on Applied Superconductivity, 13*, 2646–2650.

Cantoni, C., Christen, D. K., Kowalewski, M. M., List, F. A., Ownby, G. W., Zehner, D. M., et al. (2002). *Journal of Materials Research, 17*, 2549–2554.

Chen, Y., Selvamanickam, V., Zhang, Y., Zuev, Y., Cantoni, C., Specht, E., et al. (2009). *Applied Physics Letters, 94*, 062513.

Choi, S. M., Shin, G. M., Joo, Y. S., & Yoo, S. I. (2013). *Progress in Superconductivity and Cryogenics, 15*, 15.

Civale, L. (1997). *Superconductor Science and Technology, 10*, A11.

Cooper, S. L., & Gray, K. E. (1994). In D. M. Ginsberg (Ed.), *Physical properties of high temperature superconductors* (Vol. IV, p. 61). Singapore: World Scientific.

Cordero-Cabrera, M. C., Mouganie, T., Glowacki, B. A., Bäcker, M., Falter, M., Holzapfel, B., et al. (2007). *Journal of Materials Science, 42*, 7129.

Dechoux, N., Jimenez, C., Chaudouet, P., Rapenne, L., Sarigiannidou, E., Robau, F., et al. (2012). *Superconductor Science and Technology, 25*, 125008.

Dijkkamp, D., Venkatesan, T., Wu, X. D., Shaheen, S. A., Jisrawi, N., Min Lee, Y. H., et al. (1987). *Applied Physics Letters, 51*, 619.

Dimos, D., Chaudhari, P., & Mannhart, J. (1990). *Physical Review B, 4*(1), 4038.

Dürrschnabel, M., Aabdin, Z., Große, V., Bauer, M., Sigl, G., Prusseit, W., et al. (2012). *Physics Procedia, 36*, 1546.

Eickemeyer, J., & Holzapfel, B. (2008). Chapter 3: Cube textured nickel-based substrates for buffer/TBCO coatings. In Li-C. Liáng (Ed.), *YBCO superconductor research progress*. New York: Nova Science Publishers.

Eickemeyer, J., Huhne, R., Guth, A., Rodig, C., Gaitzsch, U., Freudenberger, J., et al. (2010). *Superconductor Science and Technology, 23*, 085012.

Fleshler, S., DeMoranville, K., Gannon, J., Jr., Li, X., Podtburg, E., Rupich, M. W., et al. (2014). *Journal of Physics: Conference Series, 507*, 022005.

Foltyn, S. R., Arendt, P. N., Dowden, C., DePaula, R. F., Groves, J. R., Coulter, J. Y., et al. (1999). *IEEE Transactions on Applied Superconductivity, 9*, 1519.

Foltyn, S. R., Wang, H., Civale, L., Jia, Q. X., Arendt, P. N., Maiorov, B., et al. (2005). *Applied Physics Letters, 87*, 162505.

Fuji, H., Teranishi, R., Kito, Y., Matsuda, J., Nakaoka, K., Izum, T., et al. (2005). *Physica C: Superconductivity, 426–431*, 938.

Fujino, K., Konishi, M., Muranaka, K., Hahakura, S., Ohmatsu, K., Hayashi, K., et al. (2003). *Physica C, 392*, 815.

http://www.fujikura.com/solutions/superconductingwire/.

Goyal, A., Lee, D. F., List, F. A., Specht, E. D., Feenstra, R., Paranthaman, M., et al. (2001). *Physica C, 351–360*, 903–913.

Goyal, A., Norton, D. P., Budai, J. D., Paranthaman, M., Specht, E. D., Kroeger, D. M., et al. (1996). *Applied Physics Letters, 69*, 1795.

Goyal, A., Paranthaman, M., & Schoop, U. (2004). *MRS Bulletin, 29*, 552–561.

Gupta, A., Cooper, E. I., Jaganathan, R., & Giess, E. A. (1988). In *Chemistry of high temperature superconductors II* (p. 265). Washington, DC: American Chemical Society.

Gupta, A., Jaganathan, R., Cooper, E. I., Giess, E. A., Landman, J. I., & Hussey, B. W. (1988). *Applied Physics Letters, 52*, 2077.

Gutierrez, J., Llordes, A., Gazquez, J., Gibert, M., Roma, N., Ricart, S., et al. (2007). *Nature Materials, 6*, 367–373.

Hanyu, S., Tashita, C., Hayashida, T., Hanada, Y., Morita, K., Sutoh, Y., et al. (2010). *Physica C, 470*, 1227.

Hasegawa, K., Fujino, K., Mukai, H., Konishi, M., Hayashi, K., Sato, K., et al. (1996). *Applied Superconductivity, 4*, 487.

Hasegawa, K., Hobara, N., Nakamura, Y., Izumi, T., & Shiohara, Y. (2001). *Physica C, 354*, 424–428.

Hazelton, D. W. (September 15–19, 2013a). In *European conf. appl. supercond*. Genova, Italy, Paper 2M-WT-08. http://www.superpower-inc.com/system/files/EUCAS2013_2M-WT-08_final.pdf

Hazelton, D. W. (November 5, 2013b). In *LTHFSW 2013*. St. Petersburg, FL.

Hensel, B., Grivel, J. C., Jeremie, A., Perin, A., Pollini, A., & Flukiger, R. (1993). *Physica C, 205*, 329.

Hikata, T., Sato, K., & Hitotsuyanagi, H. (1989). *Japanese Journal of Applied Physics, 28*, 82.

Hilgenkamp, H., & Mannhart, J. (2002). *Reviews of Modern Physics, 74*, 485–549.

Ibi, A., Fukushima, H., Kuriki, R., Miyata, S., Takahashi, K., Kobayashi, H., et al. (2006). *Physica C: Superconductivity, 445*, 525–528.

Iijima, Y., Kakimoto, K., Kimura, M., Takeda, K., & Saitoh, T. (2001). *IEEE Transactions on Applied Superconductivity, 19*, 2816.

Iijima, Y., Kakimoto, K., Takeda, K., & Saitoh, T. (2001). Honolulu, HI. In *Proc. int. workshop on superconductivity* (p. 47).

Iijima, Y., Kakimoto, K., Yamada, Y., Izumi, T., Saitoh, T., & Shiohara, Y. (2004). *MRS Bulletin, 29*, 564.

Iijima, Y., Tanabe, N., Ikeno, Y., & Kohno, O. (1991). *Physica C, 185*, 1959.

Iijima, Y., Tanabe, N., Kohno, O., & Ikeno, Y. (1992). *Applied Physics Letters, 60*, 769.

Izumi, T., Yoshizumi, M., Miura, M., Nakaoka, K., Ichikawa, Y., Sutoh, Y., et al. (2009). *Physica C, 469*, 1322.

Jia, Y., LeRoux, M., Miller, D. J., Wen, J. G., Kwok, W. K., Welp, U., et al. (2013). *Applied Physics Letters, 103*, 122601.

Jia, Y., Welp, U., Crabtree, G. W., Kwok, W. K., Malozemoff, A. P., Rupich, M. W., et al. (2011). *Journal of Applied Physics, 110*(8), 083923.

Jin, S., Sherwood, R. C., van Dover, R. B., Tiefel, T. H., & Johnson, D. W., Jr. (1987). *Applied Physics Letters, 51*, 203.

Kang, S., Leonard, K. J., Martin, P. M., Li, J., & Goyal, A. (2007). *Superconductor Science and Technology, 20*, 11−15.

Kashima, N., Shima, K., Doi, T., Kubota, S., Watanabe, T., Inoue, M., et al. (2009). *IEEE Transactions on Applied Superconductivity, 19*, 3299.

Katayama, K., Nakahata, K., Yoshizumi, M., Izumi, T., & Shiohara, Y. (2010). *Superconductor Science and Technology, 23*, 014015.

Ko, K. P., Ha, H. S., Kim, H. K., Yu, K. K., Ko, R. K., Moon, S. H., et al. (2007). *Physica C, 463 -465*, 564.

Kutami, H., Hayashida, T., Hanyu, S., Tashita, C., Igarashi, M., Fuji, H., et al. (2009). *Physica C: Superconductivity, 469*, 1290.

Lee, H. (April 26, 2013). In *Workshop on high temperature superconductors for utility applications, Beijing, China.*

Lee, B. S., Chung, K. C., Lim, S. M., Kim, H. J., Youm, D., & Park, C. (2004). *Superconductor Science and Technology, 17*, 580.

Li, C. S., Lu, Y. F., Zhang, P. X., Yu, Z. M., Tao, B. W., Feng, J. Q., et al. (2012). *Physics Procedia, 36*, 1620.

Li, X., Rupich, M. W., Thieme, C. L. H., Teplitsky, M., Sathyamurthy, S., Thompson, E., et al. (2009). *IEEE Transactions on Applied Superconductivity, 19*, 3231.

Lim, J. H., Yang, H. S., Sohn, S. H., Oh, S. R., Kim, Y. H., Lee, G. T., et al. (2012). In *Advances in. cryogenic engineering: Transactions of the cryogenic engineering conference-CEC* (Vol. 57, Vol. 1434, no. 1, p. 1657). AIP Publishing.

MacManus-Driscoll, J. L., Foltyn, S. R., Jia, Q. X., Wang, H., Serquis, A., Civale, L., et al. (2004). *Nature Materials, 3*, 439−443.

MacManus-Driscoll, J. L., Kursumovic, A., Maiorov, B., Civale, L., Jia, Q. X., Foltyn, S. R., et al. (2007). *IEEE Transactions on Applied Superconductivity, 17*, 2537.

Maguire, J., Folts, D., Yuan, J., Henderson, N., Lindsay, D., Knoll, D., et al. (2010). *AIP Conference Proceedings, 1218*, 445.

Maguire, J., Folts, D., Yuan, J., Lindsay, D., Knoll, D., Bratt, S., et al. (2009). *IEEE Transactions on Applied Superconductivity, 19*, 1740.

Maguire, J. F., Schmidt, F., Bratt, S., Welsh, T. E., & Yuan, J. (2009). *IEEE Transactions on Applied Superconductivity, 19*, 1692.

Ma, B., Koritala, R. E., Fisher, B. L., Uperty, K. K., Baurceanu, R., Dorris, S. E., et al. (2004). *Physica C, 403*, 183−190.

Ma, B., Li, M., Fisher, B. L., Koritala, R. E., & Balachandran, U. (2002). *AIP Conference Proceedings, 614*, 573.

Manasevit, H. M. (1968). *Applied Physics Letters, 12*(4), 156–159.
Matsuda, J., Nakaoka, K., Izumi, T., Yamada, Y., & Shiohara, Y. (2008). *Physica C: Superconductivity, 468*, 1017.
Matsui, H., Ogiso, H., Yamasaki, H., Kumagai, T., Sohma, M., Yamaguchi, I., et al. (2012). *Applied Physics Letters, 101*, 232601.
McIntyre, P. C., & Cima, M. J. (1994). *Journal of Materials Research, 9*, 2219.
McIntyre, P. C., Cima, M. J., & Ng, M. F. (1990). *Journal of Applied Physics, 68*, 4183.
McIntyre, P. C., Cima, M. J., & Roshko, A. (1995). *Journal of Applied Physics, 77*, 5263.
McIntyre, P. C., Cima, M. J., Smith, J. A., Hallock, R. B., Siegal, M. P., & Phillips, J. M. (1992). *Journal of Applied Physics, 71*, 1868.
Miura, M., Yoshizumi, M., Izumi, T., & Shiohara, Y. (2010). *Superconductor Science and Technology, 23*, 014013.
Moon, S. H. (July 7, 2010). In *Presentation at the Korean Superconductivity Society, Meeting 2010*. Yongpyong Resort, Gangwondo, Korea.
Moon, S. H., & Hazelton, D. W. (September 15–19, 2013). In *European conf. appl. supercond.* Genova, Italy, Paper 1M-MA-I1.
Mori, M., Watanabe, T., Kashima, N., Nagaya, S., Muroga, T., Miyata, S., et al. (2006). *Physica C, 445-448*, 515–520.
Mueller, K. A., & Deutscher, G. (1987). *Physical Review Letters, 59*, 1745.
Mukoyama, S., Yagi, M., Masuda, T., Amemiya, N., Ishiyama, A., Kashima, N., et al. (2009). *Physica C: Superconductivity, 469*, 1688–1692.
Na, J. B., Park, D. K., Yang, S. E., Kim, Y. J., Chang, K. S., Kang, H., et al. (2009). *Physica C, 469*, 1754.
Neumueller, N.-W., Schmidt, W., Kraemer, H.-P., Otto, A., Maguire, J., Jie Yuan, Y., et al. (2009). *IEEE Transactions on Applied Superconductivity, 19*, 1950.
Nomoto, S., Teranishi, R., Sato, A., Fuji, H., Aoki, Y., Izumi, T., et al. (2005). *Physica C: Superconductivity, 426–431*, 966.
Norton, D. P., Goyal, A., Budai, J. D., Christen, D. K., Kroeger, D. M., Specht, E. D., et al. (1996). *Science, 274*, 755.
Obradors, X., Puig, T., Pomar, A., Sandiumenge, F., Mestres, N., Coll, M., et al. (2006). *Superconductor Science and Technology, 19*, S13.
Obradors, X., Puig, T., Pomar, A., Sandiumenge, F., Pinol, S., Mestres, N., et al. (2004). *Superconductor Science and Technology, 17*, 1055.
Obradors, X., Puig, T., Ricart, S., Coll, M., Gazquez, J., Palau, A., et al. (2012). *Superconductor Science and Technology, 25*, 123001.
Oh, S. S., Ha, H. S., Kim, H. S., Ko, R. K., Song, K. J., Ha, D. W., et al. (2008). *Superconductor Science and Technology, 21*, 034003.
Ohmatsu, K., Hahakura, S., Hasegawa, K., & Ueyama, M. (2005). *IEEE Transactions on Applied Superconductivity, 15*, 2715.
Ohnishi, T., Huh, J.-U., & Hammond, R. H. (2004). *Journal of Materials Research, 19*, 977.
Ohya, M., Ashibe, Y., Watanabe, M., Minamino, T., Yumura, H., Masuda, T., et al. (2009). *IEEE Transactions on Applied Superconductivity, 19*, 1766.
Ohya, M., Yumura, H., Masuda, T., Nagaishi, T., Shingai, Y., & Fujiwara, N. (2010). *Journal of Physics: Conference Series, 234*, 032044.
Paranthaman, M., Aytug, T., Christen, D. K., Arendt, P. N., Foltyn, S. R., Groves, J. R., et al. (2003). *Journal of Materials Research, 18*, 2055–2059.
Paranthaman, M. P., Bhuiyan, M. S., Sathyamurthy, S., Heatherly, L., Cantoni, C., & Goyal, A. (2008). *Physica C: Superconductivity, 468*, 1587.

Prusseit, W., Hoffmann, C., Nemetschek, R., Sigl, G., Handke, J., Lumkemann, A., et al. (2006). *IEEE Transactions on Applied Superconductivity, 16*, 996–998.

Prusseit, W., Nemetschek, R., Semerad, R., Numssen, K., Metzger, R., Hoffmann, C., et al. (2003). *Physica C, 392*, 801–805.

Qiao, Y., Chen, Y., Xiong, X., Kim, S., Matias, V., Sheehan, C., et al. (2011). *IEEE Transactions on Applied Superconductivity, 21*(3), 3055.

Rey, C. M., Duckworth, R. C., Demko, J. A., Ellis, A., James, D. R., Gouge, M. J., et al. (2010). *AIP Conference Proceedings, 1218*(1), 453.

Riley, G. N., Jr., Malozemoff, A. P., Li, Q., Fleshler, S., & Holesinger, T. G. (1997). *Journal of Metals, 49*, 24.

Rupich, M. W., Li, X., Sathyamurthy, S., Thieme, C. L. H., DeMoranville, K., Gannon, J., et al. (2013). *IEEE Transactions on Applied Superconductivity, 23*, 6601205.

Rupich, M. W., Li, X., Sathyamurthy, S., Thieme, C., & Fleshler, S. (2011). *Physica C, 471*, 919.

Rupich, M. W., Li, X., Thieme, C., Sathyamurthy, S., Fleshler, S., Tucker, D., et al. (2010). *Superconductor Science and Technology, 23*, 014015.

Russo, R. E., Reade, R. P., McMillian, J. M., & Olsen, B. L. (1990). *Journal of Applied Physics, 68*, 1354.

Schmidt, W., Kraemer, H.-P., Neumueller, N.-W., Schoop, U., Verebelyi, D., & Malozemoff, A. P. (2007). *IEEE Transactions on Applied Superconductivity, 17*, 3471.

Selvamanickam, V., Carota, G., Funk, M., Vo, N., Haldar, P., Balachandran, U., et al. (2001). *IEEE Transactions on Applied Superconductivity, 11*, 3379.

Selvamanickam, V., Lee, H. G., Li, Y., Xiong, X., Qiao, Y., Reeves, J., et al. (2003). *Physica C, 392–396*, 859.

Shiohara, Y., Taneda, T., & Yoshizumi, M. (2012). *Japanese Journal of Applied Physics, 51*, 010007.

Sohn, S. H., Lim, J. H., Yang, B. M., Lee, S. K., Jang, H. M., Kim, Y. H., et al. (2010). *Physica C: Superconductivity, 470*, 1567–1571.

Solovyov, V. F., Wiesmann, H. J., & Suenaga, M. (2001). *Physica C, 353*, 14.

Specht, E. D., Goyal, A., Lee, D. F., List, F. A., Kroeger, D. M., Paranthaman, M., et al. (1998). *Superconductor Science and Technology, 11*, 945–949.

Specht, E. D., Goyal, A., Li, J., Martin, P. M., Li, X., & Rupich, M. (2006). *Applied Physics Letters, 89*, 162510.

Strickland, N. M., Long, N. J., Talantsev, E. F., Hoefakker, P., Xia, J. A., Rupich, M. W., et al. (2008). *Current Applied Physics, 8*, 372.

Strickland, N. M., Talantsev, E. F., Long, N. J., Xia, J. A., Searle, S. D., Kennedy, J., et al. (2009). *Physica C, 469*, 2060–2067.

Strickland, N. M., Talantsev, E. F., Xia, J. A., Long, N. J., Rupich, M. W., Li, X., et al. (2009). *IEEE Transactions on Applied Superconductivity, 19*, 3140.

Super3C project news release (March 17, 2009).

Tada, K., Yoshida, J., Mori, N., Yamada, K., Teranishi, R., Mukaida, M., et al. (2008a). *Physica C: Superconductivity, 468*, 1554.

Tada, K., Yoshida, J., Mori, N., Yamada, K., Teranishi, R., Mukaida, M., et al. (2008b). *Physica C, 468*, 1554–1558.

Tinkham, M. (1996). In *Introduction to superconductivity* (2nd ed.). New York, NY: McGraw-Hill.

Tixador, P., & Nguyen, N. T. (2012). *Superconductor Science and Technology, 25*, 014009.

Tokunaga, Y., Honjo, T., Izumi, T., Shiohara, Y., Iijima, Y., Saitoh, T., et al. (2004). *Cyrogenics, 44*, 817.

Ueyama, M., Hahakura, S., Hasegawa, K., & Ohmatsu, K. (2006). *IEEE Transactions on Applied Superconductivity, 16*, 999.

Usoskin, A., & Freyhardt, H. C. (2004). *MRS Bulletin, 29*, 583.

Usoskin, A., Freyhardt, H. C., Issaev, A., Dzick, J., Knoke, J., Oomen, M. P., et al. (2003). *IEEE Transactions on Applied Superconductivity, 13*, 2452.

Usoskin, A., Kirchhoff, L., Knoke, J., Prause, B., Rutt, A., Selskij, V., et al. (2007). *IEEE Transactions on Applied Superconductivity, 17*, 3235.

Verebelyi, D., Schoop, U., Thieme, C., Li, X., Zhang, W., Kodenkandath, T., et al. (2003). *Superconductor Science and Technology, 16*, L19–L22.

Wang, C. P., Do, K. B., Beasley, M. R., Geballe, T. H., & Hammond, R. H. (1997). *Applied Physics Letters, 71*, 2955.

Watanabe, T., Kashima, N., Suda, M., Mori, M., Nagaya, S., Miyata, S., et al. (2007). *IEEE Transactions on Applied Superconductivity, 17*, 3386.

Watanabe, T., Kuriki, R., Iwai, H., Muroga, T., Miyata, S., Ibi, A., et al. (2005). *IEEE Transactions on Applied Superconductivity, 150*, 2566.

Witanachchi, S., Patel, S., Zhu, Y. Z., Kwok, H. S., & Shaw, D. T. (1990). *Journal of Materials Research, 5*, 717.

Wu, X. D., Foltyn, S. R., Arendt, P. N., Blumenthal, W. R., Campbell, I. H., Cotton, J. D., et al. (1995). *Applied Physics Letters, 67*, 2397.

Xiong, X., Kim, S., Zdun, K., Sambandam, S., Rar, A., Lenseth, K. P., et al. (2009). *IEEE Transactions on Applied Superconductivity, 19*, 3319.

Yamada, Y., Ibi, A., Fukushima, H., Takahashi, K., Miyata, S., Kuriki, R., et al. (2007). *IEEE Transactions on Applied Superconductivity, 17*, 3371.

Yamada, Y., Miyata, S., Yoshizumi, M., Fukushima, H., Ibi, A., Kionoshita, A., et al. (2009). *IEEE Transactions on Applied Superconductivity, 19*, 3236.

Yazawa, T., Koyanagi, K., Takahashi, M., Toba, K., Takigami, H., Urata, M., et al. (2009). *Physica C, 469*, 1740.

Yoshida, J., Tada, K., Tanaka, T., Mori, N., Yamada, K., Teranishi, R., et al. (2008). *Physica C: Superconductivity, 468*, 1550.

Yoshizumi, M., Nakanishi, T., Matsuda, J., Nakaoka, K., Sutoh, Y., Izumi, T., et al. (2008). *Physica C: Superconductivity, 468*, 1531.

Yu, L. S., Harper, J. M. E., Cuomo, J. J., & Smith, D. A. (1985). *Applied Physics Letters, 47*, 932.

Yumura, H., Ashibe, Y., Itoh, H., Ohya, M., Watanabe, M., Masuda, T., et al. (2009). *IEEE Transactions on Applied Superconductivity, 19*, 1698.

Part Two

High-temperature superconducting (HTS) cable technology

High-temperature superconducting (HTS) AC cables for power grid applications

5

A.P. Malozemoff[1], *J. Yuan*[1], *C.M. Rey*[2]
[1]AMSC, Devens, MA, USA; [2]Energy to Power Solutions (E2P), Knoxville, TN, USA

5.1 Introduction

Power transmission and distribution at medium to high voltages have long been recognized as a significant application opportunity for superconductors in the electric power grid, and after many successful demonstrations around the world, high-temperature superconductor (HTS) AC cables are the closest among all HTS power applications to full commercial introduction.

Garwin and Matisoo (1967) made the first detailed study of the many trade-offs in superconductor cable design, recognizing that such cables based on low-temperature superconductors, operating at liquid helium temperature, would have to carry many gigawatts of DC power to be commercially competitive. The most significant demonstration of an AC cable based on low-temperature superconductors was made at Brookhaven National Laboratory in a project that started in 1972 and concluded with full power tests during the period 1982–1986. This 115 m AC cable system used liquid-helium-cooled Nb_3Sn tape and was designed to carry 1 GW of three-phase power with three such cables at 138 kV (Forsyth & Thomas, 1986). However the costs of the system and complexities of cryogenics at such a low temperature, coupled with the large cryostat diameter, prevented practical application in the power grid, especially since the dominant application opportunities were for carrying AC power, and AC losses add significantly to the cryogenic load. Only in the controlled environment of high-energy physics laboratories were low-voltage liquid helium temperature superconductor cables viable, typically operating with DC currents except during ramping.

The 1987 discovery of cuprate HTS with T_c above the liquid nitrogen (LN_2) boiling point changed the cryogenic considerations dramatically (see Chapter 1, Section 1.4.2). Operation in the LN_2 temperature range reduced cryogenic load, simplified refrigeration design and cost, and facilitated having a cryogenic fluid as part of the high-voltage dielectric system. So as soon as robust and high critical current first generation (1G) BSCCO-2223 HTS tape-shaped wires (we use wires and tapes interchangeably in what follows) became available in the mid-1990s (see Chapter 3), HTS AC cable projects were launched around the world (Hassenzahl & Tsukamoto, 2012; Okubo, 2010), most led by well-established power cable companies supported by government funding and with active utility participation (Demko et al., 2007; Honjo et al., 2003; Kim et al., 2005; Lee, Yoon, & Yang, 2011; Masuda et al.,

Superconductors in the Power Grid. http://dx.doi.org/10.1016/B978-1-78242-029-3.00005-4

2007; Maguire et al., 2011; Maguire et al., 2009; Maruyama et al., 2013; Masuda et al., 2011; Mukoyama, Hirano, Yagi, Kimura, & Kikuchi, 2003; Nassi et al., 2001; Sohn et al., 2007; Sim et al., 2008;Soika, Garcia, & Nogales, 2011; Stemmle, Merschel, Noe, & Hobl, 2013; Stovall et al., 2001; SUPER3C, 2009; Superconductive, 2012; Takahashi et al., 2005; Tonnesen et al., 2004; Volkov, Vysotsky, & Firsov, 2012; Xiao et al., 2007; Xin et al., 2005; Yumura et al., 2013), as summarized in Table 5.1. These projects typically featured a flexible 6–10″ diameter HTS AC cable, containing multiple HTS wires and a wrapped dielectric, much as in conventional AC cables, but with sophisticated terminations mating the LN_2-cooled-phase conductor and dielectric with room temperature in a high-voltage environment, along with an array of refrigeration equipment including cryocoolers, mixing tanks, pumps, and an LN_2 storage tank.

Over the next two decades these projects grew in sophistication, carrying ever greater power over ever greater lengths at ever higher voltages. Unfortunately, progress was not always steady. A major project at Detroit Edison with an HTS cable fabricated by Pirelli Cables and Systems (Norman et al., 2001) failed in 2001 because of welding problems in the vacuum jacket (not because of the superconductor wire!), and soon after, Pirelli, one of the initial industrial leaders in the field, gave up HTS cable development, sending a shock wave through the community. Nevertheless, the Japanese cable companies, particularly Sumitomo Electric Industries (SEI) and Furukawa Electric Company, and Ultera, a United States–Danish collaboration of Southwire and nkt cables, continued their programs (Honjo et al., 2003; Mukoyama et al., 2003; Tonnesen et al., 2004). They were soon joined by Nexans, which entered the major US Department of Energy (USDOE)-supported Long Island Power Authority (LIPA) cable project with American Superconductor (AMSC), successfully demonstrating the first in-grid 138 kV transmission-level-voltage HTS cable on Long Island in 2008 (Maguire et al., 2011). Active programs began in China (see Chapter 13) and Korea, with the Korean program led by LS Cable and KEPCO demonstrating one of the first in-grid cables using 2G HTS wire in 2011 (Lee et al., 2011). An active program continues in Russia as well, with projects for both AC and DC HTS cables (Volkov et al., 2012).

As of the time of this writing, the most significant recent in-grid HTS cable demonstration is the AmpaCity project in Essen, Germany, where Nexans, Karlsruhe Institute of Technology, and the local utility RWE, supported by the German Federal Ministry of Economics and Technology, have teamed to install and energize a 1-km-long 10 kV/40 MVA distribution cable, along with a standalone fault current limiter, linking two substations and replacing a 110 kV cable bringing power into the Essen city center (Stemmle et al., 2013). Another important ongoing project is the Hydra project of AMSC, Ultera, and Con Edison, supported by the US Department of Homeland Security, to install in the New York City grid a distribution-level three-phase AC cable (13.8 kV/4 kA) which is *inherently* fault-current-limiting (Maguire et al., 2009).

These demonstration projects have brought into sharper focus the benefits of HTS AC cables vis-à-vis conventional copper cables, which have been summarized previously in Chapter 1, Section 1.5.1. The initially most obvious benefit—of electrical power transmission with reduced electrical loss—has not actually been achieved in the demonstration projects so far, since the cryogenic refrigeration load arising from superconductor AC loss, thermal loss, and dielectric loss, coupled with a modest

Table 5.1 **Major worldwide HTS AC cable projects**

Year/place/country	Manufacturer	Type	Rating	Site	Wire manufacturer	Program	References
2000 Carrollton US	Southwire	CD-1 core	12.4 kV/1.25 kA/30 m/3 ϕ	Plant	BSCCO-AMSC	USDOE	Stovall et al. (2001)
2000 Milan IT	Pirelli	WD-1 core	115 kV/2 kA/50 m/1 ϕ	Test lab	BSCCO-AMSC	USDOE	Nassi et al. (2001)
2001 Copenhagen DK	nkt cables	WD-Triax	30 kV/2 kA/30 m/3 ϕ	AMK substation	BSCCO-NST	DK Dept. En.	Tonnesen et al. (2004)
2002 Ichihara JP	Furukawa	CD-1 core	77 kV/0.7 kA/30 m/1 ϕ	Test lab	BSCCO-Furukawa	Super-ACE	Mukoyama et al. (2003)
2002 Yokosuka JP	SEI	CD-3 core	66 kV/1 kA/100 m/3 ϕ	CRIEPI	BSCCO-SEI	TEPCO	Honjo et al. (2003)
2004 Anyangsi KR	LG Cable	CD-3 core	23 kV/1.26 kA/30 m/3 ϕ	LG cable	BSCCO	DAPAS	Kim et al. (2005)
2004 Yokosuka JP	Furukawa	CD-1 core	77 kV/1 kA/500 m/1 ϕ	CRIEPI	BSCCO-Furukawa	Super-ACE	Takahashi et al. (2005)
2004 Kunming CN	Innost/Innopower	WD-1 core	35 kV/2 kA/33 m/3 ϕ	Puji	BSCCO-Innost	MOST	Xin et al. (2005)
2005 Baiyin CN	ChangTong	WD-1 core	10.5 kV/1.5 kA/75 m/3 ϕ	Plant	BSCCO-AMSC	MOST/CAS	Xiao et al. (2007)

Continued

Table 5.1 Continued

Year/place/ country	Manufacturer	Type	Rating	Site	Wire manufacturer	Program	References
2006 Albany US	SEI	CD-3 core	34.5 kV/0.8 kA/ 350 m/3 ϕ	Albany- Nat'l grid	BSCCO-SEI/ YBCO-SP	USDOE	Masuda et al. (2007)
2006 Columbus US	Ultera	CD-Triax	13 kV/3 kA/200 m/ 3 ϕ	Bixby-AEP	BSCCO-AMSC	USDOE	Demko et al. (2007)
2006 Gochang KR	SEI	CD-3 core	23 kV/1.25 kA/ 100 m/3 ϕ	KEPCO	BSCCO-SEI	DAPAS	Sohn et al. (2007)
2007 Gochang KR	LS Cable	CD-3 core	23 kV/1.25 kA/ 100 m/3 ϕ	KEPCO	BSCCO-AMSC	DAPAS	Sim et al. (2008)
2008 Long Island US	Nexans	CD-1 core	138 kV/1.8 kA/ 600 m/3 ϕ	LIPA	BSCCO-AMSC	USDOE	Maguire et al. (2011)
2009 Europe	Nexans	CD-1 core	10 kV/1 kA/30 m/ 1 ϕ	LABEIN ES test lab	YBCO-BEST	Super3C, EC	SUPER3C (2009)
2010 Hannover DE	Nexans	CD-1 core	20 kV/3.2 kA/ 30 m/1 ϕ	Test lab	BSCCO-AMSC	Nexans	Soika et al. (2011)
2010 Yokohama JP	SEI	CD-3 core	66 kV/1.75 kA/ 30 m/3 ϕ	Test lab	BSCCO-SEI	METI/ TEPCO	Masuda et al. (2011)

2011 Icheon KR	LS Cable	CD-3 core	23 kV/1.25 kA/500 m/3 ϕ	Icheon substation	YBCO-AMSC	GENI	Lee et al. (2011)
2012 JP	Furukawa	CD-1 core	275 kV/3 kA/30 m/1 ϕ	Test lab	YBCO-Fujikura/SWCC	M-PACC	Maruyama et al. (2013)
2012 JP	SEI	CD-3 core	66 kV/5 kA/15 m/3 ϕ	Test lab	YBCO-SEI	M-PACC	Maruyama et al. (2013)
2012 Yokohama JP	SEI	CD-3 core	66 kV/1.75 kA/250 m/3 ϕ	Asahi substation	BSCCO-SEI	METI/TEPCO	Yumura et al. (2013)
2014 Essen DE	Nexans	CD-Triax	10 kV/2.3 kA/1000 m/3 ϕ	Delbrügge/Herkules	BSCCO-SEI	AmpaCity	Stemmle et al. (2013)
x Moscow RU	VNIIKP	CD-1 core	20 kV/2 kA/200 m/3 ϕ	Substation	BSCCO-SEI	Gov't/util.	Volkov et al. (2012)
x New York US	Ultera	CD-Triax	10 kV/4 kA/300 m/3 ϕ	Hydra-TBD	YBCO-AMSC	USDHS	Maguire et al. (2009)
x Jeju KR	LS Cable	CD-3 core	154 kV/500 m/3 ϕ	TBD	YBCO-AMSC	KEPCO	Superconductive (2012)

Year refers to year of energization. Type refers to WD, warm dielectric; CD, cold dielectric. x instead of a year indicates a project in progress, not yet energized.

coefficient of refrigeration performance (see Chapter 1, Section 1.2.2), largely nullified the lower superconductor loss advantage. There is little doubt that such a loss advantage *could* be achieved in HTS AC cables, though at significant cost in extra HTS conductor and increased thermal insulation, and new, more efficient turbo-Brayton refrigerators will also help. After all, even conventional copper cable could be designed to have arbitrarily low loss simply by adding arbitrarily more copper, but at the sacrifice of intolerable weight, diameter, and cost!

Rather, the dominant benefits of HTS AC cables emerge from the much higher current density of HTS wires compared to copper, allowing much larger power in the same cable cross-section or similar power at much lower voltage. For example, a typical conventional copper cable can carry up to 1 kA_{rms}, whereas HTS cable can carry several times as much in the same cross-sectional area. This is a major advantage, opening up the opportunity to retrofit existing underground ducts with much higher power HTS AC cables, or to avoid the complex permitting which burdens conventional transmission voltage cable projects by using lower voltage. HTS AC cables are also lighter and emit no heat or external magnetic field (EMF), making them easier to retrofit and install. The inherent sharp nonlinear increase in electrical resistance when transitioning from the superconducting to the normal (resistive) state above the critical current, a phenomenon not present in copper, enables development of inherently fault-current-limiting HTS cables. This unique functionality opens the opportunity to directly connect substations, without risk of cascading, potentially catastrophic fault currents between substations. The use of these inherently fault-current-limiting HTS AC cables connecting substations provides greater redundancy and resiliency to the local electric grid (Maguire et al., 2009). All these advantages are of particular value in high-population-density urban power grids with already crowded underground infrastructures which are facing ongoing load growth and fault currents near the rating limits of existing transformers and electrical switchgear.

Of course, HTS DC cables are a better means for achieving reduced electrical loss since they avoid AC loss, and much has been written about a SuperGrid efficiently connecting an entire continent with a DC backbone of power and bringing large quantities of clean power from remote solar and wind resources (see Chapters 6 and 7). However, since the existing power grids around the world are predominantly AC, any DC link must ultimately convert the power to AC through expensive power electronic conversion, and to amortize the cost of such AC–DC conversion stations, DC links must be long enough, usually at least 100 miles or more, to justify the AC–DC conversion expense. The vast amount of HTS wire required for such a sizeable DC link means that any such project remains quite far in the future.

All in all, the many successful in-grid demonstrations both in the United States and internationally have brought HTS AC cable technology to a significant level of technical maturity and credibility in the utility community. But the cost remains high, both for the HTS wire and for the accompanying cryogenic refrigeration, and the step to true commercialization without government support awaits opportunities where conventional cable technology is simply unable to meet application requirements. Such opportunities include space-constrained underground environments in urban centers needing increased power, or current-limited coupling between substations for increased resiliency. Once the commercial threshold is breached, one can expect ongoing cost reduction and a growing operational track record over time to drive an ever broader range of HTS AC cable installations in the power grid, including, ultimately, long-distance low-loss power transmission using DC cables.

5.2 High-temperature superconducting (HTS) AC cable design

5.2.1 Helically wound tape configuration and inductive balancing

A variety of three-phase HTS AC cable designs have been used by the different cable companies, but they typically fall into three distinct categories: (1) three separate single-phase conductors in three separate cryostats (see Figure 5.1(a)), sometimes called the "single core" design, (2) the HTS "3-core" or "3-in-1" or "triad" design which consists of three separate phase conductors held in a triad configuration in a single cryostat (see Figure 5.1(b)), or (3) the HTS Triax™ (a trademark of Southwire Company) or concentric design which has three concentric phases all contained within a single cryostat (see Figure 5.1(c)). All three HTS AC cable geometries, at their core, use one or more layers of HTS tape-shaped wires, helically wound around a flexible central former, which could be a hollow tube or bundle of copper wires for stabilization. This central assembly of HTS tapes, often called the "phase conductor," is what carries the electrical power. The HTS tapes are surrounded by a cold dielectric which is typically wrapped using an insulating tape that can survive cryogenic operating conditions. This is followed by either one or multiple HTS layers acting as screening layers or other phases, space for LN_2 flow, a surrounding cylindrical vacuum cryostat with multilayer insulation, and an outer protective jacket, as shown in Figure 5.1. An alternative design, now less used, puts a cryostat directly around the phase conductor, with a "warm dielectric" and conventional copper screening outside.

The use of helically wound HTS tapes around a flexible cylindrical former was initially prescribed by the fact that the first HTS wires available in quantity were flat and tape shaped, based on the 1G configuration using BSCCO-2223 conductor (see Chapter 3). But with the more recent development of round wires based on BSCCO-2212 or even MgB_2 (see Chapter 2), one might ask whether an alternative configuration using a cylindrical bundle of densely packed, twisted, and transposed round wires, much as in conventional copper cable, could form a competitive design for the future.

A key issue for HTS AC cable design with round wires arises from AC loss considerations, which can be understood with reference to Figure 5.2. The classic formula for hysteretic AC loss P per length (in mks units), arising when a sufficiently large AC peak magnetic induction field B_{peak} with frequency f acts on a wire or tape of critical current I_c and dimension a perpendicular to the field, is

$$P(\text{W/m}) = afB_{\text{peak}}I_c \tag{5.1}$$

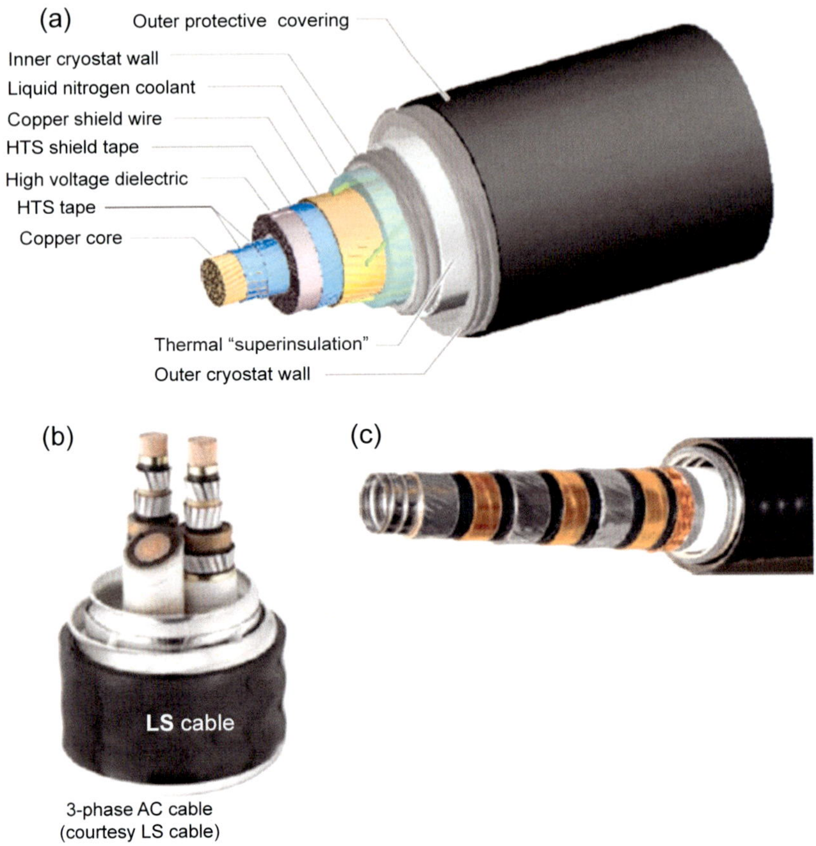

Figure 5.1 Widely used HTS cold dielectric AC cable configurations: (a) "single-core" single-phase configuration with counterwound two-HTS-layer phase conductor and screening layer, both stabilized by copper, used by Nexans for the 138 kV LIPA cable (Maguire et al., 2011). (b) "three core" or "3-in-1" or "triad" configuration with three phases, each constituted by a two-HTS-layer phase conductor and screening layer, both stabilized by copper, used by Sumitomo electric and LS cable for distribution voltage cables (Honjo et al., 2003; Sim et al., 2008). (c) "Triax" or "concentric" configuration with three phases wound sequentially on a single former, used by Ultera for the Bixby project (Demko et al., 2007), and by Nexans for the AmpaCity project (Stemmle et al., 2013). In this case, no HTS screening layer is required.

Now the magnetic induction field at radius r surrounding a cylindrical wire bundle carrying total current I is $B = \mu_0 I/2\pi r$, oriented azimuthally (circumferentially) around the bundle (see Figure 5.2). Thus, the azimuthal magnetic field at the surface of such a bundle with radius 0.4 cm, carrying 2500 A, is 125 mT, significantly higher than 33 mT at a typical radius 1.5 cm for a helically wound (flat tape) layer. In addition, typical HTS round wire diameters today are 100 μm or more, whereas the 2G tape's HTS layer has its thin dimension, typically only 1 μm, oriented perpendicular to the azimuthal field (see Figure 5.2). Then, by Eqn (5.1), the 60 Hz AC loss of each 100-μm-diameter wire with $I_c = 100$ A at the outer edge of the bundle will be

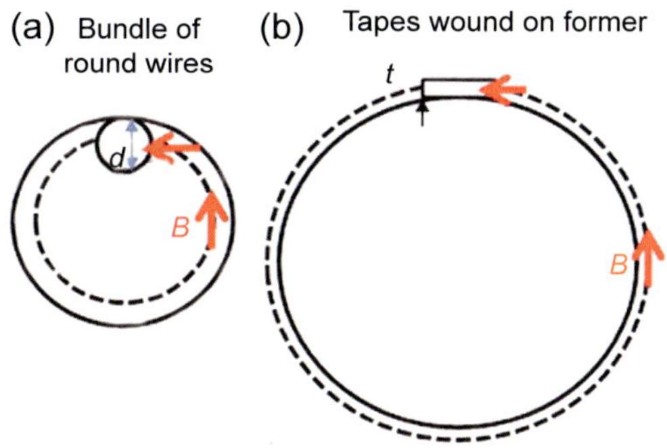

Figure 5.2 Surface magnetic field B orientation in cross-sections of (a) round wire in a compact bundle and (b) tapes wound on a former.

75 mW/m, whereas the loss of a 1-μm-thick 2G tape with the same I_c will be only 0.2 mW/m, or even less if the lower field level does not saturate the hysteresis loop. Thus, until HTS round wire diameters can be brought to submicrometer dimensions, which seems beyond hope for the foreseeable future, the helically wound HTS AC cable design with tape-shaped wires has a major AC loss advantage. Further discussion of AC loss follows in Section 5.3.

The helically wound HTS design also fulfills several other key requirements of conventional cable design. One is simply mechanical: The helical configuration allows flexibility and bending of the cable. By contrast, a cable with wires lying straight along the axis of a cylindrical former of radius 1.5 cm will buckle on the concave side of the bending, for example, when the cable is wound onto an 8-foot-diameter drum for transport, while the 1% tensile strain on the convex side will cause the HTS wires to break.

Another key advantage of the helically wound configuration is the ability to achieve a uniform current distribution among all the HTS wires (Fukui et al., 2007; Garber et al., 1976; Leghissa et al., 1999; Mukoyama et al., 1997), which is essential for avoiding potentially large AC flux-transfer losses, as discussed further in Section 5.3.3. In a single layer of helical windings where the HTS tapes lie tightly next to each other, each HTS tape is symmetric with all the others, ensuring an equivalent inductive environment and hence equal current flow in each tape. However the situation becomes more complicated if multiple layers are used. Consider, for example, a former of radius $r = 1.25$ cm, which can accommodate $2\pi r \cos\theta/w = 18$ tapes of width $w = 4$ mm, helically wound with pitch angle $\theta = 20°$. If each of these HTS tapes carries ~85 A_{rms}, a single winding layer could carry 1.5 kA_{rms}. Thus, a second winding layer would be required for a 3 kA_{rms} cable, and more layers are needed for even higher rating.

The use of multiple HTS layers creates a difficult challenge for achieving uniform current flow, since the relative positions of HTS tapes in the different layers are no longer symmetric and the number of HTS tapes that can be accommodated, or the spacing between them, increases as the diameter of the layer increases, causing a change in the inductive impedance of each layer. The most effective way to achieve such a uniform current distribution is through inductive balancing by adjusting the winding pitch angles of the different layers (Fukui et al., 2007; Garber et al., 1976; Leghissa et al.,

1999). This can be seen from straightforward application of Ampere's and Faraday's laws for the case of two winding layers (Clem & Malozemoff, 2010; Mukoyama et al., 1997), each characterized by a total AC current I_i, a radius R_i, and tangent $t_i = \tan\theta_i$, where positive and negative signs of t_i correspond to right-handed and left-handed helices, respectively. The voltages V_i per unit axial length of each layer (where subscript 1 denotes the inner and 2 the outer layer) are then

$$V_1 = L_{11}\mathrm{d}I_1/\mathrm{d}t + L_{12}\mathrm{d}I_2/\mathrm{d}t \tag{5.2}$$

$$V_2 = L_{21}\mathrm{d}I_1/\mathrm{d}t + L_{22}\mathrm{d}I_2/\mathrm{d}t \tag{5.3}$$

where the inductances per unit length L_{ij} are given by

$$L_{11} = (\mu_0/4\pi)\left[t_1^2 + 2\ln(R_2/R_1)\right] \tag{5.4}$$

$$L_{12} = L_{21} = (\mu_0/4\pi)(R_2/R_1)t_1t_2 \tag{5.5}$$

$$L_{22} = (\mu_0/4\pi)t_2^2 \tag{5.6}$$

The L_{ii}'s form an inductance matrix for this configuration. Since the voltages per unit length must be equal, one can solve for the ratio of the currents in the two layers:

$$I_1/I_2 = (L_{22} - L_{12})/(L_{11} - L_{21}) \tag{5.7}$$

Consider now the simple case where $R_1 \to R_2$ and the pitch angles are both positive and equal (e.g., both right-handed). It can immediately be seen from Eqns (5.4)–(5.7) that the current ratio in Eqn (5.7) approaches zero; that is, all current is carried on the outer HTS layer. This is unacceptable because when the total current I of the HTS AC cable reaches the total critical current of the outer layer, which occurs when $I_{\mathrm{rms}} = I_{\mathrm{c,tot}}/2\sqrt{2} = 0.35\ I_{\mathrm{c,tot}}$, the current in the outer layer becomes pinned at $I_{\mathrm{c,tot}}/2$ and further increases in applied current must be accommodated by the inner layer. This process requires flux transfer across the gap between the two layers, generating a large so-called "flux-transfer loss" (Clem & Malozemoff, 2010, 2013).

An inductively balanced configuration can be determined simply by solving for the ratio of pitch tangents for $I_1/I_2 = 1$ in Eqn (5.7). With $R_1 = R_2$, the result is $t_1/t_2 = \pm 1$. However the solution $t_1/t_2 = +1$ is excluded because this could lead to division by zero in Eqn (5.7). Thus, the inductively balanced solution $t_1/t_2 = -1$ has the pitch angles equal and opposite in the two layers. This is often called the two-layer "counter-wound" or "double helix" configuration and is widely used in both single core and multicore HTS AC cable designs (see below), as well as in the early Brookhaven Nb_3Sn cable (Forsyth & Thomas, 1986; Garber et al., 1976). Thus, it is clear that by appropriate choice of the winding pitch angles, one can equalize current flow in both layers and thereby avoid the onset of unacceptably high flux-transfer loss. Optimization of pitch angles to achieve inductively balanced current flow among all layers becomes more complex with more than two layers but follows from the same

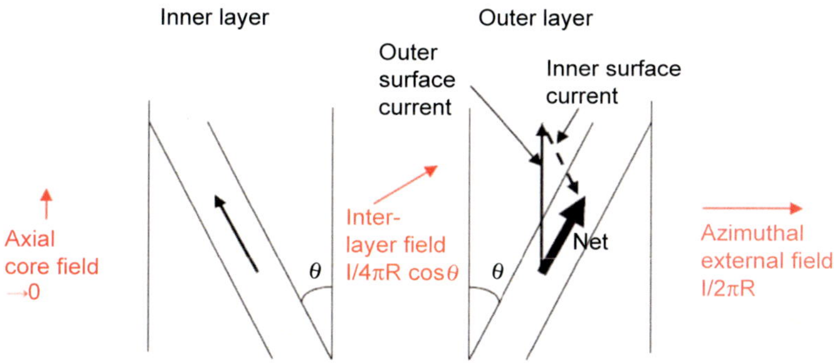

Figure 5.3 Schematic current flows (black arrows) and magnetic fields (red arrows) in a two-layer helically counterwound cable carrying a total current *I*, with radius *R* and pitch angle θ for the windings (Clem & Malozemoff, 2010). Both layers are laid flat with one tape shown in each layer. The vertical direction in the figure corresponds to the cable's axis and the horizontal direction to the azimuthal direction around the winding layer's circumference.

principles and is routinely done numerically by HTS AC cable designers (Fukui et al., 2007; Leghissa et al., 1999).

The current flow in the counterwound two-layer configuration is quite remarkable and is shown in Figure 5.3, where the two winding layers are conceptually cut open and laid out flat (Clem & Malozemoff, 2010). Since the surface supercurrent must be equal in magnitude to, and orthogonal to, the surface magnetic field, and since the outside surface magnetic field is azimuthal, the surface current on the outer surface of the outer tapes must flow along the *cable* axis. But assuming that each HTS layer carries half the total current, the *net* surface current directed along the *wire* axis must be $I/4\pi R\cos\theta$. Since the tapes lie at an angle, the axially oriented outer surface current must cross over at the tape edge to the inner surface, and the direction of the inner surface current must form an isosceles triangle with the outer surface current to give rise to the net wire current as shown in the figure. Then an interlayer field of magnitude $I/4\pi R\cos\theta$ must point perpendicular to the inner surface current of the outer layer and a corresponding surface current on the outer surface of the inner windings must flow at the equal and opposite angle from the outer windings. Thus, having the inner windings at the equal and opposite angle ensures that no surface current flows on the inner surface of the inner windings, and the axial magnetic field in the core is zero, which is useful for minimizing any eddy currents in the former.

The fact that the surface currents in the outer layer must cross over from one side to the other at the edges has stimulated studies of the perpendicular ("*c*-axis") critical current density in typical production tapes, which have shown that the perpendicular critical current density is large enough to confine the crossover region to a thin region near the tape edge (Jia et al., 2011).

5.2.2 Dielectric, screening, and multiphase configurations

A critical element of HTS AC cable design is the dielectric surrounding the phase conductor, which must withstand the full voltage of the given application, ranging from 10 kV for typical distribution cables to 138 kV and above for transmission cables, plus the much higher voltages of BIL (basic insulation or impulse level) testing including lightning impulse and switching impulse tests. An early concern was whether conventional room-temperature dielectrics would be adequate at LN_2 temperature; little was initially known about their low-temperature performance or whether special new dielectrics would be required. This led to an early proposal of a so-called "warm dielectric" design in which the phase conductor is surrounded by a cryostat, with the dielectric at room temperature outside the cryostat (Hassenzahl & Tsukamoto, 2012). Although this configuration reduces the number of HTS wires as compared to the single-core cold-dielectric design with its additional HTS screening layers, it has the disadvantage that external fields are not fully shielded, leading to higher AC loss, increased inductance, and increased EMF.

In fact, several dielectrics have proven successful at LN_2 temperature. Most are held proprietary by individual HTS cable companies, although one, polypropylene-laminated paper (PPLP), is known to have been used, for example, by Sumitomo Electric Industries (Honjo et al., 2003) and others, and its dielectric breakdown and partial discharge characteristics in LN_2 have now been intensively studied (Okubo, 2010). Pressurized LN_2 (at several bars) is sufficient for avoiding the bubble formation which could lead to dielectric breakdown due to the weakened dielectric strength of the nitrogen gas phase relative to the liquid. A further proprietary aspect in the dielectric design of HTS AC cables is the potential use of carbon-based semiconducting layers to minimize non-uniformities in the electric field distribution. Establishing a standardized procedure for high-voltage dielectric testing, including dielectric breakdown strength and BIL testing, is another requirement for utility acceptance.

In any case, as is evident from Table 5.1, now essentially all HTS cable projects use the cold dielectric approach. The simplest cold-dielectric design is shown in Figure 5.1(a). This single-core HTS AC design has a grounded screening layer or layers wound onto the cold dielectric, surrounded by a cryostat including multilayer insulation and a protective outside jacket. Three separate and parallel single-phase HTS AC cables are needed to provide three-phase power.

The cryostat walls are corrugated to ensure flexibility in the HTS cable. Open passage for LN_2 can be arranged either through a hollow central former or in a space around the screen layer. LN_2 has direct access to the dielectric and HTS layers but is largely stagnant; so any heat from AC loss must diffuse to the open channel, where it can be swept away by the flowing LN_2. The cable must be able to withstand fault currents for specified durations to ensure the HTS cable's viability until circuit breakers open and cut the current. The fault currents that these HTS cables experience can be substantial, reaching 40 kA_{rms} in distribution-level circuits and 60 kA_{rms} or more in transmission-level circuits, and lasting for a second or two in distribution circuits and several tenths of a second in a transmission circuits. Such currents will drive the superconductor into its normal state and then generate large I^2R loss which could burn up the HTS wire. The usual solution to protect the cable is to provide sufficient copper electrically connected in parallel with both the phase conductor and the screen to absorb the I^2R loss without overheating. An alternative is a fault-current-limiting design which is discussed in Section 5.6.

The single-core design is particularly advantageous for high-voltage transmission-level circuits because in this case the dielectric must be thicker, making alternative three-core and HTS Triax™ designs too bulky. Termination design is also more difficult at high voltages, and the single-phase design avoids the extra complications inherent in the 3-in-1 and concentric designs.

Another widely used design is the three-core or 3-in-1 design of Figure 5.1(b). In this case, three phases, each consisting of a former, phase conductor, dielectric, and screen, are wound together within a single cryostat. This configuration has the advantage of compactness, but because the dielectric must be thin, the design is limited to medium voltage for distribution circuits.

A third design, the so-called "HTS Triax™" or concentric design, is shown in Figure 5.1(c). In this case, all three phases are wound coaxially, with three layers of dielectric separating the phases and an external room-temperature ground. An elegant feature of this configuration is that to the extent that the three phases are perfectly balanced (current amplitudes perfectly equal), the magnetic fields H they generate at a radius r combine to precisely cancel outside the third (exterior) layer:

$$H = (I/2\pi r)e^{i\omega t}\left(1 + e^{i2\pi/3} + e^{i4\pi/3}\right) = 0$$

Thus, like the other cold-dielectric designs and as long as the three phases are perfectly balanced, there is no EMF emanating from the HTS cable, which is a significant advantage in installation and operation. A major advantage of this design vis-à-vis the other cold-dielectric designs is that because no external superconductor screen is necessary, the amount of HTS tape required in the cable is reduced by half, a substantial reduction in cost and complexity (Demko et al., 2007). Another major advantage of this configuration is its compactness; power density can reach up to 10 times that of comparably rated conventional copper cables. And vis-à-vis the single core design (each phase in a separate cryostat), only a single cryostat is required for all three phases, also a major cost saving. However, separating the phases at the terminations is complex, and so this design has been limited to lower voltage, distribution applications such as Bixby (Demko et al., 2007; Section 5.8.1) and AmpaCity (Stemmle et al., 2013; Section 5.8.5). Inductive balancing is particularly important to ensure equal current flow in all phases; otherwise EMF can appear outside the cable. The types of designs used in various demonstrations to date are listed in Table 5.1.

5.3 AC loss of HTS cables

As indicated above, minimizing AC loss is a critical consideration in HTS AC cable design because of the heavy energy penalty that must be paid for cryogenic operation. Without optimized design, superconductor AC loss can easily reach 1 W/m or more per phase at rated power, which, over a 1 km, three-phase cable, would generate 3 kW of 77 K heat to be dissipated by the cryogenic refrigeration system. To this must be added thermal loss and dielectric loss, typically in the range of 1–2 W/m. An additional 1 kW is generated at each set of three-phase terminations. One can quickly see that the refrigeration system must be designed to handle in the range of 7 kW, and with some margin, 10 kW. Since even the most efficient closed-cycle refrigeration system has a coefficient of performance of about 0.1, the total room temperature refrigeration power load approaches a substantial 100 kW. Thus, for a 1 km 40 MVA distribution cable (10 kV, 2.3 kA_{rms}), this represents a loss of 0.25%, and for 4 km, this already reaches 1%.

AC losses in superconductor cables have been studied intensively, both theoretically and experimentally. Experimental determinations of these losses are tricky because slight differences in contact resistance to different HTS tapes can cause non-uniform current flow which aggravates the loss. Typically, at least 5-m-long HTS test cables are required so that the inductive impedance exceeds the contact resistance enough to even out the current flow (Nguyen et al., 2011; Vysotksy et al., 2010). Proper grounding and control of the environment are other critical issues since even at meter-length distances from an unshielded test cable, steel beams in the floor, for example, can affect a sensitive transport measurement by absorbing power radiated from an unscreened HTS cable core. Electrical transport AC loss measurements on three-phase HTS AC cable have been particularly difficult to interpret because of the complex interaction of AC magnetic fields from adjacent phases interfering with the measured phase. But at least for single-phase measurements, such measurements are now routine and reliable. An example is shown in Figure 5.4.

Calorimetric measurements have proven to be another reliable method of determining AC loss in HTS cables as well as of benchmarking against electrical measurements (Traeholt et al., 2001). These calorimetric measurements, however, are slow and time consuming, requiring calibration with internal embedded heaters or by driving the HTS cable itself into the resistive state using DC current and measuring the corresponding temperature rise associated with the Joule heating.

From a theoretical point of view, the various mechanisms of cable AC loss can be broken up into the following categories (Clem & Malozemoff, 2010; Campbell, 1999):

1. *Eddy current loss*. This is a dominant AC loss mechanism in conventional nonsuperconducting cables, but in HTS AC cables it generally plays a negligible role. In part this is due to the

Figure 5.4 Transport AC loss measurements on 1G cable used for the Albany cable project (Masuda et al., 2007; Section 5.8.2). Courtesy of Nguyen et al. (2011).

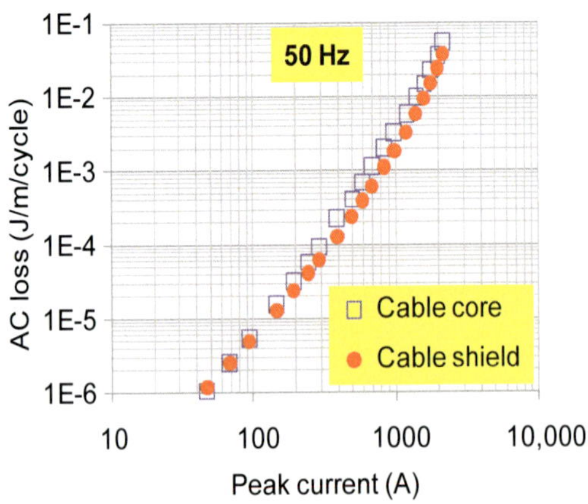

reduction in the amount of copper in the HTS cable, and in part to the elimination of AC fields outside the superconductor screen. Furthermore, in the two-layer counterwound HTS AC cable configuration, AC fields are also eliminated in the former, as explained earlier. The characteristic feature of eddy current loss is a squared dependence on frequency ($P \propto f^2$), but all HTS cable AC loss experiments show a dominant linear frequency dependence, indicating any of a variety of possible hysteretic loss mechanisms.

2. *Superconductor hysteretic loss*: *Surface Bean model loss*. A classic calculation by Norris (1970) predicts the AC hysteretic loss of a superconductor cylinder carrying an axial current I, on the basis of the well-known Bean critical state model (Bean, 1962),

$$P_L(\text{W/m}) = (\mu_0/3\pi) f I_0^3 d / I_c R (\text{mks units}) \tag{5.8}$$

where f = frequency (in Hz), I_0 is the *peak* (not rms) operating current (in Amps) in the cylinder, d is the total cylinder wall thickness (in m), I_c is the cylinder's DC critical current, and R is the cable radius (in m). Vellego and Metra (1995) proposed to use this "monoblock" model to estimate the AC loss of multilayer 1G HTS cables. This formula shows the characteristic cubic increase of AC loss with current which is observed in most experiments. It also shows the linear dependence on frequency, which characterizes all hysteretic loss phenomena and which is also seen in experiment. Even the loss magnitude has matched experiment reasonably well in a number of 1G cables (Cho et al., 2005; Xin et al., 2005). Because the fill factor of BSCCO-2223 filaments in 1G tapes is quite high, the monoblock model has some plausibility. As a rough estimate, consider an HTS cable having two HTS layers of $I_c = 100$ A HTS tapes with 20 tapes per layer, $f = 60$ Hz, $R = 1.25$ cm, a total two-layer thickness $d = 400$ μm, and a peak current $I_0 = 2000$ A (equal to only one-half of the total cable I_c). This gives a surface Bean loss of 0.51 W/m/phase, which obviously increases rapidly with increasing I_0 according to Eqn (5.8)

However, consider now a 2G cable with YBCO layer thickness of only 1 μm. It is immediately evident from the d-factor (cylinder thickness) in Eqn (5.8) that the loss is dramatically decreased. A more complete calculation including all surfaces of a two-layer counterwound 2G cable, using the surface fields shown in Figure 5.3, and assuming infinitely small gaps between neighboring tapes and tapes that conform perfectly to the cylindrical shape of the former, gives (Clem & Malozemoff, 2010)

$$P_L(\text{W/m}) = (\mu_0 d_s / 6\pi R) f I^3 (1 + 4\cos^3\theta) / I_c \cos^2\theta (\text{mks units}) \tag{5.9}$$

where now $d_s = 1$ μm is the superconductor layer thickness. Taking $\theta = 20°$ and the parameters as before, $P_L = 0.4$ mW/m, a truly negligible level. This low surface Bean model AC loss is a major advantage for 2G as compared to 1G HTS AC cables.

3. *Superconductor hysteretic loss*: *Gap and polygonal loss*. However, there are two other hysteretic AC loss contributions in HTS AC cables wound from 2G YBCO tapes that must also be minimized. These arise from gaps between the superconducting layers in neighboring tapes and from the polygonal shape when tapes do not fully conform to the cylindrical cross-section of the former, as shown schematically in Figure 5.5. These AC loss contributions have been calculated theoretically for a single layer of axially oriented 2G YBCO tapes (Amemiya et al., 2007; Malozemoff et al., 2009; Mawatari & Kajikawa, 2008). Careful winding of the 2G YBCO tapes onto the former to minimize the physical gaps between the tapes will help to minimize the gap AC loss, which scales as the square of the gap size $2g$ (Malozemoff et al., 2009; Norris, 1970),

$$P_L(\text{W/m}) = (\pi^3/48)\mu_0 f (g^2/\text{w})\, I^4 I_{c,\text{w}} / I_c^3 (\text{mks}) \tag{5.10}$$

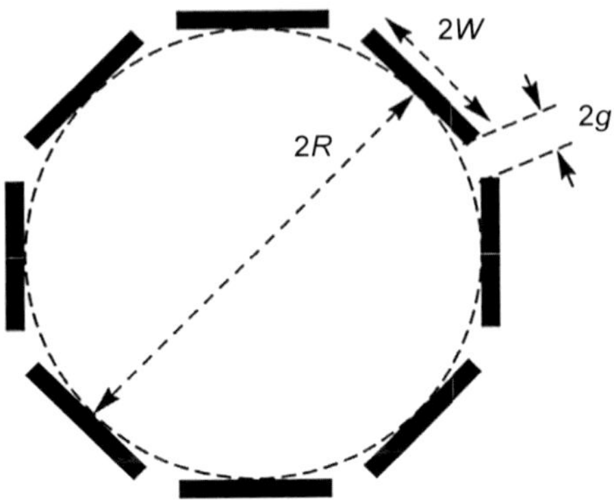

Figure 5.5 Schematic cross-section of a single-layer 2G cable with tapes indicated by black strips, with gaps between tapes of magnitude $2g$, and with polygonal corners defined by the extensions of the strips as $g \rightarrow 0$.
After Mawatari and Kajikawa (2008).

where $I_{c,w}$ is current per width of the tape. Unfortunately, all 2G tapes have a non-superconducting zone at the edges, because of fillets in laminated wires (see Chapter 4), electroplated stabilizer, or edge damage to the superconductor layer, and so the gap between edges of the fully superconducting regions is always significant. As an example, for a single layer with a gap $g = 200\ \mu m$, $w = 4$ mm, $I_{c,w} = 300$ A/cm, $I_c = 2000$ A, and $I = 1000$ A, the loss per length is $P_L = 1.8$ mW/m and double this for two layers. This is larger than the surface Bean loss estimated above, but still much smaller than the monoblock-model loss describing 1G cables.

With wider HTS tapes, fewer can be accommodated around a given circumference, and so the corresponding number of gaps will be smaller. Assuming that the gaps are kept the same size, the total gap loss will *decrease* as the HTS tape width increases in a cable of constant radius, as evident from the inverse width dependence in Eqn (5.10). On the other hand, the wider the HTS tapes, the fewer their number and, assuming that the tapes are perfectly flat and stiff, the sharper the polygonal angle between neighboring tapes shown in Figure 5.5. These polygonal corners also give rise to an AC loss, comparable to the gap loss in order of magnitude and *increasing* with tape width. Thus the competing gap and polygonal loss effects imply an optimum width for minimizing the AC losses, which for typical values falls in the range of 3–4 mm (Malozemoff et al., 2009). Analytical calculations (Amemiya et al., 2007) have indicated somewhat smaller values, as low as 2 mm, but manufacturing HTS AC cables with such fine wires is a daunting task.

Through meticulous control of the winding process to minimize all gaps, and using a large former diameter and narrow (3 mm) 2G YBCO tapes in the two-layer counterwound configuration, Chevchenko et al. (Chevtchenko et al., 2012) have demonstrated a record low loss of 0.1 W/m at 3 kA_{rms} and 50 Hz. Comparably low loss has also been achieved in the Japanese M-PACC project (Maruyama et al., 2013). Of course, industrial HTS cable manufacturing cannot be so precise, and bending of the HTS cable will also disturb the tape positions to some degree. Nevertheless, these design principles of minimizing gaps and using narrow tapes go a long way to achieving significantly reduced 2G cable AC loss.

The inverse dependence on total cable current I_c in Eqns (5.8) and (5.9) and the inverse square dependence on I_c in Eqn (5.10) imply that one of the most effective ways to reduce AC loss is simply to increase I_c. Ideally this is done by improving the I_c of the individual tapes by enhanced processing. Most simply, one can do this by adding more tapes; however, this creates a trade-off because of the increased cost of the extra tapes.

4. *Flux-transfer loss*. If the current distribution in the different layers of an HTS cable is not balanced, the layer carrying the higher current, usually the outer layer, can reach its critical current before the other layers. If the total current then increases further, the current in the outer layer will be pinned at its critical current, while further current is shunted to the inner layers. Since the increased current in the inner layers must be accompanied by increased flux according to the Bean critical state model (Bean, 1962), flux must transfer through the outer layer and the space between the two layers to reach that inner layer. A first estimate of the AC loss which arises from this process was given by Daeumling in his "duoblock" model (Daeumling, 2004), and a more general result has been derived recently (Clem & Malozemoff, 2013). Especially in 2G cables, where the spacing between layers is large compared to the thickness of individual superconductor layers, this "flux-transfer loss" can be very large, typically many watts per meter. Avoiding this large loss is a major incentive to ensure a balanced current distribution in cable design, as emphasized earlier.
5. *Ferromagnetic loss*. Textured Ni–5 at% W tapes have been used as substrates for 2G wire by some manufacturers (see Chapter 4). However, they are ferromagnetic, with a narrow hysteresis loop only a few Oersteds (several 100 A/m) wide at 77 K. Integration of the fully saturated loop area for a typical 4-mm-wide 2G wire with such a substrate gives a net loss per length of about 7 mW/m at 60 Hz (Zubko, Nosov, Polyakova, Fetisov, & Vysotsky, 2011). When combined in a two-layer counterwound cable core, 2G wires on such substrates give an AC loss shown in Figure 5.6: The presence of ferromagnetic substrates adds a "foot" to the AC loss at low currents. Figure 5.6 shows the loss decomposed into ferromagnetic substrate loss, determined from integration of the substrate's magnetic hysteresis loop, multiplied by the number of wires, and a rough estimate of superconductor loss. The calculated total lies somewhat below the measurement in this study.

   Thus, it is evident that all the ferromagnetic substrates contribute their ferromagnetic AC hysteresis loss to the total HTS cable AC loss. There is another important loss mechanism that can arise if a two-layer cable core is constructed in such a way that one or both substrates in the two layers lie *between* the superconductor layers. The ferromagnetic switching between the two HTS layers imbalances the current distribution and thereby can precipitate flux-transfer loss (Clem & Malozemoff, 2013). This problem can be avoided by orienting the substrates in the "in-out" configuration, that is, with the inner layer's substrates facing inward and the outer layer's substrates facing outward, relative to the superconductor layer (Clem & Malozemoff, 2010).

   Of course, ferromagnetic AC loss can be eliminated altogether by simply using wire with nonmagnetic substrates. For example, 2G wires using the IBAD buffer process use nonmagnetic Hastelloy or stainless steel substrates (see Chapter 4). New nonmagnetic Ni–9 at% W has also recently been developed as substrate for 2G HTS wire (Rupich et al., 2013).

5.4 Terminations

HTS cable terminations are the links between the conventional power grid at room temperature and a superconducting cable at cryogenic temperature. Inside the termination, the stranded HTS tapes in the cable are connected to the internal cold end of the termination while outside the termination, conventional power distribution cables are connected to its external warm end. In addition, the termination provides an interface

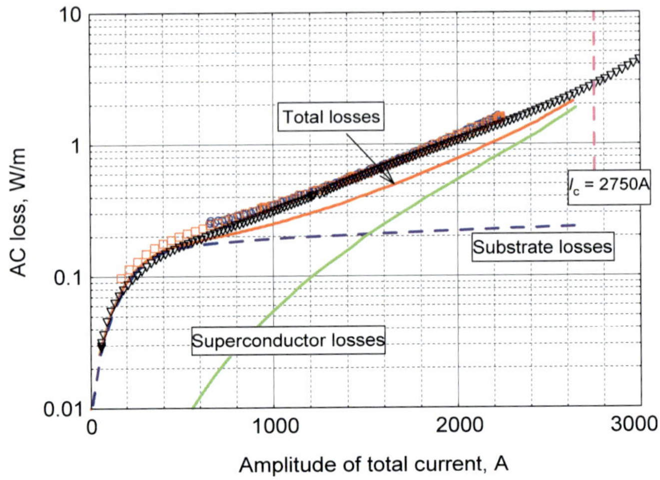

Figure 5.6 AC loss versus amplitude of AC current of a two-layer, counterwound cable core using 34 2G HTS wires with Ni–5 at% W substrates.
After Vysotksy et al. (2010).

point for the supply and return of coolant to the cable system, needed to keep the HTS cable in its superconducting state. Proper design and manufacture of the HTS cable terminations are vital for long-term reliability of the overall system. As a matter of fact, in-grid HTS cable demonstration projects show that time out of service is more frequently due to termination issues than due to problems in the HTS cable itself (Schmidt et al., 2011). Because of the complex requirements, superconducting cable terminations are more complicated than the HTS cable itself in terms of design, manufacturing and installation. Besides all the functions of conventional cable terminations, superconducting cable terminations must provide the following additional functions:

- Cooling and electrically insulating the current conductors (current leads)
- Carrying high current
- Thermally grading the conductor and insulation material from the cryogenic temperature to ambient temperature
- Thermally insulating the termination assembly in the cold region
- Managing the thermal contraction of the superconducting cable (optional)
- Smoothly grading the electrical stress from ground to line potential
- Managing transit pressure rise due to internal faults
- Interfacing the coolant stream

The design of commercial superconductor cable terminations requires the selection of a simple yet suitable configuration so that the HTS wires are easy to connect to the current leads and so that current leads as well as associated electrical insulation are easy to assemble. Two types of terminations are commonly used in demonstration projects, depending on the HTS cable design configuration. For a single-phase AC HTS cable configuration, a single-phase termination design is commonly applied where each phase has its own cryostat, while for a 3-in-1 or a Triax cable design, a three-phase termination architecture is usually adopted where all three phases share one cryostat. Typical termination configurations are shown in Figure 5.7.

The current leads used in HTS cable termination are not much different from other HTS applications. They have to carry very high current, normally in the range of a few thousand Amperes, compared to less than 1000 Amperes in conventional termination leads. To reduce the heat leak into the cold cryogen, current lead optimization is necessary. Typical optimized conduction-cooled current leads yield a heat loss of ~90 W/kA per pair working between 300 K and 77 K. So, for example, a 4 kA three-phase system will generate 1080 W, a substantial contribution to the overall refrigeration load of the system.

Besides the optimization, one should also pay attention to the different working modes of current leads. When the HTS cable is energized, the Joule heating of the current leads will maintain the top part of the termination warm. When the HTS cable is under no load or sub-optimal load conditions the termination top part must be maintained at sufficient temperature to make sure the electric insulation and seals are still working properly, because too cold a temperature might damage the seals or electrical insulation. This can be accomplished by carefully selecting the current lead material and choosing a proper ratio of length to cross-sectional area. If necessary, an external heater can be added to keep the termination top part warm when the cable is in idle or subload mode (Maguire et al., 2009).

Transition from cryogenic temperature to room temperature presents several challenges in the design of HTS cable termination bushings. First, the bushing must be capable of managing the thermal gradient from room temperature to the operating temperature. Second, the bushing has to manage high voltage for the grid connection. Besides mechanical and electric strength requirements at low temperature, the materials used for the bushing must endure significant temperature changes during the cool-down process.

Because thermal contraction of most metals between 300 K and 77 K is around 0.3%, the cold core of a 1 km HTS cable will contract by about 3 m compared to its outer, room-temperature jacket. Handling a contraction of this magnitude poses a serious issue in HTS cable design. The most common method for accommodating this large thermal contraction is based on a movable termination. In this approach, terminations are built on wheels that allow the termination to move forward and backward during cable cool down and warm-up, respectively, while an accordion-like jacket accommodates the motion (Ichikawa et al., 2005). Other methods used to manage HTS cable contraction include compensating thermal shrinkage inside the termination, introducing extra length in the HTS cable core using a serpentine configuration either at the center or at the ends of the cable (Maguire et al., 2011). In a three-core configuration, the serpentine configuration is introduced through the intertwining of the three cores (Honjo et al., 2003).

The termination cryostat is very similar to a conventional horizontal cryogen storage tank, where vacuum with multilayer superinsulation is used to reduce thermal heat invasion. Due to the large amount of energy dumped into the cryogen in an internal fault at a termination, a fast relief burst disc is often introduced for this kind of application to prevent a catastrophic overpressure event. The burst disc is normally

(a)

(b)

Figure 5.7 Typical superconducting cable termination configurations. (a) Three-phase termination for the Triax™ cable system at AEP's Bixby substation in Columbus, OH, with all three phases in a single cryostat (Demko et al., 2007). (b) Three single-phase terminations in the LIPA cable system, each phase having its own separate cryostat and connected to a separate cable (Maguire et al., 2007).

implemented on the inner volume of the termination. In case of an overpressure spike, the burst disc will open and relieve the LN_2 in a controlled manner. Typically, LN_2 is dumped and vented to an underground pit filled with rocks (Maguire et al., 2007). There are two different designs of the burst disc. One is called the cold burst disc design, where the burst disc is in direct contact with the cold cryogen, while the other is called a warm burst disc design where the disc is at room temperature and a transitional column of vapor phase is located between the room temperature burst disc and the cold cryogen. Both approaches have been demonstrated in previous cable projects (Maguire et al., 2011).

Just as in conventional cable terminations, smooth grading of electrical stress from ground to line potential must be managed properly at the high-voltage HTS cable terminals. In order to control the electric field, a stress cone is usually utilized (Ohya et al., 2013) (see Figure 5.8).

High-voltage-type tests of HTS cable terminations are similar to those of conventional cable terminations. These include partial discharge, lightning impulse (i.e. BIL), and AC withstand. Because LN_2 coolant is part of the dielectric system, HTS cable terminations must be tested under cold conditions, requiring a refrigeration system capable of providing subcooled LN_2. Further discussion of HTS AC cable terminations is given in Chapter 8.

5.5 Cryogenic refrigeration systems for HTS AC cables

5.5.1 Refrigeration system configurations

As discussed in previous sections, an HTS power cable must be cooled sufficiently below the critical temperature T_c of superconductor YBCO wires (~90 K) to maintain their superconducting state. For a variety of reasons, the real operating temperature of the HTS superconductor cable is much lower than T_c and the coolant is in its subcooled state. LN_2 has been the coolant of choice for all HTS AC cables because of its ideal temperature range sufficiently below T_c and because of its favorable dielectric properties.

The term "subcooling" refers to liquid at a temperature below its normal saturation temperature at given pressure. For example, LN_2 boils at 77 K at atmospheric pressure,

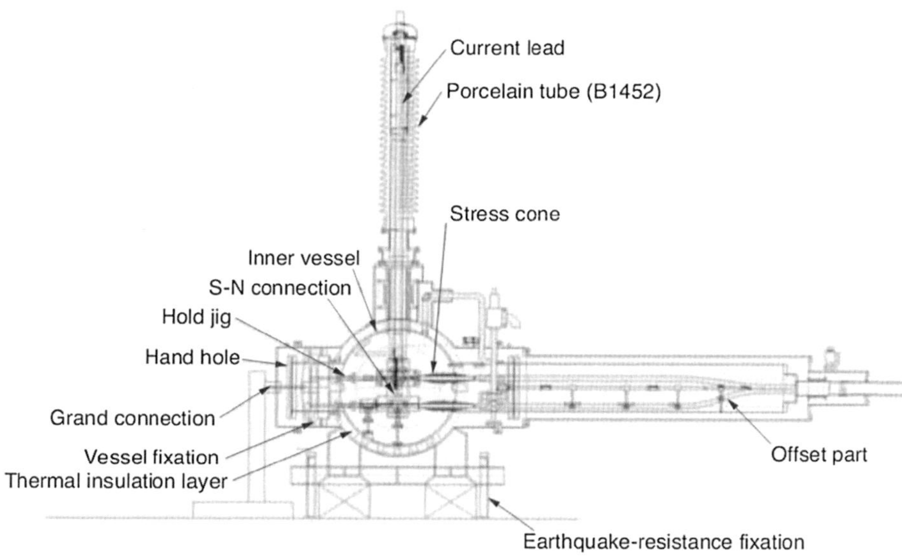

Figure 5.8 Structure of termination for 3-in-1 cable (Ohya et al., 2013).

and at 70 K and 1 atm, LN_2 is termed "subcooled" because in this state, it is 7 K below its atmospheric boiling temperature 77 K. For a cold dielectric HTS power cable design, this is necessary because the cooling fluid, which participates in the electrical insulation by impregnating the lapped dielectric material, must remain in a pure liquid phase (subcooled state) to avoid any bubble formation and so maintain the dielectric strength between the high-voltage cable core and the shield, which is at ground potential. Furthermore, the coolant must have enough temperature margin during a major fault or multiple through-faults to remain in the subcooled state. A typical operating temperature is around 70 K, 34 K below the LN_2 boiling temperature at 10 bar.

To maintain the temperature of the HTS cable and accessories within a determined range, a broadly adopted approach by most HTS cable projects is to circulate subcooled LN_2 through the cable system and its accessories, where the coolant stream removes the heat losses from the cable system and dissipates them at a refrigeration station. Figure 5.9 shows a simplified flow diagram of a typical HTS power cable cooling system. The refrigeration system is a combination of three separate, but interactive, subsystems. The primary refrigerator removes the heat from the LN_2 coolant loop under normal operating conditions. The backup system can be engaged to provide cooling capacity in the event that the primary refrigerator is partially or fully shut down. The LN_2 coolant loop is the loop through which the LN_2 flows to and from the HTS cable. It includes the LN_2 pumping and pressurization compensation systems. Finally, a heat exchanger dissipates the heat from the HTS cable system loop into the refrigeration loop. This thermal link could be a subcooled LN_2 bath as in the Bixby case (see Section 5.8.1) or a plate-fin heat exchanger as in the LIPA case (see Section 5.8.3).

Typically, an HTS cable cooling system is a combination of two separate loops. One loop is part of the refrigerator itself, which could be open or closed, and the working fluid could be helium, nitrogen, or a gas mixture, while the other loop consists of LN_2 circulated to cool the cable system, which is always a self-contained closed loop with the LN_2 fully contained inside the cable cryostat and terminations. In theory, the LN_2 coolant loop should never need to be replenished. However, in practice due to small leakage or other events such as overpressure relief venting, the inventory of LN_2 in this self-contained system will reduce over time and needs to be replenished once in a while. Normally, this kind of refill is insignificant compared to that of an open-loop refrigerator cycle, where cooling effects are produced by depressurizing and consuming LN_2, and therefore regular LN_2 refilling is part of normal operation procedures.

The LN_2 subcooled state is retained by pressurizing the coolant at low temperature at the cooling station. A variety of flow configurations have been used. For a single-phase cable design, one phase can be used as the go-flow and the other phases can be used for return-flow. For a "Triax" cable design (see Figure 5.1(c)), one can use either a once-through flow configuration with separate return line or a counterflow configuration without return line (e.g., flow one way through a hollow former and return flow in the space around the screen).

Besides maintaining coolant in the subcooled state during normal operation, the refrigeration system should also be designed to have the capability to monitor the

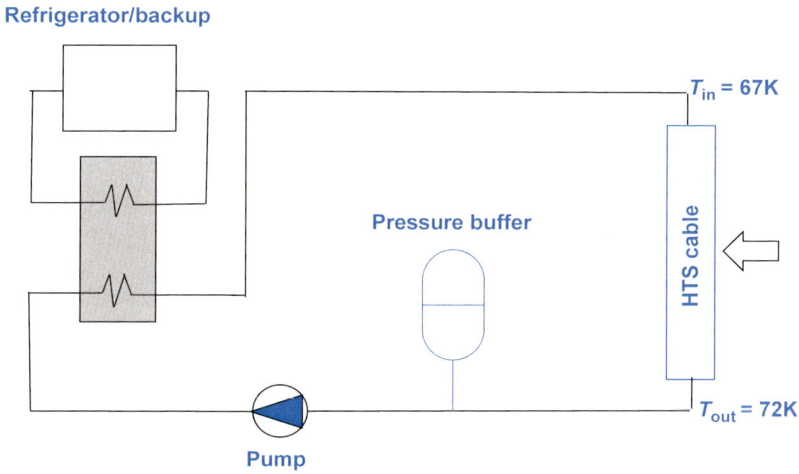

Figure 5.9 Schematic typical cooling system for an HTS AC power cable system. The shaded rectangle is a heat exchanger.

health of the refrigeration system, to initially cool down the HTS cable and to handle the cable warm-up, drying process, and a variety of fault conditions. The cool-down subsystem is used to initially cool the cable system to the operating temperature. The drying subsystem is used to bake the HTS cable core to eliminate any possible contaminating materials in the case where the HTS cable needs to be serviced. The control and monitoring system including logic processors and all associated instrumentation is used for operating the refrigerator and LN_2 cooling loop.

Since the cryogenic cooling system is absolutely critical to the operation of the HTS cable system, measures must be taken to ensure availability even during contingencies such as power failure and maintenance. The technology and overall design of the refrigeration system must ensure that loss of a single component with the highest potential of failure will not cause loss of refrigeration capability.

5.5.2 Cooling technologies

Producing low temperatures has more than a 100 year history (Barron, 1985). Various methods can be used to produce temperatures below 120 K, including phase changes, the Joule–Thomson effect, and adiabatic expansion. HTS AC cables require significant cooling powers, usually in the range of 5–10 kW of cold (~70 K) power to be removed for a 1 km, three-phase medium-voltage cable, and more for longer lengths. For long-term HTS cable operation, one should be cautious when considering phase change or Joule–Thomson effects as a primary cooling source since the cumulative operation cost of phase changes is large and the intrinsic efficiency of the Joule–Thomson effect is low. Due to its potential to be more efficient, a refrigeration system based on adiabatic expansion has been widely investigated and employed in various HTS cable projects. The cooling technologies based on adiabatic expansion include the reverse-Brayton cycle, the reverse turbo-Brayton cycle, and most mechanical cryocoolers such as Stirling, Gifford–McMahon, and pulse-tube coolers. For a utility, the selection of specific cooling technology is mainly based on a few factors including the footprint of the entire system, initial capital cost, operation cost, long-term maintenance cost, and a scheme for redundancy.

A trade-off study of three different cooling technologies is illustrated in Table 5.2. Gifford–McMahon and pulse-tube coolers are not listed in the table because of their limited unit cooling capacity at the HTS cable operating temperature. As seen, the reverse turbo-Brayton cycle has significant advantages over other technologies due to its high availability, high efficiency, and high reliability as well as small footprint. The high reliability has been achieved by using oil-free, contact-free, and hermetic system design. Assembly of the centrifugal compressor, high-speed motor, and cryogenic turbo-expander on a single shaft is a key to reach high efficiency. As shown in the table, the coefficient of performance close to 0.1 is very important to keep the overall room-temperature power requirements of the refrigeration at a reasonable level. Due to elimination of a pneumatic gas and oil removal system, the overall system footprint is small. The system will be designed as a plug-and-play module which will greatly reduce the complexity of cooling technologies employed today in cable demonstration projects. Reverse turbo-Brayton systems have not yet been used in actual HTS cable installations, but they are under development for the Japanese Yokohama cable project (see Section 5.8.4) (Yumura et al., 2013).

Table 5.2 Comparison among different cryogenic refrigeration technologies

	Reverse-Brayton	Stirling cryogenerator	Turbo-Brayton
Technology	Existing and proven technology	Proven but old technology	To be developed
Power consumption per cold watts	1 W_{cold}/23.2 $W_{electric}$	1 W_{cold}/18.3 $W_{electric}$	1 W_{cold}/11.6 $W_{electric}$
Carnot efficiency	~15%	~19%	>30%
Capital cost	High	Medium	High
Availability	High	High	Very high
Chilled water	Not required	Required	Required
Redundancy	Component level	Component level	System level
Maintenance	Medium (annual)	High (6000 h)	Low (several years)
Footprint	Medium	Large	Small

A backup refrigeration system based on an adiabatic expansion system could be an open-loop LN_2 cycle or a mechanical cooler just like the primary refrigerator. If a mechanical cooler is selected as backup, the backup scheme could be accomplished in two different approaches. One approach is component-level redundancy and the other approach is system-level redundancy. The modular approach can be used for system-level redundancy. In this approach, an extra identical modular refrigerator is installed. This approach is also referred to as an $N + 1$ redundant system. For normal operation, N units are required, and in the case of contingency, the extra one is engaged and necessary maintenance or repair can be performed on the disengaged one. It should be cautioned that if an open-loop LN_2 cycle is selected as the backup system, this option requires a large LN_2 storage tank on site, which may be prohibitive in a dense urban environment.

5.5.3 Refrigeration system capital and operation cost

Besides the initial capital cost of a cryogenic refrigeration system, operational cost is another important factor that requires special attention when selecting the cooling system for a particular HTS cable project. The operational cost can quickly add up if the system efficiency is low or an open-loop cycle is selected. Figure 5.10 shows a typical annual cost of a 10 kW system as a function of the local electricity cost, with different Carnot efficiencies, defined by

$$\eta_c = \frac{T_c}{T_h - T_c}$$

where T_c is the cold temperature and T_h is the heat rejection temperature. The annual LN_2 cost of an open-loop system for a 10 kW cooling capacity is around \$1M, assuming an average LN_2 price \$1.6/gal.

5.5.4 Cooling system design

To size the overall cooling capacity of a refrigeration system for a specific HTS AC power cable, various loss terms of the HTS cable and accessories must be taken into consideration: cryostat thermal loss, HTS cable AC loss, dielectric loss, termination loss, and loss in accessories such as transfer lines, HTS cable splices, and pumping ports.

The selection of the HTS cable operating parameters is driven by coolant properties, HTS wire characteristics, HTS cable core characteristics, and fault conditions at the installation site. As with conventional cable, HTS AC cables must be safe and reliable even when abnormal conditions, such as local and through-faults, occur in the power grid. Typically, the through-faults are those that are generated at other locations but affect the power flow on the HTS cable, while the local faults are those which happen directly at the HTS cable or related peripherals and in general require repairs, maintenance, or replacement of equipment by utility operating personnel. During a local fault, a large current many times greater than the rated current of

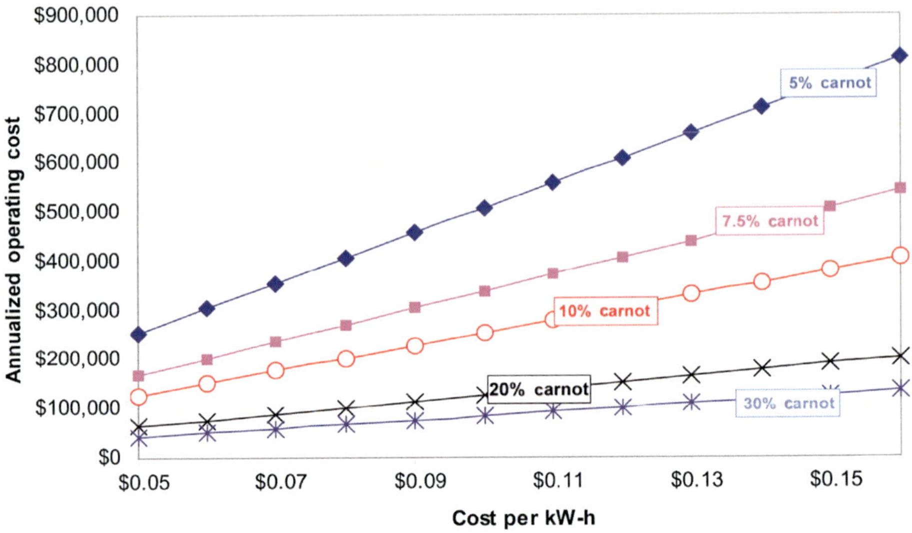

Figure 5.10 Operational cost of a 10 kW, 77 K heat load, as a function of the local electricity cost (per kWh) and percentage of Carnot efficiency.

the cable is created for a brief period of time until circuit breakers open. As a result, a tremendous amount of energy is dissipated into the HTS cable core in a relatively short time. This dissipated energy drives not only HTS cable system design but also cooling system design.

The design process usually takes several iterations due to coupling of various parameters. Various criteria have been used to select system operating parameters. One commonly accepted criterion is that LN_2 is not allowed to boil to form vapor bubbles at any location in the HTS cable system during a fault condition, so as to maintain the integrity of the HTS cable dielectric system. As a result of this restriction and knowing the solidification temperature of the coolant, maximum allowable operating temperature, minimum coolant flow rate, minimum cable cryostat diameter, and minimum operating pressure are readily defined. Since LN_2 is widely used as the coolant, the minimum operating temperature should not be below or even close to its solidification point of 63.15 K.

One important aspect of the HTS cable cooling system that can be easily overlooked when designing a refrigeration system is the cool-down scheme. Successful cool down of the HTS cable from room temperature to the operating temperature is one of the most important factors for system success. The cool-down system should be designed such that the time duration of the cool-down process is short, no more than a few days, and the axial and radial temperature gradients along the HTS cable during cool down should be insignificant. The cool-down system should be capable of varying the coolant flow rate and coolant temperature, which is critical for controlling the cool-down duration and maximum radial and axial temperature gradients. The most straightforward way to accomplish this is to use the bulk of LN_2 from a storage tank. Using the storage tank pressure as the driving force and by carefully controlling the nitrogen flow rate and inlet temperature, thermal shock to the HTS cable can be avoided. However, for a dense urban environment, where accessibility and open space are always limited, evaporating LN_2 or any approach requiring a large LN_2 tank is not an acceptable option. In this case, the cool-down system should be constructed around the normal operating LN_2 coolant loop, and preferably a closed loop using a primary refrigerator.

5.6 Principles of fault-current-limiting HTS AC cables

5.6.1 Introduction

One of the more significant opportunities for HTS AC cable is to link substations at the distribution voltage level, creating a more resilient electric grid. This is of particular relevance to high population density urban locations like New York City, where a single substation typically feeds a certain sector of the city; so if the high-voltage feeds to the substation or the HV/MV transformers are disrupted, the entire sector will go dark (Maguire et al., 2009). A link between substations at their distribution voltage busses would provide an alternative source of power in case of such disruption. Such a link also has the benefit of reducing the number of extra power transformers needed at each substation for redundancy, which can provide significant space and cost savings. However, establishing such a distribution voltage link faces many technical challenges, not easily met with conventional AC cables but addressable using HTS AC fault-current-limiting (FCL) cable:

1. For a given power, say 100 MVA, an incoming high-voltage link at 138 kV would require a cable of only $100/138x\sqrt{3}) = 0.4\ \text{kA}_{\text{rms}}$ capacity, whereas at a 13.8 kV distribution voltage, the link would require a cable of 4 kA_{rms} capacity, far above the rating of typical conventional cables. Such a current level can be easily met by an HTS AC cable (Maguire et al., 2009).
2. As discussed in Chapter 1, Section 1.1.6, any additional link in a meshed power network decreases the effective impedance of the grid and hence increases its fault current. In a city like New York, fault currents are already near the maximum rating of distribution power equipment such as transformers and circuit breakers. So, insertion of a new link between substations introduces the potential peril of excessively high fault currents cascading through the grid, possibly leading to equipment damage and system collapse. To avoid a massive and costly installation of all new equipment with higher fault rating, a new 13.8 kV distribution voltage link would require some means of fault current limiting. This can be achieved by means of a standalone FCL, as in the recent AmpaCity system in Essen, Germany (Section 5.8.5) (Stemmle et al., 2013). However, in most inner city substations, space is at a premium; so it would be far more space efficient and cost effective if the current limiting feature could be incorporated directly into an HTS AC cable, using its own HTS wires. A major project, called the Hydra project, is underway in New York City to demonstrate just such a capability (Maguire et al., 2009).
3. Many other challenges in urban cable installations are relieved through the use of an HTS AC cable: Its lack of emitted heat and EMF during normal operation, coupled with high-power capacity in the same cross-section as a conventional cable, permits simplified installation—either by retrofitting existing cable ducts or by using much smaller trenches than

allowed for conventional cable, or even by digging deep under existing subsurface infrastructure, which is not acceptable for conventional cable because its heat must dissipate to the atmosphere.

Therefore, a key to addressing this important application of linking substations at distribution voltage is the development of an effective inherently FCL HTS cable. Explaining the design principles for FCL cable is the topic of this section. Note that much of the discussion can also be directly applied to standalone FCLs (see Chapter 9).

The basic concept of resistive fault current limiting in a superconductor device, as described already in Chapter 1, Section 1.3.2, is that when a current spike caused by a short in the grid's load exceeds the critical current of a superconductor wire or cable, the superconducting elements switch nearly instantaneously from their near-zero resistivity state to their "normal" nonsuperconducting state with high resistance; that high resistance R_{FCL} in a network with fixed grid voltage V limits the current from $I_{fault} = V/Z_{grid}$ (where Z_{grid} is the net grid impedance remaining after the short) to the much lower value $I_{lim} = V/(R_{FCL} + Z_{grid})$. Thus, a superconductor is an ideal FCL device: it has nearly zero resistance during normal operation (i.e., when the grid current is below I_c) and high resistance above I_c.

If the HTS wire contains low resistance material which shunts the superconductor, such as the silver matrix of 1G HTS wire (see Chapter 3) or a laminated or electroplated copper stabilizer in 2G HTS wire (see Chapter 4), the resistance per length of the switched wire will be too low for practical implementation, leading to inordinate lengths of wire to achieve meaningful current limitation. This is why 2G HTS wire with fairly high resistance stabilizer is the preferred vehicle for resistive current limiting, whether in an FCL cable or in other applications such as standalone FCLs or FCL transformers (see Chapter 12), and we will henceforth focus here exclusively on such wire. For standalone FCLs, bulk tubes of either BSCCO or YBCO material have also been used, but these are obviously not an option for long-length HTS AC cables. And indeed, the uniformity and robustness of 2G HTS wire make it increasingly the favored HTS material for standalone FCLs as well (see Chapter 9).

5.6.2 Design requirements for fault-current-limiting high-temperature superconductor AC cable

The design of an FCL HTS AC cable starts with the definition of relevant grid parameters and the desired level of fault current reduction. Let V_{rms} be the rms system phase-to-ground voltage, with a peak voltage $V_{peak} = \sqrt{2}V_{rms}$. The rms phase-to-phase voltage is then $\sqrt{3}V_{rms}$, but for simplicity we focus here on the behavior of just one of the phases, assuming a simple single core configuration for the cable and a symmetric three-phase system with a three-phase fault. Single and double-phase faults need a more detailed analysis. For example, the phase-to-ground voltage V_{rms} for the Hydra project in New York City (Maguire et al., 2009) is 13.8 kV/$\sqrt{3}$ = 8.0 kV and V_{peak} is 11.3 kV.

The first requirement for an FCL cable linking substations is of course meeting its current rating in normal operation. For example, for the Hydra project, that rating for each phase is 4 kA_{rms} with a peak current of 5.66 kA, and to carry so much current with adequate headroom, the critical current of each cable phase should be about 8 kA at its operating temperature. In a single-phase cable core with the phase conductor wound on a 3-cm-diameter former, such a critical current could be carried by two layers of 2G HTS wire carrying about 450 A/cm-width or three layers carrying about 300 A/cm-width; these performances are indeed in the range of 2G HTS wire commercially available today, especially since the operating temperature around 70 K is below 77 K and the critical current increases substantially at lower temperature. The cable length L is also specified by the given grid installation. Although the Hydra length is only about 200 m, the typical installation in the future is expected to be in the range of 1 km.

For the purposes of this introductory analysis, we consider the HTS FCL cable to be linked to the grid environment shown in Figure 5.11. Let the net grid impedance in series with the HTS FCL cable be Z_{grid}, which has a real resistive component and an imaginary component for inductive or capacitive elements. In fact the grid impedance is usually predominantly inductive. And let the shunt impedance, arising from other links in a meshed grid or from a room-temperature shunt installed in conjunction with the HTS cable, be Z_{shunt}. As an example, we will take the value for $|Z_{grid}|$ to be 0.15 Ω at 60 Hz, and $|Z_{shunt}|$ could be 0.1 Ω, although we will also consider for comparison the extreme cases where there is no grid impedance or shunt (i.e., $Z_{grid} = 0$ and $Z_{shunt} \rightarrow \infty$). By comparison, the resistive impedance of the HTS FCL cable under usual operation is close to zero (the switch in Figure 5.11 is closed), and its inductive impedance is also very small, as can be seen from the well-known formula for the inductance of a hollow cylinder: Assuming here a single-phase configuration with an outer radius r_{screen} of the screening layer, and an inner radius r_{phase} of the phase conductor, the inductance per length is $L(\mathrm{H/m}) = (\mu_0/2\pi)\ \ln\ (r_{screen}/r_{phase})$. At 60 Hz and with $r_{screen} = 2$ cm and $r_{phase} = 1.5$ cm, this corresponds to an inductive impedance of $Z_{FCLcable} = 22\ \mu\Omega/\mathrm{m}$ or only 22 mΩ for a 1 km cable.

The above parameters determine the fault currents: The peak fault current in the absence of the FCL cable and shunt would be $I_{fault,peak} = V_{peak}/Z_{grid}$ or 75 kA for the example parameters above, well above the peak critical Hydra grid fault current of $40\sqrt{2} = 56.6$ kA, above which irreparable damage to grid equipment could occur. This would be totally unacceptable. With addition of a 0.1 Ω shunt but no superconductor cable, the peak fault current would be reduced to $I_{fault,peak} = V_{peak}/(Z_{grid} + Z_{shunt}) = 45$ kA. But a non-FCL HTS cable in the circuit of Figure 5.11, in parallel to the shunt, would cause the peak fault current to rise significantly, depending on its impedance during the fault, potentially nearing or exceeding the critical 56.6 kA limit. This example highlights the dangers which might arise, depending on grid parameters, in an HTS AC cable installation linking substations without the FCL feature.

Another key system parameter for the following analysis is the fault hold time τ, which can be as long as a couple of seconds in a distribution grid and tenths of seconds in a transmission grid. This is the time it takes for protection relays to detect a

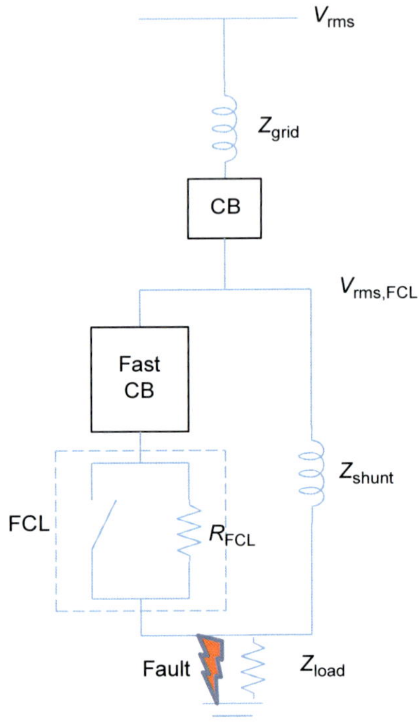

Figure 5.11 Schematic grid environment for a fault current limiting (FCL) cable, where Z_{grid} is the net grid impedance remaining after a short circuit of the load, Z_{shunt} is the shunt impedance of a parallel conventional cable or the net impedance of parallel grid links, the schematic switch in the dashed FCL box represents the superconductor which switches when the current exceeds its I_c, and R_{FCL} is the resistance of the FCL cable after the superconductor switches.

fault plus the time for the conventional circuit breakers to open. Fast-switching advanced circuit breakers can cut current in a few power-line cycles ($\sim 2 \times 16$ ms/cycle plus 1 cycle for detection) or as fast as 0.05 s, and to limit energy dumped into the FCL device, such a fast-switching circuit breaker may be deployed in series with the FCL, as shown in Figure 5.11. Thus, when an HTS FCL, cable or standalone, acts in conjunction with the fast circuit breaker, the HTS FCL limits the fault current on the first few cycles, allowing time for the circuit breaker to open. Because R_{FCL} can be designed to exceed Z_{shunt}, and as long as Z_{shunt} is smaller than Z_{grid}, most of the current is shunted to the room temperature shunt, thus significantly reducing the heat generated in the cable's cryogenic environment, and once the fast circuit breaker opens, *all* the current is shunted to the room temperature shunt, allowing the heated 2G HTS wires in the cable to cool off and reset to their superconducting state.

Now, the challenge for the HTS FCL cable is to trigger its switch to the resistive state rapidly enough to limit even the first peak of the fault current to some reduced value appropriate for the given grid. As an example, let us take 50 kA as the target value for the peak limited current; this would give a safety margin of 12% below the critical peak fault level of 56.6 kA in the distribution grid. In so doing, the HTS wire must absorb the resistive heating load during the fault hold time, and then cool off in as short as possible a recovery time (in the range of tens of seconds to minutes) after the circuit breakers open to be ready for the next fault event. Managing the resistive heating is in fact the main challenge in HTS FCL cable design, because during the fault hold time, the now-resistive superconductor wires will heat up, and the temperature rise must be limited to prevent (1) overheating, damaging, or even burning up the wire (turning it into a fuse), and (2) nucleating gas bubbles in the LN_2 which could cause a dielectric failure.

In the case of the HTS FCL cable, prevention of nucleation is the more severe requirement because, even in the high pressures (15–20 bar) used in typical HTS cables, the LN_2 boiling point is around 111–116 K, and so, assuming the initial operating temperature is about 70 K, the temperature rise ΔT must be kept below 40 K; we will take $\Delta T = 35$ K as a target. This is rather different from the case of a standalone FCL, where LN_2 boiling can be better managed (nitrogen gas bubbles can be deflected from the ground terminal) so that the temperature can rise to well above room temperature. In a cable, this is impossible because the grounded screen is located directly opposite the phase conductor across a narrow LN_2-impregnated dielectric gap.

Preventing hot spots nucleated at wire non-uniformities is another key requirement which we will address in detail below.

5.6.3 Fault-current-limiting cable design: non-uniform switching

To understand how these issues can be addressed and grid requirements met, we consider the following simplified analysis as an example. The main handles of the HTS cable and system designer are to adjust the stabilizer resistance and thickness so as to control the sheet resistivity ρ_{net}/t_{net} of the wire in its normal state (where ρ_{net} is the net resistivity of all the parallel wire constituents and t_{net} is the net wire thickness), to choose an appropriate tape critical current per width (I_c/W), to specify an appropriate in-line circuit breaker to control the fault hold time τ, and to adjust, if possible, the size of the room temperature shunt impedance Z_{shunt}. We shall see that all these parameters are of critical importance in achieving a practical HTS FCL cable design.

As described in Chapter 4, 2G HTS wire consists of a stack of tightly (we assume adiabatically) thermally coupled layers: a metallic substrate (e.g., Ni–W or Hastelloy C 276), buffer layers, the superconductor, a thin passivating silver layer, and a much thicker stabilizer, which could consist of two solder layers and laminates on top and bottom. Because the buffer layers are usually insulating, they can be ignored, and they are thin enough not to prevent close thermal coupling through the entire thickness of the tape. Then, the net sheet resistivity (resistivity per thickness in Ohms) of the stack in its normal state is

$$\rho_{net}/t_{net} = 1/\left[(t_{sub}/\rho_{sub}) + (t_{sup}/\rho_{sup}) + (t_{Ag}/\rho_{Ag}) + (t_{solder}/\rho_{solder}) + (t_{lam}/\rho_{lam})\right] \quad (5.11)$$

Of course, a detailed design will need to take into account the full temperature dependence of the various resistivities, but for our purposes here, we take them as roughly constant over the operating temperature range 70–110 K. Thus, we have substrates such as Hastelloy with $t_{sub} = 50$ μm and $\rho_{sub} \sim 1.0$ μΩm or Ni–9 at% W with $t_{sub} = 60$ μm and $\rho_{sub} \sim 0.60$ μΩm, the 1 μm superconductor layer with 1.0 μΩm resistivity, ~2 μm of silver (likely contaminated with solder) with ~0.007 μΩm, a possible 10 μm of solder with ~0.10 μΩm, and a possible stainless steel laminate with $\rho_{lam} = 1.0$ μΩρm and a thickness t_{lam} to be optimized. For the case of $t_{lam} = 100$ μm and the Ni–W substrate, for example, one finds $t_{net} = 173$ μm and $\rho_{net} = 0.29$ μΩm, corresponding to a sheet resistivity of 1.68 mΩ, and for $t_{lam} = 250$ μm, $t_{net} = 323$ μm and $\rho_{net} = 0.44$ μΩm, corresponding to a sheet resistivity of 1.36 mΩ. Because the 1 Ω sheet resistivity of the switched superconductor layer itself is so much larger than the sheet resistivity of the other layers, its contribution to the net sheet resistivity is negligible, and the transfer of current is almost complete from the superconductor to the other tape layers when the superconductor switches to its normal (resistive) state.

Now we can write the fundamental equations which govern HTS FCL wire performance. We do this first for the simple case ignoring grid impedance and the shunt. Although it is possible to model the detailed time dependence as the superconductor switches into its normal state, we also for simplicity consider the time for the superconductor to heat up above T_c and reach its normal state to be essentially instantaneous on the time scale of a power cycle (16 ms at 60 Hz). Since it takes time for heat to dissipate from the wire, we also assume that whatever region of the wire switches when the voltage over the wire during the fault reaches its peak remains normal during the ongoing power cycles, even when voltage and current drop back to zero. The switched regions are thus effectively "locked in" until, many seconds after the fast circuit breaker opens, the heat finally is dissipated and the superconductivity is restored as temperature drops back below T_c. And so, in the analysis, the behavior at the initial switching (at the voltage peak) and during the rest of the fault hold time must be carefully distinguished.

Now, if the length of tape-shaped wire that has gone normal is L_q, the resistance it generates during the rest of the fault hold time is just $R_{FCL} = \rho_{net} L_q/t_{net}W$, where W is the combined width of all the parallel superconductor films in the cable. Then the peak limited current is

$$I_{lim} = V_{peak,FCL}/R_{FCL} = V_{peak,FCL}\, t_{net}W/\rho_{net}L_q \quad (5.12)$$

But what determines the length L_q of the normal sections of the wire? A naïve assumption is that this is simply the full length of the wire, but this often-made assumption has led to significant misunderstanding of the dynamics of 2G-wire-based fault current limiting. In fact, even for highly uniform wire, there is always some degree of non-uniformity causing a range of I_c values, which leads to the possibility that only limited segments of the wire switch during a fault. Furthermore, an LN_2 temperature gradient inevitably exists across the length of a cable, and because the wire's critical current is very sensitive to temperature, this temperature gradient also creates I_c-non-uniformity along the length of the wire.

The basic physics of what happens in the presence of such non-uniformity is that as the AC grid voltage rises toward its maximum and the current approaches the overall average critical current, the lowest I_c segments of the wire switch first, and as more and more of the wire switches with increasing grid voltage, the resistance climbs, holding the total current through the FCL near I_c. Only if the voltage across the entire FCL cable rises high enough will the entire wire switch, and then the current could further climb above I_c with increasing voltage.

This fundamentally important phenomenon of non-uniform switching was first recognized by Kraemer et al. (2003) in their studies of the switching of YBCO thin film meander lines on crystal substrates, but the phenomenon applies to 2G YBCO thin film wires as well (Schmidt et al., 2007). Some of their remarkable high-speed CCD camera images of the initial quench phase in a linear YBCO strip at three different voltages are shown in Figure 5.12. In this case, the 2G YBCO film is locally driven above the LN_2 boiling point in the switched regions, generating bubbles and thus revealing them in the images. Multiple regions of different sizes are visible in some of the figures, and the length of the regions grows in proportion to the applied voltage. The surprising implication of this is that at least up to a certain voltage, the total resistance of the switched regions of the YBCO strip is not constant but increases approximately linearly with voltage! Kraemer et al. also discovered that the effective peak limiting current I_{lim} was not exactly at I_c but somewhat higher by a factor $f = I_{\text{lim}}/I_c$, which in their case was about 2, but in 2G HTS wires values closer to 1 can be deduced. That f could be greater than 1 should not be surprising given that I_c is a somewhat arbitrary value set by an arbitrary electric field criterion (see Chapter 2, Section 2.11.3.3) and not the somewhat higher current at which the entire vortex lattice finally melts into a liquid-like or flux flow state. But in what follows, we will take $f = 1$ or $I_{\text{lim}} = I_c$ for simplicity.

From Eqn (5.12) and $I_{\text{lim}} = I_c$, one has

$$L_q = V_{\text{peak,FCL}}\, t_{\text{net}}/\rho_{\text{net}}(I_c/W), \quad \text{for } V_{\text{peak,FCL}} < V_{\text{crit}}, \text{ non-uniform, and} \tag{5.13}$$

$$L_q = L\,, \quad \text{for } V_{\text{peak,FCL}} > V_{\text{crit}}, \text{ uniform} \tag{5.14}$$

where

$$V_{\text{crit}} = (I_c/W)\rho_{\text{net}}\, L/t_{\text{net}} \quad \text{or} \quad L_{\text{crit}} = V_{\text{peak,FCL}} t_{\text{net}}/(I_c/W)\rho_{\text{net}} \tag{5.15}$$

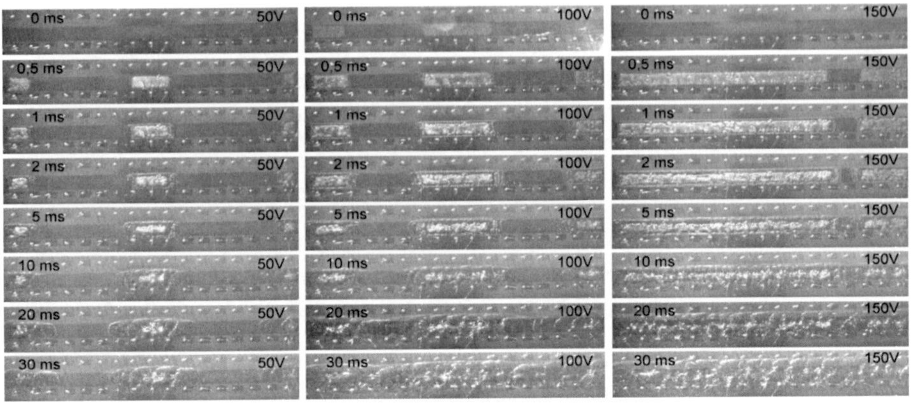

Figure 5.12 High-speed CCD camera images of quenching of a linear YBCO strip on a single crystal substrate, revealed by local LN_2 boiling (Kraemer et al., 2003).

Note that I_c/W is just the commonly reported critical current per unit width of 2G tapes, and taking the value 450 A/cm, a 100-μm-thick SS stabilizer, and other values given earlier, one finds $V_{crit}/L = 75$ V/m; so, for a 1-km cable, $V_{crit} = 75$ kV. At lower voltages, the switched wires remain in a non-uniform state. Alternatively, for a grid with $V_{peak,FCL} = 11.3$ kV, wires longer than $L_{crit} = 147$ m once again will switch into a non-uniform state.

Thus we can see that there are two regimes that must be distinguished. If the applied voltage $V_{peak,FCL}$ is large enough, the entire wire switches, and the effective resistance is a constant

$$R_{FCL} = \rho_{net}L/t_{net}W, \quad V_{peak,FCL} > V_{crit}, \quad \text{uniform} \tag{5.16}$$

The current $V_{peak,FCL}/R_{FCL}$ through the FCL device can climb in this case above I_c. This can be particularly dangerous because, as we shall see in the next section, if the current is high enough, it can overheat the *entire* FCL.

For voltages below the critical level defined in Eqn (5.15), the resistance during the ongoing power cycle remains pinned at

$$R_{FCL} = V_{peak,FCL}/I_c, \quad V_{peak,FCL} < V_{crit}, \quad \text{non-uniform} \tag{5.17}$$

because, as discussed earlier, once the segments of the wire have switched at the peak voltage, they heat up over T_c and remain normal thereafter. A successful design will require that heating in these segments not exceed the allowable limit, which in our example is $\Delta T < 35$ K. Another possible concern in this regime comes from thermal stresses between switched and unswitched segments, although this issue is negligible in an FCL cable as compared to standalone FCLs, because the designed temperature rise is constrained to be so much less to avoid LN_2 bubbling. The ideal condition for

HTS FCLs is for the parameters to be just at the crossing point $V_{\mathrm{peak,FCL}} = V_{\mathrm{crit}}$, where the full wire switches but the current remains near I_c; however, given the variability in fault conditions, this is impossible to achieve in practice. Indeed, because most HTS FCL cables are longer than L_{crit} in Eqn (5.15), they typically operate in the non-uniform regime! To understand HTS FCL behavior of systems based on 2G HTS wires, it is therefore fundamental to recognize this non-uniform switching behavior.

5.6.4 Wire heating in an HTS fault-current-limiting AC cable

The next key equation governing the behavior of resistive FCL systems determines the local temperature rise ΔT from the local heating in switched wire segments. For simplicity, we ignore the non-uniformity in heat generation caused by current flow through solder fillets at the tape edges and assume the heat spreads through the entire thickness of the YBCO tape adiabatically, uniformly across the width, but not significantly down the length of the wire. The heating occurs during a fault hold time τ until the fast circuit breaker opens; so the local heat generated per volume is

$$\rho_{\mathrm{net}} J_{\mathrm{rms}}^2 \tau = C \Delta T \tag{5.18}$$

where

$$J_{\mathrm{rms}} = V_{\mathrm{rms,FCL}} \big/ R_{\mathrm{FCL}} t_{\mathrm{net}} W \tag{5.19}$$

is the average current density in the tape and C is the average heat capacity per volume (specific heat). Recall that after the initial maximum voltage, as described earlier, R_{FCL} is "locked in" during the following cycles of voltage and current because local wire segments cannot cool down quickly; so the heating can be calculated with a constant resistance. Of course, a more complete treatment would treat the temperature-dependent specific heats of the individual stack components, but for our purposes here we can use the approximate value of $C = 2\ \mathrm{MJ/m^3K}$ in the 70–110 K temperature range.

The tape parameters must be controlled such that ΔT does not exceed about 35 K, as discussed earlier, to avoid the risk of LN_2 bubbling. This is the key to avoiding hot spots and potential damage to the wire. Combining Eqns (5.16)–(5.19), one finds

$$\Delta T = \left(V_{\mathrm{rms,FCL}} \big/ L\right)^2 \tau / C \rho_{\mathrm{net}}, \quad V_{\mathrm{peak,FCL}} > V_{\mathrm{crit}}, \quad \text{uniform} \tag{5.20}$$

$$\Delta T = \left(I_c / t_{\mathrm{net}} W\right)^2 \rho_{\mathrm{net}} \tau / 2C, \quad V_{\mathrm{peak,FCL}} < V_{\mathrm{crit}}, \quad \text{non-uniform} \tag{5.21}$$

These important equations show that the temperature rise of very short cables in the uniform switching regime drops as their length increases, until the critical length is reached, above which switching becomes non-uniform and the local temperature rise in the switched regions remains constant.

It is also interesting to use Eqn (5.20) to solve for the maximum effective electric field which can be sustained by the wire at a given temperature rise and fault hold time.

$$V_{max}/L = (C\rho_{net}\Delta T/\tau)^{1/2} \tag{5.22}$$

Interestingly, this quantity is independent of wire thickness. For a given ΔT, Eqn (5.22) has often been used to evaluate the potential of different wires for FCL applications, and high values are widely considered superior. Since heat capacities cannot be varied significantly and τ is set by circuit breaker capabilities, a focus has therefore been on increasing the resistivity ρ_{net}. Nevertheless, since most HTS FCL cables will operate in the non-uniform regime, Eqn (5.21) makes it evident that having a higher ρ_{net} is actually detrimental, causing an increase in the local ΔT!

Let us now estimate ΔT using our earlier values $\rho_{net} = 0.295$ μΩm, along with $\tau = 50$ ms and $V_{crit}/L = 75$ V/m. Let us assume a cable at the critical length of 147 m (from Eq (5.15)) for a peak FCL voltage of 11.3 kV. Then Eqn (5.20) gives for the uniform case $\Delta T = 250$ K, a value within range for a standalone HTS FCL but totally unacceptable for an HTS FCL cable. And for shorter cables or higher voltages, ΔT will be much higher. This highlights the design challenge for the HTS FCL cable. It also reveals the flaw in the idea of using a short, say 1-m-long and very high resistance 2G cable element, put in series with an HTS cable to limit it: Because of the $1/L^2$ dependence in Eqn (5.20), and unless there is an unrealistically high resistivity, the temperature rise becomes so high as to cause the element to burn up during the fault, effectively making it a fuse. Of course, with L_{crit} only 149 m at 11.3 kV peak, it is clear that most cables will in practice be longer. Then, a more realistic approach to an HTS FCL cable is to design for non-uniform switching with $L > L_{crit}$. But not surprisingly, Eqn (5.21) gives the same $\Delta T = 250$ K as we calculated before because we had chosen the critical cable length for the crossover between the two regimes.

How can this temperature rise be brought to the desired level of 35 K in an HTS FCL cable? Clearly, one of the strongest factors in Eqn (5.21) is I_c/W, the current per width of the wires; so by putting in *lower* performance and *more* wires (to achieve the same total I_c) into the cable, one can significantly reduce the heating! This is a rare case where lower wire performance is better, although it comes at the cost of more wires. So let us consider 250 A/cm wires instead of the 450 A/cm used earlier. An additional improvement, suggested by Eqn (5.21), is increasing the stabilizer thickness. Assuming a 250-μm-thick stainless-steel stabilizer, $t_{net} = 323$ μm and $\rho_{net} = 0.44$ μΩm, as we calculated earlier from Eqn (5.11). Combining these two changes, one finds $\Delta T = 33$ K, meeting our 35 K target and thus preserving the dielectric integrity of the cable. In such a system (with no grid impedance or shunt), the limited current is just $I_c = 8$ kA, far below our target of 50 kA. In fact, this limited current may be too low in the context of typical grid design because other circuit breakers in the grid may fail to trigger at such a low current.

It is worth highlighting that this treatment of non-uniform nucleation of switched segments in the wires directly addresses the concerns raised at the beginning of this section about inhomogeneities and hot spots in an FCL based on 2G wires. Equation (5.21) shows that as long as the sheet resistance of the wire is within a certain bound, say ±10%, the temperature rise will be no higher than ±10% of the calculated level.

One other basic requirement for an FCL cable must be addressed: the total heat generated in the cable during the fault hold time, because if this becomes too large, the recovery time during which LN_2 flow dissipates it could become too long. In the simple case we have been considering, without any grid impedance or shunt, the heat generated is

$$Q = V_{\mathrm{rms,FCL}}^2\ \tau / R_{\mathrm{FCL}} = V_{\mathrm{peak}} I_c\ \tau/2 \tag{5.23}$$

where we have used the non-uniform value for R_{FCL} in Eqn (5.17). With the parameters $V_{\mathrm{peak}} = 11.3$ kV, $I_c = 8000$ A, and $\tau = 0.05$ s, one finds $Q = 2.26$ MJ, a substantial amount to be dissipated by the refrigeration system.

Here is where the shunt configuration of Figure 5.11, in conjunction with the grid impedance, plays an important role. A simple analysis based on Kirchhoff's laws with the FCL current at I_c in the non-uniform regime gives

$$V_{\mathrm{peak,FCL}} = Z_{\mathrm{shunt}}\left(V_{\mathrm{peak}} - I_c\, Z_{\mathrm{grid}}\right) / \left(Z_{\mathrm{grid}} + Z_{\mathrm{shunt}}\right) \tag{5.24}$$

$$V_{\mathrm{rms,FCL}} = V_{\mathrm{peak,FCL}} / \sqrt{2} \tag{5.25}$$

Using the parameters $Z_{\mathrm{grid}} = 0.15\ \Omega$, $Z_{\mathrm{shunt}} = 0.1\ \Omega$ and treating them for simplicity as simple resistances, one finds $V_{\mathrm{peak,FCL}} = 4.0$ kV, significantly reduced from 10.1 kV in the nonshunted case, as are the calculated $V_{\mathrm{rms,FCL}} = 2.9$ kV and $Q = 0.81$ MJ. In practice, of course, the full complex impedance must be used for a complete treatment. This calculation highlights the benefit of the shunted configuration: much of the fault current is shunted to a room temperature impedance, thus substantially reducing the refrigeration requirements for the HTS FCL cable. R_{FCL} is now 0.5 Ω. Note also that the critical cable length above which non-uniform fault switching occurs (Eqn (5.15)) is now much smaller than before because $V_{\mathrm{peak,FCL}}$ is so much less than without the shunt.

In the shunted configuration, one can calculate a limited peak fault current of 48 kA, meeting our 50 kA target. After the fast circuit breaker opens, the limited peak fault current further drops to 45 kA.

A full system design for an AC HTS FCL cable requires a more detailed treatment of the time- and temperature-dependent behavior of the superconductor and its impact on the grid, as well as treatment of asymmetric and single-phase faults and the full real and reactive components of the current flow. However, the above analysis highlights the basic and often misunderstood elements of HTS FCL system design. In particular, recognizing the non-uniform nature of the superconductor-to-normal switching is essential to adequately protect the system from burnout and properly evaluate the level of current limiting that can be achieved. Clearly, the design of an AC HTS FCL cable is complex, but this analysis, and the ongoing progress in the Hydra project (Maguire et al., 2009), provides strong support for the viability of this important application. Further information on the status of the Hydra program is given in Section 5.8.6.

5.7 Inductance and capacitance

While most focus on the properties of HTS AC cables falls naturally on their low resistance, their inductance and capacitance also open unique opportunities and applications.

5.7.1 Inductance

As we have seen in Section 5.6.2, because the magnetic field in either the single core or three core (3-in-1) HTS cable design (Figure 5.1) is fully contained between the HTS screen at radius r_{screen} and the HTS core at radius r_{core}, the inductance per length of HTS AC cold-dielectric cables is given to a good approximation by the textbook formula

$$L_{HTS}(\mathrm{H/m}) = (\mu_0/2\pi)\ln(r_{screen}/r_{core}) \tag{5.26}$$

where μ_0 is the permeability of free space, 1.257×10^{-6} H/m. This is typically around six times smaller than the inductance per length of a conventional copper cable because the magnetic field around any given phase conductor in a copper cable is not screened significantly and extends to an effective "capture distance," which is approximately the distance to the other phases. Also the HTS cable inductance per length is even smaller compared to the inductance of overhead lines, which have no screening at all.

Because the effective resistance per length of the HTS cable, arising from the HTS and dielectric AC loss, is much smaller than ωL_{HTS}, its inductive impedance per length, the total inductive impedance constitutes the dominant HTS cable impedance Z_{HTS}, and its relatively low value must be taken into account in grid planning, particularly because the very low impedance of the HTS cable will essentially "attract" current of a meshed grid, and because of the increased fault current which can arise, as discussed in Section 5.6. In this context, HTS AC cables have sometimes been called very low impedance or VLI cables (Malozemoff, Kehrli, Diaz de Leon, & Kalsi, 2004).

The low impedance opens up a novel opportunity for HTS AC cable to contribute to a Smart Grid (Malozemoff et al., 2004). As pointed out already in Chapter 1, Section 1.3.1, the electric power grid is a massive network of impedances that control power flow according to Kirchhoff's laws, and readjustment of power flows is done largely by means of circuit breakers. In a Smart Grid, however, it would be desirable to more smoothly steer current away from overloaded sections of the grid or to steer it to needed loads, thus improving grid stability and addressing capacity constraints. One device that enables such control of AC power flow is the Unified Power Flow Controller, based on power electronics. Unfortunately the complexity and expense of such systems have prevented all but a very few installations so far around the world (Malozemoff et al., 2004).

The very low impedance of HTS AC cables provides a potentially cost-effective way to control current flow through the use of a phase angle regulator (PAR) in series with the HTS cable (Malozemoff et al., 2004). The PAR is a kind of tap-changing transformer that can adjust the phase angle ϕ. Now the power through the cable is given by

$$P = V_s V_r \sin\phi / Z_{HTS} \tag{5.27}$$

where V_s and V_r are the sending and receiving end voltages, respectively. Because Z_{HTS} is so low, significant power adjustment can be made with relatively small phase angle changes, which makes this concept economically attractive because the cost of a PAR is much lower when the phase angles it adjusts are much smaller. This potentially important concept still awaits demonstration.

5.7.2 Capacitance

The capacitance $C = \varepsilon A/d$ of an HTS AC cable is similar to that of a conventional cable, which is determined by the area A of the opposing surfaces carrying the phase-to-ground voltage $V_{phase\text{-}to\text{-}ground}$, the gap d between them containing a dielectric material of permittivity or dielectric constant $\varepsilon = K\varepsilon_0$, where $\varepsilon_0 = 8.85 \times 10^{-12}$ F/m is the dielectric constant of free space and K is the relative dielectric constant. K is about 3 for PPLP, whereas for LN_2 it is about 1.5. In the cylindrical geometry, the capacitance per length is (Burke, Clapp, & Ros, 1993)

$$C/L(\mu F/km) = 0.055\ K/\ln(r_{screen}/r_{core}) \tag{5.28}$$

This capacitance becomes substantial in a long cable and must be charged up two times every cycle. This leads to a charging current per length

$$\begin{aligned} I_{Charge}/L\ (A/m) &= V_{phase\text{-}to\text{-}ground}\ 2\pi f(C/L) \\ &= 0.34\ V_{phase\text{-}to\text{-}ground}\, fK/\ln(r_{screen}/r_{core}) \end{aligned} \tag{5.29}$$

where f is the grid frequency. For example, taking $K = 2.5$, $r_{screen}/r_{core} = 1.2$, and $V_{phase\text{-}to\text{-}ground} = 138$ kV, one finds $I_{Charge}/L = 39.5$ A/km.

Equation (5.29) allows one to define an effective charging length

$$L_C = I_{max}/(I_{Charge}/L) \tag{5.30}$$

where I_{max} is the maximum current which can be carried by the cable core. L_C is the length at which almost all the cable's current capability is taken up with charging rather than carrying the needed power. Clearly, for a practical cable, its length needs to be a relatively small fraction of L_C. For our numbers above and $I_{max} = 1000$ A, near the limit for a conventional copper cable, one finds $L_C = 25$ km. This charging phenomenon is one reason that conventional AC HV cables cannot be used for long runs, for example, for marine applications linking an island 25 km or more from the mainland.

Here we can see an interesting advantage of HTS AC cables. Because of their much higher current capacity, I_{max} and L_C are several times higher for HTS AC cables than for conventional AC cables of the same voltage. Also since HTS AC cables can often be operated at much lower voltages, I_{Charge}/L is correspondingly reduced and L_C further increased, allowing for single HTS AC cable runs of over 100 km! Of course, the limit in the length of HTS AC cables for marine applications is likely to be set by refrigeration rather than by the capacitive charging length because of the difficulty of arranging for refrigeration stations under the sea along the length of the cable. On land, this is not an issue, enabling long AC cable runs without intervening substations. Exploiting the large capacitive charging length of HTS AC cables is an opportunity for the future.

5.8 Some major HTS AC cable projects

Major HTS AC cable projects, both in test labs and in grid, have been conducted throughout the world, as summarized in Table 5.1. We discuss next a few significant ones in more detail.

5.8.1 AEP Bixby project

A major in-grid HTS AC cable project funded by the USDOE under its SPI (Superconductivity Partnership Initiative) program was the HTS cable installation at American Electric Power's Bixby substation located in Columbus, OH, or AEP Bixby for short (Demko et al., 2007). The AEP Bixby project consisted of team members Ultera, which is a joint venture among the Southwire Company of Carrollton, Georgia, and nkt cables of Asanes, Denmark, Praxair, and the Oak Ridge National Laboratory (ORNL). The project built on expertise developed in two important earlier HTS AC cable projects, the Carrollton project by Southwire (Stovall et al., 2001) and the Copenhagen project by nkt cables (Tonnesen et al., 2004). The HTS AC cable connected a 138/13.2 kV transformer to the distribution switchgear serving seven outgoing circuits. The cable was designed to carry 3000 A_{rms} at 13.2 kV and had an active length of 200 m not including the terminations.

The Bixby HTS AC cable used the so-called HTS Triax™ design, which, as described earlier in Section 5.2.1, consists of three concentric phases wrapped around a flexible hollow former (see Figure 5.1(c)). Each phase in the HTS Triax™ design is separated by a cold dielectric made of Cryoflex and designed for the rated phase-to-phase voltage. Copper stabilizer tape was included along with HTS tape in each phase to provide protection from overcurrent faults. Thus, the HTS cable design was that of a fault-current-tolerant HTS cable and not an FCL cable (see Section 5.6).

The AEP Bixby project consisted of two separate design and construction phases. In the first phase, a 5-m-long HTS cable prototype was design, fabricated, and tested at

ORNL. The purpose of the 5 m prototype HTS cable was to fully test the HTS cable to the extent possible prior to the fabrication and eventual installation of the final 200 m demonstration cable. Key tests performed at ORNL included: (1) a DC critical current test of each phase, (2) AC loss measurement at 3000 A_{rms} in the straight configuration and 3500 A_{rms} in the 90° bent configuration, and (3) high-voltage testing in both the straight and the 90° bent configuration. The measured critical currents were 7020, 7960, and 7450 A for phases 1, 2, and 3, respectively, at an average cable temperature of 78.5 K. The AC loss of ~4 W/m/phase was measured using a thermal technique in which the temperature rise ΔT as a function of the AC current level is measured. Finally, the high-voltage testing included: (1) AC hipot withstand tests to 39 kV on the terminations, (2) very low frequency (VLF − 0.1 Hz) AC hipot soak tests to ~20 kV, and (3) BIL testing to ±110 kV using the standard lightning impulse waveform of 1.2/50 μs (Sauers et al., 2007).

After the successful 5 m prototype testing at ORNL, a 200 m demonstration HTS cable was installed at the Bixby substation (see Fig. 5.7(a) and 5.13). HTS cable installation included both above-ground and below-ground portions to demonstrate the ability of the HTS cable to adjust to various types of contours and terrain. Another key feature of the AEP Bixby cable was the inclusion of a splice joint, shown in Figure 5.14 located in an underground vault, which is typical of underground installations for conventional copper cables.

The HTS Triax™ cable was cooled with subcooled LN_2 operating with an initial inlet temperature of about 70 K at a pressure of 7−9 bar. Subcooled LN_2 was injected in the first termination/bushing and flowed in both the central former and the annulus of the HTS cable. The now warmer 72−73 K LN_2 exited the second termination/bushing and was returned in a separate cryostat to the refrigeration system. The refrigeration system consisted of an open-cycle vacuum subcooler supplied by Praxair Inc. The Praxair system utilized a subcooled LN_2 storage tank that could provide backup cooling for up to 8 h in the event that the main system lost power. In addition, the Praxair system had provisions for including pulse-tube cryocoolers to substantially reduce the thermal load on the vacuum subcooler.

The Bixby HTS Triax™ cable system was first energized in August 2006 and ran continuously for > 30,000 h, shutting down only a few times for system maintenance and surviving faults due to the occasional squirrel serving as a short to ground. The HTS cable was officially decommissioned in October 2012, making it the longest activated HTS cable project to date.

5.8.2 Albany project

Another major project funded by the USDOE under the SPI program was the HTS cable installation between two National Grid substations in Albany, New York, or the Albany project for short (Masuda et al., 2007). The Albany project team members consisted of SuperPower Inc. (prime) of Schenectady, New York, Sumitomo Electric Industries of Osaka, Japan, and the BOC Group, now a part of the Linde Group. The HTS cable for the Albany project was an underground installation ~350 m in length (not including the terminations) and connected two substations at Menands and

Figure 5.13 HTS Triax™ cable installation at the AEP Bixby substation. The white cable on the drum is the superconductor and dielectric portion being fed into the black cryostat and outer jacket of the final cable. The cryostat and outer jacket were already placed in a below-ground trench at the substation. The cable will be connected to the above-ground terminations in the background (Demko et al., 2007).

Figure 5.14 Field splice of an HTS Triax™ cable and cryostat inside a standard concrete manhole at the Bixby station (Demko et al., 2007).

Riverside separated by a major highway. The HTS cable was a distribution class cable with a design rating of 34.5 kV and a nominal operating current of 800 A_{rms}. The HTS cable design used a triad configuration shown in Figure 5.1(b) with all three phases contained within a single cryostat. The overall outermost dimension of the HTS triad cable was ~136 mm and fit inside a corrugated flexible cryostat with an i.d. of

~152 mm. The triad design requires an HTS return shield around each of the HTS phases, which provides >95% cancellation of stray magnetic fields emanating from the HTS cable.

The HTS tapes for each phase consisted of Bi-2223 tapes manufactured by Sumitomo Electric Industries. A total of 70 km of Bi-2223 tape was used in the initial installation with each Bi-2223 tape having an average I_c ~86 A (77 K, self-field). The cryogenic dielectric used to insulate each phase conductor consisted of wrapped PPLP. In order to accommodate the thermal contraction of the HTS cable on cool down, the three separate phases were cabled very loosely in the 3-in-1 configuration (Figure 5.1(b)) so as to reduce the overall tension in the HTS cable relative to the fixed terminations/bushings. This design was built on the successful earlier 100 m demonstration by SEI in Japan (Honjo et al., 2003). One set of terminations for the system is shown in Figure 5.15. The 350 m HTS cable was purposely manufactured with a splice joint that connected a 320 m section to a 30 m section. The splice joint was manufactured in an underground vault and was later used to replace the 30 m section with 2G YBCO tape provided by SuperPower, which constituted the first in-grid test of an HTS AC cable based on 2G wire. DC critical current measurements were initially performed on short 3 m prototype 2G YBCO cables with I_c values ranging from 2660 to 2820 A for the phase conductor and 2400 to 2500 A for the shield conductor at 77 K. In addition, AC loss measurements were performed at Los Alamos National Laboratory on the short 3 m prototype 2G YBCO cables, showing AC loss of approximately 3.4 W/m/phase for an operating current of 800 A_{rms} at 60 Hz and 75 K. Although this loss per meter at rated current was high (see Section 5.3), the short length of the YBCO cable segment kept the total heat load small compared to other cable component contributions.

High-voltage tests were also successfully performed on the HTS cable per the AEIC standards. The primary high-voltage tests included: (1) DC hipot to 100 kV for 10 min, (2) AC hipot to 69 kV for 5 min, and (3) BIL ±200 kV using a 1.2/50 μs lightning impulse waveform.

The primary cooling of the HTS cable was a closed-loop mechanical (cryocooler) refrigeration that provided a minimum cooling capacity of 5 kW at 77 K and 3.7 kW at 70 K. Normal operating temperature for the HTS cable was 67–77 K, with the subcooled LN_2 operating pressures between 1 and 5 bar and cooling loop flow rates between 30 and 50 L/min. In the event of cryocooler power failure, backup cooling was provided by vaporizing bulk LN_2 operating as a low-pressure subcooler.

Highlights of the Albany project included:

- The Albany project began on November 1, 2002, and was effectively completed on December 31, 2009.
- The initial 1G Bi-2223 HTS cable system became operational on July 20, 2006, and successfully operated for more than 6720 h (280 days) before being shut down for installation of the 30 m 2G YBCO cable section on May 1, 2007.
- The manufacture of 9.7 km of 2G YBCO wire was completed in December 2006.
- The hybrid Bi-2223 (320 m)/YBCO (30 m) HTS cable became operational on January 8, 2008, and successfully operated for >2000 h before being shut down on completion of the operational phase of the project on April 30, 2008.

Figure 5.15 Three-phase termination/bushing used in the Albany project (Masuda et al., 2007). The termination skid is on rollers to accommodate the thermal contraction of the HTS core when it cools down.

The significance of the Albany project was that it was one of the first cables operating between two substations of a utility grid, using a splice and successfully demonstrating operation using both BSCCO 1G and YBCO 2G HTS wires.

5.8.3 LIPA cable project

This major project was funded under the SPI between the USDOE and industry and developed the world's first long-length transmission-voltage HTS AC power cable (Maguire et al., 2007). The HTS cable was designed for permanent installation in the LIPA grid and is able to carry 574 MVA at a voltage of 138 kV. AMSC was the prime contractor as well as manufacturer of the HTS 1G and 2G wires. Nexans provided development and manufacturing of the cable, terminations, and cryostat as well as site and installation support. Air Liquide provided the cryogenic refrigeration expertise, equipment, and installation, as well as operations monitoring and support. The host utility LIPA provided the site, civil work, controls and protection, transmission planning, and operation of the HTS cable system. The system was manufactured, installed, and successfully connected to the LIPA grid on April 22, 2008 (Maguire et al., 2007).

Three single-phase (single core) HTS cables were installed in the LIPA grid starting at the Holbrook Substation (South End), heading north for 600 m, where a new switching station (North End) was installed. This new station houses the cryogenic refrigerator, HTS cable terminations, LN_2 storage and necessary equipment for operation and control of the cable system. One set of cable terminations with cables emerging from their underground ducts is shown in Figure 5.7(b). The power transmitted by the new HTS cable system is taken from an existing 138 kV overhead circuit which will remain available in parallel as a backup.

The HTS cable used in the LIPA project has a cold dielectric design. The system contains three individual HTS cables and six terminations. The HTS cable configuration consists of a copper former, two HTS conductor layers, an insulation layer, an HTS screen layer, a copper screen stabilizer, and a cryogenic envelope, as in Figure 5.1(a). During normal operation, the HTS cable core is maintained at an operating temperature between 67 and 72 K by circulating subcooled LN_2. Each cable phase is linked to an outdoor termination at both ends, located in the substation areas that connect the HTS cable to the LIPA grid. The terminations provide both a connection to the HTS cable and the interface to the HTS cable cooling system. The HTS cable has been designed to operate reliably and to withstand fault currents up to 51 kA_{rms}, by the use of extra copper in contact with both the phase and the screen HTS conductors.

To fully qualify the cabling process and for type testing, a prototype 30 m HTS cable and two full-scale terminations were fabricated and tested by Nexans. Subcooled LN_2 was used as coolant during the entire test. The HTS cable design for this project is based on the use of hermetic first generation (1G) BSCCO 2223 superconductor wires, 4.3 mm wide, with a minimum critical current of 135 A (77 K, self-field) per wire for the phase layers, and 105 A (77 K, self-field) per wire for the shield layer. The cable manufacturing for this project was performed by Nexans. The cryostat was manufactured directly over the HTS cable core in a continuous fashion.

The refrigeration system is based on the Reverse-Brayton cycle and exploited the existing refrigeration system designed for the earlier DOE/Detroit Edison project by Linde (Norman et al., 2001). Air Liquide modified the system to meet LIPA cable specifications, by boosting cooling capacity, increasing buffer tank volume, and adding a new cool-down skid. The cooling capacity of the refrigerator was boosted from 4.3 to about 5.6 kW at 65 K. The LIPA site was completed in early 2007. The HTS cable phases were shipped separately and pulled into the already installed conduit directly from the shipping spools. Once the phase conductor was installed in the conduit, the termination connections were made. A three-phase parallel cool-down process has been used in the present project.

In 2007, a new USDOE Superconductor Power Equipment (SPE) program, called LIPA II, was awarded to the existing project team to address outstanding issues for integrating HTS cables into the utility grid (Maguire et al., 2011), and as of 2014, this project is still ongoing. The new project includes demonstration of 2G HTS wire in a transmission-level in-grid cable, a field joint, a field-repairable cryostat, and improved thermal contraction control within the HTS cable. Whereas the LIPA I project was designed to demonstrate the technology, the goal of LIPA II is to demonstrate a commercial system. Within the LIPA II project, one of the three existing phases of the 600-m-long LIPA I installation was replaced by a LIPA II phase based on 2G YBCO wire. In addition, a field-repairable cryostat is one of the items being developed in the LIPA II project to demonstrate the readiness of HTS AC power cables for grid application. Damage to an HTS cable cryostat in the field may occur for a variety of reasons (for example, through civil works), and a speedy repair process is necessary in a commercial HTS cable system. Whereas the LIPA I cable cryostats have only one vacuum space over their entire 600 m length, the LIPA II cable cryostat has been developed to include a multitude of suitably spaced vacuum barriers inside the cryostat, subdividing the vacuum space. The spacing of the barriers has been chosen so that each vacuum space can be evacuated via a single pumping port and at the same time without increasing heat load significantly. This ensures that a vacuum failure anywhere along the length of the cryostat can be repaired by having access to the location of the failure point only.

In addition, the new HTS cable system has been developed to include the ability to limit fault currents. Although this HTS cable is not long enough to provide meaningful limitation of the fault current experienced at this location, the development, fabrication, and qualification of this feature are included and tested and demonstrated in a laboratory environment for use at transmission voltages. Further discussion of FCL cables is given in Section 5.6.

5.8.4 Yokohama and other Asian HTS AC cable projects

A significant number of successful HTS AC cable demonstrations have been conducted in Asia. Japan, South Korea, and China all have active programs. The Chinese cable projects are reviewed in Chapter 13 (Xin et al., 2005; Xiao et al., 2007). In South Korea, the leading cable company developing HTS cable is LS Cable and System Ltd., one of the world's largest cable manufacturers. In developing and demonstrating HTS cable systems, they work closely with KEPCO, the electric utility of South Korea, and with the Korean Electrotechnology Research Institute and other South Korean technical institutes. Several 100 m HTS AC cable demonstrations using 1G HTS wire were conducted at KEPCO's Gochang test center, supported by the South Korean government's DAPAS program (Kim et al., 2005; Sim et al., 2008; Sohn et al., 2007), and then the world's first in-grid test of LS Cable's 500 m 23 kV, 1.25 kA cable based on AMSC 2G HTS wire was started in 2011 at KEPCO's Icheon substation in the Seoul area (Lee et al., 2011). LS Cable and KEPCO have an aggressive plan for testing and installing a series of additional HTS cables in Jeju Island, South Korea's environmental and resort showcase (Superconductive, 2012). With this aggressive program,

South Korea appears poised to achieve world leadership in commercialization of HTS AC cable technology.

The strong HTS cable program in Japan is led by SEI and Furukawa Electric Co. Ltd., with support from the Japanese government's METI/NEDO and the key involvement of TEPCO and CRIEPI. SEI demonstrated one the world's earliest HTS AC cables in 2002: a 100 m, 3-in-1 (see Figure 5.1(b)), 66 kV, 114 MVA cable (Honjo et al., 2003). In 2004, Furukawa demonstrated a single-phase, 500 m, 77 kV, 1 kA_{rms} cable (Takahashi et al., 2005). Under the METI/NEDO M-PACC program, SEI demonstrated a 15 m, 3-in-1 66 kV, 5 kA_{rms} cable with remarkably low AC loss, and Furukawa demonstrated a 30 m, single-phase 275 kV, 3 kA_{rms}, also with very low loss (see Section 5.3) (Maruyama et al., 2013). SEI has also participated in a number of international projects, including the Albany project in the United States (see Section 5.8.2) (Masuda et al., 2007) and the test of a 23 kV, 1.25 kA_{rms}, 100 m cable at the KEPCO Gochang test facility (in Korea in 2006.)

Perhaps the most significant of the Japanese HTS cable projects has been the Yokohama project, whose first phase was concluded in 2014 (Masuda et al., 2011; Mimura, Honjo, Masuda, Machida, & Hara, 2014; Yumura et al., 2013). This project brings together SEI as the cable manufacturer, Mayekawa Mfg. Co. Ltd. as the cryogenic system supplier, and TEPCO as the host utility, with Japanese government support from METI/NEDO. A 240 m, 66 kV, 1.65 kA_{rms} (200 MVA) cable has been installed in the Asahi substation of the TEPCO grid and, beginning in October 2012, it operated successfully in grid for a year, supplying electricity to 70,000 homes. The amperage, pressure, inlet temperature, and flow rate during this period are shown in Figure 5.16. The inlet temperature was controlled during the first two months but later allowed to climb slightly to test operating performance at the higher temperatures.

The HTS cable uses SEI's 3-in-1 design (see Figure 5.1(b)) with SEI's 1G BSCCO-2223 tapes, including special tapes with twisted filaments to reduce AC loss, and the cable's DC critical current is 6.5 kA. The cable achieved an AC loss of 0.9 W/m/phase at 50 Hz and 2 kA_{rms}, which met design requirements. The cable used 7-mm-thick PPLP dielectric and passed all the standard high-voltage dielectric tests. It showed no degradation from a 5-m-diameter 180° bend and tolerated a 1.3 ton maximum tension during installation. The HTS cable also is designed to withstand the 3 ton stress from thermal contraction on cooling the cable to its operating temperature. Through the use of adequate amounts of copper shunting both the phase and the screen conductors, the cable successfully handled 31.5 kA_{rms}, 2 s fault currents without damage (no fault current limitation in this project).

The closed-cycle cooling system uses six Stirling cryocoolers in a redundant arrangement. Each cryocooler had a cooling power of 1 kW at 77 K and 0.8 kW at 67 K. The inlet temperature can be regulated to 69 K. The pumps which circulate LN_2 maintained a pressure of just under 0.2 MPa at 40 L/min. The cable cool down took 3 days. The next phase of the project will introduce a turbo-Brayton refrigeration system which should further improve refrigeration efficiency. All in all, this successful in-grid demonstration with stable operating conditions is a strong confirmation of the potential for HTS AC cables in the Japanese grid.

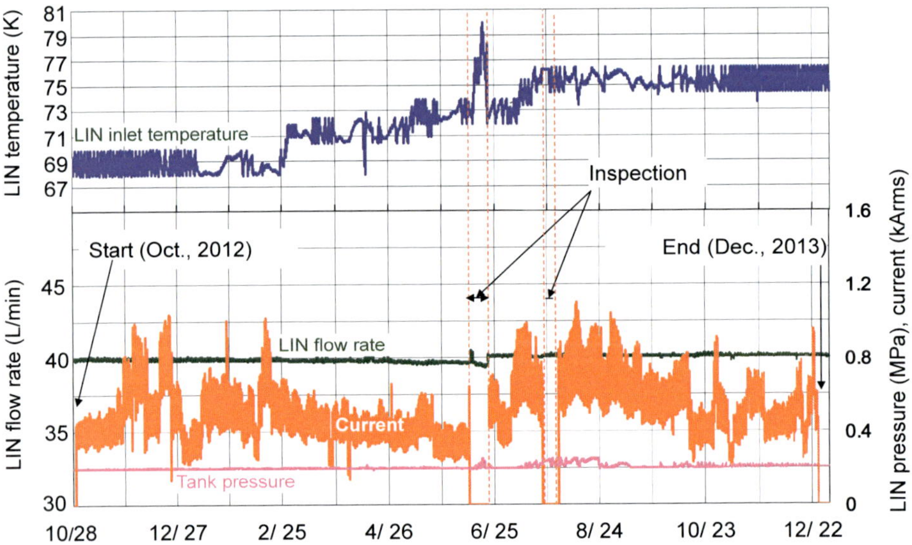

Figure 5.16 Operating parameters—inlet temperature, LN_2 flow rate, LN_2 tank pressure, and current—of the Yokohama in-grid HTS AC cable project during the period October 2012–September 2013 (Mimura et al., 2014).

5.8.5 AmpaCity project

The successful energization of the AmpaCity HTS AC cable and FCL project in Essen, Germany, was announced on May 1, 2014. This project is arguably the most significant HTS power technology project to date. It is the first in-grid installation between two utility substations combining an HTS AC cable and an FCL to protect it. At 1 km, it is also the longest HTS AC cable installation so far.

The project is a joint effort of Nexans, Karlsruhe Institute of Technology (KIT) and the German utility RWE, supported by the German Federal Ministry of Economics and Technology. In 2010 a feasibility study was conducted, led by KIT, identifying major advantages of using 10 kV HTS AC cables to replace the 110 kV cables currently used to feed the center of Essen. They found that using HTS AC cables, four out of ten 110/10 kV transformer substations in the city center could be eliminated, and that conventional AC cables were not a viable alternative because of their space requirements and high losses. Ease of HTS cable installation with no external heat load or EMF was a significant advantage in facilitating the cable route into the city center. They recognized that an FCL would also be needed in series to limit a 38 kA peak fault current, and the design target was limitation to 10 kA.

The AmpaCity project was started in 2011 to demonstrate the first 10 kV 40 MVA (2.3 kA_{rms}) cable, 1 km long, between the Herkules and Delbruegge substations of RWE's Essen grid, in conjunction with an FCL. The compact concentric cable design as in Figure 5.1(c) was chosen, using BSCCO-2223 HTS tapes from SEI, with an LN_2 inflow through a hollow former and a return flow through a space between the copper screen and the corrugated cryostat wall. The dielectric is a lapped PPLP. A cable joint is located in the middle of the 1 km length. The FCL was fabricated with YBCO 2G HTS wire (see also Chapter 9).

A successful type test on a prototype in February 2013 included a partial discharge test, 75 kV lightning impulse test, and 30 kV AC voltage withstand test, according to German standards. Component manufacturing began in March 2013, installation was completed by December 2013, and commissioning tests were completed by March 2014. With the system fully energized as of May 1, 2014, pilot operation of the system is now expected to continue for 2 years.

Figure 5.17 shows the cable wound on a drum and on-site in preparation for installation through polyethylene pipes preinstalled in the ground. Figure 5.18 shows the interior of the Herkules substation, with the HTS cable in the lower left, forming a loop for handling thermal contraction, coupled to the standalonc FCL and cooling system, which is designed for a total heat load of 4 kW with an inlet temperature of 67 K. The LN_2 is supplied to the cable with a flow of 400 g/s under a pressure of 10 bar, from a storage tank in an open-cooling arrangement. Most of the refrigeration and FCL are located at the Herkules substation, which is located outside the city center, allowing minimum space demands in the city center Delbruegge substation.

5.8.6 Hydra project

This significant project is intended to develop and demonstrate the world's first in-grid HTS AC cable with inherently fault-current-limiting functionality (Gouge et al., 2009; Ichikawa et al., 2005; Maguire et al., 2007, 2009). The project is partially funded by the US Department of Homeland Security under a cost sharing agreement with AMSC; Consolidated Edison of New York is the participating utility. The 200 m HTS cable is being provided by Southwire/Ultera. ORNL directly funded by DHS, is serving as the main test site for subscale prototype cables.

The Hydra cable is an HTS Triax™ cable design, which Southwire/Ultera has successfully employed in previous cable installations (see Figure 5.1(c)) (Demko et al., 2007). The HTS cable assembly includes a hollow flexible former, three-layered HTS phase conductors, an outer neutral copper conductor, and dielectric wrapping. The phase and neutral conductors are concentrically wound around a flexible former in a helical pattern, with a cryogenic dielectric isolating each phase and the neutral conductor. The HTS Triax™ design allows the use of substantially less HTS wire than other three-phase coaxial assemblies. As discussed in Section 5.2.2, the HTS Triax™ concentric design causes net zero EMFs when the phases are balanced; thus there is no need for an HTS shield layer. Each phase of the conductor, rated for 4000 A_{rms} at 13.8 kV, consists of two layers of helically wound HTS wire. To obtain the necessary electrical characteristics for fault current limiting, AMSC fabricated 2G YBCO-coated conductors encased in stainless-steel lamination. The design fault current is 40 kA_{rms} and the expected degree of current limiting by the HTS cable is about 17%.

An HTS cable with a built-in FCL feature takes advantage of the "switching" characteristic of superconductor wires at currents higher than their critical current

Figure 5.17 AmpaCity HTS AC cable prepared for cable pulling installation in Essen, Germany.
Courtesy of M. Stemmle, Nexans.

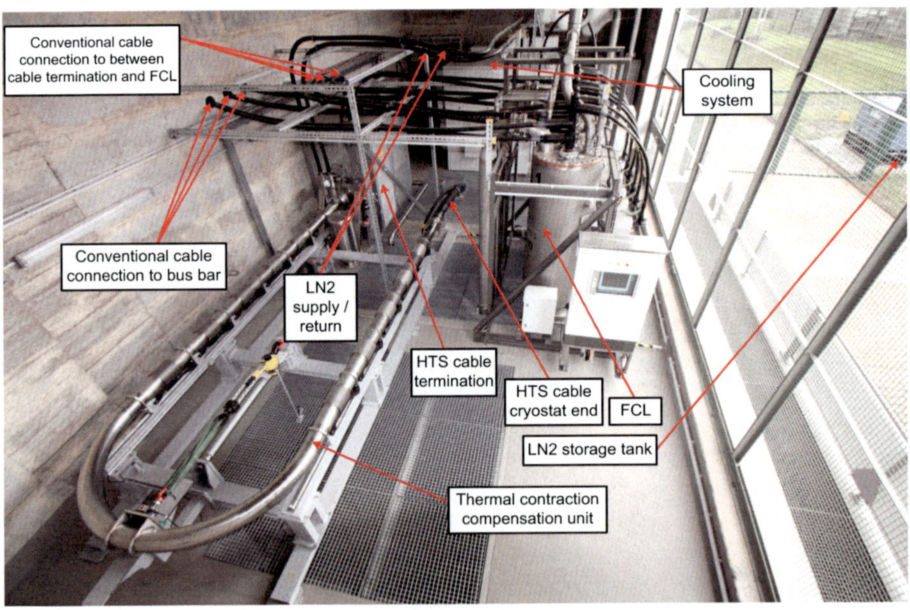

Figure 5.18 HTS system installation at the Herkules substation for the AmpaCity project, showing the fault current limiter in three individual cryostats for each phase on the right and the cryogenic cable in a loop for handling thermal contraction.
Courtesy of M. Stemmle, Nexans.

threshold, which inserts an impedance, primarily resistive, into the HTS cable, as described in Section 5.6. This change in impedance occurs within the first cycle of the fault and results in an immediate reduction in the fault current. A major system advantage of the current-limiting functionality is that substations and busses that normally could not be joined due to the increase in fault current can now be connected without increasing the fault current in either location. An additional benefit is that assets can be connected and shared, allowing, for example, a spare transformer to be shared by two substations.

To properly design an FCL HTS cable, a system-level model has been generated to study the HTS cable performance, operating parameters, and preliminary component requirements. To verify the system-level model, a subscale 25 m prototype cable has been manufactured and tested. The test results indicate that the cable has fully met the specifications as shown in Figure 5.19 and validated the model (Maguire et al., 2009).

The full-scale, 200-m-long HTS power cable system will connect two of Con Edison's substations and is scheduled to be installed and operating in 2015, in conjunction with Con Edison's construction schedule.

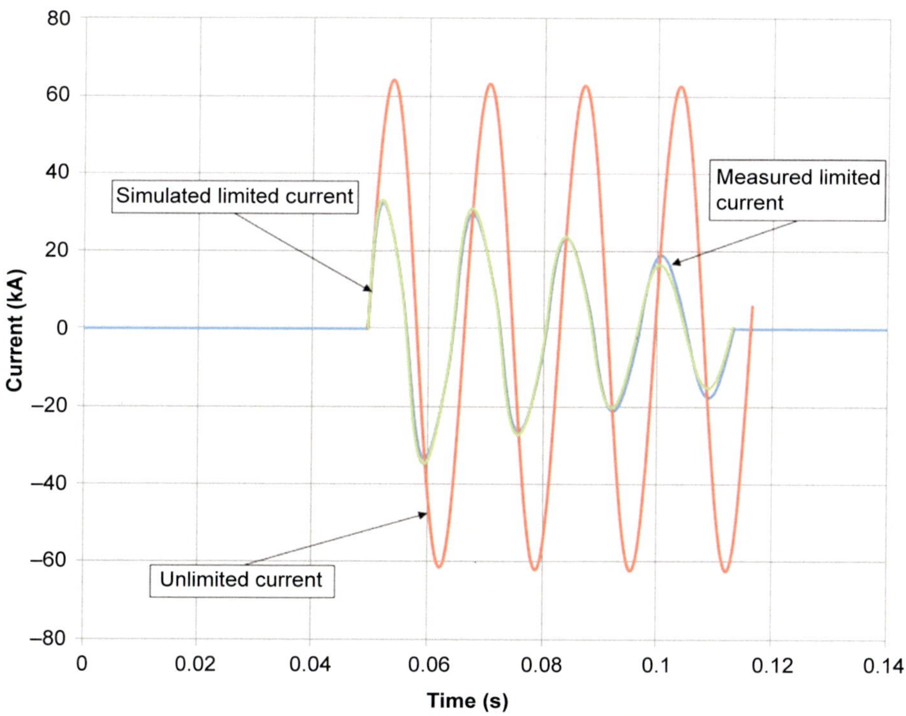

Figure 5.19 Comparison of current during an unlimited fault event to the current during a fault event on phase 2 and phase 3 of the HTS Triax™, 25-m-long cable. American Superconductor's electrothermal model of the HTS cable agrees with test results (Maguire et al., 2009).

5.9 Conclusion: commercial prospects for HTS AC cable

Up to the time of this writing, all HTS AC cable projects have been significantly supported by the governments of the given country, but now the prospects for a truly commercial, non-government-supported HTS AC cable look promising, given the many advantages described in the above sections.

One of the main obstacles remains cost, since conventional power cables cost in the range of \$1M/km or less, while HTS AC cables cost several times as much, especially when one takes into account the expensive terminations and refrigeration equipment at the terminal substations. Specific numbers depend strongly on the rating and other project specifications, of course, and in any case, actual cost projections for commercial installations are held closely proprietary by cable companies. Partially offsetting the cable and equipment costs are the installation costs, because if one can reduce permitting costs by using lower voltages and avoid tearing up city streets by threading HTS AC cables through existing ducts, one may save millions. Again, actual installation costs depend strongly on the specific site.

One cost element that can be estimated quite simply is the HTS wire cost. Up to now, costs remain above \$100/kA-m for a 4.4 mm 1G or 2G HTS wide tape, where the kA refers to the 77 K critical current. Since the critical current almost doubles when the tape is cooled to the operating temperature of 70 K, the effective tape cost is about \$50/kA-m. Now for a three-phase cable of the 1-core or 3-in-1 type carrying 4 kA_{rms}, the critical current requirement is at least 6 kA for both phase conductor and screen at operating temperature; thus, the estimated wire cost is $6 \times 50 \times 2 \times 3 = \$1800/m$ or \$1.8M for a 1 km cable! This shows the extent of cost reduction still required to bring HTS AC cable costs closer in line with conventional cable costs. This cost can be halved with the Triax design of Figure 5.1(c), as it does not require an HTS screen. Of course it can be halved again, down to \$450k, if one is dealing with a 2000 A_{rms} application. The HTS wire industry continues to work to bring down production costs further.

Another major cost element is refrigeration, because up to now, practically every refrigeration system used in an HTS AC cable project has represented a one-off design. With more efficient refrigeration, such as the turbo-Brayton cryocooler being introduced in the next phase of the Yokohama project (Masuda et al., 2011), and with more standardized production, refrigeration costs are also expected to decrease significantly.

As summarized in Chapter 1, Section 1.4, the other key challenge for utility adoption of HTS AC cable technology is the need for demonstrated reliability. This issue is being addressed through in-grid, long-term projects like Bixby, LIPA, AmpaCity, and Yokohama, making HTS AC cable one of the more promising HTS power technologies for imminent commercialization.

In a promising development for further commercialization, Commonwealth Edison and AMSC announced in July 2014 the start of a major new program, with AMSC as project lead and supported in part by the US Department of Homeland Security, to install inherently fault-current-limiting cables in the Chicago Loop to increase the reliability of this critical area of Chicago's central business district.

Acknowledgments

The authors thank Bruce Gamble, Alan Wolsky, and Wolfgang Schmidt for their careful reading and suggested modifications of Section 5.6, and David Knoll, Mark Stemmle, Pascal Tixador, Kenichi Sato, and Shoichi Hondo for their inputs on important HTS cable projects around the world.

References

Amemiya, N., Jiang, Z., Nakahata, M., Yagi, M., Mukoyama, S., Kashima, N., et al. (2007). Ac loss reduction of superconducting power transmission cables composed of coated conductors. *IEEE Transactions on Applied Superconductivity, 17*, 1712–1715.

Barron, F. R. (1985). Cryogenic systems. In *Monographs on cryogenics* (Vol. 3, pp. 9–10).

Bean, C. P. (1962). Magnetization of hard superconductors. *Physical Review Letters, 8*, 250–253.

Burke, J. J., Clapp, A. L., & Ros, W. J. (1993). Power distribution. In D. G. Fink, & H. W. Beaty (Eds.), *Standard handbook for electrical engineers* (13th ed.). (pp. 18–90). New York: McGraw-Hill.

Campbell, A. (1999). Fields and losses in high Tc conductors. *IEEE Transactions on Applied Superconductivity, 9*(2), 2137–2142.

Chevtchenko, O., Zuijderduin, R., Smit, J., Willen, D., Lentge, H., Thidemann, C., et al. (2012). Low AC loss in a 3 kA HTS cable of the Dutch project. *Physics Procedia, 36*, 1285–1289.

Cho, J., Bae, J.-H., Kim, H.-J., Sim, K.-D., Seong, K.-C., Jang, H.-M., et al. (2005). Development and testing of a 30 m HTS power transmission cable. *IEEE Transactions on Applied Superconductivity, 13*(2), 1719–1722.

Clem, J. R., & Malozemoff, A. P. (2010). Theory of AC loss in power transmission cables with 2G HTS wire. *Superconductor Science and Technology, 23*, 034014.

Clem, J. R., & Malozemoff, A. P. (2013). Flux transfer losses in helically wound superconducting power cables. *Superconductor Science and Technology, 26*, 085008.

Daeumling, M. (2004). AC loss in two AC carrying superconducting concentric tubes – the duoblock model. *Physica C, 403*, 57–59.

Demko, J. A., Sauers, I., James, D. R., Gouge, M. J., Lindsay, D., Roden, M., et al. (2007). Triaxial HTS cable for the AEP Bixby project. *IEEE Transactions on Applied Superconductivity, 17*(2), 2047–2050.

Forsyth, E. B., & Thomas, R. A. (1986). Performance summary of the Brookhaven superconducting power transmission system. *Cryogenics, 26*, 599–614.

Fukui, S., Noguchi, T., Ogawa, J., Yamaguchi, M., Sato, T., Tsukamoto, O., et al. (2007). Numerical study on AC loss minimization of multi-layer tri-axial HTS cable for 3-phase AC power transmission. *IEEE Transactions on Applied Superconductivity, 17*(2), 1700–1703.

Garber, M., Bussiere, J. F., & Morgan, G. H. (1976). Design of double helix conductors for superconducting AC power transmission. In *Magnetism and magnetic materials, AIP conference proceedings* (Vol. 34, pp. 84–87).

Garwin, R. L., & Matisoo, J. (1967). Superconducting lines for the transmission of large amounts of electric power over great distances. *Proceedings of the IEEE, 55*(4), 538–548.

Gouge, M. J., Duckworth, R. C., Demko, J. A., Rey, C. M., Thompson, J. R., Lindsay, D. L., et al. (2009). Testing of 3-meter prototype fault current limiting cables. *IEEE Transactions on Applied Superconductivity, 19*, 1744–1747.

Hassenzahl, W., & Tsukamoto, O. (2012). Superconducting cables. In H. Rogalla, & P. H. Kes (Eds.), *100 Years of superconductivity* (pp. 788–792). Boca Raton, FL: CRC Press.

Honjo, S., Shimodate, M., Takahashi, Y., Masuda, T., Yumura, H., Suzawa, C., et al. (2003). Electric properties of a 66 kV 3-core superconducting power cable system. *IEEE Transactions on Applied Superconductivity, 13*(2), 1952–1955.

Ichikawa, M., Kanegami, M., Okamoto, T., Akita, S., Yagi, M., & Kimura, A. (2005). Thermomechanical characteristics of 500-m HTS power cable. *IEEE Transactions on Applied Superconductivity, 15*, 1771–1774.

Jia, Y., Hua, J., Crabtree, G. W., Kwok, W. K., Welp, U., Malozemoff, A. P., et al. (2011). C-axis critical current density of second generation YBCO tapes. *Superconductor Science and Technology, 23*, 115017.

Kim, D.-W., Jang, H.-M., Lee, C.-H., Kim, J.-H., Ha, C.-W., Kwon, Y.-H., et al. (2005). Development of the 22.9-kV class HTS power cable in LG cable. *IEEE Transactions on Applied Superconductivity, 15*(2), 1723–1726.

Kraemer, H.-P., Schmidt, W., Utz, B., & Neumueller, H.-W. (2003). Switching behavior of YBCO thin film conductors in resistive current limiters. *IEEE Transactions on Applied Superconductivity, 13*, 2044–2047.

Lee, S., Yoon, J., & Yang, B. (2011). Modeling of a 22.9 kV 50 MVA superconducting power cable based on PSCAD/EMTDC for application to the Icheon substation in Korea. *Physica C: Superconductivity and Its Applications, 471*(21), 1283–1289.

Leghissa, M., Rieger, J., & Neumueller, H.-W. (1999). Development of HTS power transmission cables. *IEEE Transactions on Applied Superconductivity, 9*, 406–411.

Maguire, J., Folts, D., Yuan, J., Henderson, N., Lindsay, D., Knoll, D., et al. (2009). Status and progress of a fault current limiting HTS cable to be installed in the Con Edison grid. *Advances in Cryogenic Engineering, 55A*, 445–452.

Maguire, J. F., Schmidt, F., Bratt, S., Welsh, T. E., Yuan, J., Allais, A., et al. (2007). Development and demonstration of a HTS power cable to operate in the Long Island Power Authority transmission grid. *IEEE Transactions on Applied Superconductivity, 17*, 2034–2037.

Maguire, J. F., Yuan, J., Romanosky, W., Schmidt, F., Soika, R., Bratt, S., et al. (2011). Progress and status of a 2G HTS power cable to be installed in the Long Island Power Authority (LIPA) grid. *IEEE Transactions on Applied Superconductivity, 21*(3), 961–966.

Malozemoff, A. P., Kehrli, A., Diaz de Leon, J., & Kalsi, S. (2004). Superconductor technologies for a controllable and reliable high capacity grid. *PSCE04, IEEE Catalog No. 04EX943C*.

Malozemoff, A. P., Snitchler, G., & Mawatari, Y. (2009). Tape-width dependence of ac losses in HTS cables. *IEEE Transactions on Applied Superconductivity, 19*, 3115–3118.

Maruyama, O., Ohkuma, T., Masuda, T., Ashibe, Y., Mukoyama, S., Yagi, M., et al. (2013). Development of 66 kV and 275 kV class REBCO HTS power cables. *IEEE Transactions on Applied Superconductivity, 23*(3), 5401405.

Masuda, T., Yumura, H., Ohya, M., Ashibe, Y., Watanabe, M., Minamino, T., et al. (2011). Test results of a 30 m HTS cable for Yokohama project. *IEEE Transactions on Applied Superconductivity, 21*(3), 1030–1033.

Masuda, T., Yumura, H., Watanabe, M., Takigawa, H., Ashibe, Y., Suzawa, C., et al. (2007). Fabrication and installation results for Albany HTS cable. *IEEE Transactions on Applied Superconductivity, 17*(2), 1648–1651.

Mawatari, Y., & Kajikawa, K. (2008). Hysteretic ac loss of polygonally arranged superconducting strips carrying AC transport current. *Applied Physics Letters, 92*, 012504.

Mimura, T., Honjo, S., Masuda, T., Machida, A., & Hara, T. (2014). Japan's first in-grid operation of a 66 kV – 200 MVA superconducting cable system. *CIGRE SC B1: PS3 Insulated Cables in the Network of the Future*, 1–8. Paris Session.

Mukoyama, S., Hirano, H., Yagi, M., Kimura, H., & Kikuchi, A. (2003). Test results of a 30 m high-Tc superconducting power cable. *IEEE Transactions on Applied Superconductivity, 13*(2), 1926–1929.

Mukoyama, S., Miyoshi, I. K., Tsubouti, H., Mimitra, M., Uno, N., Ichtyanagi, N., et al. (1997). 50 m long HTS conductor for power cable. *IEEE Transactions on Applied Superconductivity, 7*, 1069–1072.

Nassi, M., Kelley, N., Ladie, P., Corsaro, P., Coletta, G., & Von Dollen, D. (2001). Qualification results of a 50 m – 115 kV warm dielectric cable system. *IEEE Transactions on Applied Superconductivity, 11*(1), 2355–2358.

Nguyen, D. A., Ashworth, S. P., Duckworth, R., Carter, W., & Fleshler, S. (2011). Measurements of AC losses and current distribution in superconducting cables. *IEEE Transactions on Applied Superconductivity, 21*(1), 996–1000.

Norman, S., Nassi, M., Bechis, M., Ladie, P., Kelley, N., & Wakefield, C. (2001). High temperature superconducting cable field demonstration at Detroit Edison. *Physica C, 354*, 49–54.

Norris, W. T. (1970). Calculation of hysteresis losses in hard superconductors carrying AC: isolated conductors and edges of thin sheets. *Journal of Physics D, 3*, 489.

Ohya, M., Inagaki, Y., Tatamidani, K., Ito, H., Saito, T., Ashibe, Y., et al. (2013). Japan's first live power transmission using 3-in-one superconducting cable (high-temperature superconducting cable demonstration project). *SEI Technical Review, 76*, 45–54.

Okubo, H. (Ed.). (2010). Status of development and field test experience with high temperature superconducting power equipment (pp. 21–27). *CIGRE Report #418*, (CIGRE, 21 Rue D'Artois, F-75008 Paris). ISBNL 982-2-85873-105-3.

Rupich, M. W., Li, X., Sathyamurthy, S., Thieme, C. L. H., DeMoranville, K., Gannon, J., et al. (2013). Second generation wire development at AMSC. *IEEE Transactions on Applied Superconductivity, 23*(3), 6601205. and Chapter 4 of present book.

Sauers, I., James, D. R., Ellis, A. R., Tuncer, E., Pace, M. O., Gouje, M. J., et al. (2007). High voltage testing of a 5-meter prototype triaxial HTS cable. *IEEE Transactions on Applied Superconductivity, 17*, 1734–1737.

Schmidt, W., Kraemer, H.-P., Neumueller, H.-W., Schoop, U., Verebelyi, D., & Malozemoff, A. P. (2007). Investigation of YBCO coated conductors for fault current limiter applications. *IEEE Transactions on Applied Superconductivity, 17*(2), 3471–3474.

Schmidt, F., Soika, R., Maguire, J., Yuan, J., Bratt, S., & Welsh, T. (June 20–23, 2011). Operation experience and further development of a high temperature superconducting power cable in the Long Island Power Authority grid. Versailles, France: *Jicable'11*. www.jicable.org/2011.

Sim, K., Kim, S., Cho, J., Kim, D., Kim, C., Jang, H., et al. (2008). DC critical current and AC loss measurement of the 100 m 22.9 kV/50 MVA HTS cable. *Physica C, 468*, 2018–2022.

Sohn, S. H., Lim, J. H., Yim, S. W., Hyun, O. B., Kim, H.-R., Yatsuka, K., et al. (2007). The results of installation and preliminary test of 22.9 kV, 50 MVA, 100m class HTS power cable system at KEPCO. *IEEE Transactions on Applied Superconductivity, 17*(2), 2043−2046.

Soika, R., Garcia, X. G., & Nogales, S. C. (2011). ENDESA supercable, a 3.2 kA, 138 MVA, medium voltage superconducting power cable. *IEEE Transactions on Applied Superconductivity, 21*(3), 972−975.

Stemmle, M., Merschel, F., Noe, M., & Hobl, A. (2013). AmpaCity − installation of advanced superconducting 10 kV system in city center replaces conventional 110 kV cables. *IEEE International Conference on Applied Superconductivity and Electrromagnetic Devices*, 323−3265. phys.org/news/2014-05-longest-superconducting-pcable-worldwide.html.

Stovall, J. P., Demko, J. A., Fisher, P. W., Gouge, M. J., Lue, J. W., Sinha, U. K., et al. (2001). Installation and operation of the Southwire 30-meter high temperature superconducting power cable. *IEEE Transactions on Applied Superconductivity, 11*(1), 2467−2472.

SUPER3C − Superconducting coated conductor cable.(2009). ewh.ieee.org/tc/csc/Europe/newsforum/pdf/HP18.pdf.

Superconductive cable called 'cable of dream' to replace transmission cable in Jeju.(2012). energy.korea.com/archives/37058.

Takahashi, T., Suzuki, H., Ichikawa, M., Okamoto, T., Akita, S., Mukoyama, S., et al. (2005). Demonstration and verification tests of 500 m long HTS power cable. *IEEE Transactions on Applied Superconductivity, 15*(2), 1823−1826.

Tonnesen, O., Daeumling, M., Jensen, K. H., Kvorning, S., Olsen, S. K., Traeholt, C., et al. (2004). Operation experiences with a 30 kV/100MVA high temperature superconducting cable system. *Superconductor Science and Technology, 17*, S101−S105.

Traeholt, C., Olsen, S. K., Rasmussen, C., Veje, E., & Tonnesen, O. (2001). Calorimetric measurements of losses in HTS cables. *IEEE Transactions on Applied Superconductivity, 11*(1), 1777−1780.

Vellego, G., & Metra, P. (1995). An analysis of the transport losses measured on HTSC single-phase conductor prototypes. *Superconductor Science and Technology, 8*, 476.

Volkov, E. P., Vysotsky, V. S., & Firsov, V. P. (2012). First Russian long length HTS power cable. *Physica C, 482*, 87−91.

Vysotksy, V. S., Shutov, K. A., Nosov, A. A., Polyakova, N. V., Fetisov, S. S., Zubko, V. V., et al. (2010). AC loss of a model 5m 2G HTS power cable using wires with NiW substrates. *Journal of Physics: Conference Series, 234*, 032061.

Xiao, L. Y., Dai, S. T., Lin, Y. B., Gao, Z. Y., et al. (2007). Development of HTS AC power transmission cables. *IEEE Transactions on Applied Superconductivity, 17*(2), 1652−1655.

Xin, Y., Hou, B., Bi, Y., Xi, H., et al. (2005). Introduction of China's first live grid installed HTS power cable system. *IEEE Transactions on Applied Superconductivity, 15*(2), 1814−1817.

Yumura, H., Ashibe, Y., Ohya, M., Itoh, H., et al. (2013). Update of Yokohama HTS cable project. *IEEE Transactions on Applied Superconductivity, 23*(3), 5402306.

Zubko, V. V., Nosov, A. A., Polyakova, N. V., Fetisov, S. S., & Vysotsky, V. V. (2011). Hysteresis loss in power cables made of 2G HTS wires with NiW alloy substrate. *IEEE Transactions on Applied Superconductivity, 21*(3), 988−990.

Using superconducting DC cables to improve the efficiency of electricity transmission and distribution (T&D) networks: an overview

6

C.E. Bruzek[1], *A. Allais*[2], *K. Allweins*[3], *D. Dickson*[4], *N. Lallouet*[5], *E. Marzahn*[3]
[1]Nexans France, Clichy Cedex, France; [2]Nexans France, Lyon Cedex, France; [3]Nexans Deutschland GmbH, Hannover, Germany; [4]Nexans France, Lens Cedex, France; [5]Nexans France, Calais Cedex, France

6.1 Introduction

Superconductivity was discovered 100 years ago by Dutch Physicist Kamerlingh Onnes. This discovery allowed the development of numerous applications in the field of electrical engineering, mainly for electromagnet businesses and for applications such as medical imaging. Superconductivity is also a key technology for the development of magnetically confined nuclear fusion and the physics of the high energies. These superconducting magnet technologies are mostly based on the low-temperature superconducting NbTi alloys with a superconducting transition temperature ~9.2 K. At these temperatures these types of superconducting magnets are typically cooled with liquid helium at 4.2 K. The high cost of this cooling medium makes such a technology only economically interesting for compact magnet applications.

It is a quarter of century since the discovery of superconductive materials so-called "high-temperature superconductors" (HTS). HTS materials open new and interesting perspectives for superconducting power applications. Although still requiring cryogenic cooling, this discovery was a revolution for the electrical industry and in particular for the high-power cable manufacturers. With the discovery of HTS, it was possible to operate superconducting equipment with the environmental friendly liquid nitrogen (Liq N_2), a low cost, higher heat capacity, and more commonly implemented cooling medium. Numerous scientific and industrial work has led first to the design and to the manufacture and then to the testing of these HTS cables. This chapter presents a description and the characteristic performances of the different key components required for DC superconducting cables, and the impacts of this new technology on future DC transport and distribution networks.

Superconductors in the Power Grid. http://dx.doi.org/10.1016/B978-1-78242-029-3.00006-6

6.2 Superconducting cable systems: key elements

The main interest of superconducting materials in electric power applications is that they do not show any ohmic resistance below a certain temperature T (less than $-200\,°C$), below a certain current density J, and below a certain magnetic field B, depending on the superconducting material. Figure 6.1 shows a typical voltage–current curve measured in a magnetic self-field at 77 K of a high-temperature superconducting tape. The HTS tape shows a nonlinear electrical resistance versus current. A singular point of this curve is the critical current (I_c), which is conventionally estimated at 1 μV/cm electric field along the wire. Below this critical current, electrical resistance can be neglected and the superconducting material is a perfect conductor. Above this value, the material loses its superconducting properties and becomes rapidly resistive. This transition for the superconducting state to the normal state can be expressed by a power law (Eqn (6.1)),

$$Es_{\mathrm{sc}}(T) = E_{\mathrm{c}}\left(\frac{I_{\mathrm{op}}}{I_{\mathrm{c}}\ (T)}\right)^{n} \tag{6.1}$$

where

n is the superconducting transition index, T is the absolute temperature and E_c is the electrical field defining the critical current (1 μV/cm).

For a given current, similar nonlinear current–voltage dependencies can be measured with temperature or magnetic field with a critical temperature T_c and critical magnetic field H_c, respectively. In fact, a critical surface can be drawn in the current (I)–magnetic field (H)–temperature (T) space. Below this surface the material is superconducting; above it behaves as a resistor. Most practical superconducting power applications would not perform properly at a point directly on this surface and an operating margin of approximately 25–30% below the critical current value is typically

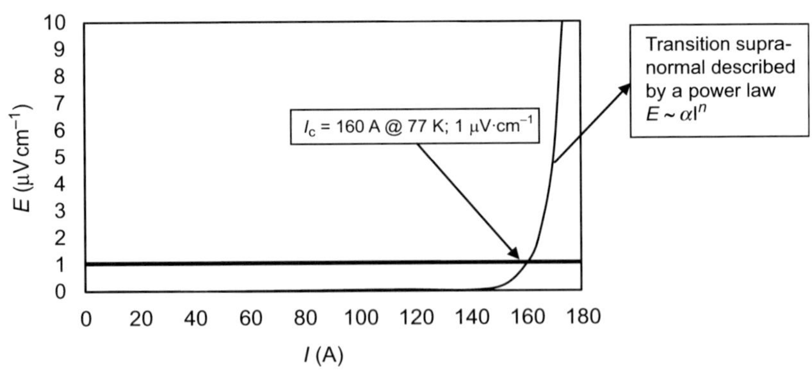

Figure 6.1 Typical voltage–current dependence of superconducting material.

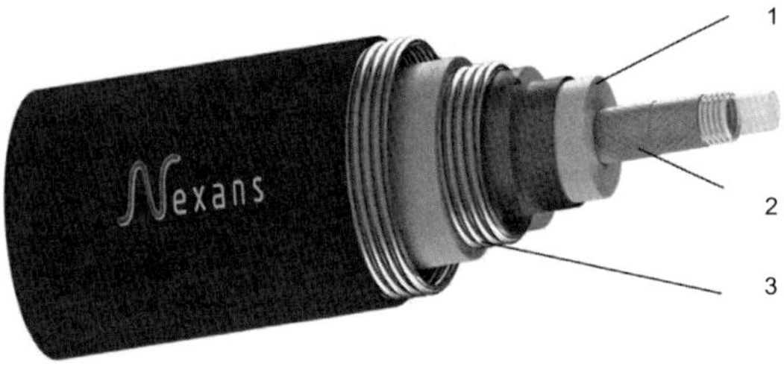

Figure 6.2 Schematic of a pole of a superconducting DC cable.

applied for the maximum current to operate the system. Consequently, a DC cable manufactured with superconducting materials and operated under these conditions is a perfect electrical conductor.

To maintain these required conditions in a superconducting cable system, a system approach including cooling and electrical systems is needed. A specific design of active and passive elements and their operation conditions depend on the application and the site requirement. However, common conceptual designs of key parts can be proposed.

Figure 6.2 shows schematically a superconducting DC cable pole. This design utilizes a "cold dielectric" and is generally considered for DC superconducting power cables by most cable companies. Its name is due to the location of the voltage insulation within the cryogenic envelope at low temperature. In Figure 6.2, four main elements of the DC superconducting cable can be distinguished: (a) the high-temperature superconducting tape or wire layers, (b) the voltage insulation, (c) the cryogenic envelope, and (d) the coolant or cooling fluid. In addition to being operational in a grid, a superconducting cable system must be connected to the grid via high-voltage terminations or bushings and must be kept at cryogenic temperature via the cooling system. Long length (several kilometers) DC superconducting cables can be assembled with joints. A detail description of all this components will be presented in the following paragraphs.

6.3 Superconducting materials

Superconducting materials are a very critical part of the superconducting cable. These materials should be available in long lengths of a few kilometers and be flexible enough to maintain their current carrying capacity after cabling, winding, and installation operations. Three kinds of HTS materials possess these rigorous characteristics. The first two are flexible tapes based on the following chemical composition: $Bi_2Sr_2Ca_2Cu_3O_{10-x}$ (Bi-2223), so-called first generation powder-in-tube (PIT) tapes with a $T_c \sim 108$ K, and $YBa_2Cu_3O_{7-x}$ (YBCO), so-called second generation or coated conductors (CCs) with a $T_c \sim 92$ K. The third one is a cylindrical wire based on $Bi_2Sr_2Ca_1Cu_1O_{10-x}$ (Bi-2212) with a $T_c \sim 85$ K. All three are superconducting below a temperature of 77 K that can be obtained with liquid nitrogen. There is fourth one that is a cylindrical wire based on the low-temperature superconductor MgB_2 that is only superconducting below 39 K, achievable typically with gaseous helium or liquid neon ($\sim$27 K). Details of the architecture of these two tapes and wires are shown in Figure 6.3 and their performances are reported in Table 6.1.

Flexible Bi-2223 PIT tape, commonly manufactured in widths of about 3–4.2 mm wide $\times$ 0.25 mm thick in cross-section, consists of several hundred filaments $\sim$10 μm in width embedded in a silver or silver alloy matrix. These PIT tapes are manufactured by successive mechanical steps of drawing and cold-rolling. The PIT fabrication method consists of filling a silver tube of several tens of millimeters diameter with a ceramic precursor to obtain a monofilament billet. The billet is drawn down to millimeter-sized hexagonal shape rods in several steps. Hundred of these rods are then inserted into a silver alloy tube to obtain a multifilamentary composite billet that is again drawn and rolled to the final Bi-2223 tape shape. The volume fraction of superconducting ceramic obtained with such a process is between 20 and 30%, corresponding to an overall cross-section of 0.2–0.3 mm^2. The superconducting section consists of several hundred electrically parallel filaments. Local defects located from place to place within a superconducting filament structure can be accepted as the current can be distributed into adjacent filaments. Consequently, thee localized defects do not dramatically affect the critical current or the wire losses. To obtain high superconducting properties, partial melting heat treatments are carried out in the range of 800–850°C under high pressure (Kato et al., 2006). During this melting/solidification heat treatment, an exchange of oxygen between the superconducting ceramic and the furnace atmosphere takes place. This exchange imposes the selection of silver as the wire matrix due to its very high oxygen diffusion coefficient in the considered temperature range. Bi-2223 PIT tapes with very high transport capacity above 400 A/cm (77 K, self-field) are commercially available on kilometric lengths

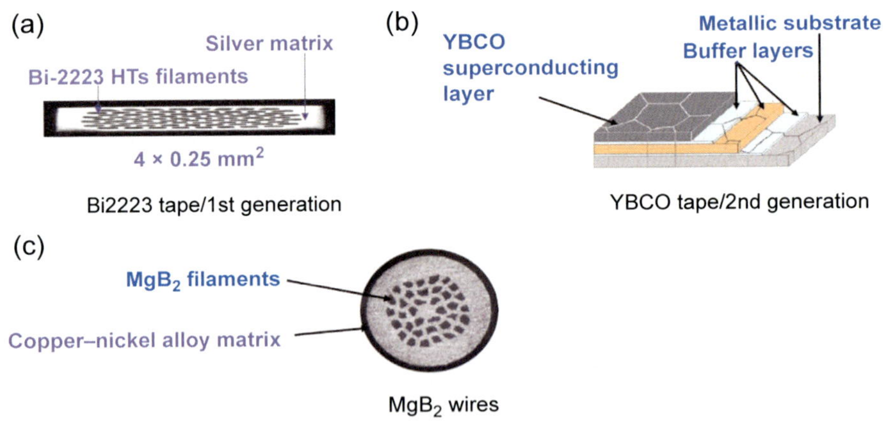

Figure 6.3 (**a–c**) HTS superconducting tapes and wires.
Reprinted with permission from Columbus SPA.

Table 6.1 **Properties of superconducting tapes and wires**

		Dimensions		Commercial performances	
	Shape	**Width (mm)**	**Thickness (mm)**	**J_e (A/mm^2) at 77 K; self-field (A/cm)**	**Length (m)**
Bi-2223	Laminated tapes	4, 5	0.3	120–150 (350–400)	<1500 m
YBCO	Laminated NiW tapes	4–12	0.2	100–150 400–500 (200–250)	<500 m
	IBAD tapes		0.05		
		Diameter (mm)		J_e (A/mm^2) at 20 K; 1 T	Length (m)
MgB_2	Cylindrical wires	0.8–1.5 mm		300	<3000 m

(http://global-sei.com/super/hts_e/index.html, 2013). Every year, this already high performance slowly increases thanks to process and filament microstructure improvements. Flexible Bi-2223 tape is now at a mature state with a fully industrialized production process where millions of kilometers have already been produced and tested in numerous electrical labs. However, the large percentage of silver matrix in the tape section (70–80%) makes it difficult to reach the cost objectives required for mass applications, even if the silver of the 1G tape is fully recyclable. This high cost justifies the R&D effort carried out on second generation (2G) CCs. The commercial market opportunity exists if an alternative raw material or manufacturing process can be used to reduce the cost and the complexity of the YBCO deposition. However, as 1G PIT tapes are available right now in large quantities with high performances, it will be used for demonstrator projects and will be kept probably for future cable applications especially if 2G tapes do not meet the cost objectives.

Flexible YBCO tapes are produced by a thin layer ($\sim$ 1–3 μm) coating processes. The CC tape structure is presented in Figure 6.3(b). The active layer has a thickness of 1–3 μm and is deposited as a covering on a suitably prepared buffered metallic substrate. As this design does not require a large cross-section of expensive material, lower costs can be predicted. The deposition techniques are numerous and derive from microelectronics (pulsed laser deposition, metal organic chemical vapor deposition, metal organic deposition). The deposition reactors have been adapted to coat kilometer lengths of flexible and textured substrates. To obtain high current in such a fine superconducting section requires a very stringent control of the chemical composition and crystallographic structure of YBCO. The manufacturing process consists in coating the metal substrate with a YBCO layer with a homogeneous composition and a very low crystallographic grain misalignment ($\Delta\theta < 5°$) over 100 m or km. This tight and very challenging specification is met by using the combined beneficial effects of two–four buffer layers that transfer or improve the texture of the metallic substrate, and act as a diffusion barrier to prevent contamination from the substrate. They also accommodate the crystallographic cell parameters to grow a YBCO monocrystalline layer. Two different metallic substrates represent the most competitive material for long length tape productions. The YBCO layer can be deposited on a buffered high strength polycrystalline substrate coated with a biaxial oriented textured ceramic using an ion bean assisted deposition (IBAD) process or on buffered softer NiW alloys textured by heavy thermo-mechanical processing. The YBCO tapes are thermally and electrically stabilized with a thin silver overlayer deposited directly on the YBCO layer with a copper shunt around the tape. For NiW substrates, the tapes can be reinforced by two soldered laminated metallic tapes. After the succession of 5–9 high-efficiency coating steps, kilometric lengths of CC with a width from 4 to 12 mm are manufactured with different architectures and produced with different coating methods; commercial 2G tapes exhibit similar characteristics, I_c/cm width = 200–250 A/cm width (http://www.amsc.com/library/BRSAMP_DS_A4_0912.pdf, http://www.bruker-est.com/ybco-tapes.html, http://www.superpower-inc.com/system/files/SP_2G+Wire+Spec+Sheet_for+web_2013FEC_v2.pdf, 2013).

These performances are still lower than those obtained with 1G tape but they have dramatically increased during recent years as manufacturing processes improve.

At the R&D level, different producers have already announced very high performance with I_c above 1000 A/cm width (Foltyn, 2007), showing the large improvement potential of the CC tapes. Today the numerous and complex deposition processes do not achieve the expected manufacturing yield or the price-to-performance metrics required for a broad market penetration. The scaling-up of the processes are still facing some difficulties but ongoing development programs should rapidly solve these issues.

Discovered in 2001, MgB_2 is technically a low-temperature superconductor with a relatively high $T_c \sim 39$ K that may offer attractive lower cost solutions. Its cross-section is presented in Figure 6.3. It is manufactured with the low-cost PIT process developed for Bi-2223 tapes. Its copper–nickel alloy matrix and composition based on the abundant materials magnesium and boron makes it about 10 times less expensive than any other HTS materials. Consequently, it is already marketed in kilometric lengths (Braccini, Nardelli, Penco, & Grasso, 2007). It offers high I_c performances either in tape or in cylindrical shapes. As a tape, it can simply replace other HTS materials or as a cylinder directly placed into cable former to give a very compact cable. However, for practical implementation as a DC superconducting power cable, it needs to be cooled to about 15–20 K with either liquid hydrogen or gaseous helium. These coolants and the associated cryogenic refrigerators are more expensive than liquid nitrogen but are significantly less expensive than liquid helium (2–3 times) and can make its use economically interesting. For high ampacity cables carrying currents above 5 kA, the additional cost for the cooling system is compensated by the very low costs of MgB_2 wires. For long length DC superconducting cables, this cost can be further mitigated with a special cryogenic envelope integrating a liquid nitrogen thermal shield circulating in parallel with the helium gas flow. Programs to introduce MgB_2 wires into cable conductor design have started with some promising results already published (Ballarino, 2010; Cheadle et al., 2013; Kostyuk et al., 2012).

6.4 Cable conductors and electrical insulation

To manufacture a cable conductor, superconducting tapes or wires are typically stranded around a central flexible former. The former or cable core usually consists of either a conductive material (Cu or Al) to accept fault current or resistive (stainless steel) to act as a "current limiter." The number of HTS tapes or wires is based on the rated current, the derating factor, and on the ampacity of a single superconducting element. Other external parameters such as the temperature rise along the cable or the magnetic field generated by the cable itself, which reduces the ampacity of the cable, must be considered in the design of a superconducting cable conductor. Considering the minimum bending radius of HTS tapes and a derating factor $I_{op}/I_c = 70–75\%$, a cable conductor of typically 30 mm diameter can transfer ampacity higher than 10 kA. For the cylindrical MgB_2 wires, the cable core can be directly filled with superconducting wires and 10 kA can be transferred in a very compact cable conductor with a diameter lower than 10 mm but at 20–25 K. When compared to conventional resistive cables, electrical power can be transported or distributed at a 5–times higher ampacity for the same voltage level without any joule losses.

Cabling is carried out on conventional stranding machines with a local arrangement of the superconducting elements to ensure that there is no damage during assembling, bending, and coiling. The superconducting tapes or wires can be laid around a flexible or hollow core that can be used to circulate the cryogenic fluid as shown in Figure 6.2. An external superconducting layer can be lapped concentrically around the electrical insulation to be used as a current return for bipolar coaxial design or as a magnetic shield blocking any stray fields.

The choice of section and materials of the core can be used to set a fault tolerance or a fault-limiting design.

In a *fault current tolerant* superconducting cable design, copper or aluminum wires are added inside the conductor and connected electrically in parallel to act as a low-resistance electrical shunt to protect the superconducting material. Under fault conditions, these highly conductive wires transport the excess current above the critical current of the superconducting conductor. The cable core exhibits a low resistance and can accept very high current fault. During a fault, the excess current flowing in the nonsuperconducting wires (e.g., Cu or Al) generates joule heating, which creates a temperature rise along the superconducting cable. The amount of temperature rise in the cable depends on the magnitude and duration of the fault current and the amount of normal (nonsuperconducting) metal in the cross-section of the cable. For fault currents of limited magnitude and duration, if the temperature rise is kept in an acceptable range, that is, does not cross the critical surface of the superconductor, after fault clearance, the transport current will eventually return back to the superconducting tapes without any intervention on the grid. For a bulk fault of larger magnitude and/or duration, the superconducting cable must be disconnected from the fault current via a bypass switch to prevent any damage to the superconducting cable. The final key parameters required to prevent dielectric breakdown or superconducting wires destruction are conductor temperature during and after fault, the available cooling power, and the temperature margin before a degradation of voltage insulation.

In a *fault current limiting* superconducting cable design, low-resistance (i.e., high conductivity) material electrically connected in parallel to the superconducting wires is purposely avoided. Instead, highly resistive (i.e., low conductivity) material (e.g., stainless steel, Hastelloy™, Ni—W) electrically connected in parallel with the superconductor is substituted in its place. The excess current during the short circuit is pushed through the highly resistive metallic matrix of the superconducting wire, which then becomes a high resistance. This quasi-instantaneous rise of electrical resistance consequently creates a temporary electrical resistive load in the damaged grid. Consequently, the increased electrical resistance "limits" the fault current level in the electric grid where the voltage is maintained constant. The electrical insulation and the superconducting material can be damaged during a fault if not adequately designed. A superconducting fault current limiting cable must always be associated with a protection device, such as a circuit breaker, to shut down the circuit if the short circuit or the overcurrent source keeps on disturbing the grid. A superconducting fault current limiting cable improves the short circuit protection level and improves the reliability of the energy delivery to customers especially in the case of subnetwork connections. Today only short laboratory cables have been manufactured and tested.

This innovative function is commercially offered for AC current using a dedicated piece of equipment known as a "superconducting fault current limiter." (http://www.nexans.de/eservice, 2013).

The electrical insulation in a superconducting cable not only must be an excellent dielectric material it must also be able to withstand operation at low temperature as well as repeated thermal cycling. Paper tapes (cellulose) or polypropylene laminated paper (PPLP) tapes or polyimide tapes are normally used to electrically insulate superconducting cables. Extruded dielectric materials (e.g., XLPE) are not typically used in superconducting cable designs; instead the preferred insulating technique is that of the wrapped or lapped conductor. The lapping process is similar to conventional resistive cable using oil-impregnated insulation. For insulation made out from lapped PPLP thin tapes (100−150 μm thick), the dielectric strength is in the range of 40−45 kV/mm with a dielectric constant of $\varepsilon_r = 2.2$. For insulation made out from lapped polyimide thin tapes (75−150 μm thick), a high dielectric strength value can be reached higher than 100 kV/mm with a dielectric constant of $\varepsilon_r = 3.5$. This material is expensive and it is generally reserved for low- to medium-voltage cable or for very compact cable. In operation, the superconducting cable insulation is impregnated with the cooling fluid (Liq N_2, Gas He, etc.). The cryogen fluid fills the gap along the side of each superconducting tape and contributes to the overall electrical strength of the insulation. The dielectric strength of Liq N_2 is similar to PPLP from 40 to 80 kV/mm with a dielectric constant of $\varepsilon_r = 1.4$, but dependent on the pressure. As the dielectric strength of cold N_2 gas is only 2−3 kV/mm when a liquid impregnated solution is selected, gas bubbles are forbidden, especially during a fault current event to avoid any dielectric breakage of the superconducting cable. To prevent their formation in the impregnated insulation material, the liquid N_2 is injected into the cable with an overpressure from 3 to 10 bar at subcooled temperatures ranging from about 65 to 70 K. These subcooled temperature and pressure conditions shift the boiling temperature of liquid N_2 up to 90−95 K and give enough thermal margin (20 K) to prevent the risk of bubble formation even under fault conditions. The subcooling of the liquid N_2 from 77 to 65−70 K also increases the I_c of the HTS tapes, which in turn reduces the amount of required HTS tape and hence the cost per meter of the superconducting cable.

To manufacture the insulation for superconducting cables, conventional paper-lapping machines are used. They can apply in line (one run) up to 200 paper layers around the HTS cable conductor in one pass. Typically the insulation thickness is between 0.5 mm for low-voltage superconducting cables below 10 kV and up to 25 mm for high-voltage superconducting cables at 400 kV.

6.5 Cable cryostat

During operation, the superconducting cable conductor must be kept below the critical temperature (T_c) of the superconducting material. For most commercial HTS wires, this temperature is below 100 K. This can be achieved using pressurized liquid nitrogen ranging from 70 to 85 K in a closed circuit or container system in absence of N_2 gas. Helium gas, which permits operating temperatures to below 20 K, is an alternative coolant when MgB_2 technology is used. It is also used at higher temperatures in a confined environment where there is an asphyxiation risk with nitrogen. The use of liquid H_2 at 20.7 K is considered for future grids (Kostyuk et al., 2012). However, as H_2 is highly flammable, safety management studies must be carried out to demonstrate to the public, utilities, and certification bodies that there is no safety risk to people.

The cooling fluid is circulated in a cryogenic envelope (the cable cryostat) that surrounds the superconducting cable with the smallest heat leak that is technically and economically achievable. As the cryogenic envelope can generate significant heat load in long DC cable systems, high thermal insulation along the cable is a key design requirement.

The design of the cryogenic envelope is based on solutions developed to connect the superconducting magnets cryostat to a cooling machine or fluid tank that allows management of the differential thermal shrinkage and expansion of the different components during cooling and reheating operations. Two physical phenomena involved in the heat transfer from a warm surface to cryogenic systems must be overcome: the heat conduction in gas and solid phases and thermal radiation.

To minimize the gas heat conduction, the solution incorporates a vacuum chamber around the equipment, like a Dewar. This vacuum chamber is built with an inner tube at cryogenic temperature and an outer tube at room or intermediate temperature. It must be hermetically sealed and also flexible enough to navigate the contours of the superconducting cable circuit. The heat input by residual gas is considered negligible when the pressure falls below about 10^{-3} mbar. This low-pressure vacuum must be kept for the lifetime or at least more than 25 years with getter materials. If needed, dynamic vacuum pumping is possible using accessible pumping ports.

Low thermal conductivity mechanical supports or spacers center the two concentric tubes and ensure the mechanical cohesion of the vacuum chamber. The spacers are manufactured with low heat conduction materials such as PTFE or glass fiber reinforced epoxy. This spacer design is a compromise between the desire to minimize the heat conduction to the inner vessel and to simultaneously maximize its mechanical strength.

Thermal losses in the vacuum cryostat due to radiation are drastically reduced by the installation of a multilayer structure of aluminized Mylar films, so-called superinsulation or MLI, in the annular space of the vacuum chamber. These aluminized films built with more than 30–40 layers are lapped around the inner tube. When properly applied, in combination these solutions have low thermal losses between 0.1 and 2 W/m at the operating temperature depending on the envelope diameter (http://www.nexans.de/eservice/, 2013). For large cryogenic envelopes and when very low losses (<0.3 W/m) are required, for example, at 20 K, a four-wall cryogenic envelope, integrating a thermal shield between the operating temperature and the room temperature, can be used. Typical performances of flexible cryogenic envelopes are given in Table 6.2.

In theory for very long and new links with limited curvatures, a possible solution is to install cryogenic envelopes made out with rigid pipes, 12–20 m long, like a gas pipeline. The different components could be transported, welded, evacuated, and sealed in the field. To date, this solution has never been field tested.

Table 6.2 Typical characteristics of flexible cryogenic envelopes

	Two walls					Four walls*		
Models	**Small**		**Medium**		**Large**	**Small**	**Medium**	**Large**
Int/Ext diameter (mm) (Shield)*	14/34	21/44	39/66	60/110	75/125	21/44 + (60/110)*	39/66 + (90/147)*	75/125 + (147/220)*
Bending radius (Several bends) (m)	0.6	0.7	1.1	1.8	2	2	2.5	3
Heat inleak at T_{op}/Shield* (W/m)	0.4	0.6	1	1.2	1.4	0.07/1.2*	0.12/1.5*	0.3/2*
Weight (kg/m)	0.5	0.8	1.7	4	6.7	5	7.5	10

*Outer thermal shield (diameters and heat losses).

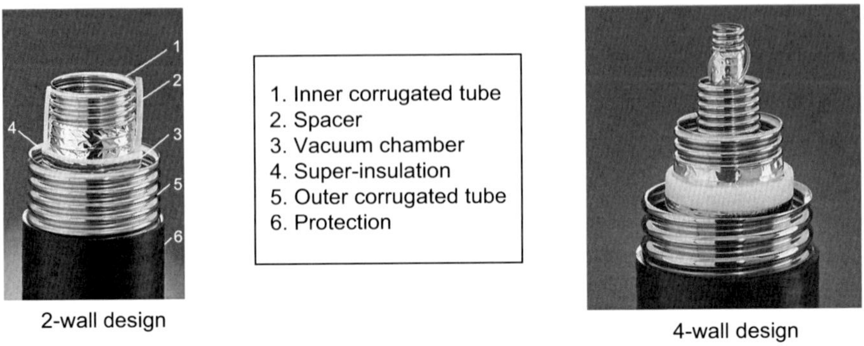

Figure 6.4 Flexible cryogenic envelopes.

Another solution is to install flexible cryogenic envelopes shown in Figure 6.4. Such a solution could be used to replace existing links or in very confined spaces such as underground, urban areas or on board ships. It can be assembled, evacuated, and tested in the manufacturers' plant. Depending on its outer diameter, long lengths over 1 km can be transported to the site on a drum. For the inner tube of the cable cryostat a stainless-steel tape is formed around the superconducting cable and welded length wise to form a plain tube. After this step, the tube is helically corrugated so it is possible to coil it onto a drum. Low thermal conductivity mechanical spacers are put around the inner corrugated tube and a multilayer superinsulation is lapped on top of these spacers. The outer tube is formed around the assembly and goes again through the welding and corrugation machine to produce the outer corrugated tube. The ends of the cable cryostat are fitted with cryostat terminations filled with getter material, which holds the vacuum for years. The space between the corrugated tubes is evacuated and sealed at the manufacturers' plant. Consequently no vacuum pump is needed on the site. The long-term vacuum tightness is guaranteed by a helium leak test as a final quality control. The cable cryostat can be pulled into cable ducts or pipes, even if the superconducting cable and cable cryostat needs to be bent. The minimum bending radius depends on the cable cryostat diameter and on the superconducting cable design. Figure 6.5 shows a cryogenic envelope drum and examples of different possible installation solutions.

Figure 6.5 Flexible cryogenic envelope installations.

6.6 Cable terminations and joints

6.6.1 Terminations

High-voltage terminations sometimes referred to as bushings connect the superconducting cable ends to the grid, to the substations, or to electrical equipment. These high-voltage terminations have a special component where very large thermal and electric gradients are located. Their design follows specific rules to both minimize the heat load on the cryogenic machine and transition the high voltage from ambient to cryogenic temperatures. Dimensions (height, diameter, etc.) and design are adapted to voltage and current parameters (see Figure 6.6(a) and (b)). Generally the cooling system inlet/outlet is connected or integrated to the termination in association with a special cryogenic circulation pump or a cryofan. Most of the heat load in the termination originates from the current leads that span from the warm ambient electrical power source to the cold cryogenic superconducting cable. The termination cryostat around the current leads is compact and thermal losses are kept as low as possible.

The current lead design is a compromise between a large cross-section for limiting joule losses and a small cross-section to reduce heat flux from thermal conduction into the system. For a superconducting cable system operating at 70–77 K, they are manufactured from conventional resistive material such as copper or aluminum alloys (brass, bronze, etc.) and connected directly to the superconducting cable conductor. For an optimized design, the current lead heat load is expected to be in the range of 45 W/kA when energized (McFee R, 1959). This heat load can be reduced to 34 W/kA if the current leads are actively cooled with cold gas circulating around the lead. When not energized the residual heat load due to thermal conduction is expected to be 28 W/kA.

For a superconducting cable system operating at a lower temperature, for example, at 20 K, the termination losses are reduced with a hybrid current lead designed in two parts where the total heat flux is shared at two temperatures. The upper (nonsuperconducting) current lead injects the current from room temperature into an intermediate chamber typically operating at $T_i \sim 77–80$ K, while the lower superconducting current lead projects it from this chamber toward the superconducting cable at 20 K, for example. Most of the heat load from the normal conducting portion of the current lead is transferred at $T_i \sim 77$ K. The design of this upper lead is similar to that for HTS cable system operating at 70–77 K. Its heat load is expected in the range of 45 W/kA when energized. Below 77 K, HTS wires can be used to suppress joule losses of the current lead. In addition, the design of the lower current lead can be optimized to significantly reduce the thermal conduction by the integration of low conduction superconducting tapes such as special Ag–Au matrix 1G tapes or stabilizer-free 2G tapes. Consequently, the current lead losses of a lower part at 20 K are expected to be in the range of 2 W/kA. However, for this design either an injection of liquid N_2 directly into the intermediate chamber or a thermal connection from the lower end of the upper current lead to a cold head of an auxiliary cooling system is required. The extra investment for this cooling machine and its associated operation costs are lower than if the heat load would be totally transferred at 20 K.

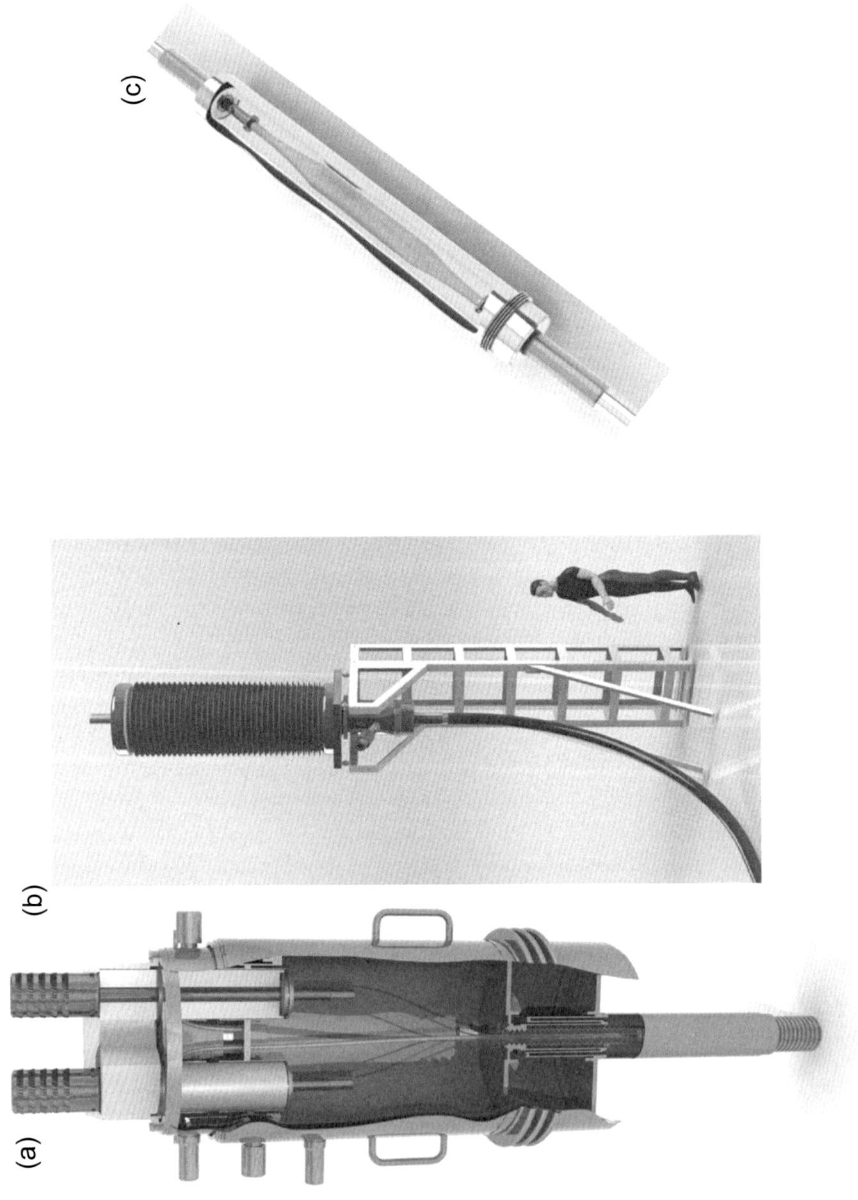

Figure 6.6 Bipolar low-voltage (a) and monopolar high-voltage (b) terminations and joints (c).

6.6.2 Joints

For multikilometric power links, the superconducting cable is produced in several shorter lengths that will need to be electrically connected during the installation. Figure 6.6(c) illustrates the concept for this connection process. Very low electrical resistance $<0.1\ \mu\Omega$ (77 K) joints or splices have been developed and consequently limits the heat load on the cooling system. The concept for the electrical field management is similar to that for conventional oil-impregnated cable joints but using cryogenic liquid or gas instead of oil. In addition, a special cryostat will have to be built around the joint to ensure the circulation of cryogenic fluid. Several prototypes have already been manufactured and tested and qualified through a type test procedure.

6.7 Cryogenic machine

A cryogenic machine is required to cool the superconducting cable system before the initial operation and to counteract the heat generation during the operation. The machine also circulates the cooling medium through the cable cryostat in order to maintain the operation temperature throughout the length of the superconducting cable. A uniform temperature is achieved using a heat exchanger connected to a cryo-refrigerator. To improve heat exchange, the cooling fluid is pressurized typically to 5−10 bar for liquid and 20 bar for gas. A cryofan for He gas or a circulation pump for liquid N_2 regulates the pressure and the circulation of the cooling fluid. The turbines can contribute significantly to the system heat losses especially for multikilometric links where the pressure drop of the cryogenic fluid can be significant.

Cryo-refrigeration is provided with maximal efficiency, when the refrigeration system is operated in the "Carnot cycle," or perfect "isentropic" cycle. Under this theoretical condition, 2.9 W at room temperature is necessary to provide 1 W of cooling at 77 K and 14 W of room temperature electrical power is necessary to provide for 1 W cooling at 20 K. However the "ideal" Carnot cycle is not technically directly applicable as an unrealistic pressure ratio exists between the warm and the cold temperature stages of the cycle. To overcome this issue, commercially available cooling systems are based on either an isobaric cycle (Ericsson cycle is implemented in Gifford McMahon refrigerators) or an isochoric cycle system (Stirling cycle is implemented in Stirling refrigerators) or a combination of isobaric and isentropic cycle systems (Brayton cycle is implemented in Brayton refrigerators). The "real" efficiency of these cooling machines increased with the cold power delivered from 8% for 50 W to 30% for 10 kW of Carnot cycle efficiency, depending on the cold power delivered. Work is underway to increase the efficiency of such machines.

Commercial cryogenic refrigerators are available with a large range of cooling powers and temperatures. Gifford McMahon and Stirling machines are commercially available for typical cooling power below 5 kW at 77 K. Brayton machines can provide for larger cooling powers (http://www.cryomech.com, 2013, http://www.stirlingcryogenics.com, 2013, http://www.airliquideadvancedtechnologies.com, 2013). The cryogenic refrigerator technology selection depends not only on the cooling power required but also on the installation constraints such as maintenance access, footprint, weight, fluid connection, etc. For an HTS cable operating at 70–77 K, one economic alternative is to use liquid N_2 delivered to storage tanks on the site as a primary source of cold power at 77 K, which is used to cool down a pressurized secondary loop of liquid nitrogen through a pumped bath heat exchanger at 65–70 K.

6.8 DC superconductive cable system configurations

Depending on the required voltage, three options are available for DC superconducting cable designs to meet the different specifications and applications. These DC superconducting cable options are:

- two coaxial poles,
- two twisted poles in one envelope,
- one pole per envelope.

The advantages of the various options are presented in Table 6.3. The power transmission efficiency of each design is discussed below.

Figure 6.7 is a schematic view of a possible installation of a superconducting DC link in close loop. This sketch presents two single pole DC transmission cables for bulk power transfer on several kilometers with one pole per cryogenic envelope including electrical joints to extend the power link. In this example, one pole (+) is used to transmit the "cold" cryogen fluid in one direction and one pole (−) is the return path to the cryogenic machine, thus forming a continuous cooling circuit. The temperature of the system is maintained by adjusting the flow rate of the cooling fluid.

For very long superconducting cable systems, several cooling stations and/or pumps could be installed from place to place along the superconducting cable to cool down and to repressurize the fluid. For a safe utilization, the maximum distance between two cryogenic stations is estimated at 5 km based on pressure drop and flow calculations.

6.9 Power dissipation sources in the superconducting system

A superconducting DC cable exhibits a virtual zero electrical resistance when operating well below the superconducting critical surface. In fact, it shows the principle electrical resistance in a range of few μΩ for the overall link associated with terminations and joints. Consequently, a high efficiency of power transmission >99% is expected. However, some energy is also consumed to operate the cryogenic system elements (turbines and cooling machines) that must be accounted for in the design.

The optimization of the overall heat load is of considerable importance for the design of cost-effective systems. The investment and operational costs (electrical

Table 6.3 Possible cable designs for DC power transport and distribution

<table>
<tr><th>Systems</th><th>Two coaxial poles</th><th>Two twisted poles in one envelope</th><th>One pole per envelope</th></tr>
<tr><td>Designs</td><td>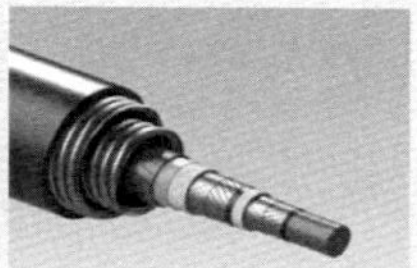</td><td>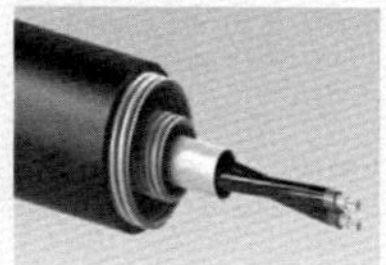</td><td></td></tr>
<tr><td>Voltage</td><td>Medium voltage <20 kV for distribution grid</td><td>Medium voltage <90 kV for distribution grid</td><td>Medium to high voltage > 90 kV</td></tr>
<tr><td></td><td colspan="2">Very compact cables</td><td>Very bulk power transferred > 5 GW</td></tr>
<tr><td>Benefits</td><td>No magnetic and thermal impacts
Common terminations and joints for both poles</td><td>No thermal impact
Common terminations and joints for both pole</td><td>No thermal impact
Maximum unit length</td></tr>
</table>

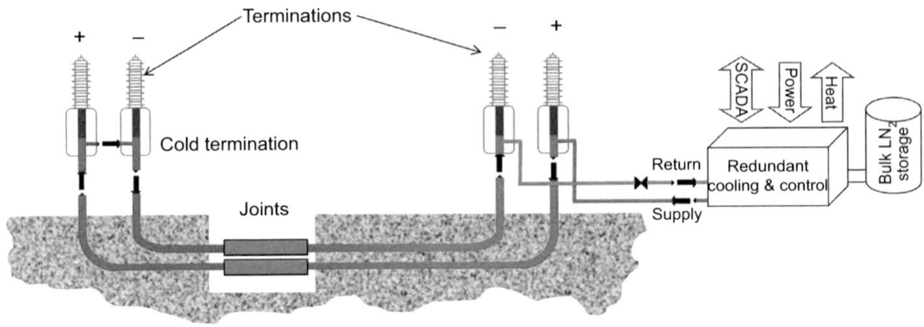

Figure 6.7 Overall DC superconducting system architecture.

and fluid consumption, maintenance operations, etc.) of the cooling machine are directly proportional to the cold power to deliver. The following paragraph analyzes the different heat sources and gives a first quantification of their respective contributions.

The overall heat load at cryogenic temperature on a superconducting DC cable system is due to four main sources:

- the heat leak through the cryogenic envelope,
- thermal conduction and joule dissipation through the current leads and joints, termination cryostats, and associated cryofans,
- heat losses due to the deviation from the nonideal superconducting material,
- AC ripples coming from the power supply.

When the DC superconducting cable is energized, the total system losses of the system can be expressed by the equation

$$W_{\mathrm{Tot}}(T_{\mathrm{op}}) = 2 \times W_{\mathrm{term}}(T_{\mathrm{op}}) + L \cdot \left(W_{\mathrm{cryost}}(T_{\mathrm{op}}) + W_{\mathrm{sc}}(T_{\mathrm{op}}) + W_{\mathrm{ACRipple}}(T_{\mathrm{op}})\right) \tag{6.2}$$

where

W_{term} is the loss through the current leads and the termination cryostat and cryofan (W);
W_{cryost} is the loss generated from the heat leak of the cable cryogenic envelope (W/m);
W_{sc} loss generated from nonideal superconducting material (W/m);
W_{ACRipple} is the loss generated by the harmonic currents and voltages (W/m);
L is the length of the cable system (m); and
T_{op} is the operating temperature (K).

The termination losses (W_{term}) are highly dependent on the rated transmission current and from the size of the cryostat built around them. The heat load from the cable cryogenic envelope (W_{cryost}) is independent of the rated current and depends only on the cryostat diameter. The superconducting residual resistance losses (W_{sc}) are dependent on the current derating factor and of the superconducting wire quality. The voltage and current AC ripple losses (W_{ACRipple}) are directly dependent on the quality of the power source.

When no power is transferred, the temperature of the overall system must be maintained to keep the DC superconducting cable system ready for power transmission. The heat inlet of the cryogenic envelope, the termination and joint cryostats, the cryofan, and the thermal conduction through the current leads must be compensated. The estimation carried out within this work considers only the maximum heat losses in operation at steady state and at maximum current.

6.9.1 Cryogenic envelope losses

As presented in the previous paragraph, heat leaks in modern flexible cryogenic envelopes depend on the cryostat diameter. The heat inlet is distributed along the cable and is dependent on the length of the cable. Table 6.2 gives the estimated heat leak in W/m of two-wall and four-wall cryogenic envelopes according to their diameters.

6.9.2 Terminations and joint losses

As terminations are only located at the ends of the DC superconducting cable system their heat inlet is not distributed along the cable. Hence, only contributions from cryostats and current leads are considered. The losses of the termination cryostat depend on its size, varying from 50 W for low-voltage systems to 100 W for very high-voltage systems. These values are far lower than what is common for three-phase AC terminations that have to contend not only with additional AC loss in the termination but also with the increased number of warm-to-cold currents leads required for all three phases and the return shield. Termination losses in three-phase AC bushings rated at >3 kA/13 kV can exceed 300−500 W at 77 K.

The contribution of the current lead is expected to be 45 W/kA at 77 K when energized at the maximum designed current. In the case of overcurrent, significantly higher losses can be expected. This situation can be accepted for a short duration and should be designed on a case by case basis. For DC superconducting cable systems operating below 20 K, it is beneficial to use hybrid current leads. In this case, the additional losses from this low-temperature segment are around 2 W/kA at 20 K when energized.

The joint losses are mainly thermal losses of the joint cryostat. They are slightly higher than the cable cryostat losses due to its larger diameter and from the connections to the cable cryostat. The expected losses are estimated to be less than 10 W per joint cryostat. According to the DC superconducting cable core design, an electrical joint resistance ranging from about 0.1 to 0.5 μΩ corresponds to a maximum heat generation of 50 W for a 10 kA superconducting cable. As there is a considerable distance between the joints their thermal impact can be considered as negligible when compared to the contribution of other parts of the DC superconducting cable system.

6.9.3 Losses from nonideal superconducting material

In practice, superconducting materials are nonideal and have microstructural defects that locally impact their current transport properties. Constant dissipation per meter of nonideal superconductors with critical current I_c, carrying a steady-state DC operating current I_{op} is expressed by the law (Eqn (6.3))

$$Ws_{sc}(T, B) = I_{op}E_c \left(\frac{I_{op}}{I_c\,(T, B)}\right)^n \tag{6.3}$$

where

n is the superconducting transition index or "n value;"
T is the absolute temperature;
B is the magnetic field experience by the superconductor; and
E_c is the electrical filed defining the critical current (100 μV/m).

The transition index or n value is most commonly found by plotting the slope of the log E versus log I curve and taking its slope near the E_c criterion. The power losses are estimated to be less than 10^{-3} W/m for a typical n value of 20 for YBCO, Bi2233 tapes, and MgB_2 and with a derating factor $I_{op}/I_c = 0.75$. It can be considered as negligible with regard to the cryogenic envelope. Note, these DC losses from nonideal superconducting behavior are far lower than what is typically experienced by AC superconducting cables, which can have losses ranging from 0.3 W up to several W/m depending on the AC superconducting cable design. Thus, the use of DC superconducting cables has a significant advantage in terms of overall efficiency over AC superconducting cables as the superconducting cable length increases.

6.10 Power losses from AC ripples

AC ripples are generated by the power source and the converters. The origin of these ripples is quick variation of current and voltage and can be generated at high frequency according to the switching frequency of the converter. They can be fully or partially eliminated with filters. Their amplitudes and frequencies are dependent and defined by the general specifications of the energized electric equipment. To quantify the AC ripples contribution to the dissipated power in the system, we have selected 3, 6, and 9% ripple full amplitudes at two frequencies of 1440–10,080 Hz (typical harmonics for multiphase generators).

6.10.1 Voltage AC ripples dielectric losses

Voltage AC ripples can generate significant power losses especially for the cold dielectric cable design where the heat generation is at cryogenic temperature. The dielectric losses per meter are expressed as Eqn (6.4)

$$W_{\mathrm{VACRipple}} = 2\pi f \tan\delta\, C\Delta U^2 \tag{6.4}$$

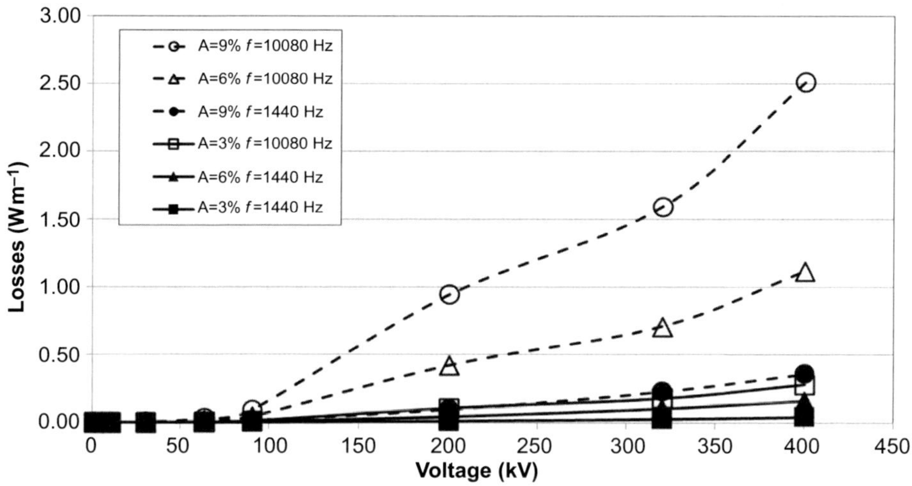

Figure 6.8 Dielectric losses from AC voltage ripples.

where

f is frequency (Hz);
ΔU is ½ full amplitude of voltage ripples at a given frequency (V);
δ is the loss angle from the insulation material (°), which can be dependent on frequency; and
C is the capacity of the cable per meter (F/m) depending on the cable design.

For impregnated lapped insulation materials the loss tangent or tan δ, which is a ratio of the "real" loss divided by the reactive power in the dielectric material, is $\sim 10^{-3}$ and is considered as constant with the frequency. The capacitance per unit length for superconducting cables ranges from about 10^{-11} F/m for low-voltage cables to 1.2 10^{-10} F/m for high-voltage cables. The dielectric losses in the insulation are plotted in Figure 6.8 as a function of rated DC voltage for different ripple amplitudes and frequencies.

Below 100 kV, the additional losses due to voltage ripples are lower than 0.1 W/m even at "high" frequencies up to 10 kHz with a large $\Delta U = 4.5\%$ corresponding to a 9% ripple full amplitude. Consequently, the contribution of voltage ripples to the overall system losses can be considered as negligible with regard to the cable cryogenic envelope for low- and medium-voltage power cables. For superconducting cables for high-voltage DC (HVDC) links rated above 100 kV, the voltage ripples should be limited to 6% full amplitude at frequencies up to 10 kHz. The use of filters is recommended.

6.10.2 Current AC ripples losses

AC ripples from the power source in turn induce AC currents in the DC superconducting cable. The AC currents in the superconducting cable generate hysteretic AC losses in the superconducting material and also eddy currents into normal metallic parts of the cable. These AC losses can be decreased but not fully eliminated by physical or geometrical means (wire and cable designs). Eddy currents in the normal metallic components are dependent on the specific cable design. Their contribution to the power dissipated is not considered in this chapter. However, it should be checked especially for high current systems that use low-quality power supplies or converters that can generate high-amplitude ripples at high frequencies.

The hysteresis AC losses in a superconducting material can be evaluated as a first approach using the Norris formula (Norris, 1970), Eqn (6.5), where the peak current does not exceed the critical current. This model is probably overestimating the power dissipation but it gives a good first approximation.

$$W_{\mathrm{IACRipple}} = \frac{\mu_0 f I_c^2(T)}{\pi} f(i) \tag{6.5}$$

Here
μ_o is the vacuum permeability (T m/A);
f is the frequency of the ripple (Hz);
$I_c(T)$ is the critical current of the cable (A);
T is the operating temperature; and
$s(i)$ is the shape factor.

For an elliptical shape this factor $s(i)$ is given by Eqn (6.6)

$$s(i) = (1 - i)\ln(1 - i) + (2 - i)\frac{i}{2} \tag{6.6}$$

where $i = \Delta i/2I_c$ and Δi is the full amplitude of the ripple in A.

The operating current I_{op} of the superconducting cable is considered at 70% of the critical current I_c. The losses in W/m at operating temperature is listed in Table 6.4 for a $\Delta I = 3$, 6, and 9% full amplitude ripples at two different frequencies 1440 and 10,080 Hz for currents from 1 to 10 kA. The current ripple losses can be considered as negligible with regard to the cryogenic envelope losses for operating currents less than 7.5 kA. For higher currents the impact of current ripples needs to be considered in the thermal analysis. Our model shows that the losses in high current systems increase dramatically when the ripple amplitude is higher than 6%. It is important to verify that the ripple amplitude is below the estimated 6% threshold for frequencies higher than 5 kHz.

A superconducting DC power link can accept AC ripples. A qualitative analysis of the losses for various applications is presented in Figure 6.9. It emphasizes the situation when special attention should be paid to the power quality and when their impact can be negligible. For high frequencies and amplitudes the contribution of AC ripples on the system heat load is acceptable with the existing cooling system. For high-frequency and amplitude AC ripples the existing cooling system should be able to account for the losses encountered. In any case, it may be beneficial to reduce

Table 6.4 Current AC ripple hysteresis losses in W/m calculated with elliptic shape Norris model

Current ripple amplitude	Frequency (Hz)	2 kA	3 kA	4 kA	5 kA	7.5 kA	10 kA
3%	1440	Negligible	Negligible	Negligible	Negligible	Negligible	Negligible
	10,080	Negligible	Negligible	Negligible	Negligible	0.01	0.02
6%	1440	Negligible	Negligible	Negligible	Negligible	0.01	0.02
	10,080	Negligible	Negligible	0.02	0.04	0.06	0.16
9%	1440	Negligible	Negligible	Negligible	0.015	0.03	0.11
	10,080	0.02	0.03	0.05	0.1	0.21	0.75

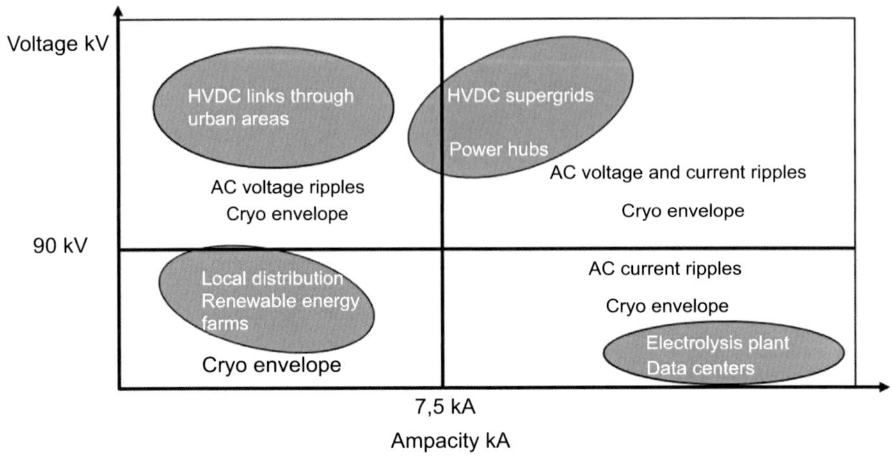

Figure 6.9 Domains of application (in blue) and main sources of losses (in black) as a function of current and voltage in superconducting cable system.

high-frequency ripples with filters so long as costs are acceptable. If not, consideration should be given to the extra investments that may be required for the cooling machines.

6.11 Comparing power dissipation in a DC superconducting system to a conventional system

Based on the proposed Eqn (6.1), this paragraph discusses and compares the power dissipation of a DC superconducting system to a resistive conventional system at room temperature for the two generic applications: a short length low/medium-voltage system and a long length high-voltage system.

In order to make an equitable comparison of the power dissipation between the DC superconducting cable system operating at cryogenic temperature and the resistive conventional system operating at room temperature, the dissipation experience by the DC superconducting cable must be translated to its room temperature equivalent. This translation is performed via the "ideal" Carnot efficiencies and "real" refrigeration cycles introduced in Section 6.7. To estimate the power consumption of the two different cable links at room temperature, the calculated heat load at cryogenic temperature of the DC superconducting cable must be multiplied by the cooling system efficiency ratio, which is dependent on the temperature and the cold power required. Based on commercially available cryogenic refrigeration technologies, the following ratios have been considered in this comparison: 25 W_{RT}/W at 70 K for DC links shorter than 1 km, 15 W_{RT}/W at 70 K for long DC links of several kilometers, and 200 W_{RT}/W at 20 K for MgB_2 DC links shorter than 1 km. Note, the enormous impact raising the temperature from just 20 to 70 K has on reducing the room temperature power consumption (i.e., nearly an order of magnitude) of the cryogenic refrigerator. This power dissipation is compared to copper cable losses at 80°C, the maximum temperature acceptable when polyethylene is used for voltage insulation. In a single pole conventional copper cable transferring 1 kA, the joule losses generated are 23 and 46 W/m for a current density of 1 and 2 A/mm^2, respectively. These joule losses are linearly dependent on the transferred current. For an easy comparison and to show the trends, the principle comparator for this study is the power dissipated per kA and per m expressed as W/kA m.

6.11.1 Low- and medium-voltage systems

The typical characteristics of superconducting DC coaxial dipole cable is given in Table 6.5 for the following configurations:

- HTS cable in a two-wall cryogenic envelope at 70 K
- MgB_2 cable in a two-wall cryogenic envelope at 20 K
- MgB_2 cable in a four-wall cryogenic envelope at 70 K (shield) and 20 K.

For the three proposed designs, the total system losses have been calculated. To calculate the overall heat load for low-voltage cables of several hundred meters long, the cryofan losses are assumed to be 10 W for both liquid N_2 (70 K) or He gas (20 K). The influences on power dissipation as a function of the cable length, of its ampacity, and of the impact of ripples have been quantified.

The most relevant trends are presented in Figures 6.10–6.12 and compared with conventional cable performances. From these figures, the following conclusions can be drawn:

- For cables length less than 1 km, the power distribution for a low-voltage superconducting cables is 2–5 times more efficient than with a conventional cable.
- Due to higher efficiency of the cooling machines, HTS cables operating at 70 K are more efficient than MgB_2 cables operating at 20 K. However, performances of MgB_2 cables with four-wall cryogenic envelopes are really close in cryogenic efficiency to those of HTS cables. Although a more complicated cryogenic system is required, this additional cost is offset by the lower cost of MgB_2 wires.
- Most of the heat load is generated in the current lead terminations that transmit the room temperature power to the cold superconducting cable. To be more efficient than a conventional resistive cable system, a superconducting system must not be too short. A DC superconducting coaxial cable system should exceed 50 m to be more efficient than a conventional system. There is a significant number of interesting business cases.
- The efficiency of superconducting DC cables increases with the cable ampacity. The very compact cable MgB_2 cable in a two-wall cryostat has a low efficiency. However when very high power transfers at low voltage are mandatory in a very limited space, for example, electrical ship or aircraft, this solution should be considered.
- For cable systems with strong AC current ripples (>6%) at high frequency (>5 kHz), a superconducting cable is more efficient than a conventional resistive cable. All the envisioned cable systems can accept current ripples. For MgB_2 cables when transfer current is very high (>7.5 kA) and strong current ripples (>6%) at high frequency (>5 kHz) are present, the ripple contribution can be up to half of the overall system losses. When strong

Table 6.5 Cryogenic envelope parameters for coaxial dipole DC cable systems for low and medium voltage including two poles and fluid return to the cooling machine

	Ampacity			
Cable parameters	2 × 3 kA	2 × 5 kA	2 × 7.5 kA	2 × 10 kA
HTS coaxial cable diameter (mm)	40	43	45	46
HTS cable two-wall cryostat diameter ID/OD^{2} (mm)	75/125			
Losses at 70 K (W/m)	1, 4			
MgB_2 coaxial cable diameter (mm)	8	11	15	16
MgB_2 cable two-wall cryostat diameter ID/OD[a] (mm)	21/44	30/55	39/66	
Losses at 20 K (W/m)	0.6	0.8	1	
MgB_2 cable four wall cryostat diameter ID/OD[a] (mm)	21/44 + 60/110	30/55 + 75/125	39/66 + 90/147	
Losses at 20 K (W/m)	0.07	0.09	0.12	
Losses at 70 K (W/m)	1.2	1.4	1.5	

[a]ID, inner diameter; and OD, outer diameter.

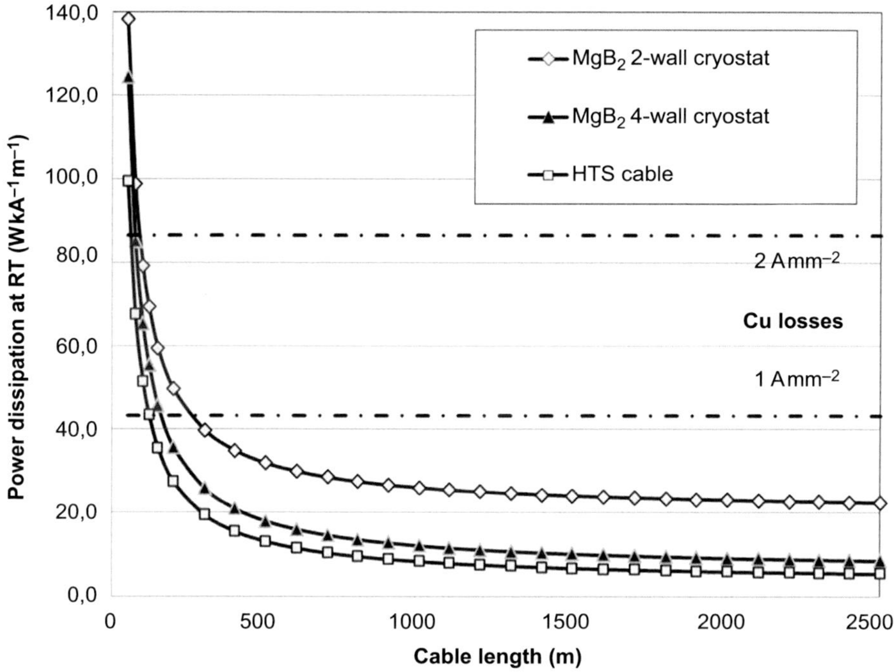

Figure 6.10 Influence of the link length on power dissipation for low- and medium-voltage coaxial systems transferring 10 kA.

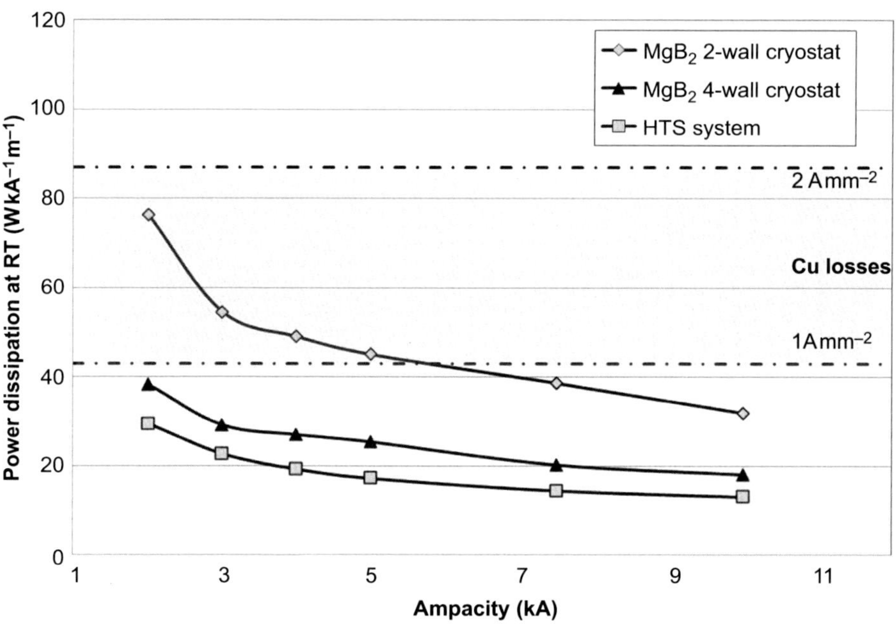

Figure 6.11 Influence of cable ampacity on power dissipation at room temperature of 500-m-long low- and medium-voltage coaxial systems.

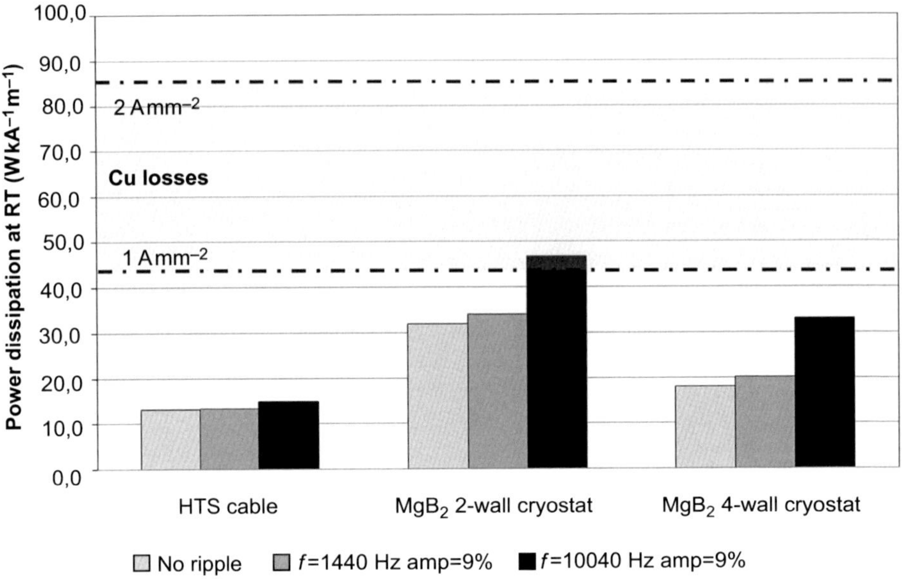

Figure 6.12 Impact of the ripples on power dissipation at room temperature for 500-m-long low/medium-voltage coaxial systems transferring 10 kA.

current ripples are present, significant investments are required for the 20 K cooling system and the cryofans. These high-frequency AC current ripples should be filtered to significantly reduce the cryogenic heat load.

6.11.2 High-voltage systems

Typical characteristics for superconducting HVDC single pole cable are given in Table 6.6 for three different DC high voltages: 200, 320, and 400 kV. As HVDC technology is used for long-range links, the results presented below are limited to superconducting cables built with HTS tapes and cooled with liquid N_2. To calculate the overall heat load of HVDC cable of several kilometers long, the cryofan losses are assumed to be 750 W at 70 K. The influences on power dissipation as a function of the cable length, ampacity, and impact of the ripples have been quantified.

Figure 6.13 shows the influence of ampacity of a DC superconducting cable on the calculated power dissipation at room temperature.

As an example a 4 kA DC system has been selected to illustrate the impact on power dissipation from the voltage and AC current ripples at two frequencies for the three HVDC systems (Figure 6.14). The conclusions can be extrapolated for other DC superconducting cable system designs and ampacities.

From these figures, the following conclusions can be drawn.

- For long length HVDC cable systems, the (room temperature equivalent) power dissipated by superconducting cables is estimated at about 2.5 W/kA m for a 10 kA system, approximately 10 times lower than conventional resistive cables where 23 W/kA m is consumed.

Table 6.6 Cryogenic envelope parameters single pole HVDC cable systems

Cable parameters	Voltage		
	200 kV	320 kV	400 kV
HTS single pole diameter for 3–10 kA transferred currents (mm)	57–60	72–74	81–84
HTS cable two-wall cryostat diameter ID/OD[a] (mm)	90/147	100/155	110/170
Losses at 70 K (W/m)	1.5	1.6	1.7

[a]ID, inner diameter; and OD, outer diameter.

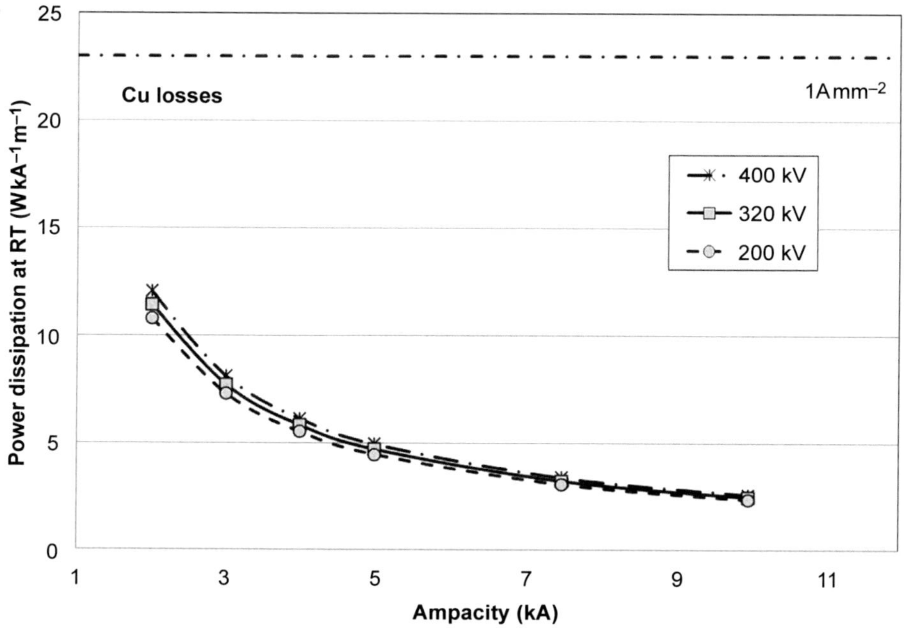

Figure 6.13 Influence of cable ampacity on power dissipation at room temperature of 5000-m-long HVDC superconducting single pole systems at 200, 320, and 400 kV (no ripples).

- Most of the power dissipation does not stem from the superconducting cable itself but instead is to compensate the cryogenic envelope heat inlet. Developments are undertaken to significantly reduce the cable cryostat losses to 0.4–0.5 W/m (Stemmle, Marzahn, West, Schmidt, & Schippl, 2012).
- Power dissipation in the DC superconducting cable systems is independent of voltage when no AC current ripple is produced by the converters
- High-frequency voltage ripples contribute significantly to total system losses especially at 400 kV where the power dissipation can increase from 2.6 to 6.5 W/kA m. For these systems, filters are recommended.

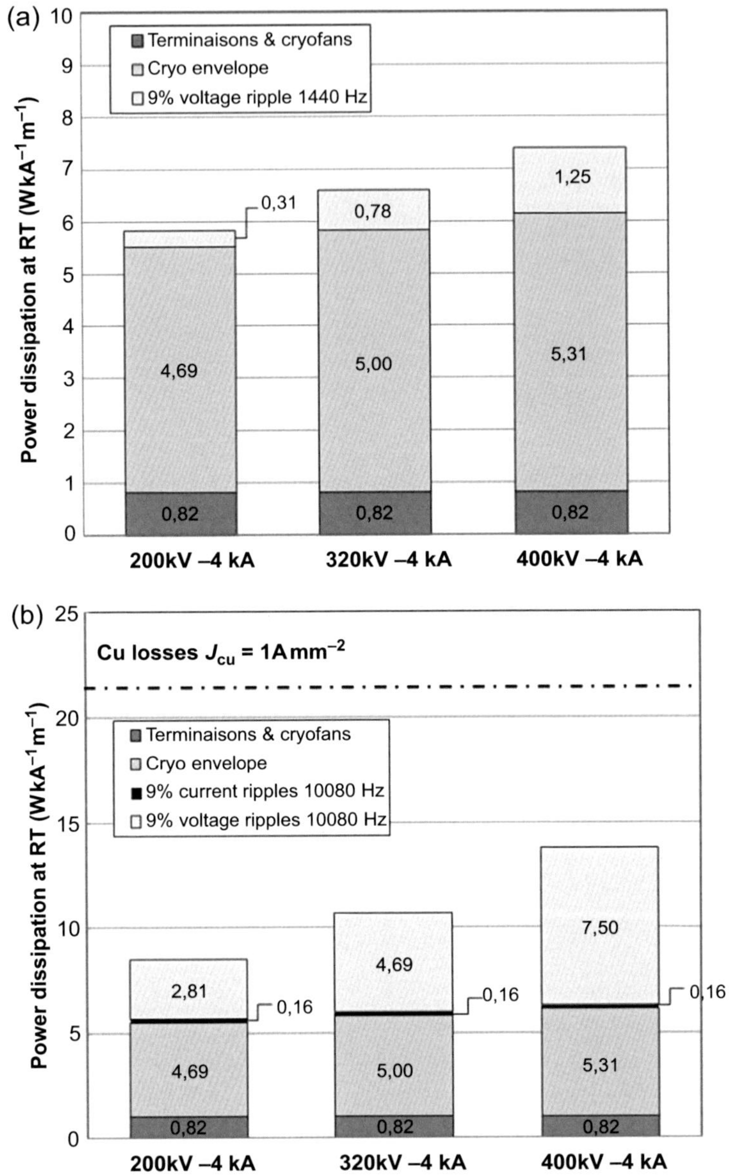

Figure 6.14 Power dissipation at room temperature in 5000-m-long HVDC superconducting single pole systems transferring 4 kA at 200, 320, and 400 kV (a) with medium frequency ripples (b) with high-frequency ripples.

6.11.3 Other environment impacts

Beside low power dissipation discussed previously, a superconducting system offers other benefits in particular to the environment. The cryogenic envelope totally eliminates heat generation outside the cable. It can be laid in confined or closed environments such as undergrounds, basements, or in hold of aircrafts or ships. No chiller or air conditioning system is needed to cool the environment around a superconducting cable to keep it operational.

This absence of heat emission combined with a very compact design (i.e., very high energy per unit volume) enables the superconducting cable system to fit into a smaller area than a conventional system. To illustrate, Figure 6.15 compares the installation space of a conventional system (left) and a superconducting (right) bipolar system transferring 2.5 GW (Stemmle et al., 2012). A picture of a HVDC superconducting cable sample is shown in Figure 6.16(a). To transfer this power only two single pole superconducting cables are needed compared to four single pole conventional cables. In addition, a space between resistive cables is required to maintain the electrical insulation under 80°C. No spacing is needed for a superconducting cable. Consequently, the trench for superconducting cables is 2.5 times smaller than for conventional cables. This can be significant in urban areas.

In addition, the electromagnetic radiation created by high ampacity conductors can be totally eliminated with a coaxial design of DC superconducting cable. No special protection is required. It can be installed very close to electrical equipment or cabinets without electromagnetic interference to the electronic components.

Lastly, the overall weight of a superconducting cable system including cryogenic envelope and cooling fluid is up to 35 times lighter than that of a conventional system. Consequently, there is a reduction of raw materials, energy, and associated greenhouse gas emissions needed for the manufacturing processes.

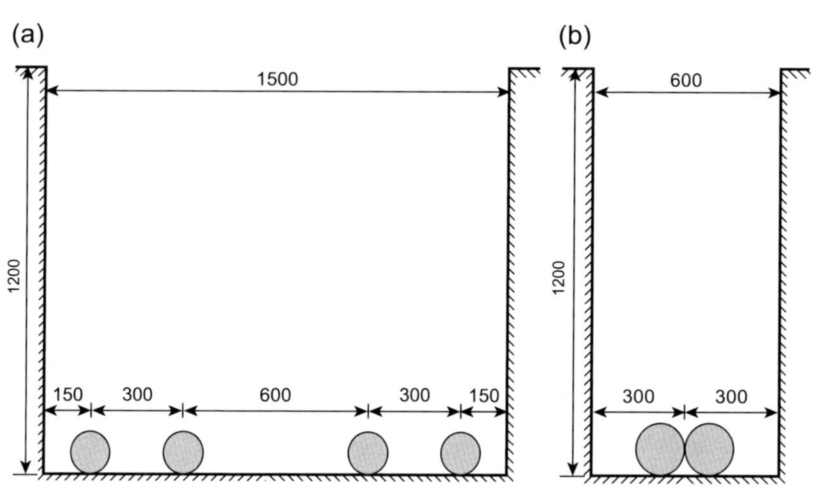

Figure 6.15 Installation footprint of a 4 kA–320 kV conventional system (a) and superconducting cable (b).

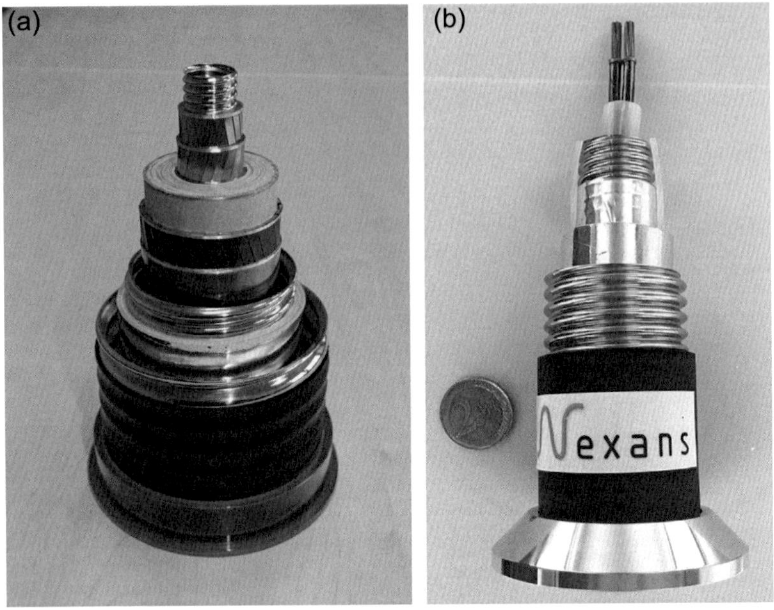

Figure 6.16 Cables for HVDC transport (a) and 20 MW LV distribution (b).

6.12 Opportunities for DC Superconducting Cables

Over the past several years there has been a significant renewal of interest for direct current cable systems. The transnational or transcontinental transport of the bulk energy in direct current, preferred for long distances, could benefit from the superconducting technologies in the future (EPRI, 2009; Xiao, 2012). Consideration is being given to distribute energy to the end user using DC technology. Furthermore, by design, photovoltaic or wind farms are required to use DC current for certain steps of their electricity generation process. DC technology is usually used in industrial and scientific infrastructures where strong or very strong currents are required for electrolysis and the reels of magnetic fields. Recently green electrical ships and aircraft designs have also been proposed using superconducting motors and generators. Up to now except for special laboratory applications, no commercial installation for superconducting HTS DC power cables has been built. However, a test platform has been built in Japan to demonstrate the potential of this technology (Yamaguchi, 2011). The most probable opportunities for the immediate future are presented below.

6.12.1 To overcome bottlenecks in power grids

Based on their many advantages, DC superconducting cable systems can be introduced to the grid to help to overcome bottlenecks. The HVDC superconducting technology can be particularly desirable in urban or suburban areas for new transmission links or for upgrading of existing grids, where the difficulties to carry out civil works or to obtain an authorization could slow or even stop the deployment of a power grid. This situation is becoming more common in densely populated countries. Furthermore, the very high ampacity of DC superconducting cables makes it possible to increase the transferred power by increasing current while maintaining the voltage. Consequently, as it may be possible to use the existing rights of way, it eases the general public's acceptance.

6.12.2 To energize power hubs

Another potential application of superconducting HVDC power cables could be its introduction to large power hubs located at key grid nodes to transfer high bulk power and to resynchronize several adjacent and asynchronous AC areas. In such a system, DC superconducting cables can transfer tens of gigawatts from and to several voltage source converters. Their underground nature also significantly enhances system security and reliability (http://www.tresamigasllc.com/technology.php, 2013).

6.12.3 For industrial and transport applications

DC Superconducting cables can be beneficial to many applications for local distribution where space is limited and where low voltage is mandatory such as off-shore oil and gas platforms, electrical ships, or future electrical aircrafts or railway grids (Tomita, Suzuki, Fukumoto, Ishihara, & Muralidhar, 2011). Superconducting DC cables are also a potential solution for reducing the environmental footprint in industrial installations such as electrolysis or smelting plants (Xiao, 2012). Finally thanks to the unique possibility of transporting huge currents with limited power dissipation, it also facilitates the local introduction of renewable energies (solar panels, wind mills, etc.) generated in direct current that can be stored directly in batteries or connected to large central substations.

Beside their low power dissipation and their reduced environmental impact, a DC superconducting cable offers a significant reduction of space and weight that makes it possible to install new electrotechnical equipment at locations not reachable with conventional technologies. It could also reduce installation costs. As an illustration of this last application, the distribution of high power at low voltage <3.3 kV in direct current can be desirable for maritime applications for propulsion. It significantly enhances electrical safety and reduces the investment required by the elimination of medium-voltage transformers and converters. For a coaxial DC superconducting cable system, it does not require any special installations such as a magnetic screening. The absence of thermal impact eliminates the need for air conditioning systems. Table 6.7 and Figures 6.16(b) and 6.17 show the typical general specifications and footprints for a 20 MW (2×10 kA/1000 V) superconducting cable system operating at 20 K made with MgB_2 wires. As a reference for an equivalent power, a conventional resistive copper system operating with a current density of 1.5 A/mm will use 50 cables (25 for each polarity) of 20 mm diameter each for an overall weight of about 20 tons for a 100-m-long cable. A superconducting system is 25 times lighter and needs 12 times less space than the conventional resistive system.

Table 6.7 Typical specification for 100-m-long 2 × 10 MW MgB_2 coaxial cable operating at 20 K

System parameters	Specified values
Diameter of cable core	16 mm
Dimensions of cryostat	ID[a] 39 mm, OD[a] 66 mm
Weight of cable	30 kg
Weight of cable cryostat	170 kg
Bending radius	1100 mm (multiple bending) and 550 mm (single bending)
Volume of coolant	100 L helium at 20 bar
Cool down time	<2 days
Cable and cryostat losses at 20 K	~120 W
Termination losses at 77 K (RT)	~2000 W
Room temperature power consumption	~74 kW
Two cable bipolar terminations	Dimension, ∅ 0.3 × 1.5 m; weight, 50 kg per termination
Cryogenic machine with redundancy	Footprint, 1 m^2; weight, 500 kg

[a]ID, inner diameter; and OD, outer diameter.

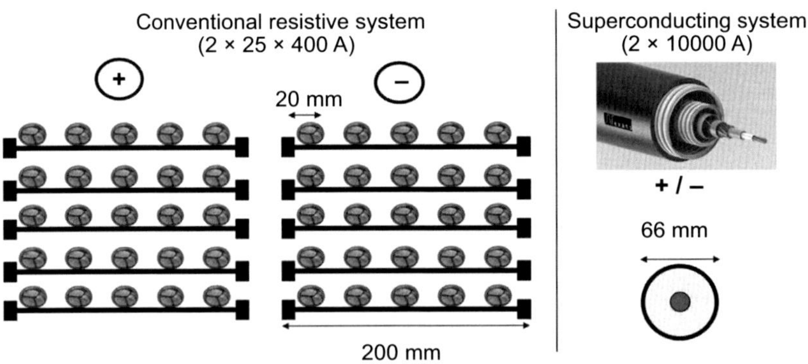

Figure 6.17 Comparison a conventional and 20 MW superconducting low-voltage coaxial DC system.

Although already profitable in some niche projects, the obstacles for the broad acceptance of this technology are still highly dependent on cost reductions for cable components (superconducting wire, cryostat, and cooling stations). R&D programs are ongoing to improve the progress toward large-scale industrialization and will soon achieve commercial objectives. In parallel to these technical tasks, other actions such as operational safety must be demonstrated before a full acceptance by end users and regulators is achieved.

6.13 Conclusions

With the existing technologies, the transmission of bulk power becomes possible with superconducting DC power cables. In contrast with conventional resistive cables a cryogenic medium is required for the superconducting system to keep the operating temperature of the system below the critical temperature of the superconducting materials. As long as this condition is met, a DC superconducting cable does not present any ohmic losses.

Superconducting tapes and wires are commercially available and can be used to build up a superconducting cable conductor (Bi-2223 or YBCO-based tapes or MgB_2-based cylindrical wire). Bi-2223 or YBCO are still expensive but they are superconducting at 77 K and therefore they can be operated with low-cost liquid nitrogen. Low-price MgB_2 must be cooled at 20–25 K in helium gas to exhibit the superconducting properties. For a high ampacity cable conductor (>5 kA), the extra cost for the cooling system is compensated by the very low price of MgB_2 wires.

The special thermal conditions required to operate superconducting cables require special accessories and unconventional pieces of equipment such as cryogenic terminations, cryogenic envelopes, cooling systems, and circulating pumps. Their specificities have been presented in this paper. Cryogenic accessories have been designed to cover a wide specification, including high currents in excess of 10 kA from the low to the high voltage.

Possible designs for low/medium- and high-voltage DC superconducting cables have been presented. They offer very flexible solutions to transfer a very high power level with high ampacity. Their very high ampacity allows the transfer of power by increasing the current while maintaining the voltage level.

Direct comparison between a conventional resistive cable system and a DC superconducting cable system demonstrated that the superconducting technology offers a high environmental benefit due to its very low power consumption. For long DC links, the calculated power dissipation in a high ampacity superconducting cable requires only 10% of the consumption of conventional resistive cable independent of the voltage level. As most of the losses are generated in the terminations, a superconducting DC link should not be too short. A DC superconducting cable system must exceed 50 m, depending on ampacity, to be more efficient than a conventional system. There is a significant number of interesting business opportunities.

AC voltage and current ripples from AC to DC converters can be accepted by DC superconducting cables in most cases. However they significantly impact the heat load on the total system when they are at high frequencies >5 kHz and amplitudes >6%.

Beside their low power consumption, DC superconducting cable designs are very compact and can be installed with a very limited footprint. Consequently, in urban areas it makes public acceptance easier. There is the possibility of using the existing rights of way or even installing within already existing ducts. For on-board distribution it offers opportunities for installations of new electrotechnical equipment at locations not reachable with conventional technologies.

Environmental friendly and in some niches already profitable, DC superconducting cables will play a key role toward the reduction of energy consumption within future local distribution grids and for long-distance power links, balancing the energy flux within future power grids.

References

Ballarino, A. (2010). Design of a MgB_2 feeder system to connects groups of superconducting magnets to remote power converters. *Journal of Physics: Conference Series, 234*.

Braccini, V., Nardelli, D., Penco, R., & Grasso, G. (2007). Development of ex situ processed MgB_2 wires and their applications to magnets. *Physica C, 456*, 209–217.

Cheadle, M., Wozniak, M., Bromberg, L., Glowacki, B., Jiang, X., Zeng, R., et al. (June 2013). DC superconducting cable using MgB_2 wires. *IEEE Transactions on Applied Superconductivity, 23*(3).

EPRI. (2009). *Program on technology innovation: a superconducting DC cable*. Palo Alto, CA: EPRI, 1020458.

Foltyn, S. R. (2007). Materials science challenges for high-temperature superconducting wire. *Nature Materials, 6*, 631.

Kato, T., Fujikami, J., Kobayashi, K., Yamasaki, K., Ayai, N., Fujino, K., et al. (June 2006). Development of drastically innovative BSSCO (DI-BSCCO) wire. *SEI Review, 62*, 10–14.

Kostyuk, V., Antyukhov, I., Blagov, E., Vysotsky, V., Katorgin, B., Nosov, A., et al. (2012). Experimental hybrid power transmission line with liquid hydrogen and MgB_2-based superconducting cable. *Technical Physics Letters, 38*(3), 279–282. Pleiades Publishing Ltd.

McFee, R. (February 1959). Optimum input leads for cryogenics apparatus. *Review of Scientific Instruments, 30*(2).

Norris, W. (1970). Calculation of hysteresis losses in hard superconductors carrying ac: isolated conductors and edges of thin sheets. *Journal of Physics. D: Applied Physics, 3*, 489–507.

Stemmle, M., Marzahn, E., West, B., Schmidt, F., & Schippl, K. (2012). Superconducting HVDC power cables for voltage. In *Cigré conference*.

Tomita, M., Suzuki, K., Fukumoto, Y., Ishihara, A., & Muralidhar, M. (2011). Next generation of prototype direct current superconducting cable for railway system. *Journal of Applied Physics, 109*(063909).

Xiao, L. (June 2012). Development of a 10 kA HTS DC power cable. *IEEE Transactions on Applied Superconductivity, 22*, 580–584.

Yamaguchi, S. (November 2011). Design and construction of 200-Meter high temperature superconducting DC power cable test facility in Chubu university. In *Twenty-third international cryogenic engineering conference* (pp. 1041–1047).

High-temperature superconducting (HTS) power cables cooled by helium gas

7

S. Pamidi, C.H. Kim, L. Graber
The Florida State University, Tallahassee, FL, USA

7.1 History of superconducting cables

Research on the application of superconductors in power transmission and distribution cables has existed since the 1960s, when commercial low-temperature superconducting (LTS) wires were becoming widely available (Baylis, 1973; Chowdhuri & Laquer, 1978; Dolgosheyev et al., 1979; Edwards, 1968; Forsyth, 1972; Forsyth et al., 1973; Forsyth & Thomas, 1986; Garber, 1979; Garwin & Matisoo, 1967; Klaudy & Gerhold, 1967, 1983; Laquer, Dean, & Chowdhuri, 1977; Meyerhoff, 1971). Superconducting power cables were thought to address the increasing demand for electrical power in urban areas while reducing environmental consequences, dealing with right-of-way challenges, maintaining urban esthetics, and addressing society's unease about perceived electromagnetic interferences (Baylis, 1973; Klaudy & Gerhold, 1983).

The daunting necessity of cooling LTS cables to <10 K with helium did not deter the superconducting cable development programs of the 1960s. A variety of competing designs were explored for both alternating current (AC) and direct current (DC) superconducting cables (Baylis, 1973; Chowdhuri & Laquer, 1978; Laquer et al., 1977). Studies were conducted to analyze and quantify the technical and economic benefits of deploying long superconducting cables in electric power grids (Klaudy & Gerhold, 1983). Prototype cables made from LTS were developed and demonstrated in the 1970s and early 1980s in the United States, Europe, and Russia (Dolgosheyev et al., 1979; Forsyth & Thomas, 1986; Garber, 1979; Klaudy & Gerhold, 1983).

Interestingly, the design approaches and technical challenges of early superconducting cable developments are almost identical to the challenges faced today by the developers of high-temperature superconducting (HTS) cables (Baylis, 1973; Chowdhuri & Laquer, 1978; Dolgosheyev et al., 1979; Edwards, 1968; Forsyth, 1972; Garwin & Matisoo, 1967; Klaudy & Gerhold, 1967, 1983; Laquer et al., 1977; Meyerhoff, 1971). The cost, complexity, and inefficiencies of cryogenic equipment and the lack of suitable high-voltage dielectric materials for cryogenic applications are a few of the common obstacles faced in the 1970s by LTS cable developers and now by contemporary HTS cable developers. Notwithstanding the high costs involved and technical challenges faced, several prototype LTS cables were successfully demonstrated in the late twentieth century, and HTS cables have been introduced in recent years. The investments responsible for rapid development

Superconductors in the Power Grid. http://dx.doi.org/10.1016/B978-1-78242-029-3.00007-8

of superconducting cable technology were a result of broad public support and government interest in the development of clean, reliable, and efficient electrical power infrastructure. The projected transition of electricity production to include distributed solar and wind energy sources typically located in remote locations will necessitate moving large quantities of power from the sources to the consumers in urban areas. HTS cables will be suitable to tackle the challenge with low losses.

The majority of HTS cable demonstrations have been in urban areas with limited space for expansion of power cables, using three-phase AC cables with varying voltage ratings (Dai et al., 2014; Landgraf, 2014; Maguire et al., 2011; Ryu et al., October 2013; Weber et al., 2007; Xiao, Dai, Lin, Zhang, & Zhan, 2013; Yumura et al., 2013). The lengths of the HTS cable segments demonstrated to date have been at or below 1 km. A detailed design of a long distance DC HTS transmission cable was undertaken by the Electric Power Research Institute (EPRI) to assess the commercial and technological viability of such a cable (Eckroad, 2009; Hassenzahl, Eckroad, Grant, Gregory, & Nilsson, 2009). Several HTS cable demonstration projects are underway to install and test in the near future. Demonstrated HTS cables (both AC and DC types) have been cooled with pressured and subcooled liquid nitrogen (LN_2) in the temperature range of 65−77 K. Gaseous helium (GHe), gaseous hydrogen, and liquid air have been investigated as alternatives to LN_2 in a DC cable design study by the EPRI (Eckroad, 2012). Liquid hydrogen is another potential option for cooling superconducting cables, if the safety concerns associated with it can effectively be addressed. A superconducting cable project using an MgB_2 superconductor and liquid hydrogen as the coolant was successfully demonstrated in Russia (Kostyuk et al., 2012). There have been a few developments in using GHe for cooling HTS cables (Ballarino, 2010; Cheadle et al., 2013; Pamidi, Kim, Kim, Crook, & Dale, 2012; Rodrigo et al., 2012). Visionary energy flow systems have proposed the combined delivery of liquid fuels such as liquid hydrogen and liquefied natural gas, which provide a cryogenic environment, with the transmission of electricity through superconducting cables (Ballarino, 2014; Grant, Starr, & Overbye, 2006; Starr, February 2002; Trevisani, Fabbri, & Negrini, 2006). The idea has been further explored by EPRI in light of high-temperature superconducting cable developments (Grant, 2005).

The discovery of many HTS materials in the late 1980s and early 1990s has spurred renewed interest in superconducting power applications, including power cables, mainly because of the higher critical temperatures (T_c) values (up to 135 K) that allow the use of inexpensive and widely available LN_2 as the coolant for maintaining the required operating temperatures (Dai et al., 2014; Landgraf, 2014; Maguire et al., 2011; Ryu et al., 2013; Weber et al., 2007; Xiao et al., 2013; Yumura et al., 2013). The lengths as well as the power ratings of the demonstrated cables have gone up since the early 2000s. The longest superconducting power cable to date is the 1-km-long HTS cable integrated into the power grid in the German city of Essen on May 12, 2014 (Landgraf, 2014; see also Chapters 5 and 6 in this book).

Various HTS cable demonstration projects, funded by federal agencies in partnership with participating utilities, have proven the technical feasibility of HTS power cables. However, a few challenges are associated with the costs of HTS conductor and cryogenic equipment, as well as the lack of long-term reliability data, which hinder the realization of widespread commercial applications of HTS cables. The demonstrations are usually terminated after a few years due to the costs involved in continuing their operation (Grant et al., 2006). Future successful implementation of HTS cables in electrical power utility infrastructure depends on technical advancements in developing simpler and more efficient cryogenic systems and low-cost superconductor manufacturing processes. The development of new cryogenic dielectric materials that will allow reliable factory testing of the cables at room temperature before their installation is also necessary (Eckroad, 2009). These developments will facilitate broad acceptance of HTS cable technologies by local utilities and electrical grid operators by reducing capital investments, reducing operational expenditures, and enhancing reliability.

7.2 Introduction to GHe-cooled superconducting cables

The interest in developing GHe-cooled superconducting power cables exists because of their potential for achieving significantly higher power densities compared to their LN_2-cooled counterparts. Besides the enhanced power densities, GHe-cooled HTS cables provide several other benefits. The advantages of GHe-cooled cables relative to the LN_2-cooled cables are discussed in this section.

7.2.1 Enhanced power density

The lowest operating temperature for an LN_2-cooled cable is 63 K, which is the freezing temperature of N_2. Hence, the operating temperature window for LN_2-cooled devices is 63–77 K. The operating temperature of GHe-cooled power cables can be as low as 4 K, but large cable systems must limit the operating temperature to 20 K due to low efficiencies of cryogenic systems below 20 K. The critical current density of HTS materials and cables depends on the operating temperature, improving significantly at lower temperatures (see Chapter 2). For example, a typical commercial second-generation (2G) HTS conductor operating at 20 K can carry six to eight times the current it can carry at 77 K in self-field (SuperPower, 2014). Thus, in principle, a GHe-cooled HTS cable operating at 20 and 50 K can support eight times and three times the power densities respectively, compared to a similar HTS cable operating at 77 K. The higher power density feature of GHe-cooled HTS cables is particularly beneficial for severely congested areas and maritime and aviation applications, where space and weight reductions translate into substantial savings in fuel consumption. The US Navy has been exploring GHe-cooled HTS power cables for an all-electric ship to take advantage of the substantially higher power densities in GHe HTS cables (Ferrara, Uva, & Nowlin, 2011). Exploratory studies conducted on the technical feasibility and benefits of GHe-cooled HTS cables for commercial ship applications concluded that they are more efficient than their copper counterparts for high-capacity and long-length cable applications (Allweins & Marzahn, 2011).

7.2.2 Power rating tunability

As discussed, the power rating of an HTS cable depends on its operating temperature. Hence, if the cryogenic system is designed appropriately, a GHe-cooled HTS cable power capacity rating can be modified by operating it at a higher or lower temperature, depending on the demand. When the power transmission demand is low, the operating temperature can be increased to still serve the loads at significantly lower operating expenses on cryogenic systems. Similarly, the power capacity rating of an HTS cable can be increased by operating it at a lower temperature to meet higher demands for transmission loads, albeit at higher operating expenses on cryogenic systems. The power rating tunability feature is beneficial for HTS cables connected in power transmission networks incorporating intermittent and seasonal power sources, such as solar and wind energy, and/or serving loads that change with the time of the day.

7.2.3 Ease of power system design optimization

The wider operating temperature window of GHe cables enables greater flexibility for a power system designer in optimizing the entire system based on a given constraint versus their LN_2-cooled counterparts. If the design is constrained by limited available space or a need to minimize superconductor material cost, a lower operating temperature could be used to obtain a higher power density. If the need is to minimize the number of cryogenic systems in a substation that houses an HTS transformer or HTS fault current limiter beside an HTS cable, a single serial cryogenic fluid loop can be used to tolerate a larger temperature gradient between the inlet and outlet gas streams of the loop. Helium gas allows a large temperature gradient without a phase change.

7.2.4 Cables that tolerate large elevation changes

Long HTS cables require large mass flow rates of cryogenic fluid at high pressures to limit the temperature increase along the cable to a desired value. If the terrain is not flat and has large elevation changes, the pressure variations and associated temperature variations in an LN_2-cooled cable could be large, requiring complex cryogenic and mechanical designs. Moreover, LN_2 could go into a two-phase flow, thus negatively affecting the thermal and dielectric performance of the cable. Pressure and temperature variations in GHe-cooled cables due to elevation changes would be significantly lower. Hence, GHe cooling is more suitable for HTS cables going through a terrain with significant elevation changes. An EPRI study on long-distance superconducting DC cables discussed the effect of terrain changes on the cryogenic systems (Eckroad, 2012).

7.2.5 Safer operation in closed spaces

In closed or confined spaces such as buildings, tunnels, and ships, LN_2-cooled cables pose a safety hazard when there is a system breach due to loss of vacuum or rupture of cryogenic vessels. Rapid evaporation of LN_2 leads to depletion of oxygen from the environment, causing an asphyxiation hazard. The much lower inventory of helium in a GHe-cooled HTS cable and its lower density than that of air significantly lower the risk of asphyxiation hazard. The potential asphyxiation hazard is one of the primary reasons that the US Navy opted to focus on GHe-cooled HTS cables and other devices instead of their LN_2-cooled counterparts (Fitzpatrick, Kephart, & Golda, 2007; Kephart et al., 2011). Similar hazards exist if an HTS cable is run through long tunnel ruptures and releases a large amount of nitrogen gas, which will stay stagnant in the tunnel. The high fraction of nitrogen in the air prevents access for quick repairs that are needed for restoring the operation of the cable system.

7.3 Potential applications of GHe cables

GHe-cooled HTS cables can be used everywhere LN_2-cooled cables are used. Because of the potential of allowing lower operating temperatures than LN_2-cooled cables, GHe HTS cables have additional superconducting material choices that operate in the 20−30 K range. Potential HTS materials for cables are discussed in a subsequent section. Considering the benefits discussed in the previous section, there are several specialized applications where GHe-cooled HTS cables have significant advantages over their LN_2-cooled counterparts. A few of the potential applications are discussed in this section. These applications use low voltages of <2 kV. To apply GHe cooling for higher voltage cables, further research on new dielectric materials and improved dielectric design is required. The limitations of GHe-cooled cables in terms of dielectrics are discussed in detail in subsequent sections.

7.3.1 *Naval and aviation applications*

The higher power densities achievable with GHe-cooled HTS cables are particularly advantageous for shipboard and aviation applications because of the limited available space and associated fuel cost, which is a major operating expense. Additionally, the confined spaces onboard ships or aircraft are not safety concerns when GHe is used as the coolant. The US Navy has been exploring GHe HTS power cables for future all-electric ships (Pamidi et al., 2012; Rodrigo et al., 2012; Rodrigo, Graber, Kwag, Crook, & Trociewitz, 2013). The interest of the US Navy has been strong since the successful sea trials of a GHe-cooled HTS degaussing system (Kephart et al., 2011). Design studies have shown that high-amperage (>5 kA) GHe-cooled HTS cables are more efficient than their copper counterparts onboard commercial ships (Allweins & Marzahn, 2011). The National Aeronautics and Space Administration and the US Air Force have conducted detailed design studies of all-electric aircraft that have HTS cables connecting an HTS motor to the onboard generator (Haugan, Long, Hampton, & Barnes, 2008; Masson, Brown, Soban, & Luongo, 2007). GHe HTS cables would be suitable and advantageous for such applications.

7.3.2 *DC cables for railway feeder applications*

Some high-traffic railway lines use DC feeding systems. The DC feeding systems using copper cables suffer from low efficiency due to the low voltages used, typically <2 kV. The large voltage drops also require a large number of substations along the line.

Studies have shown that HTS DC cables bring several benefits, including high efficiency, reduction in the number of substations, and an ability to handle much higher traffic on the same lines (Ohsaki, Lv, Sekino, & Tomita, 2012; Ohsaki, Matsushita, Koseki, & Tomita, 2014; Tomita, Fukumoto, Suzuki, & Miryala, 2010). The voltages used for the copper cables are usually low, at <1500 V. GHe-cooled HTS cables would increase efficiency and lower the infrastructure costs by eliminating the need for many substations in densely populated regions, such as the urban areas of Japan and other countries that have heavy railway traffic.

7.3.3 DC cables for data centers

A typical data center consumes up to 50 times as much electricity as standard office spaces of a similar size (Pacific Gas and Electric, 2006; Ton, Fortenbery, & Tschudi, 2008). The number of data centers and the size of a typical data center have been increasing rapidly throughout the developed world. For data security and reliability reasons, redundant electrical distribution networks are used. Enhancing the energy efficiency of data centers has been an ongoing effort. Improving the traditional technologies will bring marginal gains in efficiencies, whereas introducing HTS DC cables for powering data centers will bring significant energy savings. GHe-cooled HTS cables are suitable for this application because they are DC cables, and the cable runs tend to be indoors. Safety concerns preclude using LN_2-cooled cables inside buildings. American Superconductor Corporation has been promoting HTS cables for data center applications (AMC, 2012).

7.3.4 Cables that operate in significant elevations

LN_2 cooling is not suitable for HTS cables serving loads with significant changes in elevations that cause excessive changes in pressure and associated temperature variations along the length of the cable. The corresponding pressure and temperature variations in GHe-cooled cables would be significantly lower (Eckroad, 2012). Thus, power cables that connect large loads on mountain resorts and other such facilities would see significant energy savings after replacement with GHe-cooled HTS cables. An EPRI report discussed the 12-kA power cables at the Raccoon Pumped Hydro Facility as a potential example for cryogenic gas-cooled HTS cables (Eckroad, 2012).

7.3.5 Cables for feeder system to connect superconducting magnets

The Large Hadron Collider (LHC) facility at the European Organization for Nuclear Research (CERN) in Geneva, Switzerland plans to use power cables that supply up to approximately 150 kA of quasi-DC current from the power converters to the large magnet systems (Ballarino, 2010, 2014). Currently, a combination of copper bus bars and supercritical liquid helium (LHe)-cooled LTS cables is being used for some such loads. GHe-cooled HTS cables are being studied for the application to take advantage of the higher operating temperatures over their LTS counterparts. GHe-cooled HTS cables provide substantial energy savings compared to LHe-cooled LTS cables. The availability of 5–20 K helium gas from LHe-cooled magnet systems at the facility is an added advantage of GHe-cooled HTS cables for the application.

7.4 Technical issues pertinent to GHe-cooled high-temperature superconducting (HTS) cables

7.4.1 HTS materials

Since the discovery of superconductivity at 36 K in the La–Ba–Cu–O system (Bednorz & Muller, 1986), many new superconducting materials commonly known as HTS (or cuprates) were discovered with T_c up to 135 K. Many reviews are available on various aspects of HTS materials (Bednorz & Muller, 1986; Kalsi, 2011; Larbalestier, 1997; Larbalestier et al., November 15, 2001; Rogalla & Kes, 2012; Scanlan, Malozemoff, & Larbalestier, 2004). Only four of the materials could be developed for the commercial production of long-length conductors (see also Chapters 5 and 6). To date, all HTS electric power applications have been based on these few materials. Table 7.1 shows a list of the materials, their T_c values, and references as sources of additional information. Salient features of these practical HTS materials are discussed below as they apply to HTS cable applications. Note, as discussed previously in Chapter 2, MgB_2, although technically an LTS material based upon its electron–phonon pairing mechanism, is often included in the HTS family because of its unusually high T_c.

7.4.1.1 BSCCO-2212 conductor

BSCCO-2212 has an approximate chemical composition of $Bi_2Sr_2Ca_1Cu_2O_8$, a T_c of 85–92 K, and is made both silver alloy sheathed tape and round wire forms (see also Chapter 3). It is also available in bulk form as rods and tubes, mainly for current lead applications (Callaway, Naderi, van Le, & Schwartz, 2014; Jiang et al., 2011; Larbalestier et al., 2014; Masur et al., 2001). Processing of BSCCO-2212 involves partial melt processing; hence, the fully reacted wire does not tolerate handling due to its brittle nature. Moreover, the material exhibits useful critical current density only at temperatures below 30 K, with excellent critical current density at 4.2 K, even in large background magnetic fields. Hence, the BSCCO-2212 material is primarily being developed for high-field magnet applications operating at low temperatures (<30 K), where the coils wound from green wire can be reacted (i.e., wind then react); this eliminates the need for handling the wire after the partial melt process reaction (Callaway et al., 2014).

Figure 7.1 shows a typical cross-section of round BSCCO wire being developed for high-field magnet applications. BSCCO-2212 is not suitable for long-length power cables because of the difficulty in heat treating long lengths and its inability to tolerate any mechanical handling after the reaction. Besides high critical current density in large background magnetic fields, the round wire form allows ease of winding large magnets without having to worry about the anisotropy of superconducting properties, which is a problem in other HTS materials (Callaway et al., 2014).

Table 7.1 Practical HTS materials, their chemical compositions, T_c values, and references for additional information

Material	Common name	T_c (K)	References for additional information
MgB_2	MgB2	39	Kumakura, Matsumoto, Fujii, and Togano (2001), Soltanian et al. (2005), Tomsic et al. (2007), and Malagoli et al. (2009)
$Bi_2Sr_2Ca_1Cu_2O_8$	BSCCO; BSCCO-2223; Bi-2223	85	Larbalestier et al. (2014), Callaway et al. (2014), and Masur et al. (2001)
$REBa_2Cu_3O_7$	REBCO-123; YBCO-123; Y-123; 2G HTS	92	Rupich et al. (2013), Maguire et al. (2009), Rey et al. (2010), Shiohara et al. (2012), Marken (2011), van der Laan et al. (2011), Takayasu et al. (2012), Goldacker et al. (2014), Ballarino (2011), and Jin et al. (May 31, 2001)
$Bi_2Sr_2Ca_2Cu_3O_{10}$	BSCCO-2223; 1G HTS	110	Masur, Kellers, Feng, Fleshler, and Podtburg, 2002, Yi et al. (2004), Yamada et al. (2007), Yuan et al. (2004), Kato et al. (2004), Osabe et al. (2010), Ayai et al. (2005), and Selvamanickam et al. (2007)

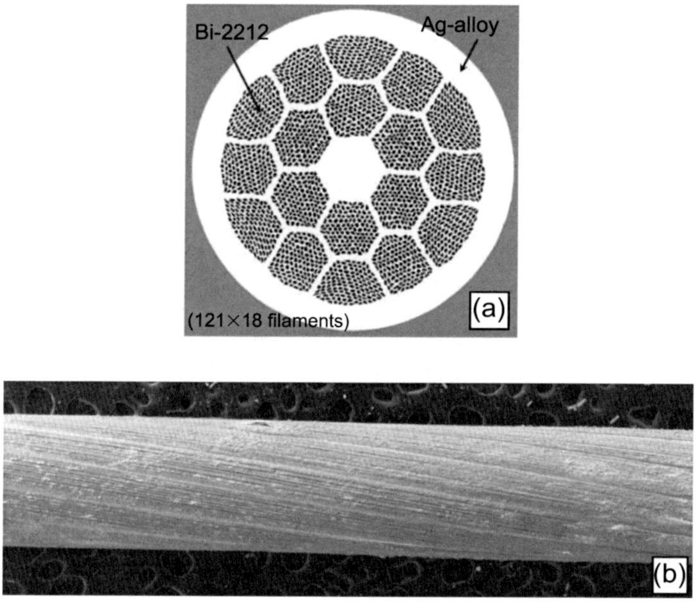

Figure 7.1 Transverse cross-section of a round BSCCO-2212 wire (a) and longitudinal cross-section of a twisted BSCCO-2212 wire after etching away the silver (b).
Reprinted with permission from the Applied Superconductivity Center, National High Magnetic Field Laboratory.

7.4.1.2 BSCCO-2223 conductor (1G HTS)

BSCCO-2223 (also known as a 1G HTS conductor) has an approximate chemical composition of $Bi_2Sr_2Ca_2Cu_3O_{10}$ with T_c of approximately 110 K; it is made as silver alloy sheathed tape. 1G HTS is the first HTS material successfully commercialized in a long-length practical conductor. Early HTS power cables (1990s and early 2000s) and other devices were made from this conductor. Because the 1G conductor is available in fully reacted form and comes with several varieties of metal-laminated stabilizer options, many large cable demonstration projects were initiated throughout the world (Masur, Kellers, Feng, Fleshler, & Podtburg, 2002; Yamada, Mogi, & Sato, 2007; Yi et al., 2004; Yuan et al., 2004; see also Chapters 5 and 6). Steady progress has been made in enhancing the performance and reliability of the conductor (Ayai, Hayashi, & Yasuda, 2005; Kato et al., 2004; Osabe et al., 2010; Selvamanickam et al., 2007).

The 1G BSCCO-2223 conductor is available in several variants depending on the requirements of the application. Of particular importance for HTS cables is the variant that has a twisted multifilamentary structure with reduced AC losses to enable AC power cables and other applications (Selvamanickam et al., 2007). Sumitomo Electric Industries is currently the largest supplier of the 1G conductor (trade name DI_BSCCO). Figure 7.2 shows the cross-section of a BSCCO-2223 (1G) tape conductor produced by Sumitomo Electric Industries. The typical commercial conductor is available as a

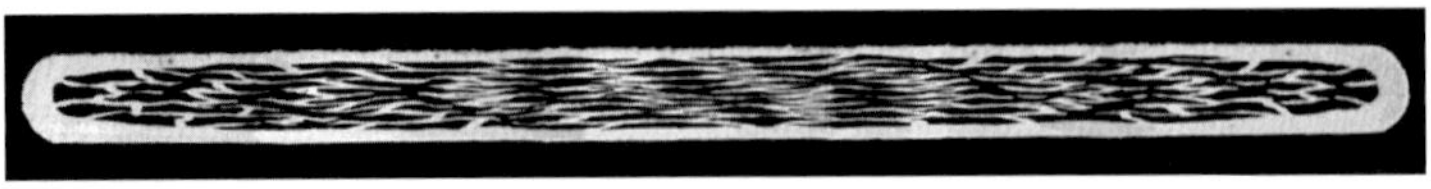

Figure 7.2 BSCCO-2223 wire cross-section reproduced with permission from Sumitomo Electric Industries, Ltd.

nominal 4-mm wide and 0.2-mm thick tape, with a self-field critical current of 170–200 A at 77 K. There has been significant progress in achieving high critical current densities in 1G conductors using the overpressure processing method (Ayai et al., 2005; Kato et al., 2004; Osabe et al., 2010).

7.4.1.3 YBCO-1223 conductor (2G HTS)

REBCO-1223, also known as the second-generation (2G) HTS conductor, has a chemical composition of $RE_1Ba_2Cu_3O_7$ (RE can be one of several rare earth elements), with a T_c of about 90 K. Many manufacturers produce long lengths (~100–300 m) of the 2G conductor as coated conductor with a thin layer (1–2 μm) of superconductor on a metal substrate with complex multilayer ceramic buffer layers (Maguire et al., 2009; Rupich et al., 2013; see also Chapter 4). Multiple intermediate buffer layers are required to achieve the required biaxial texture of the superconducting layer. The terms *tape*, *coated conductor*, and *wire* are used interchangeably throughout this chapter to refer to the same multilayer REBCO-1223 architecture.

Figure 7.3 shows the architecture of the multilayer REBCO-1223 2G tape conductor produced by SuperPower Inc. using the ion-beam assisted deposition (IBAD) process. The 2G multilayer REBCO-1223 conductor has by far the highest critical current density at 77 K and has the potential to be an inexpensive conductor if the manufacturing challenges are tackled to enhance the yield, performance, and uniformity. Many recent long-length HTS cable projects have used or are being designed

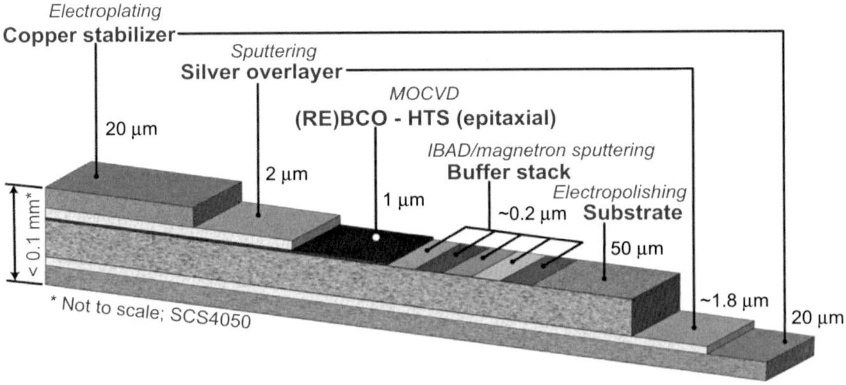

Figure 7.3 Multilayer architecture of an IBAD REBCO-1223 tape conductor manufactured by SuperPower Inc.
Courtesy of SuperPower Inc., a Furukawa Co.

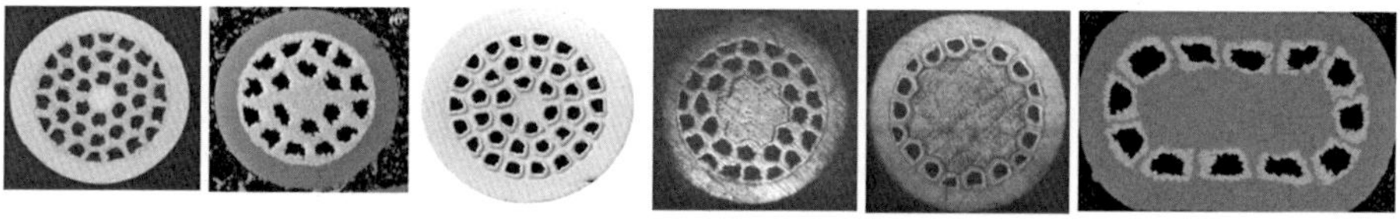

Figure 7.4 Several cross-sections of MgB_2 wire available from Hyper Tech Research.

for future installations with the 2G conductor (van der Laan, Lu, & Goodrich, 2011; Marken, 2011; Rey et al., 2010; Shiohara, Taneda, & Yoshizumi, 2012). The 2G conductor from several manufacturers is amenable to producing specialized HTS cables with small diameters and as multifilamentary and transposed cables (Ballarino, 2011; Goldacker et al., 2014; Jin et al., 2001; Takayasu, Chiesa, Bromberg, & Minervini, 2012).

7.4.1.4 MgB_2 conductor

Magnesium di-boride (MgB_2) is the most recently discovered practical superconducting wire that is available in long lengths from several manufacturers (Kumakura, Matsumoto, Fujii, & Togano, 2001; Malagoli et al., 2009; Soltanian et al., 2005; Tomsic et al., 2007). Although technically an LTS material, because of its unusually high T_c, it is also being considered for superconducting cable applications. Its simple powder-in-tube (PIT) fabrication process (see also Chapter 2) makes MgB_2 the least expensive of the LTS wires being considered for superconducting cable applications; however, it has the disadvantage of having the lowest T_c of 39 K. It is suitable for applications at operating temperatures of 25 K and lower. It is available as round wire and multifilament architecture from Hyper Tech Research Inc. and Columbus Superconductors.

Figure 7.4 shows the different cross-sections of a multifilament MgB_2 conductor produced by Hyper Tech Research. Hyper Tech Research currently supplies a conductor with a self-field engineering current density of 2 kA/mm^2 at 20 K, the expected operating temperature of MgB_2 cables, and expects to improve the conductor to 5 kA/mm^2 at 20 K in 3 years. High current cables have been demonstrated from MgB_2 wires operating at around 20 K using liquid hydrogen as the cryogenic coolant (Ballarino, 2010, 2014; Musenich, Greco, Razeti, & Tavilla, 2007; Stemmle, Merschel, Noe, & Hobl, 2013).

7.5 Dielectric design aspects of helium gas-cooled HTS cables

7.5.1 Dielectric properties of helium gas

As described previously, there are many benefits to using GHe as the coolant for HTS power cables compared to LN_2. However, the dielectric characteristics of GHe are substantially more challenging for the design of an HTS power cable compared to LN_2. This is one of the primary reasons why the HTS power cables currently installed in the distribution grid and operating at voltages greater than 10 kV use LN_2 as the coolant.

For most HTS cables (specifically, for HTS cables that are of the cold dielectric type, which will be explained in the next section), the cryogenic coolant is part of the dielectric system because it penetrates butt gaps between the layers of the lapped dielectric tape used to insulate the cable. The dielectric breakdown strength of LN_2 has been reported to be 19.6 kV/mm (Shin, Hwang, Seong, Lee, & Lee, 2012) in uniform electric fields, independent of pressure and temperature of the liquid, as long as there are no nitrogen gas bubbles present. The formation of nitrogen gas bubbles is usually minimized by subcooling and pressurizing the LN_2. However, in HTS cable applications, the formation of gas bubbles in LN_2 cannot be totally prevented. A decrease of the breakdown strength by approximately a factor of two when nitrogen gas bubbles are present has been measured experimentally (Sauers et al., 2011).

Besides the high-voltage breakdown strength, LN_2 also has a relative permittivity of approximately 1.4 and therefore is not too dissimilar from that of lapped dielectric insulating tape materials, such as PTFE (2.1) or polypropylene (2.2–2.3). When the dielectric materials insulating the HTS cable are reasonably well matched in relative permittivity, there is an advantage for the capacitive electric field in AC applications and during fast voltage transients, leading to an only moderate field enhancement in LN_2.

GHe, on the other hand, has different and more challenging dielectric characteristics than its more robust LN_2 counterpart. Gases in low electric fields are generally very good electrical insulators with low dielectric losses, low electrical conductivity, and a relative permittivity very close to unity, leading to HTS cable designs of low losses and low capacitance per unit length. However, the breakdown strength of helium gas is considerably lower than that of LN_2 or even nitrogen gas at equivalent temperature and pressure.

Breakdown strength of a particular gaseous dielectric is a function of its pressure and temperature conditions. Paschen's law describes the breakdown voltage between two electrodes as a function of the product of pressure and electrode distance. The function has a minimum breakdown voltage at a certain pressure–distance product, called the Paschen minimum. Starting from the Paschen minimum, the breakdown voltage increases rapidly toward lower pressure–distance products. This zone is important for vacuum dielectrics. The breakdown voltage also increases toward higher pressure–distance products, but only gradually. This is the zone of interest for gaseous dielectrics, such as GHe-cooled HTS devices.

The dielectric strength of a gas is also a function of temperature. The dependency on temperature is not considered in most diagrams of Paschen's law. Decreasing the temperature has a similar effect as increasing the pressure because both increase the density of the gas. Figure 7.5 shows the Paschen curve of GHe at room temperature along with the breakdown voltage of LN_2 in the same diagram. Operation at cryogenic temperature around or below 77 K and pressure levels of 1 MPa and higher allow the use of gases with low dielectric strength, such as the noble gases helium and neon, as an

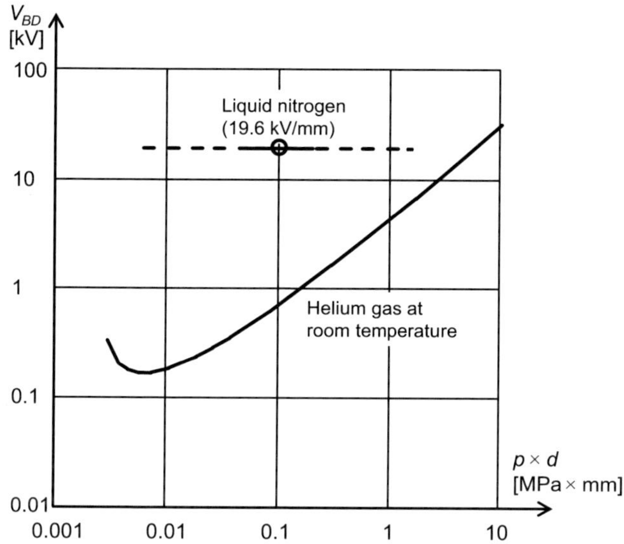

Figure 7.5 Paschen's law for helium gas at room temperature (Küchler, 2005) as well as breakdown voltage for liquid nitrogen (Shin et al., 2012).

insulation media of HTS devices. The dielectric breakdown strength of GHe at 77 K and 1.0 MPa has been reported to be 4 kV/mm (RMS, 60 Hz) in a uniform field (Rodrigo, Kwag, Graber, Trociewitz, & Pamidi, 2014). While this value is good enough to allow designing HTS cables for low- to medium-voltage applications, it is still approximately five times lower than that of LN_2.

Solely relying on Paschen's law is not sufficient for a complete assessment of dielectric properties. Corona discharges on electrode surfaces appear at voltage levels that are much lower than breakdown voltage. Because helium gas is located in butt gaps of lapped tape insulation around HTS cables, these corona discharges manifest themselves in high-frequency pulses not dissimilar to partial discharge (PD) pulses in solid dielectrics. Like PD pulses in solid dielectrics, these corona discharges can deteriorate the insulation if they occur over an extended period of time. Therefore, the maximum voltage in normal operation should never exceed the PD inception voltage. Figure 7.6 shows PD inception voltage as a function of gas pressure at a temperature of approximately 50 K.

In addition to the low corona onset voltage, GHe has a relative permittivity close to unity, which leads to considerable AC field enhancement in butt gaps. This is especially true in combination with lapped tape made of materials with high relative permittivity, such as polyimide (3.4). DC fields are also expected to be enhanced in butt gaps because GHe has a very low electrical conductivity.

Another breakdown mechanism is surface flashover on the interface of a solid insulator to helium gas. It has been reported that the temperature of the surface has a substantial impact on the voltage limit at which surface flashover occurs. This is especially important at cable terminations where the insulator body of the high-voltage bushing

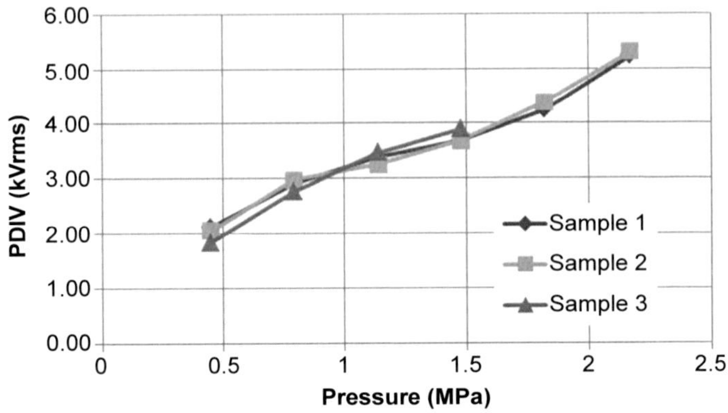

Figure 7.6 Partial discharge inception voltage as a function of helium gas pressure for three samples of a 1-m HTS model cable (Graber, Rodrigo, Salmhofer, Knoll, & Crook, 2011).

may have a temperature much higher than the helium cooling gas itself (Graber, Kim, Pamidi, Rodrigo, & Knoll, 2014).

7.5.2 Two different design philosophies: cold dielectrics and warm dielectrics

There are two different types of design of HTS power cables, categorized by the type of electrical insulation: warm dielectric and cold dielectric. In an HTS cable using a warm dielectric design, the electrical insulation is outside the HTS cable cryostat and therefore essentially at room temperature. Warm dielectric systems are typically applied using a synthetic rubber insulation material such as an ethylene−propylene polymer (EPR) layer extruded onto the outer wall of the cryostat containing the HTS cable (Demko et al., 2003). An outer armor and screen is added like in conventional nonsuperconducting power cables. The HTS cable is located inside the flexible cryostat. The HTS cable and the cryostat are on the same voltage potential; thus, no electrical insulation is required on the HTS cable. It must be noted that using the warm dielectric design, the two poles of a DC cable or the three phases of an AC cable always require separate individual cryostats. The additional volume occupied by the additional cryostat(s) reduces the effective power density compared to HTS cables using cold dielectrics, where both poles or all three phases can be contained within a single cryostat.

For an HTS cable using a cold dielectric, the high-voltage HTS cable is located inside a cryostat that is at ground potential. Therefore, the HTS cable itself must be electrically isolated from the cryostat. The cryostat might feature a protective plastic layer to simplify handling and help protect against environmental damage to the HTS cable during insertion. However, this protective layer does not need to withstand any electrical field. The HTS cable on the inside of the cryostat typically has a lapped dielectric tape insulation and a conductive return or shield layer. The return or shield layer is at ground potential and can be either made of normal metal tape (e.g., copper, aluminum) or HTS tape, depending on the anticipated magnitude of the shielding currents.

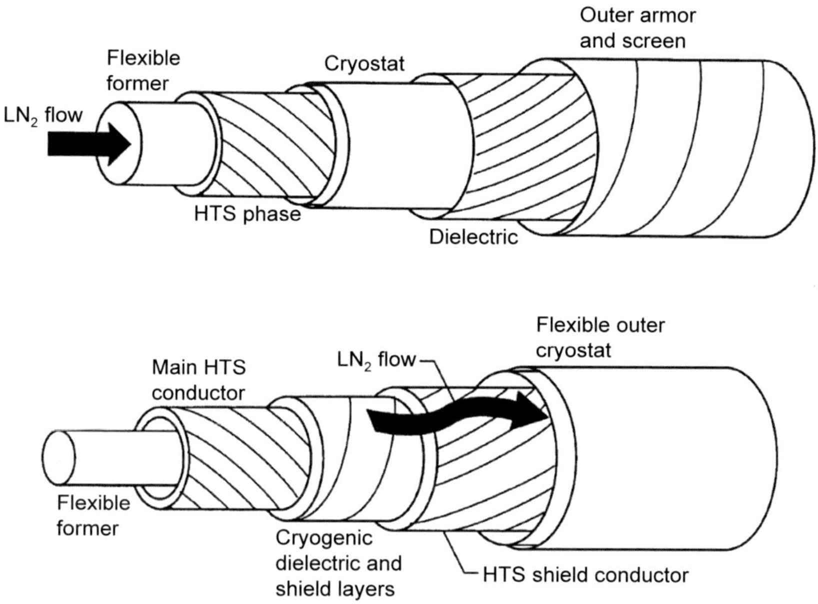

Figure 7.7 Cable layout considerations: warm dielectric (top) versus cold dielectric (bottom) (Demko et al., 2003).

There is an additional protective layer wrapped around the shield layer to reduce friction when pulling the cable into the cryostat, as well as to reduce the risk of damage to the shield layer or the inside of the cryostat. Figure 7.7 depicts the two basic designs (Demko et al., 2003).

Warm dielectric and cold dielectric HTS cables each have their advantages and disadvantages. Depending on the application, one type can be more advantageous than the other. For retrofitting existing duct or pipe systems, the warm dielectric design was reported to be more suitable (Kelley, Nassi, & Masur, 2001). Even though the power density of an HTS warm dielectric cable is lower than that of its cold dielectric counterpart, it still leads to an increase in power density by a factor of two to three compared to conventional copper or aluminum cables. However, for new systems, the cold dielectric design seems to be more favorable because it allows power densities exceeding five times that of conventional power cables (Kelley et al., 2001).

Besides the improved power density of HTS cables of cold dielectrics, they have advantages in reduced AC losses for cables carrying AC current because the HTS shield reduces the radial magnetic field of adjacent conductors. The HTS shield also prevents inductive coupling with nearby metallic materials, such as the cryostat (Kelley et al., 2001). One of the major disadvantages is the tradeoff required for the cryogenic electrical insulation, especially if GHe cooling is used. The electrical insulation needs to accommodate multiple thermal cycles without either the insulation getting damaged by cracking or the conductors getting damaged by applying excessive strain to the HTS tapes due to mismatching coefficients of thermal expansion. The thermal

Table 7.2 **Comparison of the two basic designs of HTS cables**

Characteristic	Warm dielectric	Cold dielectric
Number of cryostats	Each pole/phase in a separate cryostat	Poles/phases can be in the same or in separate cryostats
Type of shield layer	Metal at room temperature (additional losses)	HTS or metal
Magnetic field	Present	Canceled
AC losses	Higher	Lower
Dielectric losses	No heat load to coolant	Heat load to coolant
Power density	2–3× regular power cable	5× regular power cable
Complexity of insulation	Low	High
Complexity of termination	High	Low
Complexity of factory testing	Low (at room temperature)	Generally not possible

conductivity of the dielectric is also an essential feature because it allows heat conduction from the HTS conductor to the cryogenic coolant. While a large number of layers of lapped insulating tapes is beneficial to lower the dielectric stress, it is also counterproductive for thermal management. The lower complexity of warm dielectric HTS cable is simpler to manufacture and less costly overall (Grant, Gregory, & Eckroad, 2007). Table 7.2 provides a comparison of warm dielectrics and cold dielectrics.

7.5.3 Lapped tape insulation

Most conventional power cables use extruded insulation around the metallic conductor. More precisely, there is a thin inner layer of semiconductive material extruded onto the conductor, followed by the actual insulating dielectric, finished by another semiconductive layer. The extruded insulator material is typically either EPR or cross-linked polyethylene (XLPE). The extrusion process is very carefully controlled to avoid any impurities and voids that could lead to PD under high dielectric stress.

There is no evidence in the technical literature that extruding insulation on HTS conductors has been tried successfully, except for very thin layers directly extruded onto HTS tapes. The mismatch in the coefficients of thermal expansion between most dielectrics and HTS tapes may lead to significant strain and potential delamination of the extruded insulation for the HTS tape or delamination of the HTS layer from the underlying metallic substrate (Kosaki, Nagao, Mizuno, Shimizu, & Horii, 1992). Therefore, for HTS power cables with cold dielectrics, the insulation of choice has been wrapping with lapped insulating tape. Insulation materials show substantial shrinkage in cryogenic environments of around 5% or more (Kelley et al., 2001).

However, the lack of bonding directly to the surface of the HTS tape, which has a thermal shrinkage closer to 0.3%, provided by the wrapping insulation technique does not lead to damage of the HTS tapes.

To give an example, the coefficient of linear expansion of polytetrafluoroethylene (PTFE, Teflon) is approximately $140 \times 10^{-6}\,K^{-1}$, which is considerably higher than that of stainless steel ($17 \times 10^{-6}\,K^{-1}$) or copper ($16 \times 10^{-6}\,K^{-1}$) (Ekin, 2006). The mismatch in thermal expansion could lead to high levels of mechanical stress in the HTS cable, potentially damaging the superconducting material if the cable design would not take it into account. A typical solution to avoid problems associated with the mismatch in thermal expansion is to use a dielectric in tape form. The tape is helically wound around the cable along its length, allowing the insulation material to contract without causing stress on the HTS tapes. The material and thickness of the tape are chosen to allow mechanical flexibility at cryogenic temperatures. The materials proposed for lapped tape insulation of HTS cables include polyethylene (PE), polypropylene, polycarbonate, polyethyleneterephtalate, polyamide (Nomex, Nylon), polyimide (Kapton), cellulose (kraft paper), and polypropylene laminated paper (PPLP) (Rigby & Weedy, 1975).

Electric field grading is an important aspect for any power cable to reduce the electric stress in the dielectric. For this purpose, a semiconductive layer (often referred to as "semi-con") is added on the interface between conductor layers and insulation layers to reduce electric field enhancements around edges and protrusions. The semiconductive layer must be very smooth. For HTS power cables, it is most suitable to use a lapped tape (Tuncer, Zuev, Sauers, James, & Ellis, 2007). Typical examples are tapes made of fibrous graphite paper, graphite-impregnated PTFE, and semiconducting EPR. The resistivity of semi-con tape can be varied over a wide range and is also a function of frequency, temperature, and pressure. For AC power cables, the semiconductive layer has a substantial impact on the total dielectric losses of the cable. The number of layers of semiconductive tape is a tradeoff between reducing electric stress and increasing dielectric losses. The number of semiconductive layers should be kept to the necessary minimum. Ratios of semiconductor to insulation layers of between 1:40 and 1:80 have been suggested (Tuncer et al., 2007). However, this depends on factors such as the content of high-frequency components, such as power electronic switching harmonics, which can lead to increased interfacial polarization and ultimately lead to cable failure due to shield malfunction (Tuncer et al., 2007).

The HTS cable is terminated in metallic termination blocks in order to allow a low electrical resistance interface from the bushing to the HTS cable. The electrical insulation is stripped back for easy access of the terminal blocks. Without additional HTS cable design effort, the electric field would be enhanced around the HTS cable section where the insulation is stripped back to make the low electrical resistance joint. This high-stress region can be avoided by designing the appropriate stress cones. The stress cones are typically made of the same dielectric tape to avoid any mismatch in relative permittivity and electrical conductivity. A gradual increase in the number of insulating layers or an additional triangular sheet of the same material wound around the cable end result in the required conical shape of the stress cone.

7.5.4 Differences between AC cables and DC cables

There are both obvious and nonobvious differences between DC power cables and AC power cables. The most obvious difference is that a DC cable has two poles. A bipolar cable has a positive pole and a negative pole, whereas a monopole cable has one of them grounded. The two poles can be separate HTS cables each contained in its own cryostat, they can be in the same cryostat, or they can be designed as a coaxial cable.

An AC cable has three phases and a ground pole, which is sometime referred to as the "shield" or "return" phase. In rare cases, it can also have a neutral phase, which is typically on ground potential but designed to allow substantial continuous current flow. Similar to DC cable, the three phases can be in separate cryostats, they can be in the same cryostat, or they can be of tri-axial configuration in a single cryostat (see also Chapter 5).

The AC field distribution in an HTS cable is dominated by the permittivity of the dielectrics. This is called the capacitive field (Seong, Choi, Shin, Hwang, & Lee, 2013). The capacitive field can be simulated by finite element analysis. It is important to estimate the maximum electric field in butt gaps and insulator surfaces. Even though such analyses do not calculate the breakdown voltage, it helps to determine the weakest location, which could become a source of PD (Seong et al., 2013).

If the AC electric fields are too high, it can damage the insulation by constant PD activity. This is a process that can take days to months before the material breaks down. The insulator gets damaged locally at the source of PD, driving an electric tree of degraded material forward in field direction. The local electric field at the branch ends of the tree is enhanced because the tree has a higher conductivity than the surrounding undamaged insulation material, thus acting as a perturbance in the otherwise homogeneous electric field. The ultimate breakdown occurs from one of the branches, bridging the remaining distance in the pristine dielectric to the conductor.

Dielectrics under AC fields show dielectric losses. The dielectric losses typically consist of two components: polarization losses and conductive losses. Under ideal conditions, the AC current flowing in a dielectric material leads the voltage by an angle of 90°. The dielectric losses reduce this angle by a few degrees. These losses are expressed as the *loss tangent*, which is sometimes referred to as the "dissipation factor." It is defined as the ratio of the active power P_δ divided by reactive power Q_C, where $\tan\delta = P_\delta/Q_C$. The dielectric losses for a certain material can also be expressed by a complex permittivity consisting of both a real part and an imaginary part. Küchler (2005) reported that typical values for dielectric losses at room temperature range from 0.1% (PTFE, PE, quartz) to 10% (filled epoxy resins, polyamide). PPLP insulated HTS cables in LN_2 have dielectric losses between 0.2% and 1.1%, depending on the applied electric stress (Kwag et al., 2005).

Power system aspects, such as the maximum capacitive load, need to be considered when designing AC systems. The capacitive load added by power HTS cables can be considerable. There are indications that HTS power cables could have a higher capacitive load than regular XLPE or EPR power cables (Graber et al., 2014). This would limit the maximum length of such cables to avoid problems with grid stability. Details on cable capacitance and grid stability can be found in the power systems literature (see also Chapter 5).

Dielectrics in DC cables are different than their AC counterparts. The dielectric withstand capabilities of an insulator are typically substantially higher for DC electric fields than AC electric fields: values of up to three times higher withstand voltage have been reported for regular high voltage power cables (Fenger, 2010). That being said, the electric field distribution is much more complex than for AC fields and is a function of time.

Figure 7.8 shows the time-dependent voltage $v(t)$ and current $i(t)$ versus time commonly observed when charging an insulator with a time-varying voltage $\mathrm{d}v/\mathrm{d}t$. The time-dependent behavior can be characterized by four distinct regions or phases. In the first phase (phase I in Figure 7.8), during which the voltage is ramped up, the field distribution is of the capacitive type, which is dominated by the relative permittivities of the materials. The charging current is mainly a capacitive current $i_C(t) = C \times \mathrm{d}v(t)/\mathrm{d}t$ with the total effective capacitance C. In phase II of Figure 7.8, the voltage $v(t)$ is constant and the current is dominated by a polarization current $i_p(t)$. The polarization current is responsible for the accumulation of space charges on the interfaces of the insulating tapes, as well as within the material of the insulating tapes. In phase III of Figure 7.8, a purely resistive field is established that depends on the specific resistivities of the insulating materials. The residual current that remains after the transients decay is a constant resistive leakage current. After switching off the voltage

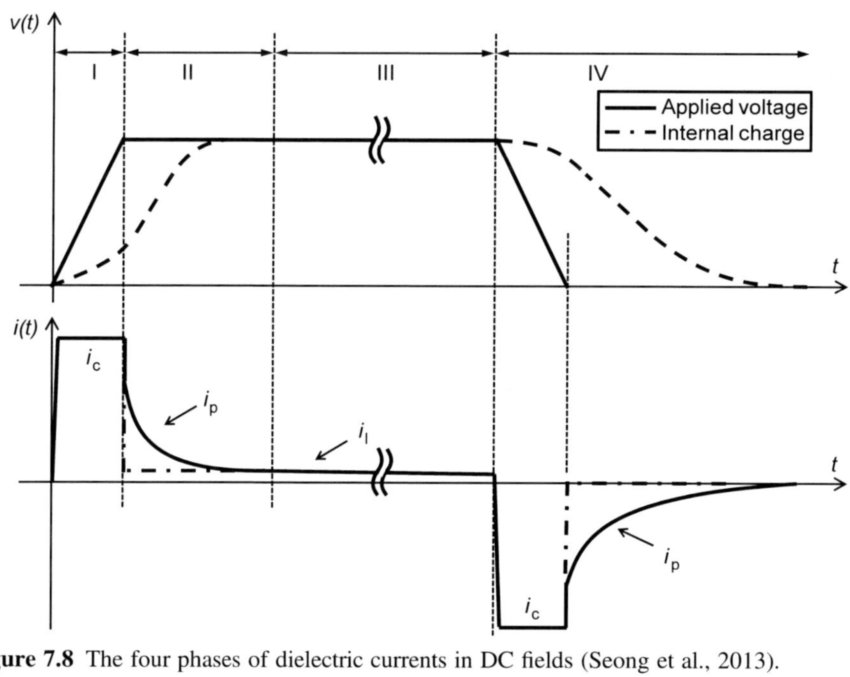

Figure 7.8 The four phases of dielectric currents in DC fields (Seong et al., 2013).

in phase IV of Figure 7.8, the local surface charges and space charges remain for a certain moment of time. This can be a slow process with a time constant similar to the polarization time constant in phase II (Seong et al., 2013).

Under ideal conditions, DC cables have considerably lower dielectric losses than AC cables. However, this assumes that the supply voltage is pure DC. In most practical applications, there are power electronic rectifiers generating the DC current. This leads to a certain level of voltage ripple in case of a line commutated diode or thyristor rectifiers. In case of a switched-mode IGBT rectifier, high-frequency harmonics of the switching frequency should be expected to be present, leading to parasitic dielectric losses. The amplitude and frequency of the voltage ripple and switching harmonics need to be carefully considered for the dielectric design. Hence, dielectric losses can still occur under these circumstances.

7.6 Design aspects for GHe-cooled HTS cable terminations

The terminations of an HTS power cable have the primary function of interfacing the room temperature power source hardware with the superconducting cable operating at cryogenic temperature. Besides the current leads, the terminations must also include helium gas inlet and outlet ports as well as instrumentation to monitor the HTS cable system (Pamidi et al., 2012). The large temperature gradient, along with Joule heating in nonsuperconducting feed-throughs, result in a substantial heat load to the cryogenic cooling system. For short power cables, the heat load from the terminations can exceed all other heat loads to the cryogenic system. Kim, Kim, Graber, and Pamidi (2014) reported a total heat load from ambient of 25 W for each termination for a 3000-A rated HTS cable. The heat load from the flexible cable cryostat is approximately 1−2 W/m, depending upon the size (i.e., diameter) of the cryostat. The lower heat capacity of GHe compared to LN_2 makes the thermal design of the terminations particularly challenging. An optimization of the conductor cross-section in the current leads is most important and usually performed using McFee optimization (McFee, 1959).

The dielectric design of terminations for GHe-cooled cables is also demanding. Electric fields are problematic in helium gas at elevated temperatures and electric fields in proximity to warm insulator surfaces because of lower density of helium at such spots. Such a situation can be prevented by maintaining low temperatures at the insulator bodies of the feed-throughs. The requirement for cold insulator surfaces in helium gas led to the development of hybrid HTS cable terminations (Figure 7.9; Crook, Graber, & Pamidi, 2014; Graber et al., 2014; Suttel et al., submitted for publication). Hybrid terminations include a volume of LN_2 to keep the feed-through into the GHe section at temperatures close to 77 K and thus increase the density and withstand the voltage of helium gas in proximity to the feed through. This design also reduces the heat load to the GHe system. A disadvantage of a hybrid termination is the requirement of a second feed-through from ambient into the LN_2 chamber.

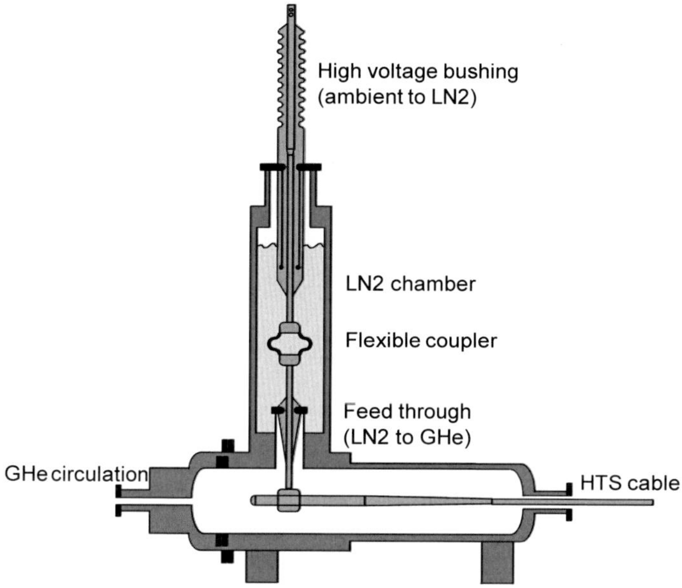

Figure 7.9 Hybrid HTS cable termination combining LN_2 cooling of the power leads with GHe cooling of the HTS cable (Suttel et al., submitted for publication).

7.7 Cryogenic helium circulation systems

A cryogenic refrigerator is a critical and integral part of any superconducting cable system. Generally, a maintenance-free and high-efficiency cryogenic refrigerator is the hallmark of a successful and practical superconducting cable system. For LN_2-cooled HTS cables, two general design options exist for providing the required cryogenic environment. The first option is an open-loop system where LN_2 is supplied to the cable from a large reservoir tank and evaporated LN_2 is periodically replenished. The supply from the tank is pumped on, pressurized, and circulated through the cable system. Some HTS cable demonstration systems preferred open-loop systems because of their simplicity and low maintenance, as well as to take advantage of the higher efficiency of large-scale centralized refrigerators (Lee, Dada, & Ringo, 2005; Stemmle, Merschel, Noe, & Hobl, 2014). The second option is to use a closed loop system in which a cryo-refrigerator removes the heat from the cryogen between the outlet and inlet to maintain the operating temperature. Some HTS cables preferred closed-loop cryogenic systems to avoid the need for trucking LN_2 to the site and the associated uncertainty of supply interruptions (Cryomech, Inc; Demko et al., 2007; Weber, Reis, Dada, Masuda, & Moscovic, 2005).

For GHe-cooled HTS cables, the only viable cooling option is a closed-loop circulation of cryogenic helium gas. There are a few methods available to produce cryogenic helium gas. Depending on the operating temperature, a single-stage or two-stage cryogenic refrigerator can be used (Linde Cryogenics; Stirling Cryogenics).

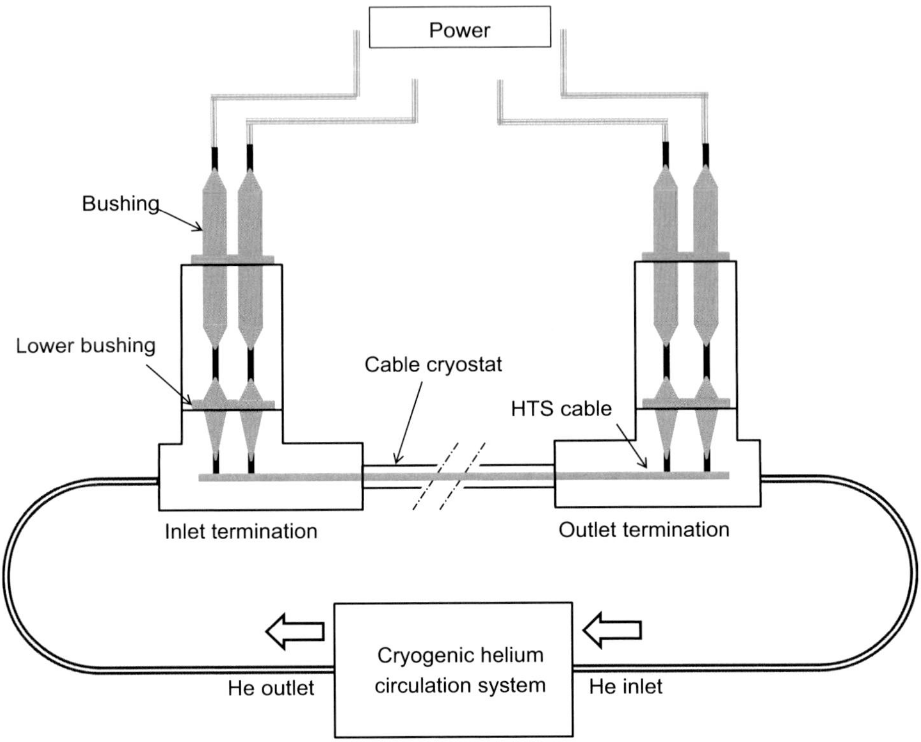

Figure 7.10 A schematic of a GHe-cooled HTS DC cable system with a cryogenic gas circulation system, terminations, cable cryostat, and HTS cable.

Standard LHe refrigerators can also be modified to obtain cold helium gas (Air Liquide; Vermeuleun, 2013). Circulation of helium gas through the cable can be accomplished using cryogenic circulator fans or a compressor—either the same compressor as for the working gas of the cryocooler or a separate compressor unit (Cheadle et al., 2013; Cryomech, Inc; Pamidi et al., 2012; Radebaugh, 2009). The cryogenic helium gas loop could be either the primary loop of the working gas of the cryogenic refrigerator or a secondary loop that is a separate fluid loop, but thermally connected to the primary loop through a heat exchanger that transfers heat from the secondary loop to the primary working gas loop. Some of the options available to produce cryogenic helium gas and to circulate it through an HTS cable system are described here.

Figure 7.10 shows a schematic of an HTS cable system cooled by GHe circulation. The main components of a helium circulation system are identified in Figure 7.10. As shown in Figure 7.10, a cryocooler cold head is represented as the source of cryogenic cooling power, but many kinds of cryocoolers or helium refrigerators can serve this purpose. Typically, large capacity (>300 W at 77 K) single-stage Gifford McMahon (GM) type cryocoolers are used because the operating temperatures of HTS cables are in the range of 50–75 K (Stirling Cryogenics). Two-stage cryocoolers have

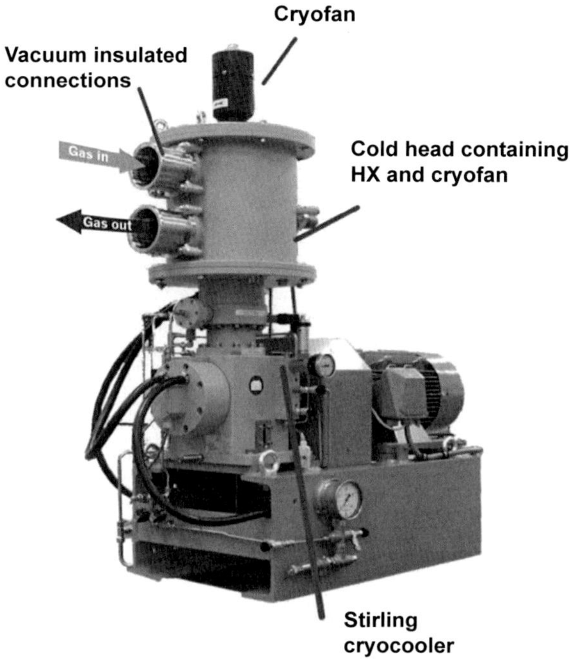

Figure 7.11 A picture of a cryogenic helium circulation system (Stirling Cryogenics GPC-1 with integrated CryoFan) similar to the one used at the Center for Advanced Power Systems in a demonstration of an HTS DC power cable.

been used for cables involving MgB_2 superconductors at operating temperatures of 20 K (Linde Cryogenics). Single-stage GM cryocoolers are available with large capacities of up to 600 W at 77 K, but they are generally inefficient. Comprehensive reviews of many available types of cryocoolers and their operating principles are a good source of information for anyone considering a cryocooler-based helium circulation system (Boman & Doty, 2002; Radebaugh, 2004).

Figure 7.11 shows the helium circulation system used at the Center for Advanced Power Systems to demonstrate an HTS DC power cable system. The helium circulation system was supplied by DH Industries and uses a single-stage Stirling unit (SPC-1, 650 W at 60 K) with a helium circulator manufactured by Cryozone (Bohmwind CryoFan). Large-capacity helium liquefiers are also useful as sources of cooling power, particularly if the operating temperatures required are at 20 K or higher. Reverse Turbo-Brayton type cryogenic systems with large capacities and high efficiencies are available from Air Liquide for HTS cable applications (Vermeuleun, 2013). Standard helium refrigerators up to 1 kW of cooling power at 20 K are available from Linde (models LR1420/LR1620) (Air Liquide).

The second major component in cryogenic helium circulation systems is the gas circulator. Unlike some applications, such as small superconducting magnet systems, HTS cables cannot use the primary helium loop of a refrigerator's working helium gas and the cooling medium for the application. The primary reason for this difference is that there is a risk of contamination of the working fluid that could deteriorate the refrigerator's efficiency or damage it. Hence, typical GHe circulation systems use a secondary forced-flow cooling loop that is thermally anchored to the cold source through a heat exchanger. The secondary helium loop requires a cryogenic circulator(s) to push the gas through the HTS cable system. The circulators come in a range of capacities (Cryomech, Inc) and are simply high-speed centrifugal pumps. The cryogenic circulators are specially designed for the required capacity and operating temperature range to minimize the static heat load. Unlike with liquid cryogens, obtaining large mass flow rates in a helium gas circulation system is difficult because of its low density. Operating pressures up to 20 bar are used to increase the density needed to obtain the required GHe mass flow rates. Typical GHe flow rates obtained are in the range of 10–50 g/s, depending on the operating pressure and temperature. Several firms offer cryogenic gas circulators with a range of capacities. The specifications of a cryocirculator that need to be considered for a given application are volume flow rate and the corresponding pressure drop that can be handled. The performance characteristics of a circulator are usually represented in a plot of volume flow rate (m^3/h) versus pressure head (m). As with other cryogenic equipment, the larger the fan capacity is, the higher the static heat load is.

Cryozone, Linde, Air Liquide, Barber Nicholes, and R&D Dynamics are some examples of commercial sources of helium circulators. Figure 7.12 presents some cryogenic circulators that are commercially available from Cryozone. The selection of a circulator should consider several design factors, such as required mass flow rates, operating temperature, and static heat load limits. A review of the basic principles and design process of cryogenic gas circulators has been published recently (Cryomech, Inc).

Another critical component of a cryogenic helium circulation system is the heat exchanger that makes a thermal link between the source of cooling power, such as

Figure 7.12 Pictures of commercial helium circulators available from Cryozone.

the cold head of a cryocooler, and the secondary helium loop that collects the heat from the application and transfers to the cold head. The efficiency of the heat exchanger is important to be able to obtain the secondary loop temperature as close to the cold head temperature as possible. For large mass flow rates, it is not easy to design the heat exchanger that has the required heat transfer capability while keeping the pressure drop across the heat exchanges within the capability of the helium circulator being used. The heat exchangers are usually made of copper, either as helically wrapped finned tubing or plate-fin heat exchangers (Barron, 1999; Gallego & Klett, 2003; Straatman, Gallegob, Thompsonc, & Hangan, 2006). There have been new developments in lightweight metal and carbon foam heat exchangers with large surface areas (Barron, 1999; Gallego & Klett, 2003; Kuang, Kim, Xu, & Lu, 2012; Straatman et al., 2006; Venkatarathnam & Sarangi, 1990). The objective of the developments is to enhance heat transfer efficiency without increasing the pressure drop generated. Low pressure drop is particularly important for cryogenic helium circulation systems due to the limited pressure head available with helium circulators.

To achieve large cryogenic helium gas flow rates, multiple circulators were used in parallel configuration with the individual gas streams joining at the inlet and outlet. Test results of a large helium circulation system with four circulators, each serving a path with a cryocooler and associated heat exchanger, have been described (Pamidi et al., 2012). Figure 7.13 shows a schematic of the GHe circulation system with four circulators in a parallel configuration.

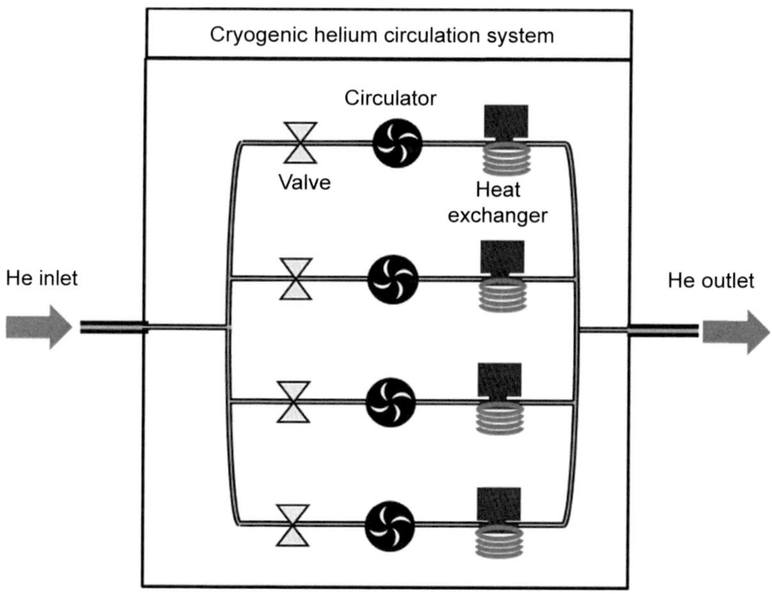

Figure 7.13 A schematic of the GHe circulation system with four circulators in parallel configuration that was used at Florida State University Advanced Power Systems.

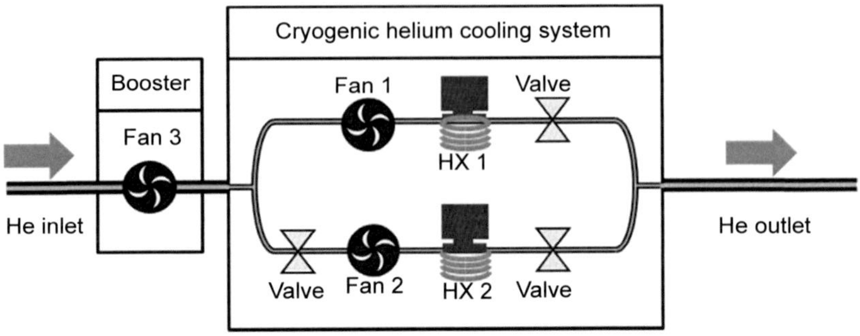

Figure 7.14 A schematic of a helium circulation system with multiple circulators in series and parallel configuration.

Multiple circulators in series configuration have been used to counter the pressure drop across the application (Kim, Kim, & Pamidi, 2014). Figure 7.14 shows a schematic of a cryogenic helium circulation system with circulator fans both in series and in parallel configurations. These types of helium circulation systems were used to cool a 30-m long HTS DC cable system carrying up to 3 kA current. The temperature gradient across the cable system was as low as 3 K for a mass flow rate of approximately 10 g/s at 50 K inlet temperature (Kim, Kim, Graber, et al., 2014; Pamidi et al., 2012).

The total useful cooling power of a cryogenic helium circulation system depends upon the mass flow rate ($\dot{m}$) and the temperature gradient (ΔT) between the inlet and outlet of the circulation system, as given by

$$Q = \dot{m} C_{\mathrm{p}} \Delta T \tag{7.1}$$

Here, $\dot{m}$ is the mass flow rate and C_{p} is heat capacity (approximately 5 J/g K between 20 and 80 K), and temperature gradient ΔT. As an example, a circulation system with a mass flow rate of 10 g/s and 5 K temperature gradient can provide useful cooling power of approximately 250 W. A typical 100-m long HTS cable requires about 250 W of cooling power if designed well. Heat load from commercial cable cryostats is approximately 1 W/m. Heat leak from cable terminations usually add 25–30 W each (Kim, Kim, Graber et al., 2014). The heat load from an HTS cable depends on its design and whether it is an AC or DC cable. HTS DC cables have essentially no heat load, except when there is significant AC ripple. AC cables will have AC losses in the cable and dielectric losses in the insulation if the cable uses a cold dielectric design.

7.8 Ongoing GHe-cooled HTS cable projects

There are a few ongoing research and development activities on GHe-cooled HTS power cables (Del Rosso, 2014; EPRI-Minervini, 2013; EPRI-Pamidi, 2013; Florida State University, 2013; Kim, Kim, & Pamidi, 2014). Florida State University Advanced Power Systems (FSU-CAPS) has been working on a GHe-cooled HTS

DC cable project. FSU-CAPS has developed GHe circulation systems and conducted detailed cryogenic thermal studies on emulated cables using typical cryostats used for HTS cables (Pamidi et al., 2012). In parallel to cryogenic thermal studies, FSU-CAPS studies many design variations of HTS DC cables cooled with GHe to understand the thermal and dielectric aspects of GHe-cooled cables. Many experimental tools were established for characterizing short design cables and 30-m long prototype cables. At present, the US Navy is the primary supporter of the work at FSU-CAPS on GHe-cooled HTS cables and other power devices. The US Navy is funding many small businesses to develop various components of GHe-cooled HTS devices, including cryogenic equipment and HTS cables and insulation materials. The funding from the US Navy has contributed to significant progress in understanding the challenges involved in GHe-cooled HTS power cables and the development of innovative solutions that address the challenges. Recently, FSU-CAPS has successfully demonstrated a 30-m DC cable made with 2G HTS by operating at 3 kA at 60 K in GHe (Florida State University, 2013; EPRI-Pamidi, 2013). The dielectric design of the GHe-cooled HTS cable has been validated up to 3.5 kV through high-voltage DC soak tests that preceded and were followed by PD tests (Graber et al., 2014). Ultera, a joint venture of Southwire Company and NKT Cables, fabricated the 30-m cable for the FSU-CAPS project. Figure 7.15 shows a photograph of the GHe cable demonstration at FSU-CAPS. The picture shows the helium circulation system, the two cable terminations, and part of the cable cryostat that houses the HTS cable.

Figure 7.16 shows the measured temperature increase across the prototype cable system demonstrated at FSU-CAPS at three different GHe mass flow rates. The cable system consists of the inlet and outlet terminations (described in Section 7.6) and the 30-m long superconducting cable in a Nexans Cryoflex cryostat. As seen in the figure, the temperature rise was just 3 K for a current of 3 kA at GHe mass flow rate of 8.6 g/s. Figure 7.17 shows a plot of the voltage across the 30-m superconducting cable as a function of time while the cable is carrying 3 kA. The data in Figures 7.16 and 7.17

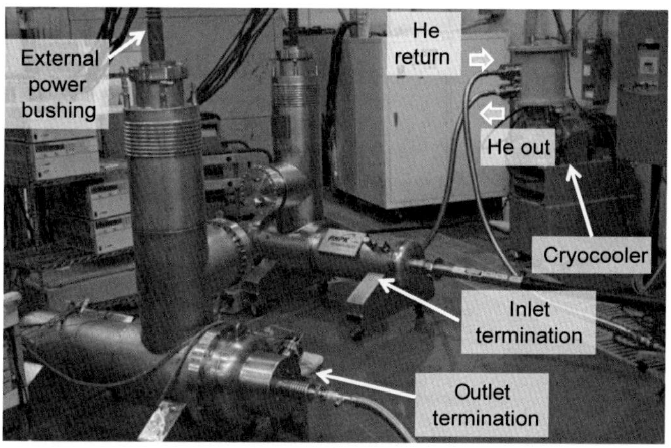

Figure 7.15 The GHe-cooled HTS cable test facility at FSU-CAPS.

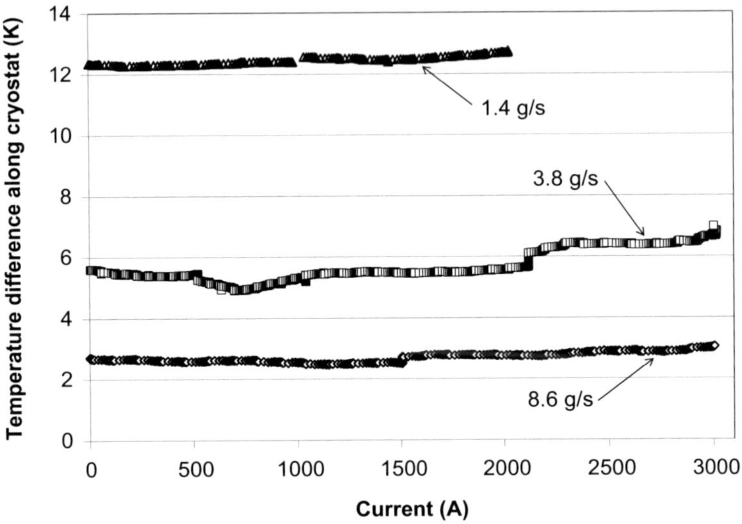

Figure 7.16 Temperature rise across the 30-m long HTS cable system at 3 kA current for three different GHe mass flow rates.

show that GHe is a viable cooling option for HTS cable systems and that the behavior of the cables depends on the operating temperature, not the cryogenic media used for obtaining the operating conditions.

The Massachusetts Institute of Technology (MIT), in collaboration with Tsinghua University (China) and University of Cambridge (UK), has been working on the design of a GHe-cooled HTS DC cable made of MgB_2 superconducting wire for microgrid applications (Cheadle et al., 2013; EPRI-Pamidi, 2013). The design

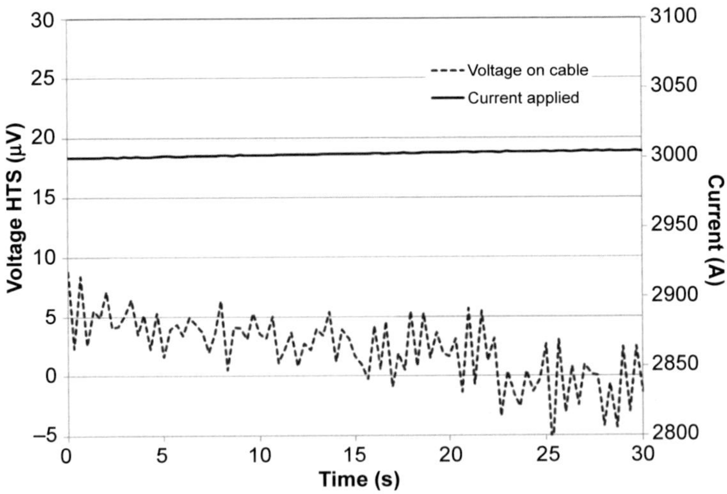

Figure 7.17 Current versus the electric field across the 30-m HTS cable during the GHe-cooled demonstration at FSU-CAPS.

incorporates two-stage current leads: GN_2-cooled copper leads and GHe-cooled HTS. Preliminary designs were rated for 1-kV and 1-kA cable, with the ultimate goal of a 5-kA cable to be installed and tested at Tsinghua University in Beijing. The design and thermal models for the cable's cryogenic system have been published (Cheadle et al., 2013). The HTS cable is designed to be cooled with forced-flow GHe at 20–25 K. Both poles of the DC HTS cable are in separate tubular cryostats contained in one common vacuum envelope. Besides the GHE-cooled MgB_2 cable, MIT is also developing high-current 2G HTS cables for superconducting power transmission for directed energy and data center applications (EPRI-Minervini, 2013).

Under the framework of High Luminosity LHC project, CERN has been working on designs and prototype HTS cables for high-current applications (Ballarino, 2010, 2011, 2014). CERN demonstrated a 20-m long HTS cable made of MgB_2 superconductor for power transportation (Del Rosso, 2014). The MgB_2 cable was operated at 24 K under a forced-flow helium environment and carried a world-record 20 kA. The project supported the development of round MgB_2 wires at Columbus Superconductors. The ultimate goal of the project is to use similar MgB_2 cables for connecting power converters that supply current to superconducting magnets in the tunnel of future LHC. It is necessary to move the power converters from the tunnel to the surface or other radiation-free underground areas away from the magnets.

7.9 Summary

In recent years, there has been a significant progress in GHe-cooled HTS cable technology. High-current GHe-cooled HTS cables have been successfully demonstrated by multiple institutions. Many challenges of GHe HTS cables have been addressed through innovative solutions that resulted from systematic research and development studies in many research centers. The results of the studies suggest that GHe-cooled cables are attractive for some high-current cable applications. However, designers of GHe-cooled HTS power cable systems have to keep in mind the limitations of GHe when compared with LN_2. Detailed studies on the cryogenic thermal aspects of quench initiation, propagation, and their implications on GHe systems need to be studied through detailed modeling and experimentation. Such studies are in progress at many research institutes.

References

Air Liquide. "Air liquide" [Online]. Available http://airliquideadvancedtechnologies.com/file/otherelementcontent/pj/f8/a4/5a/58/turbo-brayton_sd1997969771067587692.pdf. Accessed September 2014.

Allweins, K., & Marzahn, E. (2011). *Feasibility of HTS DC cables on board a ship* [Online]. Available http://mydocs.epri.com/docs/publicmeetingmaterials/1110/7TNRSL46577/Meeting_Materials.pdf. Accessed September 2014.

AMSC. (2012). *Superconductor LVDC data Center power cables* [Online]. Available http://www.amsc.com/documents/superconductor-lvdc-data-center-power-cables. Accessed September 2014.

Ayai, N., Hayashi, K., & Yasuda, K. (2005). Development of Bi-2223 superconducting wires for AC applications. *IEEE Transactions on Applied Superconductivity, 15*(2), 2510–2513.

Ballarino, A. (2010). Design of an MgB2 feeder system to connect groups of superconducting magnets to remote power converters. *Journal of Physics: Conference Series, 234*, 032003.

Ballarino, A. (2011). Alternative design concepts for multi-circuit HTS link systems. *IEEE Transactions on Applied Superconductivity, 21*(3), 980–983.

Ballarino, A. (2014). Development of superconducting links for the large hadron collider machine. *Superconductor Science and Technology, 27*(4), 044024.

Barron, R. (1999). *Cryogenic heat transfer*. Taylor & Francis.

Baylis, J. A. (1973). Superconducting cables for a.c. and d.c. power transmission. *Philosophical Transactions of the Royal Society of London. Series A, Mathematical and Physical Sciences, 275*(1248), 205–224.

Bednorz, J. G., & Muller, K. A. (1986). Possible high T_c superconductivity in the Ba–La–Cu–O system. *Z. Physik B – Condensed Matter, 64*(2), 189–193.

Boman, A., & Doty, D. (2002). Design and manufacture of ultra-low-mass, cryogenic heat exchangers. *Cryogenics, 41*, 797–803.

Callaway, E. B., Naderi, G., van Le, Q., & Schwartz, J. (2014). Statistical analysis of the relationship between electrical transport and filament microstructure in multifilamentary Bi2Sr2CaCu2Ox/Ag/Ag?Mg round wires. *Superconductor Science and Technology, 27*, 044020.

Cheadle, M. J., Woźniak, M., Bromberg, L., Glowacki, B. A., Jiang, X., Zeng, R., et al. (2013). DC superconducting cable using MgB2 wires. *IEEE Transactions on Applied Superconductivity, 23*(3), 6200805.

Chowdhuri, P., & Laquer, H. L. (1978). Some electrical characteristics of a DR superconducting cable. *IEEE Transactions on Power Apparatus and Systems, PAS-97*(2), 399–408.

Crook, D. G., Graber, L., & Pamidi, S. V. (2014). *Cryogenic heat sink for gas cooled superconducting power devices*. US Patent 8809679 B1.

Cryomech, Inc. Cryorefrigerators, [Online]. Available http://www.cryomech.com/products/cryorefrigerators/. Accessed September 2014.

Cryomech, Inc. "Cold helium circulation systems at cryomech" [Online]. Available http://www.cryomech.com/products/cold-helium-circulation/. Accessed September 2014.

Dai, S., Xiao, L., Zhang, H., Teng, Y., Liang, X., Song, N., et al. (2014). Testing and demonstration of a 10-kA HTS DC power cable. *IEEE Transactions on Applied Superconductivity, 24*(2), 5400104.

Del Rosso, A. (2014). *World-record current in a superconductor* [Online]. Available http://home.web.cern.ch/about/updates/2014/04/world-record-current-superconductor.

Demko, J. A., Lue, J. W., Gouge, M. J., Linsay, D., Roden, M., Willen, D., et al. (2003). Cryostat vacuum thermal considerations for HTS power transmission cable systems. *IEEE Transactions on Applied Superconductivity, 13*(2), 1930–1933.

Demko, J. A., Sauers, I., James, D. R., Gouge, M. J., Lindsay, D., Roden, M., et al. (2007). Triaxial HTS cable for the AEP bixby project. *IEEE Transactions on Applied Superconductivity, 17*(2), 2047–2050.

Dolgosheyev, P. I., Peshkov, I. B., Svalov, G. G., Bortnik, I. M., Karapazyuk, V. L., Kubarev, L. P., et al. (1979). Design and first stage test of 50-meter flexible superconducting cable. *IEEE Transactions on Magnetics, MAG-15*(1), 150–154.

Eckroad, S. (2009). *A superconducting DC cable*. Palo Alto, CA: Electric Power Research Institute.

Eckroad, S. (2012). *Novel approaches and alternative cryogens for cooling a superconducting cable: A comparison of liquid air with other fluids*. Palo Alto, CA: Electric Power Research Institute.

Edwards, D. R. (1968). *Electrical and electronics technician engineer* (vol. 1).

Ekin, J. W. (2006). *Experimental techniques for low-temperature measurements* (pp. 234–235). Oxford University Press.

EPRI-Minervini. (2013). *Development of HTS high current cables and joints for DC power and high field magnet applications* [Online]. Available http://mydocs.epri.com/docs/PublicMeetingMaterials/MRNYPKPLTGV/15-MINERVINI-HTS_Cables_for_DC_Apps.pdf.

EPRI-Pamidi. (2013). *Installation and testing of helium gas cooled superconducting DC cable at FSU-CAPS* [Online]. Available http://mydocs.epri.com/docs/PublicMeetingMaterials/MRNYPKPLTGV/19-PAMIDI-Helium_Cooled_DC_Cable.pdf. Accessed September 2014.

Fenger, M. (2010). *Partial discharge in cables: Lowering the risk of unforeseen cable insulation failures*. San Diego, CA, USA.

Ferrara, P. J., Uva, M. A., & Nowlin, J. (2011). Naval ship-to-shore high temperature superconducting power. *IEEE Transactions on Applied Superconductivity, 21*(3), 984–987.

Fitzpatrick, B. K., Kephart, J. T., & Golda, E. M. (2007). Characterization of gaseous helium flow cryogen in a flexible cryostat for naval applications of high temperature superconductors. *IEEE Transactions on Applied Superconductivity, 17*(2), 1752–1755.

Florida State University. (2013). [Online]. Available http://news.fsu.edu/More-FSU-News/24-7-News-Archive/2013/December/Florida-State-Center-for-Advanced-Power-Systems-researchers-achieve-major-breakthrough. Accessed September 2014.

Forsyth, E. B. (1972). *Underground power transmission by superconducting cable*. Upton, NY: Brookhaven National Lab.

Forsyth, E. B., Blewett, J. P., Britton, R. B., Garber, M., Gurinsky, D. H., Hendrie, J. M., et al. (1973). Flexible superconducting power cables. *IEEE Transactions on Power Apparatus and Systems, PAS-92*(2), 494–505.

Forsyth, E. B., & Thomas, R. A. (1986). Performance summary of the brookhaven superconducting power transmission systems. *Cryogenics, 26*(11), 599–614.

Gallego, N. C., & Klett, J. W. (2003). Carbon foams for thermal management. *Carbon, 41*(7), 1461–1466.

Garber, M. (1979). A 10 m Nb3Sn Cable for 60 Hz power transmission. *IEEE Transactions on Magnetics, MAG-15*(1), 155–158.

Garwin, R. L., & Matisoo, J. (1967). Superconducting lines for the transmission of large amounts of electrical power over great distances. *Proceedings of the IEEE, 55*(4), 538–548.

Goldacker, W., Grilli, F., Pardo, E., Kario, A., Schlachter, S. K., & Michal Vojenčiak, M. (2014). Roebel cables from REBCO coated conductors: a one-century-old concept for the superconductivity of the future. *Superconductor Science and Technology, 27*, 093001.

Graber, L., Kim, C. H., Pamidi, S. V., Rodrigo, H., & Knoll, D. (2014). *Dielectric design validation of a helium gas cooled superconducting DC power Cable*. Philadelphia, PA, USA.

Graber, L., Rodrigo, H., Salmhofer, F., Knoll, D., & Crook, D. G. (2011). *Experimental determination of dielectric properties of cryoflex in high pressure cold helium gas*. Hannover, Germany.

Grant, P. M. (2005). The SuperCable: dual delivery of chemical and electric power. *IEEE Transactions on Applied Superconductivity, 15*(2), 1810–1813.

Grant, P. M., Gregory, W. V., & Eckroad, S. W. (2007). Design criteria for warm temperature dielectric superconducting DC cable: impact of co-pole magnetic fields. In *Eighth European conference on applied superconductivity*.
Grant, P. M., Starr, C., & Overbye, T. J. (2006). A power grid for the hydrogen economy. *Scientific American, 295*, 76–83.
Hassenzahl, W. V., Eckroad, S. E., Grant, P. M., Gregory, B., & Nilsson, S. (2009). A high-power superconducting DC cable. *IEEE Transactions on Applied Superconductivity, 19*(3), 1756–1761.
Haugan, T. J., Long, J. D., Hampton, L. A., & Barnes, P. N. (2008). Design of compact, lightweight power transmission devices for specialized high power applications. *SAE International Journal of Aerospace, 1*, 1088.
Jiang, J., Starch, W. L., Hannion, M., Kametani, F., Trociewitz, U. P., Hellstrom, E. E., et al. (2011). Doubled critical current density in Bi-2212 round wires by reduction of the residual bubble density. *Superconductor Science and Technology, 24*, 082001.
Jin, S., Mavoori, H., Bower, C., & van Dover, R. B. (May 31, 2001). High critical currents in iron-clad superconducting MgB2 wires. *Nature, 411*, 563–565.
Kalsi, S. S. (2011). *Applications of high temperature superconductors to electric power equipment* (1st ed.). Hoboken, NJ: Wiley-IEEE Press.
Kato, T., Kobayashi, S., Yamazaki, K., Ohkura, K., Ueyama, M., Ayai, N., et al. (2004). Development of high performance Ag sheathed Bi2223 wire. *Physica C: Superconductivity, 412-414*(2), 1066–1072.
Kelley, N., Nassi, M., & Masur, L. (2001). *Application of HTS wire and cables to power transmission: State of the art and opportunities*. Columbus, OH, USA.
Kephart, J. T., Fitzpatrick, B. K., Ferrara, P., Pyryt, M., Penkos, J., & Golda, E. M. (2011). High temperature superconducting degaussing from feasibility study to fleet adoption. *IEEE Transactions on Applied Superconductivity, 21*(32), 2229–2232.
Kim, C. H., Kim, S.-K., Graber, L., & Pamidi, S. V. (2014). *Cryogenic thermal studies on terminations for helium gas cooled superconducting cables*. Enschede, The Netherlands.
Kim, C. H., Kim, J.-G., & Pamidi, S. V. (2014). *Cryogenic thermal studies on cryocooler based helium circulation system for gas cooled superconducting power devices*. Syracuse, NY.
Klaudy, P. A., & Gerhold, J. (1967). *Advances in Cryogeninc Engineering, 11*, 684.
Klaudy, P. A., & Gerhold, J. (1983). "Practical conclusions from field trials of a superconducting cable. *IEEE Transactions on Magnetics, MAG-19*(3), 656–661.
Kosaki, M., Nagao, M., Mizuno, Y., Shimizu, N., & Horii, K. (1992). Development of extruded polymer insulated superconducting cable. *Crogenics, 32*(10), 885–894.
Kostyuk, V. V., Antyukhov, I. V., Blagov, E. V., Vysotsky, V. S., Katorgin, B. I., Nosov, A. A., et al. (2012). Experimental hybrid power transmission line with liquid hydrogen and MgB2-based superconducting cable. *Technical Physics Letters, 38*(3), 279–282.
Kuang, J. J., Kim, T., Xu, M. L., & Lu, T. J. (2012). Ultra-lightweight compact heat sinks with metal foams under axial fan flow impingement. *Heat Transfer Engineering, 33*(7), 642–650.
Küchler, A. (2005). *Hochspannungstechnik*. Heidelberg: Springer.
Kumakura, H., Matsumoto, A., Fujii, H., & Togano, K. (2001). High transport critical current density obtained for powder-in-tube-processed MgB2 tapes and wires using stainless steel and Cu-Ni tubes. *Applied Physics Letters, 79*, 2435–2437.
Kwag, D. S., Nguyen, V. D., Baek, S. M., Kim, H. J., Cho, J. W., & Kim, S. H. (2005). A study on the composite dielectric properties for an HTS cable. *IEEE Transactions on Applied Superconductivity, 15*(2), 1731–1734.

van der Laan, D. C., Lu, X. F., & Goodrich, L. F. (2011). Compact GdBa2Cu3O7-δ coated conductor cables for electric power transmission and magnet applications. *Superconductor Science and Technology, 24*, 042001.

Landgraf, M. (2014). *Operation of longest superconducting cable worldwide started* [Online]. Available http://phys.org/news/2014-05-longest-superconducting-cable-worldwide.html. Accessed 03.09.14.

Laquer, H. L., Dean, J. W., & Chowdhuri, P. (1977). Electrical, cryogenic and system design of a DC superconducting power transmission line. *IEEE Transactions on Magnetics, MAG-13*(1), 182–187.

Larbalestier, D. C. (1997). The road to conductors of high temperature superconductors: 10 years do make a difference! *IEEE Transactions on Applied Superconductivity, 7*(2), 90–97.

Larbalestier, D., Gurevich, A., Feldmann, D. M., & Polyanskii, A. (November 15, 2001). High-T_c superconducting materials for electric power applications. *Nature, 414*, 368–377.

Larbalestier, D. C., Jiang, J., Trociewitz, U. P., Kametani, F., Scheuerlein, C., Dalban-Canassy, M., et al. (2014). Isotropic round-wire multifilament cuprate superconductor for generation of magnetic fields above 30 t. *Nature Material, 13*(4), 375–381.

Lee, R. C., Dada, A., & Ringo, S. M. (2005). Cryogenic refrigeration system for HTS cables. *IEEE Transactions on Applied Superconductivity, 15*(2), 1798–1801.

Linde Cryogenics, "Linde cryogenics," [Online]. Available http://www.linde-kryotechnik.ch/public/datenblaetter/helium_eng_einzeln.pdf. Accessed September 2014.

Maguire, J., Folts, D., Yuan, J., Lindsay, D., Knoll, D., Bratt, S., et al. (2009). Development and demonstration of a fault current limiting HTS Cable to be installed in the Con Edison grid. *IEEE Transactions on Applied Superconductivity, 19*(3), 1740–1743.

Maguire, J. F., Yuan, J., Romanosky, W., Schmidt, F., Soika, R., Bratt, S., et al. (2011). Progress and status of a 2G HTS power Cable to Be installed in the Long Island power authority (LIPA) grid. *IEEE Transactions on Applied Superconductivity, 21*(3), 961–966.

Malagoli, A., Bernini, C., Braccini, V., Fanciulli, C., Romano, G., & Vignolo, M. (2009). Fabrication and superconducting properties of multifilamentary MgB2 conductors for AC purposes: twisted tapes and wires with very thin filaments. *Superconductor Science and Technology, 22*, 105017.

Marken, K. R. (2011). *Status of coated conductor activities in the United States. No. LA-UR-11-06650.* Los Alamos: Los Alamos National Laboratory.

Masson, P. J., Brown, G. V., Soban, D. S., & Luongo, C. A. (2007). All electric aircraft (update). *Superconductor Science and Technology, 20*, 748–756.

Masur, L. J., Kellers, J., Feng, L., Fleshler, S., & Podtburg, E. R. (2002). Industrial high temperature superconductors: perspectives and milestones. *IEEE Transactions on Applied Superconductivity, 12*(1), 1145–1150.

Masur, L., Parker, D., Tanner, M., Podtburg, E., Buczek, D., Scudiere, J., et al. (2001). Long length manufacturing of high performance BSCCO-2223 tape for the Detroit Edison power cable project. *IEEE Transactions on Applied Superconductivity, 11*(1), 3256–3260.

McFee, R. (1959). Optimum input leads for cryogenic apparatus. *The Review of Scientific Instruments, 30*(2), 98–102.

Meyerhoff, R. W. (1971). Superconducting power transmission. *Cryogenics, 11*(2), 91–101.

Musenich, R., Greco, M., Razeti, M., & Tavilla, G. (2007). Electrical characterization of a multi-strand MgB2 cable. *Superconductor Science and Technology, 20*(3), 235–238.

Ohsaki, H., Lv, Z., Sekino, M., & Tomita, M. (2012). Application of superconducting power cables to DC electric railway systems. *Physics Procedia, 36*, 908–913.

Ohsaki, H., Matsushita, N., Koseki, T., & Tomita, A. (2014). Characteristic analysis of DC electric railway systems with superconducting power cables connecting power substations.

In *Journal of Physics: Conference series, vol. 507, no. 11th European conference on applied superconductivity* (p. 032035).

Osabe, G., Ayai, N., Kikuchi, M., Tatamidani, K., Nakashima, T., Fujikami, J., et al. (2010). Recent progress of high performance Ag-sheathed Bi2223 wire. *Physica C: Superconductivity, 470*(20), 1365–1368.

Pacific Gas and Electric Company. (2006). *High performance data centers : A design guidelines sourcebook*. Pacific Gas and Electric Company.

Pamidi, S. V., Kim, C. H., Kim, J-h., Crook, D., & Dale, S. (2012). Cryogenic helium gas circulation system for advanced characterization of superconducting cables and other devices. *Cryogenics, 52*(4–6), 315.

Radebaugh, R. (2004). Refrigeration for superconductors (invited). *Proceedings of the IEEE, 92*(10), 1719.

Radebaugh, R. (2009). Cryocoolers: the state of the art and recent developments. *Journal of Physics: Condensed Matter, 21*, 164219.

Rey, C. M., Duckworth, R. C., Demko, J. A., Ellis, A., James, D. R., Gouge, M. J., et al. (2010). Test results for a 25 meter prototype fault current limiting HTS Cable for project Hydra. *Advances in Cryogenic Engineering: Transactions of the Cryogenic Engineering Conference, 55*, 453–460.

Rigby, S. J., & Weedy, B. M. (1975). Liquid nitrogen-impregnated tape insulation for cryoresistive cable. *IEEE Transactions on Electrical Insulation, 10*(1), 1–9.

Rodrigo, H., Graber, L., Kwag, D. S., Crook, D. G., & Trociewitz, B. (2013). Comparative study of high voltage bushing designs suitable for apparatus containing cryogenic helium gas. *Cryogenics, 57*, 12–17.

Rodrigo, H., Kwag, D., Graber, L., Trociewitz, B., & Pamidi, S. (2014). AC flashover voltages along epoxy surfaces in gaseous helium compared to liquid nitrogen and transformer oil. *IEEE Transactions on Applied Superconductivity, 24*(3), 7700506.

Rodrigo, H., Salmhofer, F., Kwag, D. S., Pamidi, S. V., Graber, L., Crook, D., et al. (2012). Electrical and thermal characterization of a novel high pressure gas cooled DC power cable. *Cryogenics, 52*, 310.

Rogalla, H., & Kes, P. H. (2012). *100 years of superconductivity*. Boca Raton: CRC Press.

Rupich, M. W., Li, X., Sathyamurthy, S., Thieme, C. L. H., DeMoranville, K., Gannon, J., et al. (2013). Second generation wire development at AMSC. *IEEE Transactions on Applied Superconductivity, 23*(3), 6601205.

Ryu, C., Jang, H., Choi, C., Kim, Y., & Kim, H. (October 2013). Current status of demonstration and comercialization of HTS Cable system in grid in Korea. In *Proceedings of 2013 IEEE international conference on applied superconductivity and electromagnetic devices* (pp. 539–542).

Sauers, I., James, R., Ellis, A., Tuncer, E., Polizos, G., & Pace, M. (2011). Effect of bubbles on liquid nitrogen breakdown in plane-plane electrode geometry from 100-250 kPa. *IEEE Transactions on Applied Superconductivity, 21*(3), 1892–1895.

Scanlan, R. M., Malozemoff, A. P., & Larbalestier, D. C. (2004). Superconducting materials for large scale applications (invited). *Proceedings of the IEEE, 92*(10), 1639–1654.

Selvamanickam, V., Chen, Y., Xiong, X., Xie, Y., Reeves, J., Zhang, X., et al. (2007). Recent progress in second-generation HTS conductor scale-up at SuperPower. *IEEE Transactions on Applied Superconductivity, 17*(2), 3231–3234.

Seong, J.-K., Choi, W., Shin, W.-J., Hwang, J.-S., & Lee, B.-W. (2013). Experimental and analytical study on DC breakdown characteristics of butt gap condition in LN2/PPLP composite system. *IEEE Transactions on Applied Superconductivity, 23*(3), 5401604.

Shin, W.-J., Hwang, J.-S., Seong, J.-K., Lee, J.-G., & Lee, B.-W. (2012). *Breakdown characteristics of liquid nitrogen for transmission-class superconducting electric equipment*. Bali, Indonesia.

Shiohara, Y., Taneda, T., & Yoshizumi, M. (2012). Overview of materials and power applications of coated conductors project. *Japanese Journal of Applied Physics, 51*, 010007.

Soltanian, S., Wang, X. L., Horvat, J., Dou, S. X., Sumption, M. D., Bhatia, M., et al. (2005). High transport critical current density and large Hc2 and Hirr in nanoscale SiC doped MgB2 wires sintered at low temperature. *Superconductor Science and Technology, 18*, 658–666.

Starr, C. (February 2002). National energy planning for the century: the continental SuperGrid. *Nuclear News*, 31–35.

Stemmle, M., Merschel, F., Noe, M., & Hobl, A. (2013). *Ampacity project – Worldwide first superconducting cable and fault current limiter installation in a German city center*. Stockholm.

Stemmle, M., Merschel, F., Noe, M., & Hobl, A. (2014). *AmpaCity – Advanced superconducting medium voltage system for urban area power supply*. Chicago.

Stirling Cryogenics, "Stirling cryogenics," [Online]. Available http://www.stirlingcryogenics.com/. Accessed September 2014.

Straatman, A. G., Gallegob, N. C., Thompsonc, B. E., & Hangan, H. (2006). Thermal characterization of porous carbon foam-convection in parallel flow. *International Journal of Heat and Mass Transfer, 49*(11–12), 1991–1998.

SuperPower. (2014). *SuperPower 2G HTS wire specifications* [Online]. Available http://www.superpower-inc.com/system/files/SP_2G+Wire+Spec+Sheet_2014_web_v1.pdf. Accessed September 2014.

Suttel, N., Kim, C. H., Ordonez, J., Shah, D., Graber, L., & Pamidi, S. Thermal modeling of gaseous helium as a cryogen for high temperature superconducting cable components. *IEEE Transactions on Applied Superconductivity*, in print.

Takayasu, M., Chiesa, L., Bromberg, L., & Minervini, J. V. (2012). HTS twisted stacked-tape cable conductor. *Superconductor Science and Technology, 25*, 014011.

Tomita, M., Fukumoto, Y., Suzuki, K., & Miryala, M. (2010). Development of prototype DC superconducting cable for railway system. *Physica C, 470*, s1007–s1008.

Tomsic, M., Rindfleisch, M., Yue, J., McFadden, K., Phillips, J., Sumption, M. D., et al. (2007). Overview of MgB2 superconductor applications. *International Journal of Applied Ceramic Technology, 4*(3), 250–259.

Ton, M., Fortenbery, B., & Tschudi, W. (2008). *DC power for improved data center efficiency*. Berkeley, CA: Lawrence Berkeley National Laboratory.

Trevisani, L., Fabbri, M., & Negrini, F. (2006). Long-term scenarios for energy and environment: Energy from the desert with very large solar plants using liquid hydrogen and superconducting technologies. *Fuel Processing Technology, 87*(2), 157–161.

Tuncer, E., Zuev, Y. L., Sauers, I., James, D. R., & Ellis, A. R. (2007). Electrical properties of semiconducting tapes used in HTS power cables. *IEEE Transactions on Applied Superconductivity, 17*(2), 1497–1500.

Venkatarathnam, G., & Sarangi, S. (1990). Matrix heat exchangers and their application in cryogenic systems. *Cryogenics, 30*(11), 907–908.

Vermeuleun, H. (2013). Cryogenic circulators: the solution for cooling problems? *Cold Facts, 29*, 46–48.

Weber, C. S., Lee, R., Ringo, S., Masuda, T., Yumura, H., & Moscovic, J. (2007). Testing and demonstration results of the 350 m long HTS cable system installed in Albany, NY. *IEEE Transactions on Applied Superconductivity, 17*(2), 2038–2042.

Weber, C. S., Reis, C. T., Dada, A., Masuda, T., & Moscovic, J. (2005). Overview of the underground 34.5 kV HTS power cable program in albany, NY. *IEEE Transactions on Applied Superconductivity, 15*(2), 1793–1797.

Xiao, L., Dai, S., Lin, L., Zhang, Z., & Zhan, J. (2013). HTS power technology for future DC power grid (Invited). *IEEE Transactions on Applied Superconductivity, 23*(3), 5401506.

Yamada, Y., Mogi, M., & Sato, K. (2007). Examples and future prospects of high-temperature superconductor products. *SEI Technical Review, 65*, 51–59.

Yi, H. P., Han, Z., Zhang, J. S., Liu, T., Liu, L., Li, M. Y., et al. (2004). Research status of the manufacturing technology and application properties of Bi-2223/Ag tapes at Innost. *Physica C: Superconductivity, 412–414*(2), 1073–1078.

Yuan, Y., Jiang, J., Cai, X. Y., Larbalestier, D. C., Hellstrom, E. E., Huang, Y., et al. (2004). Significantly enhanced critical current density in Ag-sheathed (Bi,Pb)2Sr2Ca2Cu3Ox composite conductors prepared by overpressure processing in final heat treatment. *Applied Physics Letters, 84*(12), 2127–2129.

Yumura, H., Ashibe, Y., Ohya, M., Itoh, H., Watanabe, M., Masuda, T., et al. (2013). Update of YOKOHAMA HTS cable project (invited). *IEEE Transactions on Applied Superconductivity, 23*(3), 5402306.

High-temperature superconducting cable cooling systems for power grid applications

8

J.A. Demko
LeTourneau University, Longview, TX, USA

8.1 Introduction

The thermal management of high-temperature superconducting (HTS) cable systems remains an important topic in order for these systems to eventually gain acceptance as part of the electricity grid. To deploy an HTS power cable, it must be maintained at an operating temperature that typically ranges between 60 and 80 K. The minimum allowable operating temperature is a balance between superconductor performance, which increases as the temperature is lowered, and the increase in cost to produce refrigeration at lower temperatures. Most often, the lowest operating temperature is determined by the cooling medium selected. Liquid nitrogen (LN_2) is the most common cooling medium selected for HTS cable operation because of its many excellent properties, including excellent dielectric properties at high voltages, a large operating temperature range (particularly at high pressures), high heat capacity, and low cost. With LN_2 as the coolant, the lowest temperature in the cable system would be about 65 K. As the refrigeration operating temperature of the HTS cable moves to lower temperatures, more input power is required by the refrigerator, making it more expensive to operate the HTS cable. The technology of cryogenic refrigeration is a very broad subject, with many applications in research and commercial applications. An overview of current cryogenic refrigeration technology was presented by Ganni and Fesmire (2012). They showed that the size of the cryogenic refrigeration systems required are within existing capabilities. However, specific equipment needs to be developed, which has been optimized for HTS cable operation.

The required capacity of the refrigeration system drives the cost of the unit. To determine the required capacity of the refrigeration unit, an accurate estimation of the minimum heat load of the HTS cable and its accompanying terminations under full load is required. Minimizing the thermal loads is accomplished by both proper HTS cable design and by thermally isolating the HTS cable from the ambient surroundings and external heat loads by a vacuum-jacketed cryostat. The HTS cable is electrically connected at the ends to an electrical supply and load using a termination or high-voltage bushing. The termination also provides for ambient temperature electrical connections to the HTS cable. The termination serves many functions and must

Superconductors in the Power Grid. http://dx.doi.org/10.1016/B978-1-78242-029-3.00008-X

be specifically designed to accommodate large currents and high voltages, as well as to minimize heat loads from the ambient electrical grid to the HTS cable operating at cryogenic temperatures. For direct current (dc) operation below the critical current (I_c) of the HTS cable, there would be minimal heat generation that would require cooling of the cable. The dominant heat load during dc operation stems from both the terminations and the vacuum-jacketed cryostat. Under alternating current (ac) operation, there is additional heat generation from the ac losses of the HTS wire/tapes and dielectric losses from the electrical insulation, which contribute to the overall refrigeration load of the HTS cable system.

8.2 Thermal loads

To understand the design and operation of a refrigeration system for an HTS cable, it is necessary to understand the thermal loads that are present not only by magnitude but also by how they are distributed in the cable. The thermal loads that determine the required refrigeration capacity come from three primary sources: termination or bushings, the HTS cable, and the vacuum-jacketed cryostat. First, there is a heat load from the interfaces where the room-temperature connections are made and the transition to the cold superconducting cable takes place. This junction is commonly referred to as the termination or bushing. A compact termination for a three-phase ac triaxial HTS cable fabricated by the Southwire Company is shown in Figure 8.1.

Second, the HTS cable itself has electrical losses that arise from the ac currents flowing in each phase, and for HTS cable designs possessing cold dielectric insulation between each phase. The dielectric losses in the cable electrical insulation increase as

Figure 8.1 Southwire Co. triaxial high-temperature superconducting cable termination installed at the American electric power Bixby substation in Columbus, OH.
Photo courtesy of Southwire Co.

the voltage level increases. The third source is background thermal load through the vacuum-jacketed cryostat that thermally isolates the low-temperature cable from the ambient environment and also serves as a pressure barrier for the circulating cooling stream of the HTS cable.

8.2.1 Terminations

The terminations located at each end of the HTS power transmission cable must provide a room-temperature connection to the electrical network as well as the cryogenic liquid couplings that provide "supply" and "return" locations for the LN_2 coolant to the HTS cable. These terminations have many stringent design and operational requirements. They must operate at both high voltages and high currents, with the exterior exposed to the harshest elements of the environment; meanwhile, the interior must transition the electrical connection from a room-temperature electrical network down to the operating temperature of the HTS cable. All of this must be accomplished while minimizing the thermal load to the cryogenic refrigeration system.

For a high-current connection, McFee (1959) demonstrated that the minimum heat load between room temperature (300 K) down to LN_2 temperature (77 K) is around 45 W/kA for a conduction-cooled copper conductor in a vacuum (no external heat sources) at full current. In a shielded HTS cable, two current connections are required at each end. If, for example, 2 kA is the desired current rating of an HTS cable operating at 77 K, this results in at least 180 W per termination of thermal load from the current leads alone. To cool the HTS cable with a cryogenic fluid and circulate that fluid through the HTS cable and eventually return the fluid back to the refrigeration system, both a supply and return connection from the refrigeration system to the HTS cable must be employed. These types of connections also introduce parasitic thermal loads on the refrigeration system. A common connection of this

Figure 8.2 A standard bayonet connection used to join sections of cryostat.
Photo courtesy of PHPK Inc., Columbus, OH.

type is the standard bayonet connection, as shown in Figure 8.2. A standard bayonet connection to the HTS cable end provides a thermal load of ~20 W at 77 K for an inner diameter of ~75 mm and up to 90 W for a 200-mm inner diameter. Additional heat loads are present because of the finite size of the termination. Previously measured heat loads on a typical termination were around 300 W (Demko et al., 2000) for a termination rated at 13.8 kV and 1.2 kA_{rms} with an HTS cable operating at 77 K. The heat load for a termination on one end of the HTS cable system can be estimated by Eqn (8.1), where N_e is the electrical connections, I_e is the capacity of the connection in kilo-amperes, with an additional heat load from the environment of Q_{env}:

$$Q_{end} = 45 \times N_e \times I_e + Q_{env} \tag{8.1}$$

The environment heat load depends on the physical size of the termination and could be as much as 200 W. The physical size of the termination is determined by several factors, including the number of cable phases contained in the termination, the current and voltage rating, and the operating pressures.

8.2.2 Cryostat thermal loads

The static thermal load from the environmental heating depends on the thermal performance of the cryostat used in the system. The HTS cable cryostat typically is a flexible double-wall construction with vacuum-jacketed multilayer superinsulation (MLI). The ambient sink temperature, T_∞, could range from 245 to 320 K, depending on the geographical location of the HTS cable installation and time of the year. The performance of several thermal insulation systems for both rigid and flexible piping, or cryostats, for use with HTS is addressed in two papers by Fesmire, Augustynowicz, and Demko (2002a,b). The first paper compares the thermal performance of rigid and flexible cryostats with different insulation systems. Fesmire et al. indicated that the flexible cryostat will generally have higher heat loads than a rigid cryostat due in part to the corrugated geometry of the tube walls. In the second paper, Fesmire et al. investigated the additional heat load from simulated bending of a cryostat. The typical commercially available vacuum-insulated flexible cryostats have an effective or actual field-installation thermal conductivity, k_{oafi}, of 0.0008 W/m/K. The local heat transfer through the cryostat per unit length can be calculated using Eqn (8.2) and depends on the local LN_2 temperature, T_v, and the cryostat inner and outer tube diameters, D_{ci} and D_{co}. The temperature difference driving this heat-transfer term is typically more than 220 K for the outer cryostat:

$$Q = \frac{2\pi k_{eff} L (T_\infty - T_v)}{\ln(D_{co}/D_{ci})} \tag{8.2}$$

8.2.3 Electrical losses

The AC and dielectric losses are distributed along the length of the cable and must be removed locally in order to maintain the cable within its operating temperature. The sum of the ac loss and dielectric loss can vary from 0.1 to 5 W/m/phase, depending on the operating current and the design critical current of the cable. The ac losses in the HTS cable will vary over time depending on the instantaneous operating current. A great deal of time and care must go into the design of HTS cable in order to minimize the ac losses, and several papers have been written to address these issues (Amemiya et al., 2007; Hamajima, et al., 2007; Rostila et al., 2007). The ac loss, P'_{ac}, in watts per meter, may be simply estimated using the monoblock model (Vellego & Metra, 1995):

$$P'_{ac} = \frac{\mu_0 f I_c^2}{2\pi h^2}\{(2 - Fh)Fh + 2(1 - Fh)\ln(1 - Fh)\} \tag{8.3}$$

where $F = I_p/I_c$ is the ratio between the peak current in the ac cycle and the critical current of the superconductor, f is the frequency, and $h = (D_o^2 - D_i^2)/D_o^2$. The terms D_i and D_o are the inner and outer diameters of the superconductor. The dielectric losses are typically modest, around 0.05 W/m/phase, and depend on the operating voltage of the cable.

The total refrigeration heat load, Q_R, is the sum of the contributions from the terminations, cryostat, and electrical losses. For example, the total heat load of a 1000-m long, three-phase ac HTS cable system with a rating of 2000 A_{rms} and 15 kV is found to be between $Q_R \sim 8$ and 12 kW at 77 K. A more precise heat load estimate will depend on the HTS cable configuration and operating temperature.

8.2.4 Recovery after fault

The cryogenic refrigeration system for HTS cables must also be designed for fault events. This is a scenario for when an HTS cable takes an overcurrent, which can be many times the design operating current (10–20× for a number of ac electrical cycles (typically 50 or 60 Hz) until the protection system (typically a circuit breaker) cuts off the power. A discussion of the refrigeration response to these types of fault events was provided by Demko and Duckworth (2009). It was shown that this type of upset condition provides a temporary increase in heat load until the HTS cable can be brought back to its design operating temperature. This would result in a temporary higher refrigeration temperature as the refrigeration system compensates for the increase in heat load.

8.3 Topology of high-temperature superconducting (HTS) cable cooling circuits

Some early studies regarding refrigeration installations for low-temperature superconducting (LTS) power cables concerning cooling schemes (Morgan & Jensen, 1977), arrangement of refrigeration systems (Longsworth & Schoch, 1980), and reliability (Kadi & Longsworth, 1977) provide an excellent background for more

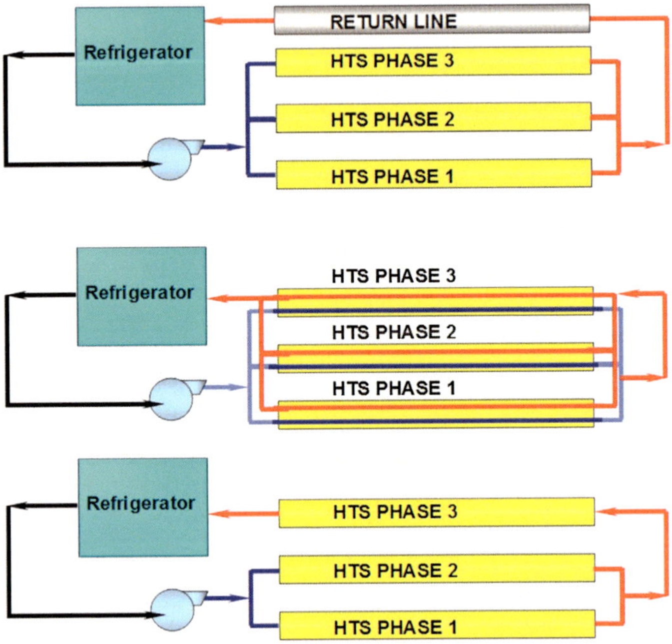

Figure 8.3 Illustration of different cooling schemes used for HTS cables.

recent HTS cable cooling circuit designs. Cooling schemes and refrigeration concepts for HTS cables depend on the cable configuration. Some cable designs and cooling arrangements that can be used are listed in Table 8.1. An illustration of different coolant flow arrangements is shown in Figure 8.3.

8.4 Coolant selection

To distribute cooling throughout the HTS cable, a coolant must be circulated. The choice of coolant used directly impacts the design of the HTS cable system. Typically, HTS cables are cooled using LN_2 because it is a liquid in the temperature range of interest for HTS power applications with a triple point of 63.2 K. LN_2 is relatively inexpensive and readily available from many suppliers. The pressure–volume diagram for nitrogen is provided in Figure 8.4, which shows that the nitrogen will remain liquid at the required operating temperatures for typical operating pressures above 0.5 MPa.

Table 8.1 Summary of cooling topologies for HTS cables

Cooling layout	3-Separate	Triad	Triaxial
Counter flow	Yes	Yes	Yes
Parallel flow	Yes	No	No
Parallel flow, separate return	Yes	Yes	Yes

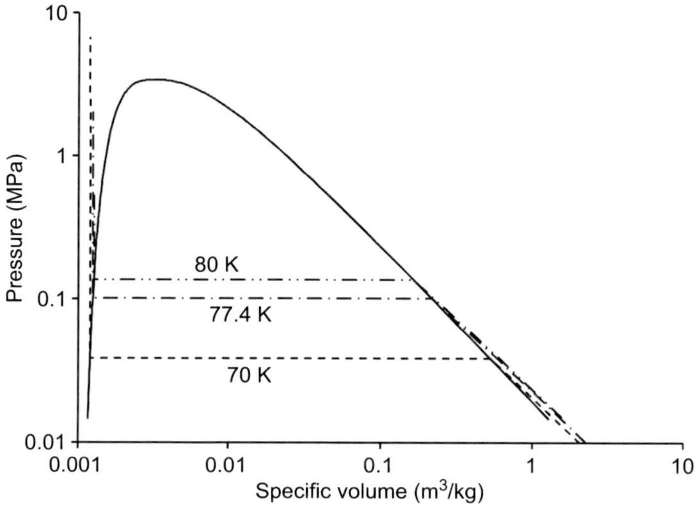

Figure 8.4 Pressure—volume diagram for nitrogen.

Alternative coolants, such as gaseous helium and hydrogen, have been proposed (Demko & Hassenzahl, 2011, 2012). Gaseous helium has been proposed as a coolant for HTS degaussing cables (Kephart et al., 2011). Gaseous coolants will be used in HTS cable installations where there are large changes in elevation along the cable because of the head pressures developed as a result of the change in elevation. For example, an elevation change of 200 m in LN_2 results in a 1.57 MPa differential pressure in the nitrogen from the elevation change. This is substantial considering that most existing systems are designed to operate at pressures of 1.0—1.5 MPa. As the length of HTS cable systems become longer, studies indicate that the refrigeration systems will operate at pressures greater than 3.0 MPa. The elevation head is a significant portion of the total system pressure that would require additional increases in the system pressure rating and in turn increase the cost of the system. To keep the system pressure ratings low, a gaseous coolant will be necessary.

The two most likely candidates for gaseous coolants are helium and hydrogen. Table 8.2 compares the properties of these coolants at 3.0 MPa and 66 K. These gases

Table 8.2 Fluid properties at an assumed supply condition of 3.0 MPa and 66 K

Fluid	Density (kg/m^3)	Specific heat (kJ/kg/K)	Dielectric constant
Liquid nitrogen	860.9	1.98	1.4645
Gaseous helium	20.675	5.31	1.0366
Gaseous hydrogen	11.98	13.51	1.0088

have high specific heats, making them very suitable for cooling mediums. The densities of the gaseous coolants are significantly lower than LN_2. The pumping work required for the circulation of gases is significantly higher for gaseous coolants compared with liquid coolants as a result of their lower density. This combined with their much lower dielectric strength means that gaseous coolants will most likely be used only when necessary.

8.5 Refrigeration system overview

Although there is no official classification, cryogenic refrigeration systems can be subdivided into three general categories: small, medium, or large scale, depending on the size of the refrigeration load. For the purposes of HTS cables, the refrigeration temperature will generally be no lower than 50 K for gaseous-cooled cables and 64 K for LN_2-cooled cables. Small-scale refrigeration would be in the 0.1–100 W range. Commercially available units would be of the Gifford McMahon or pulse-tube cryocooler type (Cryomech, Sumitomo). These units have low refrigeration capacities. These smaller capacity cryocoolers are not normally considered for use in large-scale HTS cable systems. Some of the larger pulse-tube cryocoolers, such as the Cryomech AL330, have been applied to experiments related to ship degaussing cables (Fitzpatrick, Kephart, & Golda, 2007).

Medium-scale refrigeration systems would be in the range of 100 W to 1 kW. These machines are typically based on reciprocating engines or Stirling cycle machines. This scale machine has been proposed but would require using a bank of machines that could be switched on or off depending on the HTS cable load, which can vary by both the time of the year and the time of day. It is generally not a preferred option due to reliability issues with having several small coolers.

Large-scale refrigeration systems are typically turbo machinery based with capacities >1 kW. These systems are typically based on a reversed turbo-Brayton cycle or a Claude cycle. For HTS cables, a simplified open-cycle refrigeration scheme has been employed at test beds, such as at Oak Ridge National Laboratory (Demko et al., 2000, Rey, Duckworth, Demko, James, & Gouge, 2010) and multiple installation sites such as the Southwire Co., Carrollton GA, and AEP Bixby demonstration projects (Demko et al., 2002, 2007). The advantages of the open-cycle system are its low cost, high reliability, and simple maintenance. The main disadvantage is that a constant supply of LN_2 is required to provide the refrigeration. Typical refrigeration systems that can be applied to HTS cables are discussed below.

8.5.1 Refrigeration system thermodynamics

Refrigeration systems operate on a series of consecutive thermodynamic processes that form a cycle that returns the working substance to the same state. A thermodynamic refrigeration cycle removes heat from a low-temperature object (refrigeration) and rejects it at a higher temperature. This cyclic process requires that work be input so that the second law of thermodynamics is not violated (Wark, 1983). The ideal refrigeration system is based on the concept of the reversible Carnot cycle. The Carnot cycle is a theoretical concept because it is ideal, which means that it is lossless, frictionless, and there are no temperature differences in heat exchange processes. None of these ideals can ever be achieved in practice. The Carnot cycle therefore provides a reference standard against which the performance of all other refrigeration cycles can be compared.

8.5.2 Refrigeration basics

The refrigeration load Q_R (discussed in Section 8.2.3) must be removed from the HTS cable using a refrigeration system that requires some amount of input power P. The ideal amount of input power P_{Carnot} can be determined from the Carnot cycle. The ratio of the refrigeration load to the input power supplied is known as the coefficient of performance (COP). For the ideal case, this is given by Eqn (8.4) and can be shown to be related to the temperatures at which the compression is done (room temperature or T_H) and the temperature that the refrigeration is supplied, T_L:

$$\mathrm{COP}_{\mathrm{Carnot}} = \frac{Q_R}{P_{\mathrm{Carnot}}} = \frac{T_L}{T_H - T_L} \tag{8.4}$$

The ideal compressor work per unit mass ($w_{c,ideal}$) of the working fluid in the refrigerator is based on the principle that the minimum compressor work is accomplished using an isothermal process. The isothermal compression can be shown to be determined from the high-end temperature and the change of entropy that is a function of temperature and pressure by the following:

$$w_{c,ideal} = T_H\left(s\left(p_{out}, T_H\right) - s\left(p_{in}, T_H\right)\right) \tag{8.5}$$

$$\eta_c = \frac{w_{c,ideal}}{w_{c,actual}} \tag{8.6}$$

$$P_{c,ideal} = m_f w_{c,ideal} \tag{8.7}$$

The actual compressor work, $w_{c,actual}$, is the actual electrical power needed to run the compressor per unit mass. It is determined from the vendor-supplied compressor performance parameters, input power, $P_{c,actual}$ and mass flow, m_{actual}:

$$w_{c,act} = \frac{P_{c,actual}}{m_{actual}} \tag{8.8}$$

$$P_{c,actual} = m_f w_{c,act} \tag{8.9}$$

$$\mathrm{COP}_{\mathrm{actual}} = \frac{Q_R}{P_{\mathrm{actual}}} \tag{8.10}$$

When discussing cryogenic refrigeration, the specific power consumption, SPC, is frequently used to compare the performance of different systems. The SPC is simply the inverse of the COP as given in Eqn (8.11). The SPC is a measure of the amount of power required to produce a unit of refrigeration, with units of Watts per Watt.

$$SPC = \frac{1}{COP_{actual}} = \frac{P_{actual}}{Q_R} \tag{8.11}$$

A figure of merit that is commonly used is the %Carnot, which can be determined from:

$$\%Carnot = \frac{COP_{actual}}{COP_{Carnot}} \times 100\% \tag{8.12}$$

8.5.3 Functional requirements

The cryogenic refrigeration system is comprised of different subsystems. Each subsystem performs a necessary function during all operating modes of the overall cryogenic refrigeration system.

Typically, the four subsystem divisions are:

1. *Refrigeration skid*: Transfers heat from low-temperature HTS cable to warm ambient surroundings.
2. *Pump station*: Provides pressurized flow of coolant for cooldown and warm-up, circulation during normal operation, or rapid recovery from a fault condition.
3. *Coolant storage*: Tanks to maintain sufficient inventory of coolant (e.g., LN_2). These may be used for backup refrigeration in the event that the refrigeration system shuts down.
4. *Inventory management*: Minimizing losses of inventory in the system due to background losses and fault conditions.

8.6 Types of refrigeration systems

Refrigeration for HTS cable systems can be divided into three categories. In open cycle, the refrigerant is vaporized and then vented as illustrated in Figure 8.5. The open system is the least expensive option in initial capital cost, but it requires a constant replenishment of the refrigerant. This is the LN_2 vacuum-pumped system that is most often used in demonstration and prototype systems. A second category is the mechanical refrigerator, which has typically been based on a reversed Brayton cycle, Claude cycle, or Stirling cycle (Stirling Cryogenics and Refrigeration BV), and in the future may be a pulse tube system (Fleck, Vogel, & Ziegler, 2002). This is illustrated in Figure 8.6. The third category, which is most likely to be applied in practical HTS cable installations, is the hybrid that uses a mechanical refrigerator as the primary source for refrigeration, and an open-cycle refrigeration system as a back-up when the primary system is down for maintenance or repair.

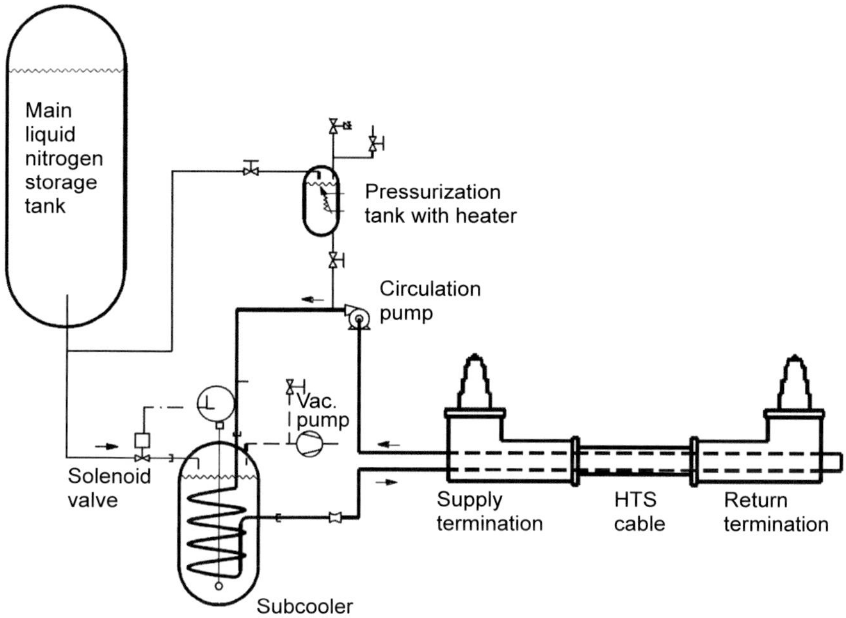

Figure 8.5 Vacuum-pumped subcooler system.

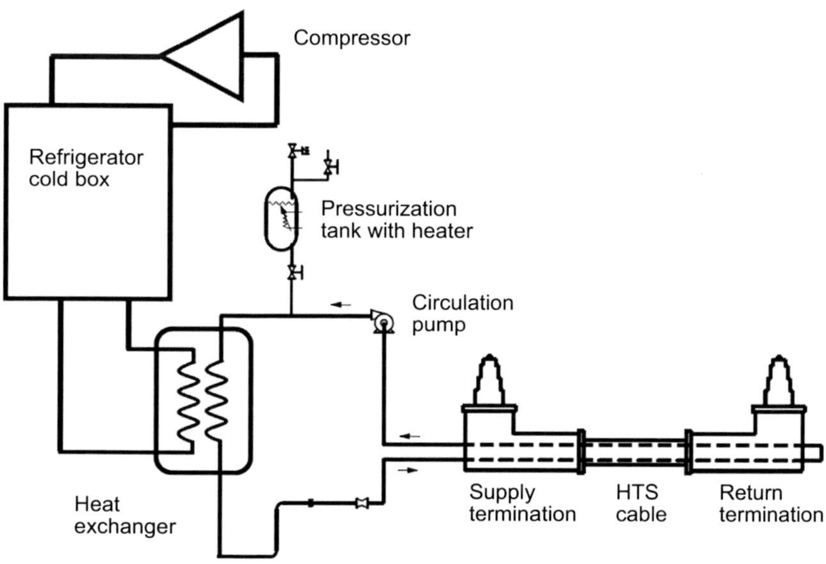

Figure 8.6 Integration of a mechanical refrigeration system in the cooling of an HTS cable.

8.6.1 Open-cycle refrigeration

The use of an open-bath subcooler is one means of providing refrigeration to an HTS cable; it has been widely used in many installations around the world. This system is often used as a backup to the mechanical refrigerator. The system consists of a heat exchanger that has high-pressure, LN_2 stream circulating the cooling stream on the tube side that is cooled by pool boiling heat transfer to a subatmospheric bath on the shell side of the heat exchanger. This is illustrated in Figure 8.7. Refrigeration is accomplished by the boiling of LN_2 so that a level controller is used to open a valve, shown as 0−1, from a supply tank to maintain the liquid level in the subcooler heat exchanger. The simplest control uses a solenoid or on−off valve, but this can produce slight temperature oscillations from the change in pressure when the solenoid valve opens or closes. A more common practice is to use a control valve that gradually opens and closes to maintain the level close to the set point. This system has the ability to quickly respond to changes in load as the current applied to the cable changes.

Vacuum pumps are used to reduce the pressure of the LN_2 in the subcooler heat exchanger bath, in turn lowering the temperature. At 65 K, the vapor pressure of nitrogen is around 0.174 bar. The triple-point temperature is 63.149 K and pressure is 0.126 bar, so the subcooler bath pressure must be carefully controlled to prevent the bath from freezing. Most vacuum pumps operate at or above room temperature. The cold vapor leaving the subcooler must be heated prior to entering the pump, as illustrated in the figure.

The subcooler system can be modeled by direct application of the integral form of conservation of mass and the first law of thermodynamics for a constant volume process. Treating the cable cooling stream as a heat addition, the supply stream as the only inlet stream, the vacuum pump, and possibly the relief valve as the outlet streams, the following equations are derived. For conservation of mass:

$$\frac{\partial \rho_{cv}}{\partial t} = \left(\frac{1}{V_{Tank}}\right)\left(\sum \dot{m}_{in} - \sum \dot{m}_{out}\right) \tag{8.13}$$

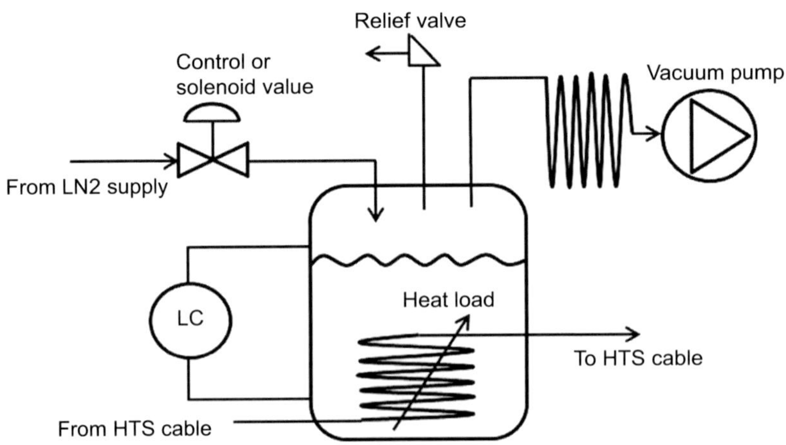

Figure 8.7 Diagram of the open cycle subcooler.

In this equation, ρ_{cv} is the average density of the nitrogen inside the heat exchanger volume, V_{Tank}. The remaining term is the difference in mass flow into and out of the heat exchanger volume.

For conservation of energy, the equation that models the subcooler is

$$\frac{\partial u_{cv}}{\partial t} = \left(\frac{1}{\rho_{cv} V_{Tank}}\right)\left(Q_{Load} + \sum \dot{m}h_{in} - \sum \dot{m}h_{out} + V_{tot}u_{cv}\frac{\partial \rho_{cv}}{\partial t}\right) \quad (8.14)$$

where u_{cv} is the specific internal energy of the contents of the subcooler heat exchanger and Q_R is the total heat load from the cable system. This will also include any pump work that is provided to circulate cooling flow through the HTS cable system. The remaining enthalpy terms represent the net energy changes that occur as the supply stream is evaporated and leaves as vapor.

A pressure-specific volume (p—v) diagram has been given in Figure 8.4. The approximate region in which the subcooler should operate most of the time is to the left of the critical point on the diagram, very near to the liquid saturation line.

8.6.2 Reversed turbo-Brayton cycle

A schematic of the simplest form of a reversed Brayton cycle with one expander is shown in Figure 8.8. The cycle consists of a room-temperature compressor, a counter-current heat exchanger, an expander, and the load heat exchanger. Possible working fluids for refrigeration in the 60—80 K range are helium, hydrogen, or neon (Saji et al., 2001).

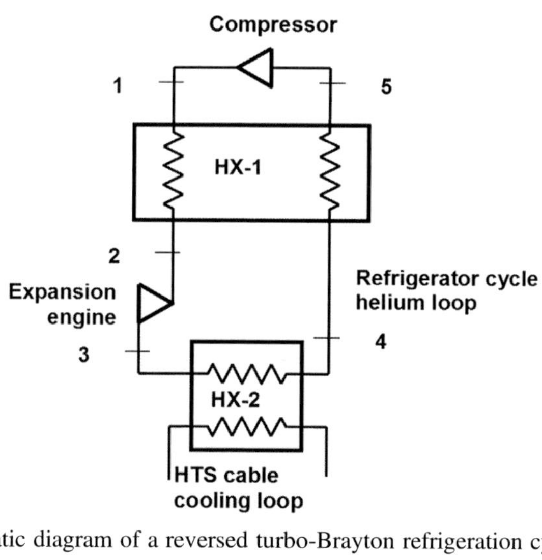

Figure 8.8 Schematic diagram of a reversed turbo-Brayton refrigeration cycle.

8.7 Recent installations

One of the first HTS power transmission cable demonstration projects was installed by Southwire Co. in Carrollton, GA as shown in Figure 8.9. It was a three-phase power transmission system designed for 1250 A_{rms} and 12.4 kV alternating current. The 30-m installation supplied all the electricity for three large manufacturing facilities and operated for more than 4 years. Each HTS cable was a co-axial design that was counterflow cooled. Each of the three separate phases was housed in its own separate cryostat. The refrigeration was supplied by an open bath subcooler built by PHPK Technologies in Columbus, OH. The cooling system developed 4 kW of refrigeration at 72 K during acceptance tests (Stovall et al, 2001; Demko et al, 2002).

Ultera installed a single 200-m long HTS three-phase triaxial cable at the American Electric Power Bixby Substation delivering power to Columbus, OH (Demko et al., 2007). The cable connects a 138-kV/13.2-kV transformer to the distribution switchgear serving several circuits. It is designed for 3000 A_{rms} operation and is cooled in a parallel flow arrangement with its own separate return cryostat. Refrigeration was supplied using a system patterned after the Carrollton, GA installation that was an open bath subcooler, but it was upgraded to use incorporate two pulse-tube cryocoolers, as shown in Figure 8.10, that would remove part of the heat load and reduce LN_2 consumption. It was also designed to provide 8 h of subcooled liquid back-up in the event that the cryogenic system had an outage. The subcooled backup would be provided by storing a supply of subcooled LN_2. In the event of a loss of the pumping system, the subcooled storage could be pressurized and supply subcooled LN_2 to the cable.

Figure 8.9 The Southwire Co. 30-m HTS power transmission cable and refrigeration system. Photo courtesy of Southwire Co. Carrollton, GA.

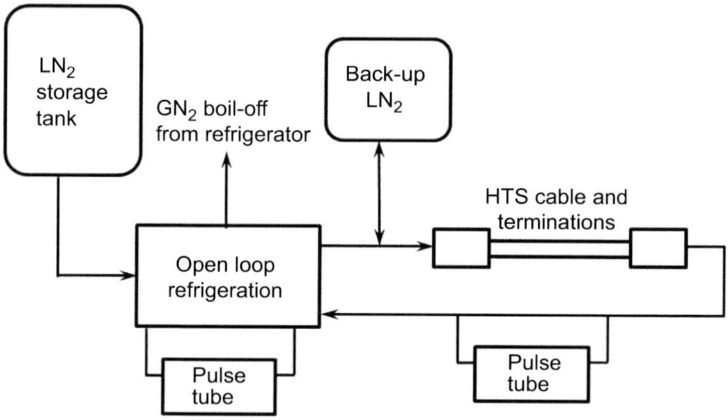

Figure 8.10 Hybrid open-loop/pulse tube refrigeration system diagram for AEP Bixby triaxial cable.

The U.S. Department of Energy has sponsored the demon stration of a 574-MVA, 138-kV HTS cable with the team of Nexans, Air Liquide, Long Island Power Authority (LIPA), and American Superconductor (AMSC). The LIPA HTS cable uses a closed-loop helium refrigerator with an open-loop backup (Maguire, Schmidt, Hamber, & Welsh, 2005, Maguire et al., 2007, 2011). The main helium refrigerator is a closed-loop refrigeration process based on a Brayton cycle. A screw compressor provides the input power to circulate the helium gas through the closed-loop system. The helium gas flow is cooled down, respectively warmed up, in a countercurrent heat exchanger. The high-pressure cold helium gas is expanded with two turbines in series down to the compressor suction pressure level. The refrigeration capacity after the expansion is used to cool down the LN_2. Two pumps circulate the LN_2 through its nitrogen closed-loop system. In a two-expander cycle arrangement, as shown in Figure 8.11, a second expander is used to re-cool the refrigeration stream after removing part of the heat from the cable cooling stream, then the refrigeration stream passes through a second expander to provide the remaining cooling to the HTS cable.

The nitrogen backup refrigerator is an open-loop refrigeration process. LN_2 is supplied in a vaporizer operating at subatmospheric conditions. The system uses a vacuum screw compressor, which provides the low pressure required in the subcooler and compresses the nitrogen vapor up to ambient pressure.

The Albany HTS cable demonstration used a combination of open and closed refrigeration in a novel arrangement as a part of the Cryogenic Refrigeration System (Lee, Dada, & Ringo et al., 2005; Weber et al., 2007). This hybrid arrangement has the advantage of being able to provide the high reliability required for in-grid operation, while not introducing excessive costs through cryocooler redundancy. The design uses a thermosyphon as depicted in Figure 8.12. Inside the thermosyphon unit is a fixed quantity of LN_2 under normal operating conditions. The LN_2 is in contact with a heat exchanger that removes heat from the high-pressure cooling stream circulating through the HTS cable by vaporizing the liquid. A second heat exchanger,

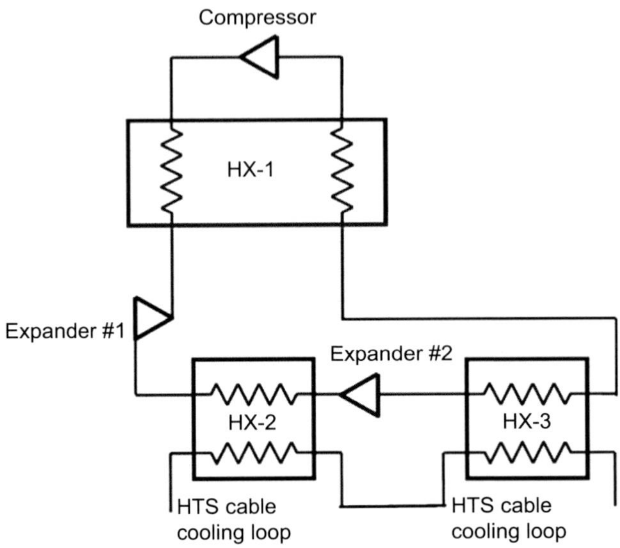

Figure 8.11 Schematic of the mechanical refrigeration system diagram for LIPA HTS cable project.

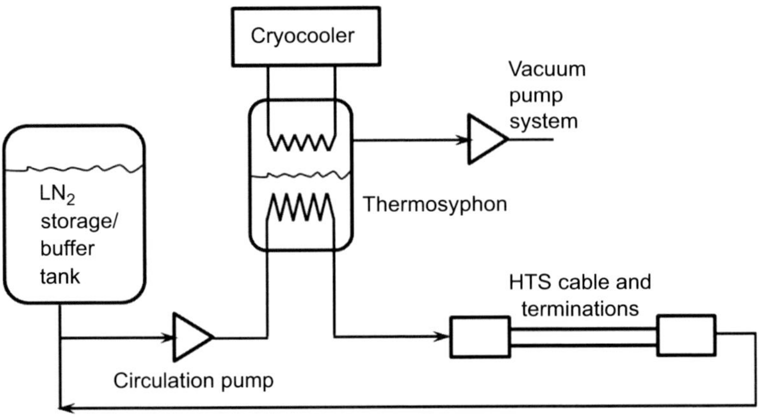

Figure 8.12 Schematic of the thermosyphon refrigeration system used in the Albany New York HTS cable project.

which is connected to a cryocooler, is located at the top of the thermosyphon in the nitrogen vapor. This second heat exchanger condenses the nitrogen vapor replenishing the liquid. As can be seen in the *p*–*v* diagram in Figure 8.2, the LN_2 contained in the thermosyphon is subatmospheric in order to provide refrigeration at temperatures near 70 K. The boiling could produce an increase of the bath pressure if the refrigeration cannot follow the load. The Albany Project demonstrated the feasibility of the operation of this system and the ability to maintain stable temperatures with a step change of

1 kW of refrigeration load. As a backup, should the mechanical refrigerator go down for repair or maintenance, the refrigeration can be supplied by a supply of LN_2 to the thermosyphon and a vacuum pump to remove the vapor boiled-off as an open-cycle refrigeration system.

8.8 Future trends

Discussion of refrigeration system designs for future HTS cable installations has been reported by several authors. Kim et al. (2012) and Lim et al. (2012) discussed the installation and performance test for the Incheon, South Korea HTS cable project. The thermodynamic design of a 10-kW Brayton cryocooler has been described by Chang et al. (2012), and results from testing of a hybrid cooling system are discussed by Lim et al. (2010).

The use of liquid air (LAir) as a coolant for HTS cable systems has been proposed as a coolant because it clearly offers some potential benefits over its LN_2 counterpart (Demko & Hassenzahl, 2012). First, LAir has lower freezing and higher boiling points than nitrogen and can thus provide cooling at a lower temperature (Lemmon, Jacobsen, Penoncello, & Friend, 2000). The triple point and normal boiling point temperatures for LN_2 are 63.15 and 77.36 K, versus 59.75 and 78.90 K for LAir. The significance of operating at a lower temperature is that the critical current of HTS wires increases as the temperature decreases. Second, from a safety perspective, a leak of LAir into an enclosed space will not result in a potential hazard caused by the reduced concentration of oxygen. Third, the use of LAir does not require fractional distillation; thus, it is a simpler process and can be done locally.

8.9 Conclusions

Assessments that have been reported (Gouge & Demko, 2013) provide recommendations that are being addressed for HTS cable systems. One main thrust for research and development efforts to support the commercial application of HTS power cables is to improve existing technology so that the target efficiency of 30% Carnot can be attained. A primary focus for achieving the higher efficiencies is to reduce the input work (i.e., significantly increase the compressor performance). The performance of other system components is high and, although small gains in performance may be achieved, the largest amount of lost work or irreversibility is in the compressor system of a cryogenic refrigerator.

There are difficulties associated with applying refrigerators using helium as the working fluid. Refrigeration cycles based on gas mixtures or other gases such as nitrogen show some promise. Process simulation is needed to study the different refrigerator system configurations. In addition, research that reduces system heat loads should be an objective. The largest sources of heat load to the HTS cable system are static load in the flexible cryostat, followed by reductions in the dynamic load from ac losses.

Informal discussions that took place with different individuals from utility companies provided common features such as minimized footprint, increased reliability, and favorable cost of the system to the benefits provided. Due to the space limitations that exist in some systems, the ability to provide backup stored LN_2 may not be practical.

Standardization was described by different utility personnel as a key factor in reducing costs and providing reliability. It is recommended that the use of standard components (compressors, pumps, cold box) be part of future installations. Redundant systems are required for reliability. For example, if the utility replaces a single component (e.g., the cable) with two components (i.e., the cable and refrigerator or cryocooler), the net result is that both must be more reliable than the cable that was replaced.

Utilities continue to move toward unmanned operation. Thus, a key component is instrumentation and needed experience with unattended operation. Lower cost cryogenic instrumentation is a must. The required minimum instrumentation for safe, unmanned operation needs to be determined. The industrial gas suppliers (Lee et al, 2005) already provide remote monitoring and control of large plants, but the integration of this with the utility grid needs to be worked out.

References

Amemiya, N., Jiang, Z., Nakahata, M., Yagi, M., Mukoyama, S., Kashima, N., et al. (June 2007). AC loss reduction of superconducting power transmission cables composed of coated conductors. *IEEE Transactions on Applied Superconductivity, 17*, 1712–1717.

Chang, H. M., Park, C. W., Yang, H. S., Sohn, S. H., Lim, J. H., Oh, S. R., et al. (2012). Thermodynamic design of 10 kw Brayton cryocooler for HTS cable. In *Advances in cryogenic engineering, AIP conference proceedings* (Vol. 1434, pp. 1664–1671).

Cryomech, Inc., 113 Falso Drive, Syracuse, NY, USA 13211. http://www.cryomech.com/.

Demko, J. A., & Duckworth, R. C. (June 2009). Cooling configuration design considerations for long-length HTS cables. *IEEE Transactions on Applied Superconductivity, 19*(3), 1752–1755.

Demko, J. A., Lue, J. W., Gouge, M. J., Stovall, J. P., Martin, R., Sinha, U., et al. (2000). Cryogenic system for a high temperature superconducting power transmission cable. In *Advances in cryogenic engineering* (Vol. 45). New York: Plenum Press.

Demko, J. A., Lue, J. W., Gouge, M. J., Stovall, J. P., Butterworth, Z., Hughey, R. L., et al. (2002). Acceptance test and operation of the Southwire company 30-m high-temperature superconducting cable cryogenic system. In *Advances in cryogenic engineering* (Vol. 47A, pp. 180–187). American Institute of Physics.

Demko, J. A., & Hassenzahl, W. V. (2011). Thermal management of long-length HTS cable systems. *IEEE Transactions on Applied Superconductivity, 21*(3), 957–960.

Demko, J. A., & Hassenzahl, W. V. (2012). Liquid air as a coolant for thermal management of long-length HTS cable systems. In *Advances in cryogenic engineering: transactions of the cryogenic engineering conference—CEC* (Vol. 57A, pp. 273–289). American Institute of Physics.

Demko, J. A., Sauers, I., James, D. R., Gouge, M. J., Lindsay, D., Roden, M., et al. (2007). Triaxial HTS cable for the AEP Bixby project. *IEEE Transactions on Applied Superconductivity, 17*, 2047–2050.

Fesmire, J. E., Augustynowicz, S. D., & Demko, J. A. (2002a). Thermal insulation performance of flexible piping for use in HTS power cables. In *Advances in cryogenic engineering* (Vol. 47A, pp. 1533–1540). American Institute of Physics.

Fesmire, J. E., Augustynowicz, S. D., & Demko, J. A. (2002b). Overall thermal performance of flexible piping under simulated bending conditions. In *Advances in cryogenic engineering* (Vol. 47A, pp. 1525–1532). American Institute of Physics.

Fitzpatrick, B. T., Kephart, J. T., & Golda, M. (June 2007). Characterization of gaseous helium flow cryogen in a flexible cryostat for naval applications of high temperature superconductors. *IEEE Transactions on Applied Superconductivity, 17*(2).

Fleck, U., Vogel, D., & Ziegler, B. (2002). Cooling of HTS applications in the temperature range of 66 K to 80 K. In *Advances in cryogenic engineering: Proceedings of the cryogenic engineering conference* (Vol. 47, pp. 188–195).

Ganni, V., & Fesmire, J. E. (2012). Cryogenics for superconductors: refrigeration, delivery, and preservation of the cold. In *Advances in cryogenic engineering AIP conference proceedings* (Vol. 1434, pp. 15–27).

Gouge, M. J., & Demko, J. A. *Cryogenics assessment report.* Oak Ridge National Laboratory, Oak Ridge, TN. http://web.ornl.gov/sci/htsc/documents/pdf/CryoAssessRpt.pdf. Accessed 13.09.13.

Hamajima, T., Tsuda, M., Yagai, T., Monma, S., Satoh, H., & Shimoyama, K. (June 2007). Analysis of ac losses in a tri-axial superconducting Cable. *IEEE Transactions on Applied Superconductivity, 17*, 1692–1695.

Kadi, F. J., & Longsworth, R. C. (1977). Optimization for helium refrigerators for superconducting power transmission lines in terms of cost and reliability. *ASME Journal of Engineering for Industry, 99*, 551–557.

Kephart, J. T., Fitzpatrick, B. K., Ferrara, P., Pyryt, M., Pienkos, J., & Golda, E. M. (2011). High temperature superconducting degaussing from feasibility to fleet adoption. *IEEE Transactions on Applied Superconductivity, 21*(3), 2229–2232.

Kim, Y., Lee, S., Jang, H., Kim, Y., Lee, K., Choi, C., et al. (2012). The application of the cryogenic system on the HTS power cable circuit in actual grid. *Cryogenics, 52*, 661–666.

Lee, R. C., Dada, A., & Ringo, S. M. (2005). Cryogenic refrigeration system for HTS cables. *IEEE Transactions on Applied Superconductivity, 15*, 1798–1801.

Lemmon, E. W., Jacobsen, R. T., Penoncello, S. G., & Friend, D. G. (2000). Thermodynamic properties of air and mixtures of nitrogen, argon, and oxygen from 60 to 2000 K at pressures to 2000 MPa. *Journal of Physical and Chemical Reference Data, 29*(3), 331.

Lim, J. H., Sohn, S. H., Hwang, H. S., Kim, D. L., Ryoo, H. S., & Choi, H. O. (2010). Results of the KEPCO HTS cable system tests and design of hybrid cryogenic system. *Physica C, 470*, 1597–1600.

Lim, J. H., Yang, H. S., Sohn, S. H., Oh, S. R., Kim, Y. H., Lee, K. T., et al. (2012). Installation and performance test of cooling system for HTS cable in Icheon substation. In *Advances in cryogenic engineering, AIP conference proceedings* (Vol. 1434, pp. 1657–1663).

Longsworth, R. C., & Schoch, K. F. (1980). A study of refrigeration for liquid nitrogen cooled power transmission cables. In *Advances in cryogenic engineering* (Vol. 25, p. 585). New York: Plenum Press.

Maguire, J. F., Schmidt, F., Bratt, S., Welsh, T. E., Yuan, J., Allais, A., et al. (2007). Development and demonstration of a HTS power cable to operate in the long island power authority transmission grid. *IEEE Transactions on Applied Superconductivity, 17*, 2034–2037.

Maguire, J. F., Schmidt, F., Hamber, F., & Welsh, T. E. (June 2005). Development and demonstration of a long length HTS cable to operate in the Long Island power authority transmission grid. *IEEE Transactions on Applied Superconductivity, 15*, 1787–1792.

Maguire, J. F., Yuan, J., Romanosky, W., Schmidt, F., Soika, R., Bratt, S., et al. (2011). Progress and status of a 2G HTS power cable to be installed in the long island power authority (LIPA) grid. *IEEE Transactions on Applied Superconductivity, 21*, 961–966.

McFee, R. (1959). Optimum input leads for cryogenic apparatus. *Review of Scientific Instruments, 30*(2).

Morgan, G. H., & Jensen, J. E. (May 1977). Counter-flow cooling of a transmission line by supercritical helium. *Cryogenics*, 259–267.

Rey, C. M., Duckworth, R. C., Demko, J. A., James, D. R., & Gouge, M. J. (2010). Test results of a 25-m prototype fault current limiting HTS cable for project hydra. In *Advances in cryogenic engineering: Transactions of the cryogenic engineering conference–CEC* (Vol. 55A, pp. 453–460). American Institute of Physics.

Rostila, L. J., Lehtonen, J. R., Masti, M., Mikkonen, R., Gömöry, F., Melíšek, T., et al. (June 2007). AC losses and current sharing in an YBCO Cable. *IEEE Transactions on Applied Superconductivity, 17*, 1688–1691.

Saji, N., Asakura, H., Yoshinaga, S., Ishizawa, T., Miyake, A., Obata, M., et al. (July 17–20, 2001). Design of oil-free simple turbo type 65 K/6 kW helium and neon mixture gas refrigerator for high temperature superconducting power cable cooling. Paper C-11E-10. In *Cryogenic engineering conference*. Madison, WI.

Stirling Cryogenics & Refrigeration BV, Science Park Eindhoven 5003, 5692 EB Son, The Netherlands.

Stovall, J. P., Demko, J. A., Fisher, P. W., Gouge, M. J., Lue, J. W., Sinha, U. K., et al. (2001). Installation and operation of the Southwire 30-m high temperature superconducting power cable. *IEEE Transactions on Applied Superconductivity, 11*, 2467–2472.

Sumitomo (SHI) Cryogenics of America, Inc., 1833 Vultee Street, Allentown, PA 18103. http://www.shicryogenics.com/.

Vellego, G., & Metra, P. (1995). An analysis of the transport losses measured on HTSC single-phase conductor prototype. *Superconductor Science and Technology, 8*, 476.

Wark, K. (1983). *Thermodynamics*. McGraw-Hill.

Weber, C. S., Lee, R., Ringo, S., Masuda, T., Yumura, H., & Moscovic, J. (June 2007). Testing and demonstration results of the 350 m long HTS cable system installed in Albany, NY. *IEEE Transactions on Applied Superconductivity, 17*, 2038–2042.

Part Three

Applications

High-temperature superconducting fault current limiters (FCLs) for power grid applications

9

V. Meerovich, V. Sokolovsky
Ben-Gurion University of the Negev, Be'er Sheva, Israel

9.1 Introduction

Electric power systems are designed such that the impedances between generation sources and loads are relatively low. This configuration assists in the maintenance of a stable, fixed system voltage on loads. Due to a strong connection of many parallel power sources, the short-circuit impedance of networks can be relatively low ($<2-5\%$ of the load impedance), resulting in high levels of fault currents (dozens of times of the nominal current).

Ongoing growth of an electric power grid is accompanied by the further growth of the fault current level. Adverse effects on the electric power equipment also increase as the fault current level increases. As the fault current level increases so do the requirements to electrodynamic and thermal stability of the power devices, generators, transformers, and circuit breakers, which leads to a significant increase in their size, weight, and cost. Another adverse effect of fault conditions on the electric power grid is the disturbance of parallel operation of generators and electric machines. To prevent damage to the existing power-system equipment and to reduce customer downtime from fault currents, protection engineers and utility planners have developed various measures including: (1) splitting a grid into smaller electrically independent subsystems or buses, (2) increasing transformer impedance, and (3) inserting shunt reactors. These measures lead to reduced system efficiency and can impair static and transient stability. When these measures are exhausted, more expensive electrical equipment with higher fault-withstand ratings must be installed.

One of the more promising solutions to the fault current problem is the application of fast-operating nonlinear fault current limiters (FCLs) with the capability of rapidly increasing their impedance when the current exceeds a given value, and thus limiting high fault currents. Superconducting fault current limiters (SFCLs) exploit two remarkable properties of superconducting materials (see Chapter 2): (1) near perfect conductivity (i.e., zero electrical resistance) below a critical current, and (2) a rapid nonlinear phase transition from the superconducting state to the normal conducting state (S—N transition) above the critical current, or at sufficiently high magnetic field and temperature. An SFCL is one of the more attractive applications of superconductors in power systems, and has no classical equivalent.

Superconductors in the Power Grid. http://dx.doi.org/10.1016/B978-1-78242-029-3.00009-1

The chapter is devoted to the consideration of various types of SFCLs with particular emphasis on high-temperature superconducting (HTS) SFCLs. The primary focus is given to the various designs and operation principles, and a comparison of the various types and their possible application in power systems. Recent achievements in the manufacture and implementation of the devices are presented, and the future trends are also discussed.

The general requirements to FCLs are formulated in Section 9.2. Section 9.3 is devoted to various developed types of SFCLs and their operation principles and characteristics. The present status of the development is described in Section 9.4. Comparisons of the advantages and disadvantages of various FCL types and designs as well as the analysis of their applicability in power networks are performed in Section 9.5. Review of the SFCL applicability in power systems is done in Section 9.6. Section 9.7 contains a discussion of future trends and Section 9.8 provides sources of further information about SFCLs. A closely related description of fault-current-limiting behavior in cables is also given in Chapter 5, Section 5.6.

9.2 Utility requirements for fault-current-limiting parameters

Electric utility requirements to a FCL have been discussed in many publications (Lopez-Roldan, Price, De La Rosa, & Moriconi, 2011; Raju, Parton, & Bartram, 1982; Slade et al., 1992; Sokolovsky, Meerovich, Vajda, & Beilin, 2004). The primary requirements are the following:

- limitation of the first peak of a fault current (transient current);
- limitation of the steady-state fault current;
- low impedance under nominal operation conditions of the protected circuit;
- low power loss under nominal conditions;
- quick and automatic return to the low impedance (initial) state after fault is interrupted by other devices;
- FCL operation should not increase the risk of transient instability of the power system;
- high reliability and long lifetime similar to a conventional power transformer;
- no overvoltages related to the FCL operation;
- ensuring the normal operation of the system automatics and circuit breakers;
- fault-current-limiting performance shall not be degraded significantly as a result of sequence of operation of downstream devices;
- capability of many operations (not less than 50 (Slade et al., 1992)) in the current-limiting mode without maintenance;
- suitable consumption of coolant or power for cooling under nominal and fault conditions;
- no switching under sufficiently small fault currents (through-faults).

An ideal FCL has zero impedance under the normal conditions of the circuit to be protected and changes its impedance quickly (before the first peak of fault current) from zero to a reasonably large value required for the limitation of a fault current. In reality, it is enough that the voltage drop across a FCL under the normal conditions of a circuit is less than several percent (usually 5%) of the rated circuit voltage. In this case, the FCL does not disturb the static stability of the power system operation and does not influence handling properties of the lines (Sokolovsky, Meerovich, Györe, & Vajda, 2007).

The usual value of power losses in a transformer is about 1–2% of its nominal power (Fassbinder). The total losses in a FCL must be much less than the transformer loss: in the range of 0.01–0.25% of the nominal power of the protected circuit (Slade et al., 1992; Sokolovsky et al., 2004). In an SFCL, the total losses are the sum of the AC loss in the superconducting elements, the losses in the normal conducting elements of the construction, and the energy consumption needed for cooling.

Fast activation of a FCL is required to limit the first peak of a fault current. On the other hand, a very fast increase of the impedance leads to the appearance of dangerous overvoltages in the circuit. Estimates (Hoshino, Nishikawa, Salim, Muta, & Nakamura, 2001; Hoshino, Salim, Nishikawa, Muta, & Nakamura, 2001; Meerovich, Sokolovsky, Jung, & Goren, 1995a,b) show that, for power systems operating at typical power frequencies of 50–60 Hz, the time of impedance change of 2–4 ms is enough for the limitation of the first fault current peak and does not cause system overvoltages across the system components.

The basic "external" parameters when designing a FCL device are: (1) impedance values under normal operation and fault conditions, (2) activation current, (3) device activation time, and (4) device recovery time. The acceptable values of these parameters lie in the limits prescribed by characteristics of the normal operating and fault regimes of the power network. The normal operating regime is realized when a current in the circuit is less than the maximum current (I_{nm}) of the normal operating regime (Figure 9.1). If a current is higher than this maximum value, but lower than twice its value, there is the overload regime. This regime continues not more than several minutes. A fault regime arises if the current is more than double the maximum of the rated current in the normal regime. A fault event can occur at any point of the circuit and the magnitude and duration of the fault current depend on the location of a fault. The minimum of the steady-state fault current is $2I_{nm}$; its maximum (I_{fm}) is

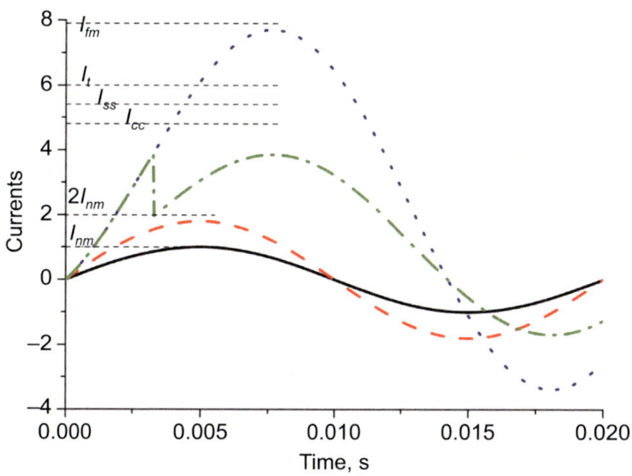

Figure 9.1 Diagram of currents in a power network.

achieved at a fault event close to the buses. All the electric power equipment installed in a system is designed for well-defined values of the admissible transient fault current I_t (admissible dynamic influence) and of the admissible steady-state fault current I_{ss} (admissible heating). In any case, the fault current must be less than these values and also less than the current capability of the circuit breaker (I_{cc}). The implementation of a FCL allows one to install the power equipment with much lower admissible I_{cc}, I_t, I_{ss}. The installation of a FCL into existing power systems allows one to use the system configurations that, without FCLs, would lead to an inadmissible high fault current. A limiter must limit any fault current below the minimal value of the admissible currents, on the one hand, but, on the other hand, the limited current should be higher than $2I_{nm}$ to provide successful operation of the system's automatic equipment.

To provide successful operation of a circuit breaker, a FCL must limit transient and steady-state fault current below the admissible magnitudes but must also provide crossing zero by the limited current at least once within about 0.1 s and the admissible transient recovery voltage (the rate of rise of transient recovery voltage and its peak (Lopez-Roldan et al., 2011)).

9.3 Designs and operation principles of various types of superconducting fault current limiters (SFCLs)

Many various designs of SFCL were proposed: resistive, inductive, saturated core, and bridge. Various "hybrid" designs were also considered with different modifications, based on different superconducting materials (Giese, 1995; Giese & Runde, 1993; Leung, 1997; Meerovich et al., 1995a,b; Sokolovsky et al., 2004; Sokolovsky, Meerovich, Beilin, & Vajda, 2003). Several prototypes of SFCLs have been built and successfully tested showing the feasibility of various proposed concepts for application in electric power systems (Gromoll et al., 1999; Meerovich et al., 1995a,b; Noudem, Barbut, Sanchez, Tixador, & Tournier, 1999, Paul et al., 1997; Witzmann, Schmidt, & Volkmar, 2001). Several devices have been brought to the commercial level and installed in power systems.

Below the principles of prospective designs and the main achievements in their implementation are described.

9.3.1 Resistive superconducting fault current limiters

The earliest concept proposed for a FCL is the "resistive FCL" (first proposed in 1936 by Kalashnikov (Neklepaev, 1978)), based on the intrinsic property of a superconductor (SC) to increase its electrical resistance from zero in the SC state to a some value in the normal state (this transition is usually called "quench") when current exceeds its critical value (see also Chapter 2). The simplest resistive FCL contains an active SC element that is connected in series with the circuit to be protected. When a fault occurs, the current increases and causes the superconductor to quench, thereby increasing its electrical resistance. Typical results of the limitation are demonstrated in Figure 9.2.

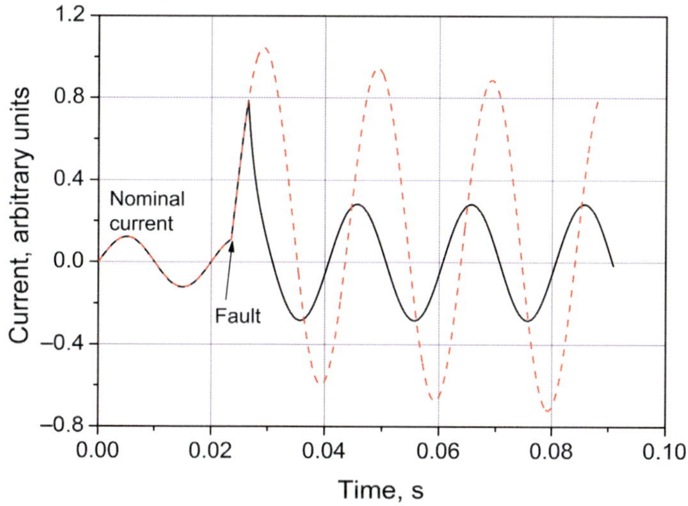

Figure 9.2 Typical traces of the current in a circuit with (solid line) and without (dashed line) a FCL.

Despite the apparent simplicity of the resistive approach, this concept faces a number of challenges: The AC current flows throughout the SC elements, resulting in possibly significant AC losses in a normal operating regime; the operation time in the limitation regime should be small to minimize heating of the SC elements in their resistive state, and to provide short recovery time required for the FCL to recover into initial "ready state." Design of terminations is especially challenging for high-voltage operation, especially to avoid dielectric breakdown when the SC elements heat up and boil the cryogenic fluid (nitrogen for HTS devices). The principal design of a resistive-type SFCL is shown in Figure 9.3, where the superconductor is connected in parallel with a resistive or an inductive shunt. The shunt allows one to solve some intrinsic problems of a resistive FCL. It is needed to adjust the impedance of the line in the limitation regime, which is essentially governed by the shunt element. In addition, because most current is transferred from the SC element to the shunt during the fault, the shunt helps to protect the superconductor from hot spots and overheating during the quench, and it provides quick recovery of the element into the SC state after the fault is cleared by a circuit breaker. The superconductor is covered by a normal metal layer to prevent hot spots (thermal stabilization). A more detailed discussion of handling hot spots is given in Chapter 5, Section 5.6.4.

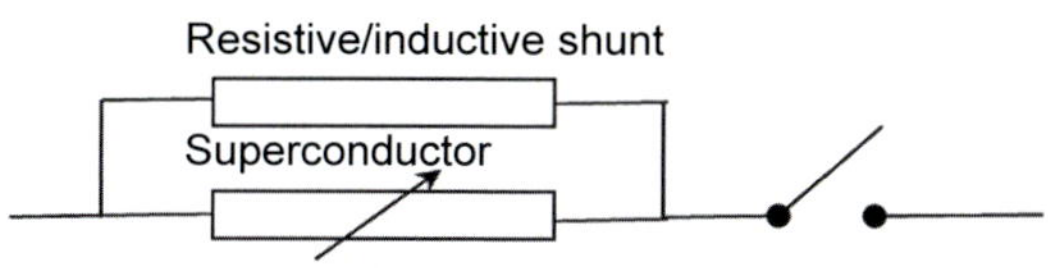

Figure 9.3 Schematic view of a resistive FCL. The superconductor is parallel with the resistive and/or inductive shunt inserted in series into a circuit.

Resistive-type FCLs are studied and developed by many research groups and companies (Gromoll et al., 1999; Noe & Steurer, 2007; Paul et al., 1997; Witzmann et al., 2001). The main directions of the R&D work have been AC loss reduction, solution of the hot spot problem, and providing quick recovery of the initial state. The first prototypes of resistive HTS FCLs were fabricated using bulk and thin film SC elements. Currently, the main emphasis is done to the designs based on coated conductors. AC losses have been minimized by using noninductively (bifilar) wound pancake coils of coated conductors (Clem, 2008). Such coils are wound so that current reverses direction in every second layer of the winding, and AC losses decrease as the tape width to spacing between the tapes decreases. The hot spot problem has been solved by quality control of the superconductor, most easily achieved in coated conductors, and also by careful design to keep the local temperature rise $\Delta T = J_{\text{lim}}^2 \rho_{\text{net}} \tau / C$ below a defined safety level for the material. Here J_{lim} is the rms limited current density of the entire wire (including its stabilizer), ρ_{net} is the net resistivity of the entire thickness of the local region of the wire being heated, τ is the fault hold time, and C is the volume heat capacity.

To provide a more homogeneous and faster quench, the "resistive with magnetic field assisted quench" design has been proposed (Elschner, Breuer, Walter, & Bock, 2006). In this design, a shunted coil is arranged coaxially outside a superconductor. When the quench starts, part of the current passes to the coil and produces a magnetic field at the surface of the superconductor, reducing its critical current.

The time it takes the superconductor to recover from the resistive state and return back to the superconducting state, that is, the recovery time, depends very much on the design, cooling conditions, and the superconducting material. The larger the dissipative energy, the higher the temperature of the heating and the more time is required for the recovery. An important task in FCL design is to provide a recovery time of about 1 s (Sokolovsky et al., 2004). Reduction of the dissipated energy can be achieved by application of fast-operating breakers. However, the conventional backup breakers take about 3–10 cycles of AC current to achieve current interruption, and it is very difficult to interrupt a fault current within 0.5 cycle.

9.3.2 Hybrid designs

In order to overcome the intrinsic disadvantages of resistive FCLs, to decrease the heating of the superconductor and to decrease the recovery time, so-called "hybrid" designs have been proposed (Endo et al., 2009; Hori et al., 2006; Lee et al., 2008; Lee, Yoon, Yang, & Lee, 2012; Na et al., 2012; Park et al., 2009; Yim et al., 2013). In these designs an active superconducting element is connected in series with a fast vacuum interrupter (VI) (Hori et al., 2006). An electromechanical linkage that opens a fast switch during a quench can also be used (Park et al., 2009).

In the devices developed by a Japanese group (Endo et al., 2009; Hori et al., 2006), the superconductor and VI are shunted by a normal coil (Figure 9.4). Under the nominal regime of the protected circuit, the superconductor is in the superconducting state, the VI is closed, and the whole current flows through the superconductor/VI branch. At a fault event, the superconductor passes into the dissipative state and a part of the

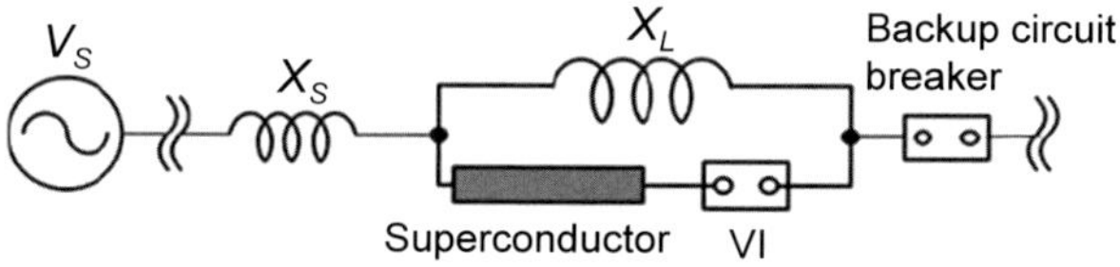

Figure 9.4 Schema of a hybrid FCL (Endo et al., 2009): V_S and X_S are the equivalent generator and its impedance, respectively, X_L is the coil impedance, VI is the vacuum interrupter. Reprinted with Permission of the Institute of Engineering Technology (IET).

current is diverted to the coil. The magnetic field of the coil pushes on an electromagnetic repulsion plate (ERP), which separates the VI contacts. There is an arc between the VI contacts till the current reaches zero. The coil carries out two functions: limitation of the fault current and providing high-speed switching.

Prototypes of the hybrid device were based on three types of superconductors: Bi-2223 silver sheath wires, quench-melt-growth (QMG) bulk elements, and YBCO thin films. As an example, Figure 9.5 presents the experimental results for the SFCL based on an YBCO thin film and using a 24 kV VI driven by ERP (Endo et al., 2008, 2009). The device was tested in an LC resonance circuit with a large capacitor which was used as a power supply. Above 500 A, the YBCO thin film starts to quench; the current passes to the parallel coil and the fault current is limited from 8000 to 1000 A_{peak}. The ERP begins to move at about 3 ms after the quench and the VI interrupts the current flowing through the superconductor at about 20 ms. The maximum VI stroke is 4 mm and the stopper fixes a 3.5-mm distance between the VI contacts. In this design, the superconductor plays two roles: a trigger for the ERP and a limiter of a current flowing through the VI.

Intensive research and development of hybrid FCLs are carried out in Korea (Hori et al., 2006; Lee et al., 2008; Lee et al., 2012; Park et al., 2009; Yim et al., 2013). Initially, the S−N transition in a superconducting element was used only for triggering

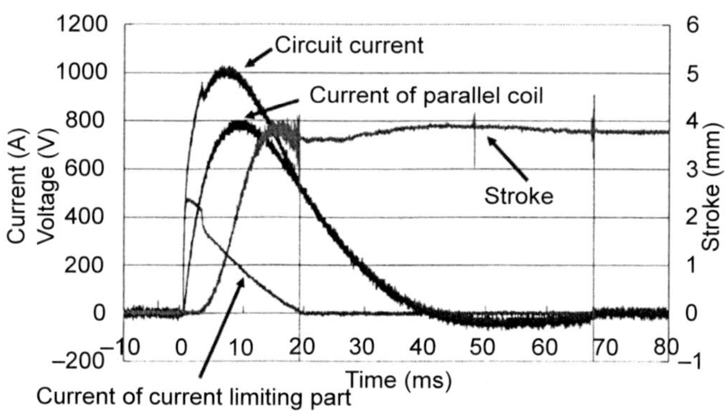

Figure 9.5 Experimental results for the YBCO film-based FCL using the electromagnetic repulsion plate (ERP) (Endo et al., 2008). © 2008 IOP Publishing.

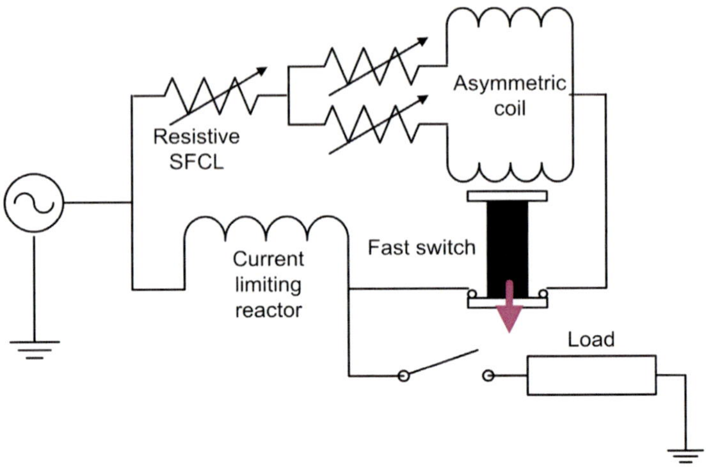

Figure 9.6 Schema of a hybrid FCL with an asymmetric HTS coil.

fast switch interrupting fault current during a half of the AC period. Recently (Lee et al., 2008; Park et al., 2009) a more advanced design consisting of a fast switch, a resistive FCL, and an asymmetric HTS coil was proposed (Figure 9.6). The asymmetric coil was composed from two kinds of single pancake coils wound in opposite directions so that, under normal conditions, the total magnetic field at the moving contacts of the switch was zero. The pancake coils were fabricated from HTS materials (AMSC 344 Amperium and AMSC Bi-2223) with strongly different resistances in the normal conducting state. At a fault event, the coils quench in sequence providing, on the one hand, operation of fast switch and, on the other hand, limitation of the first peak of the fault current. To increase the FCL impedance and to achieve deeper limitation of the first peak, a resistive FCL could be connected into the branch of an asymmetric SC coil and a fast switch in series.

9.3.3 *Inductive or shielded core design*

One of the first SFCL designs developed for grid deployment was the inductive, named also "shielded-core" or "transformer-type," design, which is based on magnetic coupling between a circuit containing a superconducting element and a protected circuit. This design allows the cryogenic environment to remain mechanically isolated from the rest of the circuit. Basically, the device resembles a transformer with a short-circuited secondary coil that can be fabricated completely from a superconducting material or as a coil containing a superconducting switching element (see Figure 9.7) (Bashkirov, Fleishman, Patsayeva, Soboler, & Vdovin, 1991; Giese, 1995; Vershinin, Meerovich, Naumkin, Novikov, & Sokolovsky, 1989). The primary coil of this transformer is connected in series to the protected circuit. Under normal operating conditions in the circuit, the secondary coil (or superconducting switch) is in the superconducting state. The magnetic flux produced by the primary coil is compensated by the flux

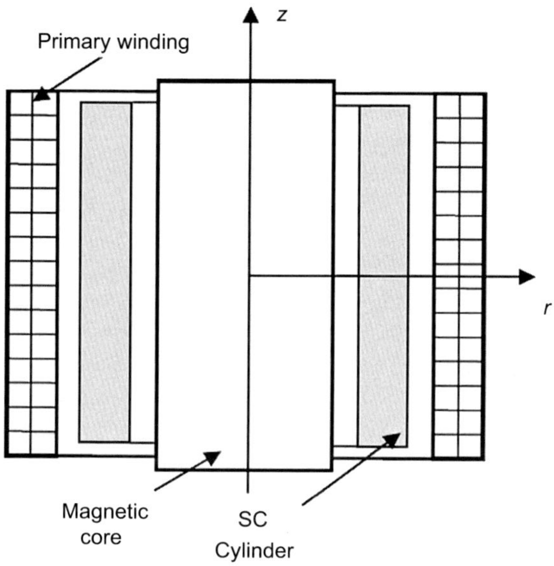

Figure 9.7 Principal shielded-core design.

originated from induced screen currents in the short-circuited superconducting coil. The device impedance is determined only by the leakage flux in an air gap between the primary and secondary coils. Under short-circuit conditions in the circuit, the increased current in the secondary coil exceeds the critical value and the coil or its part made from a superconductor passes into the resistive state. If the resistance due to the dissipation becomes much greater than the inductive reactance of the coil, the induced current in the secondary coil is sharply reduced and the magnetic flux of the primary coil is compensated no more. Thus, the FCL impedance increases and limits the fault current.

The main advantages of the shielded core design are that there are no current leads to cryogenic temperatures and the superconducting element is subjected to large currents but low voltages.

In early designs based on low-temperature (LTS) materials (Vershinin et al., 1989), it was proposed that a secondary coil from a well-stabilized superconducting wire be made and then short circuit it by a special superconducting switching element. In this case the switching element only undertakes the S−N transition under a fault while the secondary coil remains in the superconducting state. The same design was also considered in the end of the 1980s for inductive FCLs based on HTS (Vershinin et al., 1989). Progress in the fabrication technology of large diameter HTS rings and cylinders with high critical current density stimulated the building of secondary coils in the form of a set of rings or cylinders (Cave, Willen, Brissette, Richer, & Aresteanu, 1994; Elschner, Bock, & Gauss, 1995; Leung, 1997; Meerovich et al., 1995a,b; Meerovich et al., 1999; Paul et al., 1997; Paul, Baumann, Rhyner, & Platter, 1995). Now this is the typical design of an inductive HTS FCL. The popularity of this design is explained by the technological difficulties of the fabrication of multiple-turn HTS windings and switching elements as well as a low-resistance connection of these elements. The main difference between various inductive FCL designs consists in the arrangement of superconducting rings or in the configuration of a magnetic core. It was shown that the open-core design (Figure 9.7) is the most effective: large-scale devices are based on this design. The position of the rings with respect to the primary coil, inside or outside, is of no principal importance (Sokolovsky, Meerovich, Chubraeva, & Vajda, 2010).

With the progress in the technology of HTS materials, other designs have been considered for FCLs. One recent example of the return to the initial design of the inductive FCL is the device that combines a normal metal secondary coil with an YBCO bulk switching element (Granados et al., 2002). However, it was shown that the most powerful inductive FCLs were built on the base of a stack of rings made from coated conductor (CC) tape (Kaltenborn, Mumford, Usoskin, Schmidt, & Janetschek, 2011; Klein, 2012).

The basic features of the resistive, transformer-type, and saturated-core designs are used in so-called flux-lock and flux-coupling limiters, which are developed for South Korean power grids (Lim, Choi, & Han, 2004; Jung et al., 2011, 2013; Ko, Lim, & Han, 2013). The flux-coupling-type SFCL combines transformers and superconducting elements. Different schemes of these devices were proposed: with series or parallel connection of the coils, with one or two triggering levels, one- and three-phase designs. Comprehensive study of their characteristics and operation modes can be found in Jung et al. (2013) and Ko et al. (2013).

9.3.4 Saturated-core type

Saturated core FCLs utilize the dynamic magnetic characteristic of iron to change the inductive reactance on the protected line. The basic saturated FCL concept of a one-phase device is shown in Figure 9.8. There are two rectangular iron cores arranged side by side. A small DC power supply energizes the superconducting coil with a DC bias current to create a very strong DC magnetic field that magnetizes the ferromagnetic cores into a fully saturated state. The DC superconducting coil is used here for

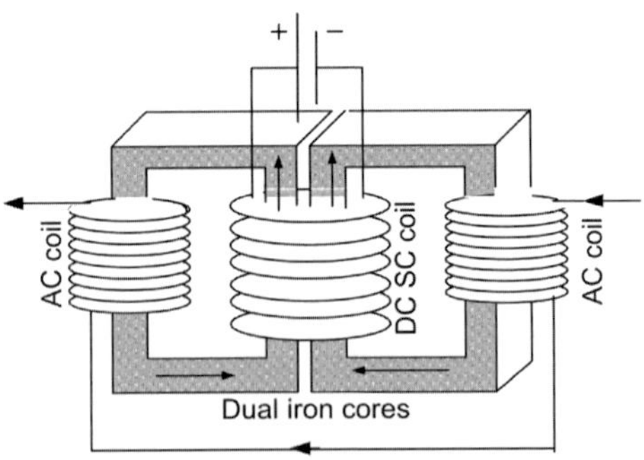

Figure 9.8 The basic saturated FCL concept.

its ability to supply permanent ampere turns to magnetize the iron cores almost without losses. Copper AC coils wound on the outer limbs of the iron cores are connected in series to the circuit that is to be protected. These AC coils are wound in opposite magnetic "senses," so that using this arrangement a single-phase device can be made in which each of the rectangular iron cores acts independently during each positive and negative half-cycle of the AC line current.

When the protected circuit is operating in the nominal regime, the core iron is highly saturated (a sum of the DC and AC magnetic field saturates the iron) and has low relative permeability. The saturated iron acts like air, so the impedance of the AC coils is small. When a fault occurs, an increasing fault current forces the iron core out from saturation, increasing the AC coil impedance during part of each half-cycle. The result is a considerable reduction in fault current. For application in power systems, all six cores for a three-phase device link a common DC loop in one cooling tank or there could be three single-phase units forming a set. In a three-phase device six iron-core frames are usually conjugated in a hexagonal structure. Each frame comprises a top yoke, a bottom yoke, and two vertical limbs. Six vertical limbs, one of each frame, are closely arranged together to form the central column at the center of the hexagon. The superconducting DC coil is installed inside the cryostat and is entwined with the central column. The edge vertical limb of each iron-core frame is surrounded by an AC coil (Gong et al., 2008, Moriconi, De La Rosa, Darmann, Nelson, & Masur, 2011; Raju et al., 1982; Xin et al., 2012, 2007, 2009, 2011).

There are two disadvantages of major concern for a saturated FCL:

- use of massive iron (resulting in a large size, heavy weight, and high cost). Two AC coils are connected in each phase but only one of these coils participates in current limitation during a half-period. The iron-core frame of the other coil is saturated.
- high voltage can be induced in the DC superconducting coil as a fault takes place due to the magnetic coupling. This can cause damage of the DC source, isolation in the DC circuit, and quenching of the superconducting coil due to a rapid increase of current.

In several papers (Gong et al., 2008; Xin et al., 2012, 2009) a method has been proposed to deal with these two problems by rapidly breaking the DC circuit. To do this, a saturated-core FCL needs a system to register a fault event in the protected circuit, a high-speed switch and the energy release unit in the superconducting circuit. A fault in the protected circuit (line) is detected in its early stage (a few milliseconds from it starts) and the DC circuit is opened within another few milliseconds. The energy stored in the DC coil is passed into the energy release unit. After current limitation is completed, the DC source should remagnetize the iron core in about 800 ms according the utility requirements. The main benefits of breaking the DC circuit are:

- Two AC coils in one phase are connected such that the magnetic flux created by the phase current is equal in magnitude but in opposite direction at the central column. Thus the induced voltage in the DC coil during normal and fault operation can be greatly minimized if not eliminated.
- Both AC coils in one phase "participate" in current limitation during the whole period. Thus their diameter and hence weight and volume are reduced.

Other technical solutions to decrease the weight and size of saturated-core FCLs were proposed (Nikulshin, Wolfus, Friedman, & Yeshurun, 2013; Wolfus, Friedman, Yeshurun, Rozenstein, & Bar-Haim, 2013). A small experimental model of the device was tested (Nikulshin, Wolfus, Friedman, & Yeshurun, 2014).

9.3.5 Bridge-type superconducting fault current limiters

The bridge-type limiters, sometimes called DC reactors, rectifier-type FCLs, or DC FCLs, use diodes or thyristors arranged as a full bridge rectifier, with a superconducting coil and a voltage source. Figure 9.9 shows the electrical circuit of the FCL proposed initially by Boenig and Paice (Boenig & Paice, 1983). The principal operation is as follows: A DC current I_0 is supplied by the DC current source. Under the nominal regime of the protected circuit, the AC current amplitude is lower than the DC current and all the diodes operate in the conducting regime; the AC current flows through the diodes. The total voltage drop and losses are practically determined by the power electronics. At a fault event the AC current i_{ac} exceeds the DC current I_0 in the positive or negative half-cycle of i_{ac}, the diodes D_3 and D_4 or D_1 and D_2, respectively, close, and the fault current is limited by the value of the DC current. In real schemes, the DC current source is replaced by a DC voltage source U_b and a reactor with an inductance L, and the fault current is limited by the inductance. To decrease losses in the reactor, it was proposed that the reactor be fabricated using superconducting wires. A current flowing through the reactor is close to a DC current under both the nominal and the fault regimes and the limitation is achieved without the transition of the superconducting reactor into the non-superconducting dissipating state. This allows one to use, along with coated conductors, low-temperature superconducting, MgB_2 and bismuth wires.

The idea has been developed and realized in several models and prototypes (see, e.g., Hoshino, Nishikawa, et al., 2001; Hoshino, Salim, et al., 2001; Hui et al., 2006; Morandi & Imparato, 2009; Noe & Steurer, 2007; Salim, Muta, Hoshino, Nakamura, & Yamada, 2004). In one of the designs, it was proposed that the DC

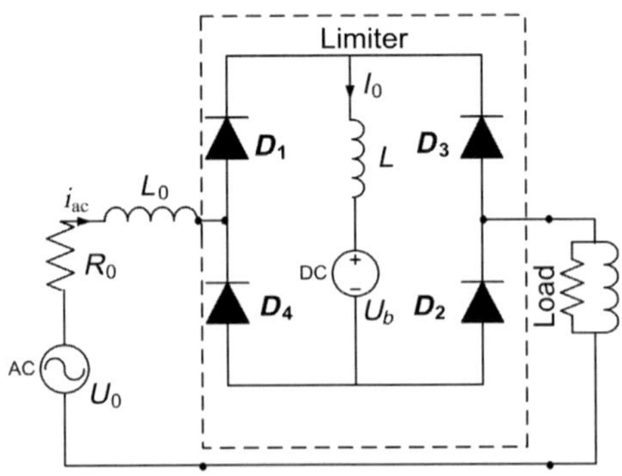

Figure 9.9 Electrical circuit of a diode bridge-type FCL (Noe & Steurer, 2007). © 2007 IOP Publishing.

voltage source be removed (Hoshino, Nishikawa, et al., 2001; Hoshino, Salim, et al., 2001; Hui et al., 2006). In this scheme the current i_{ac} flows through the reactor in both regimes. In a first half-cycle the current flows through, for example, the diodes D_1 and D_2; the diodes D_3 and D_4 are closed. During the next half-cycle, the diodes D_3 and D_4 operate in the conducting regime; the diodes D_1 and D_2 close. Under the nominal regime, the current I_0 is mainly a DC current equal to the i_{ac} amplitude with low fluctuations that increase the voltage drop across the reactor and, hence, the FCL. Under a fault event, the DC current in the inductance cannot increase suddenly; the current and, hence, an amplitude of the fault current increase slowly with the characteristic time determined by the ratio of inductance L and the active resistance in the protected circuit. The limitation is achieved without the transition of the superconducting reactor into the nonsuperconducting dissipating state. The circuit breaker S_1 should open earlier than the current in the reactor achieves the critical value or exceeds the circuit breaker capacity.

The three primary disadvantages of this design are:

- inductance L makes it difficult to control the nominal regime with a fast change of the current;
- a voltage drop appears across the reactor due to fluctuations of the current I_0 under the nominal regime;
- circuit breaker S_1 must be fast (limited fault current is increasing with time).

To overcome these disadvantages, several other designs have been proposed such as the hybrid bridge type in which two or four diodes are replaced by thyristors to decrease the limitation time (Hoshino, Nishikawa, et al., 2001; Hoshino, Salim, et al., 2001). An advantage of this hybrid bridge device in comparison with the application of only thyristors is that the power electronics interrupts the nominal current. To decrease the inductance L under the nominal regime, a saturated DC reactor-type FCL was proposed in which the reactor is saturated with the help of an additional control coil and a current source similar to saturated-core FCLs (Hoshino, Nishikawa, et al., 2001; Hoshino, Salim, et al., 2001).

Another type of bridge-type FCL is the so-called "noninductive reactor." It consists of two superconducting coils that are connected in antiparallel: a trigger coil and a limiting coil (Salim et al., 2004). These coils are magnetically coupled and have the same number of turns. Their total inductance is small under the nominal circuit regime. When a fault current appears, a current in the trigger coil exceeds the critical value of the superconducting coil, the superconducting coil passes into the normal state with a finite resistance, and the current in this coil is limited and does not equal the current in the limiting coil. The flux in the reactor increases; the equivalent impedance of these coils increases and limits the fault current.

9.3.6 Superconducting CLTs

One of the most widely used methods to limit the magnitude fault currents in the electrical grid is through the use of conventional (normal conducting) fixed impedance transformers. These transformers are specifically designed to have specified impedance in the event of faults. The downside of using these conventional devices is that their fixed impedance exists even under normal operating conditions, and therefore increases system losses and voltage drop, and so lowers overall electrical energy efficiency.

As an efficient and low cost solution to the fault current problem in power systems, special superconducting transformers with changing impedance—low, under normal operating conditions, and high, under fault conditions—were proposed. This change in impedance can be realized by fabricating the secondary winding, or at least part of it, from SC material (see Chapter 12). There are two basic approaches to a SFCL transformer. The first SFCL transformer design utilizes the resistive FCL approach, where under fault conditions the secondary SC coil, or at least part of it, undergoes the S−N transition and limits the current due to the increased resistance (Fault Current Limiting Superconducting Transformer; Hayakawa, Chigusa, Kashima, Nagaya, & Okubo, 2000, 2011; Tomioka et al., 2013).

The second SFCL transformer approach is based on the idea of the transformer-type FCL (Bashkirov, Yakimets, Fleishman, & Narovlyanskii, 1995; Meerovich & Sokolovsky, 2005; Meerovich, Sokolovsky, & Vajda, 2007; Vajda et al., 2007, 2011). One design is a current-limiting transformer (CLT) (it is also known as a transformer circuit breaker) that is capable of limiting fault currents in the circuit connected to its secondary winding to a very low level (Bashkirov et al., 1995). While the FCL action is based on the insertion of an additional impedance into the circuit, the limiting property of a CLT is provided by the compensation of the voltage in the secondary circuit.

The same principle was also used for the development of multifunctional fast-operating systems combining the functions of a conventional transformer with the functions of power redistribution and fault protection, providing a fast switching of power flow to a standby line during faults. The so called "current-limiting transformer" (Meerovich and Sokolovsky, 2005) constitutes a single-phase transformer with three additional windings located on the same leg of the magnetic circuit (Figure 9.10): one of the windings, w_5, is an SC short-circuited winding, the second winding, w_3, is counterconnected in series with the main secondary winding (w_2) so that the voltages induced in the windings are of opposite signs but equal in magnitude. When the winding w_5 is in the SC state, the electromotive forces of the windings w_3 and w_4 are zero, and the device operates as a conventional transformer. Under fault conditions, the winding w_5 passes into the normal state and the electromotive forces in w_3 and w_4 appear. The ratio of turn numbers in the windings w_2 and w_3 can be chosen so that with an opened SC winding their electromotive forces are compensated. The current in line 1 is limited and the standby line 2 is switched on.

As a result of this feature, limiting impedance due to quench of superconducting windings is utilized and leakage impedance of a transformer can be reduced, compared with that of conventional transformers. The reduced leakage impedance of the superconducting transformer enhances the static stability and transmission capacity of a power system under normal operating conditions.

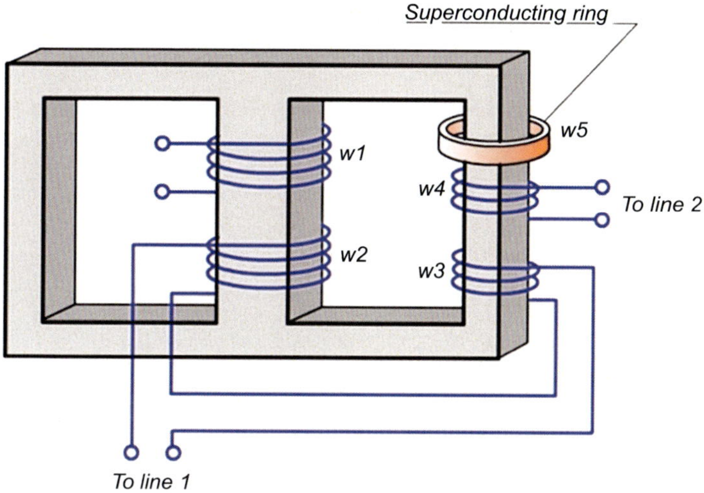

Figure 9.10 A current-limiting transformer (Meerovich & Sokolovsky, 2005). © 2005 Published by Elsevier B. V.
Reprinted with Permission from Elsevier.

9.3.7 Fault-current-limiting high-temperature superconducting cable

Development in the fabrication technology of the HTS wires, and especially coated conductors, allows one to create a superconducting power cable operating at temperatures of 66–77 K (see Chapter 5). The fault-current-limiting HTS cable combines the properties of a power cable and a resistive FCL. Introduction of fault-current-limiting functionality directly into a superconducting cable can have the benefit of avoiding separate standalone FCLs, resulting in savings in space and cost. This idea was proposed in the framework of the European SUPERPOLI project (Paasi, Lehtonen, Verhaege, & Herrmann, 2001). It was proposed that either Bi-2212 bulk conductor or YBCO 2G tapes be used.

A three-phase HTS cable with a built-in fault-current-limiting feature is also being developed in a project funded by the US Department of Homeland Security under a cost sharing agreement with American Superconductor (Maguire et al., 2009; Maguire & Yuan, 2009). The full-scale cable is designed to connect two of Con Edison's Manhattan substations. The cable is being provided by Southwire/Ultera; the subscale prototypes are tested at Oak Ridge National Laboratory. The simplified one line diagram of the cable installation in the system and schematic utility grid are shown in Figures 9.11 and 9.12, respectively. The cable system, called Secure Super Grids (Maguire et al., 2009), is designed with the key characteristics: length of a 300 m, rated nominal AC voltage/current of 13.8 kV/4000 A, and limiting a prospective fault current of 40 kA by 17% on the first quarter cycle. Single-phase 3-m-long

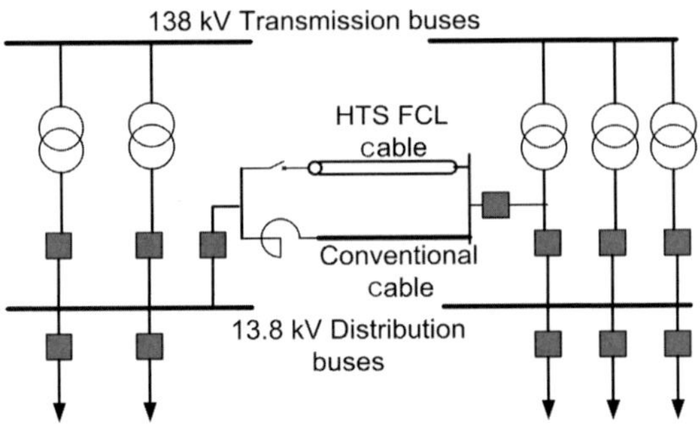

Figure 9.11 Simplified diagram of the grid with a HTS cable.

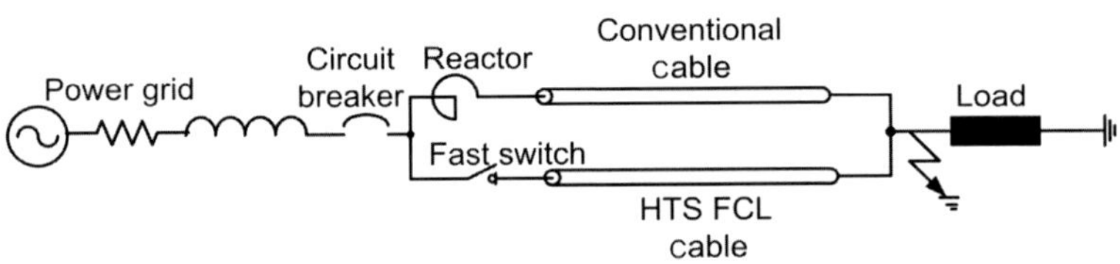

Figure 9.12 Schematic utility grid including a fault-current-limiting cable in series with a fast switch and in parallel with a conventional cable and a reactor.

and 25-m three-phase prototypes have been tested (Gouge et al., 2009; Maguire et al., 2009; Maguire & Yuan, 2009; Yuan et al., 2011).

The Secure Super Grids system consists of a low impedance, high-capacity superconducting cable shunted by a conventional lower capacity cable (in parallel). Optionally, a reactor can be included in series with the conventional cable to increase the total impedance of this branch to the value Z_R, providing limitation of a fault current to the required level.

In the normal operating regime, the impedance Z_F of the superconducting branch is about 1/6 Z_R or less and the dominant portion of the nominal current flows through the superconducting cable.

When a fault event occurs, a current exceeds the critical value and wires of the superconducting cable pass into the resistive state, limiting the fault current, just as in a standalone FCL. The superconducting cable is designed so that its resistance in this state is larger than Z_R. The main portion of the fault current flows through the conventional cable.

The fast switch in series with the superconducting cable is designed to open in about four cycles, limiting the energy dumped into the superconducting cable (an operation time of conventional circuit breaker is about three times larger). The fast-operating switch allows the superconducting cable to recover to its initial state in a few minutes. During this time, the parallel conventional cable carries power based on its short-time overload capability.

9.4 Status of fault current limiters development and implementation

9.4.1 Superconducting materials for fault current limiters

The history of the FCL design reflects the progress in technology of superconductor materials and wires. The first prototypes of superconducting FCLs were based on LTS and they were known long before the discovery of HTS. The first FCL designers proposed the use of type I superconductors (Neklepaev, 1978). Later several power models and prototypes of various types of FCL based on LTS were developed and tested (see, e.g., Ito et al., 1992; Raju et al., 1982; Ulbricht, 1979; Verhaege et al., 1993; Vershinin et al., 1989 and references therein). The highest power devices (Ito et al., 1992; Raju et al., 1982; Ulbricht, 1979; Verhaege et al., 1993) were built using Nb–Ti wires, which have been available for decades at competitive costs. Nb–Ti wires are largely used in MRI magnets and operate under DC conditions other than when being charged and discharged. However, the very low operating temperature, ~4 K, requires expensive helium cryogenic equipment, which makes use of these wires for FCLs impractical.

At present, the main attention for FCLs is on designs employing HTS materials that can operate conveniently at the liquid nitrogen temperature (66–77 K). Operating FCLs at liquid nitrogen temperatures has many advantages over liquid helium operation including: (1) lower cost cryogenic fluid, (2) higher heat capacity of the cryogenic fluid, (3) superior dielectric properties, and (4) larger dynamic temperature and pressure range to avoid unwanted bubble formation. The first power HTS FCL prototypes were inductive devices built using melt-cast processed BSCCO hollow cylinders (Meerovich et al., 1995a, 1995b; Paul et al., 1997). Progress in the HTS–normal metal contact technology allowed designing a power-resistive FCL based on HTS bulk materials (Bock, Bludau, et al., 2011; Bock, Hobl, et al., 2011). Application of the HTS films coated by stabilizing Au or Ag layers allows one to overcome the well-known drawbacks of the resistive FCLs (Gromoll et al., 1999).

The first generation (1G) HTS wires have also been available for over a decade and have been applied in FCL designs (Oda et al., 2011). They consist of BSCCO filaments immersed in a silver or Ag–Au alloy matrix. Nevertheless the cost of this material is high due to the silver and/or gold, and their use is problematic due to the large amount of material needed (which implies both high operating costs due to AC losses and high capital costs). The second generation (2G) HTS wires based on YBCO material (coated conductors) are under development and fabrication worldwide (Malozemoff et al., 2008). During the last several years, rapid progress has been made in the development of 2G HTS wire, which is now displacing 1G wire for most if not all applications. The engineering critical current density of these wires matches or exceeds that of 1G wire, and the mechanical properties are also superior. Scale-up of manufacturing is proceeding rapidly, with several companies already supplying hundreds of kilometers for grid projects worldwide. One especially attractive application, which relies on the specific properties of 2G HTS wire, is fault current limitation. By incorporating a high resistivity stabilizer in the coated conductor, one can achieve high resistance in a quenched state during a fault event and at the same time provide significant heat capacity to limit the temperature rise. The trade-offs in optimizing the stabilizer are described in detail in Chapter 5, Section 5.6.

YBCO material used in 2G HTS wires is preferred over Bi-2223 and Bi-2212 because of its sharper S—N transition. The last is explained by higher n value in the power-law dependence of the electric field E on the current density J: $E = E_0(J/J_0)^n$. Another material that is also being considered for FCL is magnesium diboride (MgB_2). This compound has a critical temperature of 39 K in zero magnetic field and shows good performance at 25 K and below. Experiments with FCL designed using bulk (Prikhna et al., 2011) and wire (Morandi, Brisigotti, Grasso, & Marabotto, 2013; Smith, Oliver, Pei, Husband, & Rindfleisch, 2012) MgB_2 superconductors confirmed the possible applicability of these materials in FCLs. The relatively low operating temperature makes the use of MgB_2 for FCLs costly due AC losses and the coefficient-of-performance penalty. Nevertheless, some encouraging prospects for this material can be achieved in devices in which a superconductor operates under DC conditions, for example, in saturated-core or in bridge-type (DC resistive) (Dalessandro, Bocchi, Rossi, & Martini, 2007) FCLs.

9.4.2 Development of resistive and hybrid SFCLs

Many prototypes of the resistive FCL based on LTS and HTS materials have been built and successfully tested, and a number are currently in active use in the power grid in Europe. A number of the designs were developed based on YBCO-based thin film technology (Ainslie et al., 2009; Hyun et al., 2007; Meerovich et al., 1997; Noguchi, Tanikawa, & Igarashi, 2007). The 100 kVA and 1 MVA distribution-level functional prototypes were built at Siemens (Gromoll et al., 1999; Witzmann et al., 2001).

Two different superconducting material options are currently available for manufacturing resistive SFCL systems: BSCCO-2212 bulk material and 2G YBCO-coated conductor tapes. The resistive FCL prototypes employing BSCCO-2212 rods were built by Nexans SC (Elschner, Bock, Brommer, & Cowey, 2000; Verhaege et al., 2001), and several prototypes were also built using YBCO-coated conductor tubes (Verhaege et al., 2001) and YBCO bulk meanders (Tixador et al., 2001).

Nexans has designed, built, and installed into different power grids a number of resistive SFCLs (Hobl et al., 2012; Tixador, 2010). Nexans adapts two technologies for their superconducting elements. The first one is the technology of mono- or bifilar coils made from bulk melt-cast processed BSCCO-2212 tubes (typically 50-mm-outer, 42-mm-inner diameter, 370 mm length), which are machined into a circular coil geometry to reach a length of 3.6 m with a current capacity of 470 A at 73 K (Figure 9.13). A shunt is soldered on the Bi-2212. Then these tubes are parallel-connected in modules (Figure 9.14).

Four large projects based on BSCCO-2212 bulk superconducting elements led by Nexans were completed to 2012. Two 12 kV FCLs with nominal currents 100 and

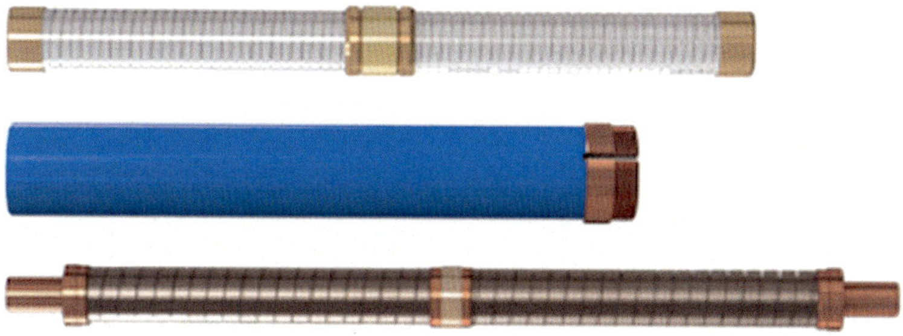

Figure 9.13 Nexans' components for three FCLs: 12 kV/100 A, 12 kV/400 A, 12 kV/800 A (Bock, Hobl et al., 2011).
Published with Permission from Nexans Super Conductors GmbH.

Figure 9.14 A module of Bi-2212 elements (Nexans picture) (Tixador, 2010). © 2010 Published by Elsevier B. V.
Reprinted with Permission from Elsevier.

400 A have been installed as busbar couplers in medium-voltage power stations in the United Kingdom. One more 12 kV system with the current of 800 A has been the first FCL installed in the house load of a thermal power plant. With progress in fabrication of YBCO-coated conductor tapes, which have critical current J_c (77 K) > 1 MA/cm^2 in 1- to 2-μm-thick superconductor layers, further projects of Nexans, the German ENSYSTROB (SFCL 12−800), and the European Community ECCOFLOW (SFCL 24−1000) are based on the second technology, 2G REBCO tapes with lower AC losses and more nonlinear S−N transitions due to higher n value (Bock, Bludau, et al., 2011; Bock, Hobl, et al., 2011; Elschner et al., 2011). The projects unite the efforts of five European electric utilities to install an SFCL in two different locations: busbar coupling and transformer feeder.

A good longitudinal uniformity in critical current and resistivity was achieved in coated conductor tapes produced by American Superconductor Corporation (AMSC) (Schmidt et al., 2010). These coated conductor tapes are used by other companies for building superconducting elements. The joint project of AMSC and Siemens aimed to investigate the fault-current-limiting performance of the modules fabricated on the basis of 2G YBCO-coated conductor stabilized with stainless-steel laminates. This development goes to the commercial level for distribution systems: two Nexans-resistive SFCLs will be installed in power distribution networks in Birmingham, United Kingdom (Nexans' press release from March 31, 2014).

Within a DOE-funded joint project led by AMSC and including Siemens, Nexans, and Southern California Edison, one electrical phase of a resistive superconducting FCL for the transmission voltage level of 115 kV has been designed, manufactured, and tested (AMSC; Kraemer et al., 2012). The system successfully passed 30 power switching tests at voltages up to 30 kV and prospective currents up to 60 kA.

This SFCL device consists of power modules formed by bifilar coils (Figure 9.15) made from coated conductor tape connected in series and parallel to reduce the inductance and the AC losses.

Another distribution-voltage SFCL operating in combination with a three-phase superconducting cable (see Chapter 5, Section 5.8.5) with a nominal current of 2300 A was funded by the German government in 2012 in the framework of the project AmpaCity. After commissioning of the system at the end of 2013 a field test period until at least the end of 2015 is foreseen.

Several FCLs were installed in the United Kingdom grid under the leadership of Applied Superconductor Ltd, a United Kingdom company based in Blyth, Northumberland (recently closed). In the beginning of 2009 Nexans has supplied a SFCL to this company for a pilot installation in the United Kingdom. The device was installed in

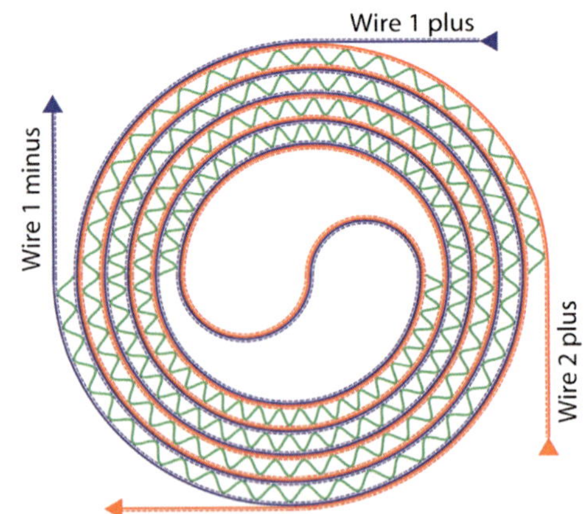

Figure 9.15 A bifilar coil wound from a CC tape (Siemens) (Schmidt et al., 2010), © 2010 IOP Publishing.

networks of Electricity North West in Bamber Bridge 11 kV Primary Substation near Preston in 2009. An 11 kV SFCL was successfully tested in August 2012. In 2010, Applied Superconductor Ltd announced the receipt of the world's first commercial order for a 33 kV SFCL for the distribution network of CE Electric UK (Northern Powergrid). In 2013, the project was closed down and is scheduled to be reopened in 2014 (Goodhand, 2013).

SuperPower, Inc. and its research partners, including American Electric Power, have been working on a program to develop a high-voltage three-phase modular SFCL to meet the needs of the utilities for Distribution and Transmission Power, particularly, recovery under load. They have performed a successful test of a distribution class SCFL, 11−15 kV phase, 800−2000 A assembled with three SFCL modules and a transmission class 138 kV device with 14 modules built on the base of 2G HTS wires (Xie et al., 2009).

The Italian SFCL program (Martini et al., 2012) has been conducted in two steps. Preliminary work concerned one- and three-phase resistive SFCL with BSCCO tapes noninductively wound on solenoids. A three-phase prototype of 1.2 MVA was successfully tested at the end of 2005. Since 2009 Ricerca sul Sistema Energetico S.p.A. (RSE), the second largest Italian utility, has been involved in the development of a resistive-type SFCL to be installed in the A2A Reti Elettriche S.p.A. distribution grid in the Milano area. A 9 kV/3.4 MVA SFCL device, based on 1G BSCCO tapes, has now been permanently installed as a single-feeder fault protection. The next step of the R&D activity will consist in the design of a three-phase SFCL demonstrator with the parameters 9 kV/15.6 MVA to be installed at the same substation in an incoming feeder to limit the contribution of this feeder to the total short-circuit current of the network (Martini et al., 2013).

To reduce the cost of superconducting material in a current limiter, limiting properties of single-strand round MgB_2 wires fabricated by Hyper Tech were intensively studied (Ye et al., 2007). A prototype of the MgB_2-based resistive SFCL was successfully tested (Smith et al., 2012). A consortium, led by Applied Superconductor Ltd., was established to develop a resistive FCL based on a MgB_2 superconductor. Funding of the project was provided by the Energy Technologies Institute. The consortium has planned to design and install an 11 kV/1250 A MgB_2-based demonstrator in Loughborough substation, but this program was suspended.

LS Industrial Systems Co., Ltd. (LSIS) and the Korea Electric Power Research Institute (KEPRI) have jointly developed a 22.9 kV/630 A hybrid-resistive SFCL through the Development of Advanced Power System by Applied Superconductivity Technologies (DAPAS) program. This device was installed in the Green Superconducting Electric Power Network (GENI project) at the Icheon substation of the Korea Electric Power Corporation (KEPCO) and operates on an incoming feeder connected with the first main transformer of the substation since August 2011 (Kim et al., 2012, 2013).

9.4.3 Inductive shielded type

The largest European FCL project ongoing in recent years was based on a long-lasting research cooperation of Schneider Electric and ALSTOM Grid (former AREVA T&D,

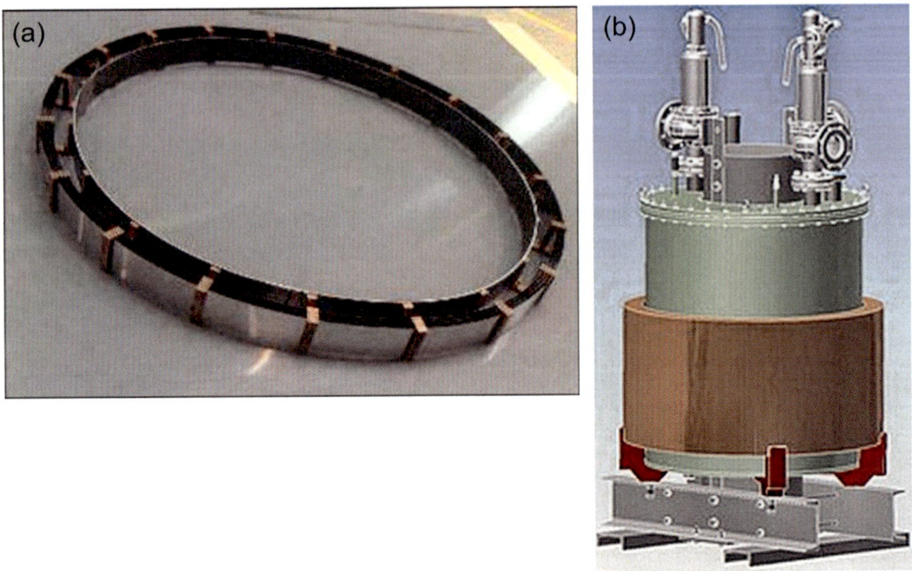

Figure 9.16 A 10 kV/15 MVA inductive, shielded core, FCL: (a) a HTS module and (b) one-phase device. © 2013 Bruker EST.
Published with Permission of Director R&D of Bruker HTS GmbH, Dr Alexander Usoskin.

leading industrial partner) with Bruker HTS. The project team planned to install its inductive SFCL (iSFCL), a first-of-its-kind prototype, between a substation of Stadtwerke Augsburg and a nearby factory, owned by MTU Onsite Energy GmbH, where its products are extensively tested (Usoskin et al., 2011, 2009). The iSFCL was intended to limit faults on a three-phase, 10 kV, 0.8 kA line. A prospective 25 kA fault is to be limited to 5 kA on the first peak and soon thereafter to 2 kA. The iSFCL was designed to initiate limitation very quickly, in 0.1 ms, and to hold the fault for 0.5 s (Usoskin et al., 2009). The device was built from 12 2G HTS modules stacked vertically (Figure 9.16).

9.4.4 Saturated-core type

The 3 kV/556 A prototype based on a low-temperature superconducting composite (Niomax A61/50, having 61 strands of niobium–titanium) was successfully tested in 1982 (Raju et al., 1982).

Currently, several teams are making progress toward commercial saturated-core HTS FCLs. Since 2006, Zenergy Power, Inc., has been developing saturated-core HTS FCLs for electric power grid applications (Moriconi et al., 2011), although the company has recently been closed down, and already earlier it withdrew from using superconductors in its system. Four full-scale prototypes were tested at Powertech Laboratories in 2009. These prototypes had the same nominal voltage of 15 kV but the nominal AC current ranged from 1250 to 2500 A, and the targeted AC fault current reduction levels ranged from about 30% to more than 50%. These prototypes were built using the concept known as the: "Compact FCL" with the financial support from the US Department of Energy and the Consolidated Edison Company of New York. The Compact FCL prototypes were built using standard "oil-filled" liquid dielectric transformer construction techniques and "dry-type" cryogenics to conductively cool the HTS coil without the use of liquid cryogens based on conduction cooling. This allowed reducing sufficiently the size and weight of the device. The dry-type cryogenics allowed the operating temperature of the HTS coils to be reduced below the freezing temperature of liquid nitrogen, enabling further increases in the critical current and using the 1G HTS in the DC bias magnet system. The use of conduction cooling also removes potential utility concerns about having large volumes of liquid cryogens in confined spaces and potential pressure vessel overpressurization, rupture, and venting concerns. As a result of the intensive testing since 2007, Zenergy Power has received contracts for the installation of a 15-kV-class device in the United Kingdom and a 138-kV-class device in the United States (Ohio) (Moriconi et al., 2011). These devices were installed in 2010 and 2011, respectively. In 2012, with the participation of Applied Superconductor Ltd., a 12 kV/1250 A device based on a saturated-core design was installed in the Avanti distribution circuit of Southern California Edison networks (as was noted above, these companies have changed their plans).

A 35 kV/90 MVA superconducting FCL was developed and installed in a transmission network at Puji substation of China Southern Power Grid in 2007 (Gong et al., 2008; Xin et al., 2012, 2007, 2009) and its performance was fully tested under artificially imposed short conditions in 2009 (Xin et al., 2011). The device's current-limiting capacity has been convincingly demonstrated. The reaction time and recovery time are also satisfactory to meet the requirements from grid routine protective procedures. The results of the fail-safe test demonstrate the reliability and robustness of this device. Using the experience in the development of a 35 kV FCL, a 220 kV/300 MVA three-phase SCFL (Figure 9.17) has been designed, successfully manufactured, and extensively tested by the Chinese group of the specialists of Innopower Superconductor Cable Co., Ltd, Tianjin University, and Tianjin Machinery and Electric Industry Holding Group Company Co. (Xin et al., 2012). The FCL has been installed and has passed all field tests at Shigezhuang substation in Tianjin, China (Xin et al., 2013).

Now, a 500 kV/1800 MVA single-phase prototype is under design and manufacturing. Factory tests are planned in early 2015 (Gong et al., 2013).

9.4.5 Bridge-type fault current limiters, CLTs, and current-limiting cables

A 10.5 kV/1.5 kA three-phase rectifier-type FCL was developed by the Institute of Electrical Engineering, Chinese Academy of Sciences, and installed in Gaoxi substation located in Hunan, China (Hui et al., 2006).

Figure 9.17 A photograph of the 220 kV saturated-core FCL (Xin, Gong et al., 2012). © 2012 IOP Publishing.

CLTs with SC windings were intensively developed in Japan, Korea, and the United States (Fault Current Limiting Superconducting Transformer; Hayakawa et al., 2000, 2011; Tomioka et al., 2013). Several prototypes of the transformers were described in Hayakawa et al. (2000, 2011). The work led by Nagoya University, Japan, was carried out since 1998, resulting in the development in 2010 of a 22 kV/6.6 kV, two MVA transformer with current-limiting functionality. Another Japanese team led by Fuji Electric Co. and Kyushu Electric Power Co. has developed and built a 6.9 kV/2.3 kV 400-kVA-class model transformer, with the further target to build a three-phase 66 kV/6.9 kV 20 MVA transformer for a distribution power grid (Tomioka et al., 2013).

Using 2G YBCO tapes fabricated by SuperPower, Waukesha Electric Systems (WES) and its partners designed and developed a smart-grid-compatible fault-current-limiting superconducting transformer on a live grid utility host site (Fault Current Limiting Superconducting Transformer). The 28 MVA three-phase Medium-Power Utility transformer (69 kV/12.47 kV class) is planned to be installed in 2015 in Southern California Edison's (SCE) MacArthur substation in Irvine, California.

In the framework of project HYDRA funded by the US Department of Homeland Security (DHS), a consortium led by AMSC has developed a HTS cable system with inherent fault-current-limiting capability for application in New York City, as described above and also in Chapter 5, Sections 5.6 and 5.8.6 (Maguire & Yuan, 2009).

9.4.6 Summary of the current status

Main achievements in the implementation of different SFCLs are summarized in Table 9.1.

Table 9.1 Main SFCLs built and installed in 2008–2013

Leader company/project	Country/year	Type	Data	SC material
Nexans SC	UK/2010	Resistive	12 kV/400 A	BSCCO tubes
Nexans SC/Vattenfall	Germany/2010	Resistive	12 kV/800 A	BSCCO tubes
Nexans SC/ENSYSTROB	Germany/2011	Resistive	12 kV/800 A	2G tapes
Nexans SC/ECCOFLOW	EU/2013	Resistive	24 kV/1000 A	2G tapes
AMSC&Siemens/Nexans	USA/2013	Resistive	138 kV/1 kA	2G tapes
RSE	Italy/2011	Resistive	9 kV/1 kA	2G tapes
AMSC&Siemens	USA, Germany/2012	Resistive	66 kV/1.2 kA	2G tapes
Zenergy Power[a]	USA/2009	Saturated iron core	12 kV/800 A	BSCCO tapes
Zenergy Power/Avanti[a]	USA/2010	Saturated iron core	15 kV/1250 A	BSCCO tapes
Zenergy Power[a]	USA/2012	Saturated iron core	138 kV/1200 A	BSCCO tapes
LSIS, KEPRI	Korea/2011	Hybrid-resistive	22.9 kV/630 A	2G tapes
HyperTech/ASL		Resistive	11 kV/1250 A	MgB_2 wires
Innopower	China/2010	Saturated iron core	220 kV/300 MVA	BSCCO tapes
SuperPower	USA/2012	Resistive	15 kV/2300 A	2G tapes
SuperPower	USA/2012	Resistive	138 kV/1840 A	2G tapes

[a]The activity of this company was stopped. Applied Superconductor in July 2012 acquired the FCL projects and key personnel from Zenergy. The FCL projects in Applied Superconductor Ltd are scheduled to be reopened in 2014 (Goodhand, 2013).

9.5 Comparison of different fault current limiters

Comparisons of the characteristics, advantages, and disadvantages of various types of FCLs are given in numerous publications (see, e.g., Hassenzahl & Tsukamoto, 2011; Lei, Ruifeng, Wanrong, & Xiaoping, 2011; Sokolovsky et al., 2004). According to their operation principle all FCLs can be divided into two main groups:

- current limitation is achieved by quenching an active superconducting element;
- current limitation occurs without quenching.

Under the nominal operation of the protected circuit, the superconductors of the first group (resistive and inductive) are subjected to AC currents or/and AC magnetic fields, while superconductors in the second group (saturated core, bridge type) carry DC current. AC losses and problems of thermal stabilization limit the applicability of cheaper low-temperature Nb–Ti and MgB_2 superconductors in the FCLs of the first group.

The common problems of the devices of the first group are to provide sufficient quenching of the SC element, minimize AC losses, avoid hot spots, and recover rapidly back to the initial SC state after the fault event.

One other intrinsic disadvantage of the saturated-core FCL is the resistive loss and voltage drop over the primary circuit. Both are higher than the low loss and voltage drop achieved in resistive FCLs. On the other hand, a significant advantage is that the system circuit closely resembles a transformer, and so well-established transformer technology can be applied, in particular for handling transmission voltage, something that is hard (though possible (Kraemer et al., 2012)) to do in the resistive FCL.

The resistive FCL offers significant advantages compared to the saturated-core one in terms of compactness, simplicity, and performance during the fault. However, in the resistive design the superconductor is directly connected in series with the circuit being protected. AC losses and the heat transferred through current leads are the main intrinsic problems of the resistive design. Moreover, at the fault event, normal resistance limits the fault current and all the circuit supply voltage is developed across the SFCL terminals unless it is reduced by means of a shunt.

To overcome the problems of inhomogeneous quenching, heating, and recovery and to improve the performance, the superconductor of a resistive FCL is usually shunted by a resistor or an electrical reactor. On the other hand, an inductive FCL meets these requirements without additional measures. Therefore, the resistive FCL advantage, lower weight in comparison with the inductive one, is not apparent especially in comparison with the inductive devices with an open magnetic system or magnetic coupled air coils. The active superconducting element of an inductive FCL, as in a resistive device element, can be fabricated using bulk, thin film, or wire superconductors (Hassenzahl & Tsukamoto, 2011; Lei et al., 2011; Paul et al., 1997; Sokolovsky et al., 2004, 2003; Usoskin et al., 2011). It operates at a lower voltage and can be shorter than that in a resistive FCL. This last feature facilitates realization of simultaneous quenching of the whole superconductor. An advantage of the inductive FCL is also the physical separation of cryogenic and normal parts of the device; there are no current leads into cryogenic cryostat.

However, since the normal metal primary circuit is in series with the grid connection, there is an inevitable resistive loss as well as inductive voltage drop during standard operation of the inductive FCL. Those inductive systems that include a magnetic yoke are also bulky, although the advantage in size vis-à-vis resistive systems comes back to the inductive FCLs at transmission voltages because of the size of the terminations and cryostats to maintain dielectric integrity in the liquid nitrogen bath. An advantage of resistive devices is also their modularity: multiple noninductive superconductor pancake coils can be connected in series or parallel to meet specific grid requirements of voltage and current (see more detailed comparisons, for example, in Sokolovsky et al. (2004)).

Despite the apparent advantages of the inductive approach, most inductive FCL projects have been terminated, and at this point, the greatest commercial strides are being made by resistive FCLs.

The initial hybrid schemes use the superconductor quenching as a trigger to a fast operation switch or an interrupter. On the one hand, this design leads to reduction of a superconductor volume, AC losses, and losses in the superconductor during a current limitation and, hence, to shortening the recovery time. On the other hand, development of a fast operation interrupter opening within a half of the current cycle is required, which is technically difficult, and the cost grows with the voltage class of the circuit; also the first, and potentially most damaging, peak of the fault current is not limited. This means that the power devices in the system are not protected from dynamic influence of this peak, which can achieve double the amplitude of the steady fault current. To overcome this obstacle, the application of a superconductor as both a trigger and a resistive device with a shorter time of current limitation was proposed. This solution increases the superconductor volume and complicates the design.

Devices that combine functions of the energy transmission and voltage transformation with fault current limitation, such as superconducting CLTs and cables, reduce the weight, volume, and footprint in comparison with systems requiring a separate device for the fault current limitation.

The advantage of the bridge and saturated FCLs is that the current limitation is realized without quenching; superconductors operate under DC conditions and, hence, losses in the superconductors are very low under both the nominal and the fault regimes of the circuit. This opens a way to apply MgB_2 or even low-temperature superconductors, which are much cheaper than high-temperature ones. The recovery time is very low and determined by the operation rate of the electronic systems. However the bridge-type limiter depends on operation of power electronics and so is not fail-safe, in comparison to systems in which the superconductor directly transitions into its resistive state, which is a fail-safe process.

Advantages and disadvantages of various types of FCLs are summarized in Table 9.2.

Table 9.2 Comparison of various FCL types

FCL type	Advantages	Disadvantages
Resistive	Low volume and weight; self-triggered; easy way to increase rated current and voltage.	Need for current leads; high-voltage drop across SC; quenching with heat dissipation; need for shunting resistor or reactor; prolonged recovery time.
Inductive	No need for current leads; low voltage drop across SC; self-triggered; physical separation of cryogenic and normal parts; no interruption of the main current path during failure of the HTS or cooling system.	Quenching; high volume and weight; prolonged recovery time.
Hybrid I (SC as a trigger)	Self-triggered; low volume and cost of superconductor; short recovery time; low AC losses.	Need in fast operation interrupter; the first peak of a fault current is not limited.
Hybrid II (SC as a trigger and resistive FCL)	Self-triggered; low volume and cost of superconductor; short recovery time; low AC losses.	Need in fast operation interrupter; superconductor volume, losses, recovery times are larger than those in Hybrid I; more complicated design.
Bridge	No quenching; DC operation of SC; low AC losses; short recovery time.	Use of power electronic devices (diodes, thyristors); difficult control of nominal regimes of the circuit; limited current increases with time; need in DC source and special control system (in some designs).
Saturated	No quenching and no need for triggering; DC operation of SC;	Large volume and weight; need in a DC source and a control system for DC

Table 9.2 Continued

FCL type	Advantages	Disadvantages
	low losses; immediate full recovery after a fault; Fail-safe operation.	source (in some designs); high voltages induced in the saturating coil under nonsymmetrical regimes.
Current-limiting transformers and cables	Reducing weight, volume, and footstep square.	Requirement of special measures to meet the system requirements.

9.6 Applicability of superconducting fault current limiters in power systems

This is no limitation in principle on the rated voltage and current of SFCLs and they can be used at all voltage classes of power and industrial systems, as well as in autonomous systems such as large ships (Blair et al., 2011; Hassenzahl & Tsukamoto, 2011; Jo, Joo, & Lee, 2013; Morandi, 2013; Noe, 2011; Noe & Steurer, 2007; Park, Sung, & Park, 2010; Sokolovsky et al., 2004; Noe & Oswald, 1999), although the size and expense climb rapidly with increasing voltage. Figure 9.18 presents possible FCL locations in distribution networks. FCLs can also be located similarly in a transmission network. The main possible locations can be classified as coupling (3, 4, 5, 9, and 11), feeder (2, 6, 7, 8, 12, 13, and 14), and generator circuit locations (1, 2, and 10). The influence of a FCL on the fault currents in a system depends strongly on the location. FCLs protect the devices placed "upstream." However, the FCL actuation can lead to a decrease in a fault current in the whole system, including "downstream"; for example, FCL 13 in Figure 9.18 not only protects the transmission network but also can decrease a fault current in the distribution networks. The main advantage of the application of a FCL in coupling is that a device can protect two subnetworks from a fault in any branch of one of the subnetworks. However, a FCL for the coupling location should possess larger impedance and be designed for higher rated power than devices for a feeder location, where the device protects the system from only a fault in a single line.

Application of FCLs allows one to avoid the traditional expensive measures required by an increasing fault current level due to power system expansion, such as:

- replacement of circuit breakers by breakers with higher current interruption capability;
- development of power devices for a higher fault current;
- stationary separation of networks;
- designing and using transformers with specially increased short-circuit impedance;
- using connections with rated voltage higher than that determined from economical considerations.

The three last measures are necessary, in the absence of FCLs, to decrease static and transient stabilities of a power system.

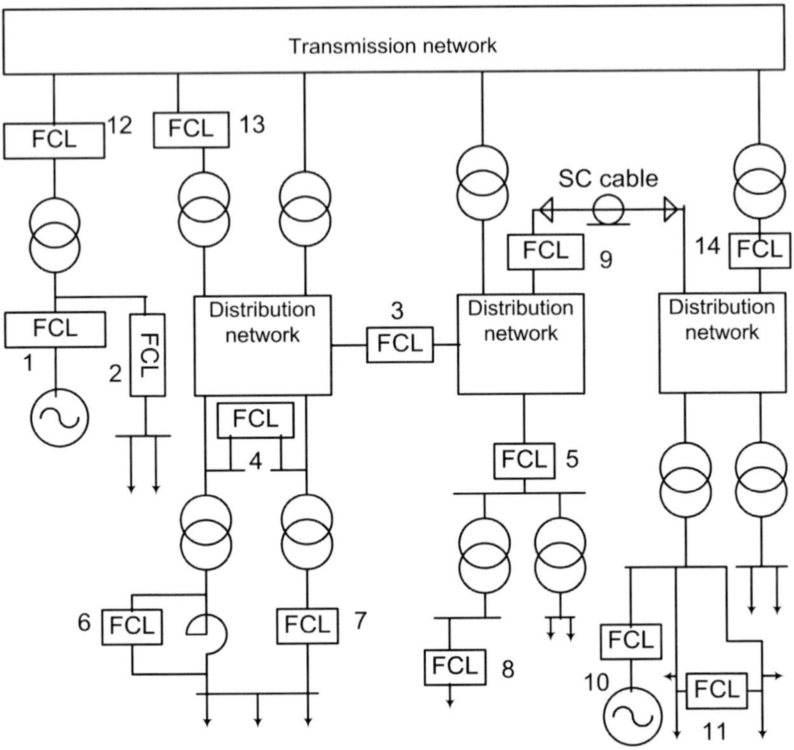

Figure 9.18 Possible locations of FCLs in a power system.

The gain of the fault current limitation by FCLs is in the reduction of weight, size, and cost of electric power equipment, the possibility of using circuit breakers with lower current-interrupting capabilities, and more effective operating schemes of the electric network.

As an example, we consider a distribution network with the configuration and parameters shown in Figure 9.19. The network consists of two separate substations: the first supplies an industrial zone and the second a residential area. The development of the regions requires building new lines of high voltage and new substations. However, there is another way to solve the problem of rising electric energy demand: to connect the buses of the substations. This solution is possible because the maximum power consumptions in both areas do not coincide. In the industrial zone, the maximum of consumption is during a workday and in the residential area in morning and evening hours. However, the connection of the substations increases the fault current to a value more than the current-interrupting capabilities of the circuit breakers installed in this network. Thus, the connection of the substations leads to the replacement of all the breakers and other equipment of the substations. FCLs can solve such problems of fault currents in an interconnected system.

Numerous studies (Almeida, Rocha, Dente, & Costa Branco, 2009; Lee, Lee, Ko, & Hyun, 2001; Sjostrom, Cheraoui, & Dutoit, 1999; Sokolovsky et al., 2004, 2007) have

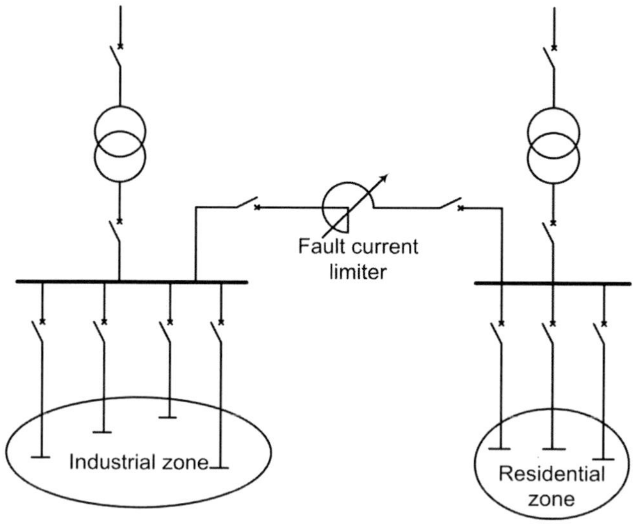

Figure 9.19 Schema of two distribution substations with connecting tie.

shown that superconducting FCLs can enhance power system transient stability. The problem of transient stability increases with the introduction of superconducting generators and motors. In comparison with the traditional generators, the inertia constant of these generators is lower by 50–60% for 3000 rpm and by 60–70% for 1500 rpm and the inertia constant of the turbine–generator block is 1.2–1.5 times less (Glebov, Danilevich, & Shachtarin, 1981). This may lead to the disturbance of the synchronous operation of superconducting generators under fault conditions in those cases where the synchronous operation of traditional generators is stable. As an example, Figure 9.20 presents simulation results for superconducting generators of 300 MW connected with a power system through two high-voltage lines; FCLs are introduced in the lines. The calculation was carried out for a three-phase fault event in one of the high-voltage lines.

The fault current problem increases with use of other superconducting devices such as transformers and cables; in some cases nominal currents of these devices are comparable with the current-interrupting capability of the existing circuit breakers. Joint application of superconducting FCLs with other superconducting devices allows one to create power complexes with optimized levels of fault currents as well as to share cryogenic equipment.

The joint application of various superconducting devices can reduce requirements on device parameters and control systems. For example, joint application of superconducting magnetic energy storage (SMES) and SFCLs can effectively increase the transient stability of a power system and sufficiently reduce the requirements of the parameters of both devices in comparison with application of each device separately (Ngamroo & Vachirasricirikul, 2012, 2013; Sokolovsky, Meerovich, & Chubraeva, 2009). The only HTS power substation in the world up to now was successfully developed in 2011 in Baiyin, in the Gansu Province of China (Zhu et al., 2013) (see also

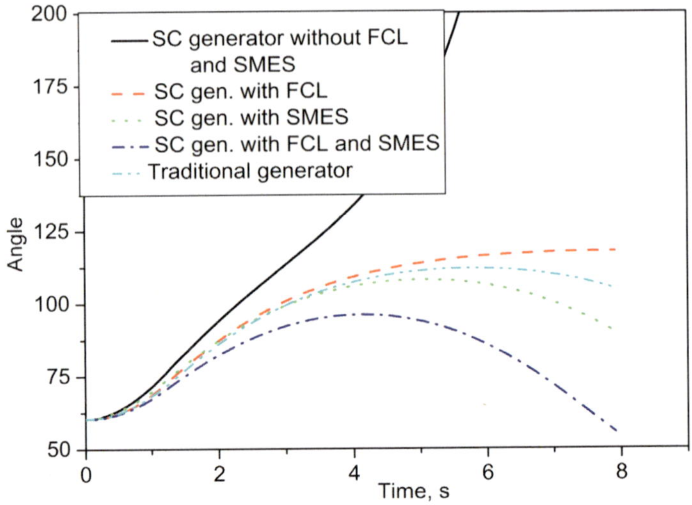

Figure 9.20 Influence of FCL and superconducting magnetic energy storage (SMES) on transient stability of power system: relative angle of the generator rotors versus time.

Chapter 12). This substation integrates a 630 kVA/10 kV/0.4 kV HTS transformer, a 380 V/1.5 kA HTS AC cable of 75 m length, a 10 kV/1.5 kA FCL, and a 1 MJ/0.5 MVA SMES. The cooling system provides liquid nitrogen for these four HTS devices. The HTS power substation is controlled automatically by a real-time measuring and control system.

9.7 Future trends

The fault current limitation is an important problem for the future growth and expansion of the electrical grid. Today only ABB offers commercial off-the-shelf fault current limitation equipment—its "Is-Limiter." The Is-Limiter, is a one-time-use fuse mechanism that is suitable for lines at voltages not exceeding 40.5 kV, but it cannot restore the line to operation after the fault has ended. Instead, ABB's Is-Limiter must be reset or replaced after each use. Therefore, because of the higher voltage and resettable nature of superconducting FCLs, they are devices highly desired for electric power utilities. Due to their features, FCLs are key components for future smart grids (Hamidi, Smith, & Wilson, 2010; Neumann; Tixador, 2010). Their introduction in the power system will provide a high level of network interconnections, thus making possible a flexible link among electric power producers, distribution operators, and ultimately consumers, which is required in the modern electric market. According to a report by Morgan Stanley on the Smart Grid (Kim et al., 2013), the potential market of FCLs may reach five billion dollars per year. Because of their unique properties, SFCLs are intensively being developed throughout the world. Despite the tremendous efforts and impressive funding for the work, these devices are still far from wide commercial introduction. However, in recent years, several SFCLs have been installed into power networks. They are mostly of the resistive and saturated types and they are intended for demonstration of the feasibility and testing of long-time operation. Nevertheless, improvements in superconducting materials since 1994 have driven the FCL technology to the forefront. Efforts to incorporate SFCLs into the power grid are underway worldwide. A promising option is to integrate several superconducting devices in the same power system. A recent example of such an approach is the AmpaCity project in Essen, German (SFCL 12–2300), headed by Nexans SC and funded by the German government. A three-phase superconducting cable at a distribution level has been combined with a resistive SFCL, which protects the cable against high fault currents (see Chapter 5, Section 5.8.5).

9.8 Sources of further information

Various types of superconducting FCLs have being developed in the United States, Japan, Germany, Spain, South Korea, China, United Kingdom, Israel, and other countries. The design, development, and application of the FCLs are discussed in numerous publications. The electronic data base INSPECT notes about 3000 publications related to these FCLs. Here we briefly presented the main results achieved. Additional information can be found in reviews given below and references therein, as well as in the specialized magazines such as IEEE Transaction on Applied Superconductivity, Superconductor Science and Technology, and Physica C. Many papers on the development of superconducting FCLs are presented at the Applied Superconductivity Conferences (USA) and European Conference on Applied Superconductivity (Eucas) and are published in IEEE Transaction on Applied Superconductivity and the Institute of Physics Publishing Conference Series, respectively.

References

Ainslie, M. D., Baba, J., Salvucci, V., Nitta, T., Fukunaga, T., Shibuya, M., et al. (2009). Superconducting fault current limiter design using parallel-connected YBCO thin films. *IEEE Transactions on Applied Superconductivity, 19*(3), 1918–1921.

Almeida, M. E., Rocha, C. S., Dente, J. A., & Costa Branco, P. J. (2009). Enhancement of power system transient stability and power quality using superconducting fault current limiters. In *Proceedings of POWERENG 2009,* 18-20 March, Lisobon, Portugal (pp. 425–430).

Bashkirov, Yu. A., Fleishman, L. S., Patsayeva, T. Yu., Soboler, A. N., & Vdovin, A. B. (1991). Current-limiting reactor based on high-Tc superconductors. *IEEE Transaction on Magnetics, 27*, 1089–1092.

Bashkirov, Yu. A., Yakimets, I. V., Fleishman, L. S., & Narovlyanskii, V. G. (1995). Application of superconducting shields in current-limiting and special-purpose transformers. *IEEE Transactions on Applied Superconductivity, 5*(2), 1075–1078.

Blair, S. M., Booth, C. D., Elders, I. M., Singh, N. K., Burt, G. M., & McCarthy, J. (2011). Superconducting fault current limiter application in a power-dense marina electrical system. *IET Electrical Systems in Transportation, 1*, 93–102.

Bock, J., Bludau, M., Dommerque, R., Hobl, A., Kraemer, S., Rikel, M. O., et al. (2011). HTS fault current limiters-first commercial devices for distribution level grids in Europe. *IEEE Transactions on Applied Superconductivity, 21*, 1202–1205.

Bock, J., Hobl, A., Kramer, S., Bludau, M., Schramm, J., Janke, C., et al. (6–9 June 2011). Nexan's superconducting fault current limiter for medium voltage applications – status and prospects. In *proceedings of the 21st international conference on electricity distribution (CIRED2011)*. Frankfurt, Germany (Paper 0352).

Boenig, H. J., & Paice, D. A. (1983). Fault current limiter using a superconducting coil. *IEEE transactions on magnetics, 19*, 1051–1053.

Cave, R. A., Willen, D. W., Brissette, Y., Richer, C., & Aresteanu, V. A. (1994). Test results for laboratory scale inductive high-Tc superconducting fault current limiters. *IEEE Transactions on Magnetics, 30*, 1895–1898.

Clem, J. R. (2008). Field and current distributions and ac losses in a bifilar stack of superconducting strips. *Physical Review B, 77*, 134506.

Dalessandro, R. B., Bocchi, M., Rossi, V., & Martini, L. F. (2007). Test results on 500 kVA-class MgB2- based fault current limiter prototypes. *IEEE Transactions on Applied Superconductivity, 17*, 1776–1779.

Elschner, S., Bock, J., Brommer, G., & Cowey, L. (2000). Study of MCP BSCCO 2212 bulk material with respect to application in a resistive current limiter. In *Institute of physics conference series: vol. 167. Applied superconductivity 1999* (pp. 1029–1032).

Elschner, S., Bock, J., & Gauss, S. (1995). Development of HTSC-rings for inductive fault current limiters. In *Proceedings of the seventh international symposium on superconductivity (ISS'94): vol. 2. Advances in superconductivity VII* (pp. 1211–1214).

Elschner, S., Breuer, F., Walter, H., & Bock, J. (2006). Magnetic field assisted quench propagation as a new concept for resistive current limiting devices. *Journal of Physics: Conference Series, 43*(1), 917–920.

Elschner, S., Kudymow, A., Fink, S., Goldacker, W., Grilli, F., Schacherer, C., et al. (2011). ENSYSTROB - resistive fault current limiter based on coated conductors for medium voltage application. *IEEE Transactions on Applied Superconductivity, 21*, 1209–1212.

Endo, M., Hori, T., Koyama, T., Kaiho, K., Yamaguchi, I., Arai, K., et al. (2009). Development of a superconducting fault current limiter using various high-speed circuit breakers. *IET Electric Power Applications, 3*, 363–370.

Endo, M., Hori, T., Koyama, K., Yamaguchi, I., Arai, K., Kaiho, K., et al. (2008). Operating characteristics of superconducting fault current limiter using 24 kV vacuum interrupter driven by electromagnetic repulsion switch. *Journal of Physics: Conference Series, 97*, 012068 (5 pp.).

Fassbinder, S., Application notes efficiency and loss evaluation of large power transformers, Electric Cooper Institute (ECI) Publication No Cu0144. Available from http://www.leonardo-energy.org/sites/leonardo-energy/files/documents-and-links/Cu0144_Efficiency%20and%20Loss%20Evaluation%20of%20Large%20Power%20Transformers_v1.pdf. Accessed 08.07.11.

Giese, R. F., & Runde, M. (1993). Assessment study of superconducting fault-current limiters operating at 77 K. *IEEE Transactions on Power Delivery, 8*(7), 1138–1147.

Giese, R. F. (March 1995). *Fault current limiter – A second look*. Rep. for International Energy Agency (IEA). Argonne Nat. Lab.

Glebov, I. A., Danilevich, Y. B., & Shachtarin, B. N. (1981). *Turbogenerators with use of superconductivity*. Moscow: Nauka.

Gong, W. Z., Xin, Y., Hong, H., Zhang, J. Y., Niu, X. Y., Yu Wei Sun, Y. W., et al. (October 25–27, 2013). Plenary Talk – practical HTS FCL development - updates of Innopower's SFCL R&D projects. In *Proceedings of 2013 IEEE international conference on applied superconductivity and electromagnetic devices, Beijing, China*. ID3091.

Gong, W. Z., Zhang, J. Y., Cao, Z. J., Hong, H., Tian, B., Wang, Y., et al. (2008). HTS dc bias coil for 35 kV/90 MVA saturated iron-core fault current limiter. *Physica C, 468*, 2050–2053.

Goodhand, C. (December 2013). CET 1001/LNCF-T1-001 Interim Closedown Report. *33 kV superconducting fault current limiter* https://www.ofgem.gov.uk/ofgem-publications/87432/33kvsfclthreeyearclosedownreport.pdf.

Gouge, M. J., Duckworth, R. C., Demko, J. A., Rey, C. M., Thompson, J. R., Lindsay, D. L., et al. (2009). Testing of 3-meter prototype fault current limiting cables. *IEEE Transactions on Applied Superconductivity, 19*, 1744–1747.

Granados, X., Puig, T., Obradors, X., Mendoza, E., Teva, J., Calleja, A., et al. (2002). Design, building up and testing of a 400 kVA hybrid FCL. *Physica C*, 372–376, 1680–1683.

Gromoll, B., Ries, G., Schmidt, W., Kraemer, H.-P., Seebacher, B., Utz, B., et al. (1999). Resistive fault current limiters with YBCO films-100 kVA functional model. *IEEE Transactions on Applied Superconductivity, 9*(6), 656–659.

Hamidi, V., Smith, K. S., & Wilson, R. C. (2010). Smart grid technology review within the transmission and distribution sector. *Proceedings of 2010 IEEE PES innovative smart grid technologies conference Europe (ISGT Europe)*, 8.

Hassenzahl, W., & Tsukamoto, O. (2011). Electric power application of superconductivity. In H. Rogalla & P. H. Kes (Eds.), *100 years of superconductivity* (pp. 769–797). Taylor & Francis Group.

Hayakawa, N., Chigusa, S., Kashima, N., Nagaya, S., & Okubo, H. (2000). Feasibility study of superconducting fault current limiting transformer (SFCLT). *Cryogenics, 40*, 325–331.

Hayakawa, N., Kojima, H., Hanai, M., & Okubo, H. (2011). Progress in development of superconducting fault current limiting transformer (SFCLT). *IEEE Transactions on Applied Superconductivity, 21*(3), 1397–1400.

Hobl, A., Elschner, S., Bock, J., Kramer, S., Janke, C., & Schramm, J. (2012). Superconducting fault current limiters: a new tool for the "grid of the future" (4 pp.). In *CIRED 2012 workshop: Integration of renewables into the distribution grid* (p. 0296). Available from http://www.cired.net/publications/workshop2012/pdfs/CIREDWS2012_0296_final.pdf. Accessed 23.05.2013.

Hori, T., Otani, A., Kaiho, K., Yamaguchi, I., Morita, M., & Yanabu, S. (2006). Study of superconducting fault current limiter using vacuum interrupter driven by electromagnetic repulsion force for commutating switch. *IEEE Transactions on Applied Superconductivity, 6*, 1999–2004.

Hoshino, T., Nishikawa, M., Salim, K. M., Muta, I., & Nakamura, T. (2001). Preliminary studies on characteristics of series-connected resistive type superconducting fault current limiter for system design. *Physica C, 354*, 120–124.

Hoshino, T., Salim, K. M., Nishikawa, M., Muta, I., & Nakamura, T. (2001). Proposal of saturated DC reactor type superconducting fault current limiter (SFCL). *Cryogenics, 41*, 469–474.

Hui, D., Wang, Z. K., Zhang, J. Y., Zhang, D., Dai, S. T., Zhao, C. H., et al. (2006). Development and test of 10.5 kV/1.5 kA HTS fault current limiter. *Superconductor Science and Technology, 16*, 687–690.

Hyun, O. B., Sim, J., Park, K. B., Kim, H. R., Kang, J. S., Lee, B. W., et al. (2007). 14 kV single-phase superconducting fault current limiter based on YBCO films. *Cryogenics, 47*(4), 183–188.

Ito, D., Yoneda, E. S., Tsurunaga, K., Tada, T., Hara, T., Ohkuma, T., et al. (1992). 6.6 kV/1.5 kA class superconducting fault current limiter development. *IEEE Transactions on Magnetism, 28*, 438–441.

Jo, H.-C., Joo, S.-K., & Lee, K. (2013). Optimal placement of superconducting fault current limiters (FCLs) for protection of an electric power system with distributed generations (DGs). *IEEE Transactions on Applied Superconductivity, 23*, 5600304.

Jung, B. I., Cho, Y. S., Choi, H. S., Ha, K. H., Choi, S. G., Chul, D. C., et al. (2011). Reclosing operation characteristics of the flux-coupling type SFCL in a single-line-to ground fault. *Physica C, 471*, 1327–1331.

Jung, B. I., Cho, Y. S., Park, H. M., Chung, D. C., & Choi, H. S. (2013). Comparison of the quench and fault current limiting characteristics of the flux-coupling type SFCL with single and three-phase transformer. *Physica C, 484*, 258–262.

Kaltenborn, U., Mumford, F., Usoskin, A., Schmidt, S., & Janetschek, T. (6-9 June, 2011). Iinductive shielded superconducting fault current limiter-an enabler of smarter grids. In *C I R E D 21st international conference on electricity distribution Frankfurt*. Paper 0955. Available from http://www.cired.net/publications/cired2011/part1/papers/CIRED2011_0955_final.pdf. Accessed 23.05.13.

Kim, H.-R., Yang, S.-E., Yu, S.-D., Kim, H., Kim, W.-S., Park, K., et al. (2012). Installation and testing of SFCLs. *IEEE Transactions on Applied Superconductivity, 22*(3), 5602704.

Kim, M. J., Choe, W., Bang, S.-H., Park, H. Y., Lee, G.-H., Sim, J., et al. (2013). Development of a fault current limiter for 22.9 kV distribution power line. In *2013 second international conference on electric power equipment - switching technology, ICEPE-st 2013, 2013 second international conference on electric power equipment – switching technology, ICEPE-st 2013*.

Klein, H.-U. (9–11 May 2012). *Superconducting fault current limiter (iSFCL)*. Presentation at IEA ExCO Meeting. Heidelberg. Available from http://www.superconductivityiea.org/document/Klein Bruker_SchneiderElectric iSFCL.pdf. Accessed 23.05.13.

Ko, S., Lim, S.-H., & Han, T.-H. (2013). Analysis of fault current limiting and recovery characteristics of a flux-lock type SFCL with an isolated transformer. *Physica C, 484*, 263–266.

Kraemer, H.-P., Schmidt, W., Cai, H., Gamble, B., et al. (2012). Superconducting fault current limiter for transmission voltage. *Physics Procedia, 36*, 921–926.

Lee, S., Lee, C., Ko, T. K., & Hyun, O. (2001). Stability analysis of power system with superconducting fault current limiter installed. *IEEE Transactions on Applied Superconductivity, 11*, 2098–2101.

Lee, B. W., Park, K. B., Sim, J., Oh, I. S., Lee, H. G., Kim, H. R., et al. (2008). Design and experiments of novel hybrid type superconducting fault current limiters. *IEEE Transactions on Applied Superconductivity, 18*, 624–627.

Lee, S. R., Yoon, J.-Y., Yang, B., & Lee, B. (2012). Specifications of the 22.9 kV SFCL considering Protection systems in korean power distribution system. *IEEE Transactions on Applied Superconductivity, 22*, 5602404.

Lei, F., Ruifeng, G., Wanrong, Z., & Xiaoping, Y. (23-27 October, 2011). Type and characteristic analysis of fault current limiter. In *Proc. of first international conference on electric power equipment – switching technology (ICEPE-ST), Xi'an, China* (pp. 356–361).

Leung, E. (1997). Surge protection for power grids. *IEEE Spectrum, 34*, 26–30.

Lim, S.-H., Choi, H.-S., & Han, B.-S. (2004). The improved hysteresis characteristics of flux-lock type SFCL using third winding. *Physica C, 406*, 37–45.

Lopez-Roldan, J., Price, A. C., De La Rosa, F., & Moriconi, F. (24–29 July 2011). Analysis of the effect of a saturable-core HTS fault current limiter on the circuit breaker transient recovery voltage. In *Proceedings of power and energy society general meeting, 2011 IEEE*. Detroit, MI, USA.

Maguire, J., Folts, D., Yuan, J., Lindsay, D., Knoll, D., Bratt, S., et al. (2009). Development and demonstration of a fault current limiting HTS cable to be installed in the Con Edison Grid. *IEEE Transactions on Applied Superconductivity, 19*, 1740–1743.

Maguire, J., & Yuan, J. (2009). Status of high temperature superconductor cable and fault limiter projects at American Superconductor. *Physica C, 469*, 874–880.

Malozemoff, A. P., Fleshler, S., Rupich, M., Thieme, C., Li, X., Zhang, W., et al. (2008). Progress in high temperature superconductor coated conductors and their applications. *Superconductor Science and Technology, 21*, 034005–034007.

Martini, L., Bocchi, M., Ascade, M., Valzasina, A., Rossi, V., Ravetta, C., et al. (2012). Live-grid installation and field testing of the first Italian superconducting fault current limiter. *IEEE/CSC ESAS Europe Superconductivity News Forum, 22*.

Martini, L., Bocchi, M., Ascade, M., Valzasina, A., Rossi, V., Ravetta, C., et al. (2013). Live-grid installation and field testing of the first Italian superconducting fault current limiter. *IEEE Transactions on Applied Superconductivity, 23*(3), 5602504.

Meerovich, V., & Sokolovsky, V. (2005). Experimental study of a transformer with superconducting elements for fault current limitation and energy redistribution. *Cryogenics, 45*(8), 572–577.

Meerovich, V., Sokolovsky, V., Jung, G., & Goren, S. (1995a). Inductive superconducting current limiter: state of art and prospects. In *Institute of Physics Conference Series: vol. 148. Applied superconductivity* (pp. 603–606).

Meerovich, V., Sokolovsky, V., Jung, G., & Goren, S. (1995b). High-Tc superconducting inductive current limiter for 1 kV/25 A performance. *IEEE Transactions on Applied Superconductivity, 5*(6), 1044–1048.

Meerovich, V., Sokolovsky, V., Goren, S., Kozyrev, A. B., Osadchy, V. N., & Holmann, E. K. (1997). Operation of hybrid current limiter based on high-Tc superconducting thin film. *IEEE Transactions on Applied Superconductivity, 7*, 3783–3790.

Meerovich, V., Sokolovsky, V., Bock, J., Gauss, S., Goren, S., & Jung, G. (1999). Performance of an inductive fault current limiter employing BSCCO superconducting cylinders. *IEEE Transactions on Applied Superconductivity, 9*(4), 4666–4676.

Meerovich, V., Sokolovsky, V., & Vajda, I. (2007). Comparison of a self-limiting transformer and a transformer type FCL with HTS elements. *IEEE Transactions on Applied Superconductivity, 17*, 1911–1914.

Morandi, A. (2013). State of the art of superconducting fault current limiters and their application to the electric power systems. *Physica C, 484*, 242–247.

Morandi, A., Brisigotti, S., Grasso, G., & Marabotto, R. (2013). Conduction cooling and fast recovery in MgB2-based DC resistive SFCL. *IEEE Transactions on Applied Superconductivity, 23*(5), 5604409.

Morandi, A., & Imparato, S. (2009). A DC-operating resistive-type superconducting fault current limiter for AC applications. *Superconductor Science and Technology, 22*, 045002.

Moriconi, F., De La Rosa, F., Darmann, F., Nelson, A., & Masur, L. (2011). Development and deployment of saturated-core fault current limiters in distribution and transmission substations. *IEEE Transactions on Applied Superconductivity, 21*, 1288–1293.

Na, J. B., Kim, Y. J., Jang, J. Y., Ryu, K. S., Hwang, Y. J., Choi, S., et al. (2012). Design and tests of prototype hybrid superconducting fault current limiter with fast switch. *IEEE Transactions on Applied Superconductivity, 22*, 5602604.

Neklepaev, B. N. (1978). *Coordination and optimization of fault current level, in power systems (in Russian)*. Moscow: Energya.

Neumann, C., Superconducting fault current limiter (SFCL) in the medium and high voltage grid. In *Proceedings of 2006 IEEE Power Engineering Society General Meeting* (p. 6).

Nexans and AMSC Introduce fault current limiter for North American utilities, Press Release of AMSC. Available from http://ir.amsc.com/releasedetail.cfm?ReleaseID=725019. Accessed 23.05.13.

Ngamroo, I., & Vachirasricirikul, S. (2012). Coordinated control of optimized SFCL and SMES for improvement of power system transient stability. *IEEE Transactions on Applied Superconductivity, 22*, 5600805.

Ngamroo, I., & Vachirasricirikul, S. (2013). Optimized SFCL and SMES units for multimachine transient stabilization based on kinetic energy control. *IEEE Transactions on Applied Superconductivity, 23*, 5000309.

Nikulshin, Y., Wolfus, Y., Friedman, A., & Yeshurun, Y. (2013). Dynamic core length in saturated core fault current limiters. *Superconductor Science and Technology, 26*, 095013.

Nikulshin, Y., Wolfus, Y., Friedman, A., & Yeshurun, Y. (2014). Dynamic inductance in saturated core fault current limiters. *Journal of Superconductivity and Novel Magnetism* (accepted for publication).

Noe, M., & Oswald, B. R. (1999). Technical and economical benefits of superconducting fault current limiters in power systems. *IEEE Transactions on Applied Superconductivity, 9*, 1347–1350.

Noe, M., & Steurer, M. (2007). High-temperature superconductor fault current limiters: concepts, applications, and development status. *Superconductor Science and Technology, 20*, R15–R29.

Noe, M. (25 February, 2011). Superconducting power applications and their potential to increase energy efficiency. In *Report at symposium on superconducting devices for wind energy, Barcelona*.

Noguchi, S., Tanikawa, T., & Igarashi, H. (2007). Operating property analysis of parallelized resistive fault current limiter using YBCO thin films. *IEEE Transactions on Applied Superconductivity, 17*(2), 1835–1838.

Noudem, J., Barbut, J. M., Sanchez, O., Tixador, P., & Tournier, R. (1999). Current limitation at 1080 A under 1100 V with bulk Bi-2223. *IEEE Transactions on Applied Superconductivity, 9*(6), 664–667.

Oda, S., Noda, S., Nishioka, H., Mori, M., Baba, J., & Shirai, Y. (2011). Current limiting experiment of transformer type superconducting fault current limiter with rewound structure using BSCCO wire in small model power system. *IEEE Transactions on Applied Superconductivity, 21*, 1307–1310.

Paasi, J., Lehtonen, J., Verhaege, T., & Herrmann, P. F. (2001). Design performance of a superconducting power link. *IEEE Transactions on Applied Superconductivity, 11*, 1928–1931.

Park, D. K., Chang, K. S., Yang, S. E., Kim, Y. J., Ahn, M. C., Yoon, Y. S., et al. (2009). Analytical and experimental studies on the hybrid fault current limiter employing asymmetric non-inductive coil and fast switch. *IEEE Transactions on Applied Superconductivity, 19*, 1896–1899.

Paul, W., Baumann, Th., Rhyner, J., & Platter, F. (1995). Tests of 100 kW high Tc superconducting fault current limiter. *IEEE Transactions on Applied Superconductivity, 5*, 1059–1062.

Paul, W., Lakner, M., Rhyner, J., Unternahrer, P., Baumann, Th, Chen, M., et al. (1997). Test of a 1.2 MVA high-Tc superconducting fault current limiter. *Institute of Physics Conference Series, 158*, 1173–1178.

Park, W.-J., Sung, B. C., & Park, J.-W. (2010). The effect of SFCL on electric power grid with wind-turbine generation system. *IEEE Transactions on Applied Superconductivity, 20*, 1177–1181.

Prikhna, T. A., Gawalek, W., Goldacker, W., Savchuk, Y. M., Noudem, J., Soldatov, A., et al. (2011). High-pressure synthesized nanostructural MgB_2 materials with high performance of superconductivity, suitable for fault current limitation and other applications. *IEEE Transactions on Applied Superconductivity, 21*, 2694–2697.

Raju, B. P., Parton, K. C., & Bartram, T. C. (1982). A current limiting device using superconducting D.C. bias: application and prospects. *IEEE Transactions on Power Apparatus and Systems, PAS-101*, 3173–3177.

Salim, K. M., Muta, I., Hoshino, T., Nakamura, T., & Yamada, Y. (2004). Proposal of rectifier type superconducting fault current limiter with non-inductive reactor (SFCL). *Cryogenics, 44*, 171–176.

Schmidt, W., Gamble, B., Kraemer, H.-P., Madura, D., Otto, A., & Romanosky, W. (2010). Design and test of current limiting modules using YBCO-coated conductors. *Superconductor Science and Technology, 23*, 014024.

Sjostrom, M., Cheraoui, R., & Dutoit, B. (1999). Enhancement of power system transient stability using superconducting fault current limiters. *IEEE Transactions on Applied Superconductivity, 9*, 1328–1330.

Slade, P. G., Wu, J. L., Stacey, E. J., Stubler, W. F., Voshall, R. E., Bonk, J. J., et al. (1992). The utility requirements for a distribution fault current limiter. *IEEE Transactions on Power Delivery, 7*, 507–515.

Smith, A. C., Oliver, A., Pei, X., Husband, M., & Rindfleisch, M. (2012). Experimental testing and modelling of a resistive type superconducting fault current limiter using MgB2 wire. *Superconductor Science and Technology, 25*, 125018.

Sokolovsky, V., Meerovich, V., Beilin, V., & Vajda, I. (2003). Application of an HTS thin film switching element in the inductive current limiter. *Physica C, 386*, 480–484.

Sokolovsky, V., Meerovich, V., & Chubraeva, L. I. (2009). Effect of SMES and current limiter on transient stability of a power system. In *Zawalishinskie chteniya 09* (pp. 115–120). St. Petersburg.

Sokolovsky, V., Meerovich, V., Chubraeva, L. I., & Vajda, I. (2010). An improved design of an inductive fault current limiter based on a superconducting cylinder. *Superconductor Science and Technology, 23*, 085007.

Sokolovsky, V., Meerovich, V., Györe, A., & Vajda, I. (2007). Transient stability of a power system with superconducting fault current limiters. *Periodica Polytechnica, Electrical Engineering, 51*, 3–9.

Sokolovsky, V., Meerovich, V., Vajda, I., & Beilin, V. (2004). Superconducting FCL: design and application. *IEEE Transactions on Applied Superconductivity, 14*, 1990–2000.

Tixador, P. (2010). Development of superconducting power devices in Europe. *Physica C, 470*, 971–979.

Tixador, P., Porcar, L., Floch, E., Buzon, D., Isfort, D., Bourgault, D., et al. (2001). Current limitation with bulk Y-Ba-Cu-O. *IEEE Transactions on Applied Superconductivity, 11*, 2034–2037.

Tomioka, A., Bohno, T., Kakami, S., Isozaki, M., Watanabe, K., Toyama, K., et al. (2013). The short-circuit test results of 6.9 kV/2.3 kV 400 kVA-class YBCO model transformer with fault current limiting function. *Physica C, 484*, 239–241.

Ulbricht, A. (1979). Test results of resistive SC power switch for 40 MW switching power at a voltage of 47 kV. *IEEE Transactions on Magnetism, 15*, 172–175.

Usoskin, A., Mumford, F., Dietrich, R., Handaze, A., Prause, B., Rutt, A., et al. (2009). Inductive fault current limiters: kinetics of quenching and recovery. *IEEE Transactions on Applied Superconductivity, 19*(3), 1859–1865.

Usoskin, A., Rutt, A., Withnell, T., Schlenga, K., Steurer, M., Pamidi, S., et al. (October 11, 2011). Physical principles and applications of inductive HTS shielded fault current limiting devices with suppressed cryogenic losses. In *10th EPRI conf, TLH*. Talk #2, BEST.

Vajda, I., Gyore, A., Semperger, S., Baker, A. E., Chong, E. F. H., Mumford, F. J., et al. (2007). Investigation of high temperature superconducting self-limiting transformer with YBCO cylinder. *IEEE Transactions on Applied Superconductivity, 17*(20), 1887–1890.

Vajda, I., Hyde, A., Gyore, A., Nador, G., Troller, T., Sailer, B., et al. (2011). Slimformer-self-limiting transformer pre-prototype and pilot plant design, construction, and tests. *IEEE Transactions on Applied Superconductivity, 21*, 1298–1302.

Verhaege, T., Herrmann, P. F., Cottevieille, C., Bock, J., Wolff, A., Moulaert, G., et al. (2001). HTS materials for ac current transport and fault current limitation. *IEEE Transactions on Applied Superconductivity, 11*, 2503–2506.

Verhaege, T., Tavergnier, J. P., Agnoux, C., Cottervieille, C., Laumond, Y., Bekhaled, M., et al. (1993). Experimental 7.2 kVrms/1 kArms/3 kApeak current limiter system. *IEEE Transactions on Applied Superconductivity, 3*, 574–577.

Vershinin, Y. N., Meerovich, V. M., Naumkin, I. E., Novikov, N. L., & Sokolovsky, V. L. (1989). A comparative analysis of nonlinear reactors with shields from low-temperature and high-temperature (about 90 K) superconductors. *Electric Technology USSR (UK), 1*, 1–9.

Waukesha Electric Systems Inc. (fault current limiting superconducting transformer). Available from http://www.smartgrid.gov/project/waukesha_electric_systems_inc_fault_current_limiting_superconducting_transformer. Accessed 31.08.14.

Witzmann, R., Schmidt, W., & Volkmar, R.-R. (2001). Resistive HTSL-strombegrenzer. In *Proceedings of Internationaler ETG-Kongress. Energietechnik für die Zukunft, Int. ETG-Kongress, Nuremberg, Germany, ETG Fachberichte Band 85* (pp. 23–24). VDE-Verlag.

Wolfus, S., Friedman, A., Yeshurun, Y., Rozenstein, V., & Bar-Haim, Z. (2013). *Fault current limiters (FCL) with the cores saturated by superconducting coils*. Patent US 8351167 B2.

Xie, Y., Selvamanickam, V., Marchevsky, M., Chen, Y., Xiong, X., Rar, A., et al. (2009). Second-generation HTS wire manufacturing and technology advancement at superpower. In *2009 International conference on applied superconductivity and electromagnetic devices (ASEMD)* (pp. 398–402).

Xin, Y., Gong, W. Z., Hong, H., Gao, Y. Q., Niu, X. Y., Zhang, J. Y., et al. (2012). Development of a 220 kV/300 MVA superconductive fault current limiter. *Superconductor Science and Technology, 25*, 105011.

Xin, Y., Gong, W., Niu, X., Cao, Z., Zhang, J., Tian, B., et al. (2007). Development of saturated iron core HTS fault current limiters. *IEEE Transactions on Applied Superconductivity, 17*, 1760–1763.

Xin, Y., Gong, W. Z., Niu, X. Y., Gao, Y. Q., Guo, Q. Q., Xiao, L. X., et al. (2009). Manufacturing and test of a 35 kV/90 MVA saturated iron-core type superconductive fault current limiter for live-grid operation. *IEEE Transactions on Applied Superconductivity, 19*, 1934–1937.

Xin, Y., Gong, W. Z., Sun, Y. W., Cui, J. B., Hong, H., Niu, X. Y., et al. (2013). Factory and field tests of a 220 kV/300 MVA saturated iron-core superconducting fault current limiter. *IEEE Transactions on Applied Superconductivity, 23*(3), 5602305.

Xin, Y., Hong, H., Wang, J. Z., Gong, W. Z., Zhang, J. Y., Ren, A. L., et al. (2011). Performance of the 35 kV/90 MVA SFCL in live-grid fault current limiting tests. *IEEE Transactions on Applied Superconductivity, 21*, 1294–1297.

Ye, L., Majoros, M., Campbell, A. M., Coombs, T., Astill, D., Harrison, S., et al. (2007). Experimental studies of the quench behaviour of MgB_2 superconducting wires for fault current limiter applications. *Superconductor Science and Technology, 20*(7), 621–628.

Yim, S.-W., Park, B.-C., Jeong, Y.-T., Kim, Y.-J., Yang, S.-E., Kim, W.-S., et al. (2013). Improvement in operational characteristics of KEPCO's line-commutation-type superconducting hybrid fault current limiter. *Physica C, 484*, 267–271.

Yuan, J., Maguire, J., Folts, D., Henderson, N., Knoll, D., Gouge, M., et al. (October 11-13, 2011). *Status and progress of a fault current limiting HTS cable to be installed in the consolidated Edison grid.* Report at Tenth EPRI Superconductivity Conference. Tallahassee, FL. http://mydocs.epri.com/docs/publicmeetingmaterials/1110/7TNRSL46577/19-Status and Progress of a Fault Current Limiting HTS Cable to be Installed in the ConEd Grid (Yuan).pdf. Accessed 23.05.13.

Zhu, Z., Dai, S., Xiao, L., Zhang, J., Teng, Y., Guo, W., et al. (2013). A real-time measuring and control system for the world's first HTS power substation. *IEEE Transactions on Applied Superconductivity, 23*, 5000404.

Further reading

Blair, S. M., Booth, C. D., Elders, I. M., Singh, N. K., Burt, G. M., & McCarthy, J. (2011). Superconducting fault current limiter application in power-dense marine electrical system. *IET Electrical Systems in Transportation, 1*, 93–102.

Malozemoff, A. P., Fleshler, S., Rupich, M., Thieme, C., Li, X., Zhang, W., et al. (2008). Progress in high temperature superconductor coated conductors and their applications. *Superconductor Science and Technology, 21*, 034005–034007.

Morandi, A. (2013). State of the art of superconducting fault current limiters and their application to the electric power system. *Physica C, 484*, 242–247.

Noe, M., & Steurer, M. (2007). High-temperature superconductor fault current limiters: concepts, applications, and development status. *Superconductor Science and Technology, 20*, R15–R29.

Orpe, S., & Nair, N.-K. C. (2009). State of art of fault current limiters and their impact on overcurrent protection. http://dx.doi.org/10.1109/PES.2009.5275284 Conference: Power & Energy Society General Meeting, 2009. PES '09. IEEE.

Paul, W., Chen, M., Lakner, M., Rhyner, J., Braun, D., & Lanz, W. (2001). Fault current limiter based on high temperature superconductors – different concepts, test results, simulations, applications. *Physica C, 354*, 27–33.

Pham, V. D., Laumond, Y., Verhaege, T., Fevrier, A., Collet, M., & Bekhaled, M. (1991). Towards the superconducting fault current limiter. *IEEE Transactions on Power Delivery, 6*, 801–808.

Sokolovsky, V., Meerovich, V., Vajda, I., & Beilin, V. (2004). Superconducting FCL: design and application. *IEEE Transaction on Applied Superconductivity, 14*, 1990–2000.

Tixador, P. (2010). Development of superconducting power devices in Europe. *Physica C, 470*, 971–979.

Verhaege, T., & Laumond, Y. (1998). Fault current limiters. In B. Seeber (Ed.), *Handbook of applied superconductivity* (vol. 2, pp. 1691–1702). IOP Publishing Ltd.

Xin, Y., Hui, H., Gong, W. Z., Wang, J. Z., Tian, B., Ren, A. L., et al. (September 25–27, 2009). Superconducting cable and superconducting fault current limiter at Puji Substation. In *Proc 2009 IEEE intern conf applied superconductivity and electromagnetic devices, Chengdu, China* (pp. 392–397).

High-temperature superconducting motors and generators for power grid applications

10

J.W. Bray
GE Global Research, Niskayuna, NY, USA

10.1 Introduction

According to the International Energy Agency (Waide & Brunner, 2011), motors account for 43–46% of all global electricity consumption, which is about twice that of the second place, lighting. About 23% of this consumption is by large motors, which they define as greater than 375 kW output power. It is this number that is more relevant to potential superconducting (SC) motors, since, as I shall discuss, the cryogenic costs will limit SC applications to large motors.

The world generated about 20 trillion kWh of electrical energy in 2009 (International Energy Agency, 2009), and almost all of that was generated by mechanical-to-electric conversion via rotating generators. Photovoltaics, the prime example of nonmechanical-to-electric conversion, supplied only about 20 billion kWh, or ~0.1%. By far most electrical generation is from large plants (Tidball, Bluestein, Rodriguez, & Knoke, 2010), and therefore large electrical generators, >~50 MW, dominate generation globally, potentially allowing SC versions by accommodating their cryogenic costs.

Given the size of these markets for new units, retrofits, and replacement, it is not surprising that motors and generators are often noted as prime targets for SC versions. Given the fact that SC can carry much larger currents than copper without ohmic losses, the usually emphasized benefits of such machines are higher electrical efficiency and smaller size and weight for the same power. Despite these potential benefits and large markets, there are at present no commercial SC motors or generators. In this chapter, we will explore the reasons for this, which are primarily economic. I will begin with a description of these machines and then some discussion of prototypes that have been built. I will then discuss the particular constraints that SC wires and their required cryogenics place on these machines. I will close with a look at the current status and outlook.

Superconductors in the Power Grid. http://dx.doi.org/10.1016/B978-1-78242-029-3.00010-8

10.2 Principles of superconducting (SC) motors and generators

10.2.1 Design and construction principles

Electrical motors and generators are very old topics in electrical engineering and physics, and there are many good textbooks and references on the subject (e.g., Boldea, 2005; Emadi & Andreas, 2005; Hughes, 2006). There are also various references that pay particular attention to the engineering of SC motors and generators (Kalsi, 2011). It is beyond the scope of this chapter to explain the design and operating principles of motors and generators in general. Instead we will emphasize and discuss those design principles and characteristics that are peculiar to SC machines. We also restrict ourselves to electromagnetic (EM) motors and generators, since they are relevant to SC, while acknowledging other non-SC types, such as electric field (capacitive), magnetohydrodynamic, and biological machines.

The important precepts for this chapter are:

- Motors and generators are for most practical purposes the same machines. Essentially all motors can be run backward as generators and vice versa. Therefore, one can describe the application of SC generally to both. Specific applications require specific machine designs to achieve optimum performance, but the discussion of application-specific designs is beyond the scope and purpose of this chapter.
- While not trying to explain the origin of superconductivity itself, Maxwell's equations can be successfully applied to the design of SC motors and generators, just as they can be for conventional copper-wire or permanent magnet generators. Special requirements of SC will be discussed in this chapter.
- EM motors and generators have two basic magnetic field sources: the armature (usually attached to the stator, which is the stationary part of the machine) and the field (usually attached to the rotor). The field may be an EM coil/winding (always, if SC) or a permanent magnet, while the armature is always a coil/winding. Figure 10.1 shows the usual layout. The field produces a mostly static magnetic induction B, which is forced through the armature windings to produce an electric field E in the armature (and thereby a voltage) during generator mode or is pulled through the armature by E (i.e., a voltage) applied to the armature during motor mode, all in accord with Maxwell's equations, especially the third (Faraday's law):

$$\nabla X\mathrm{E} + \partial B/\partial t = 0 \tag{10.1}$$

  (SI—system international—units used). The partial derivative in time (t) in Eqn (10.1) is caused by the mechanical motion of the field, usually rotational, and results in the conversion of mechanical to electrical energy (in a generator) or vice versa (in a motor). In the motor mode, the time-varying E (or voltage) supplied to the motor armature produces an armature current that in turn produces a time-varying magnetic field that attracts and moves the B of the field, with some back reaction from the field's $\partial B/\partial t$. In the generator mode, $\partial B/\partial t$ from the mechanically driven field produces most of the E (and hence the current and voltage) in the armature, with some back reaction from the circuit (load) to which the generator armature is attached.
- I will not explain the basic properties of SC here, as they are treated in Chapter 2 and in many excellent references (Orlando & Delin, 1991).

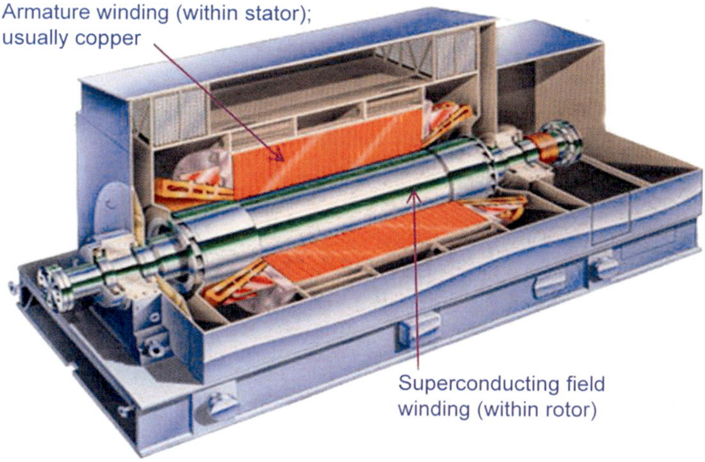

Figure 10.1 General picture of an ac motor or generator showing the usual locations of the superconducting field winding and armature winding.

A first and obvious design difference between SC and conventional motors and generators, which utilize copper or aluminum windings, is the need to cool the SC windings to cryogenic temperatures. Copper windings in conventional machines usually require cooling by air or water, but they operate well above ambient at temperatures usually limited by the electrical insulation of the windings. LTS (low-temperature SC) machines operate at about 4 K, and HTS (high-temperature SC) ones usually operate at around 20–40 K. LTS and HTS wires have considerably higher transition temperatures (T_c) than these, but engineering safety margins and the high magnetic fields in these machines usually limit the operating temperature to roughly half of the zero-field T_c. The requirement to cryocool the SC windings is difficult and places considerable restrictions on the overall machine design. The machine must be designed to protect the SC winding and the cryogen from excessive heating by hot machine parts or the ambient environment. This usually involves reducing thermal conduction to a minimum by using vacuum insulation and reducing radiative heating by using low emissivity coatings on the hot areas with a view of the cryogenic sections. Both of these requirements are difficult and expensive to implement and can lead to potential reliability issues. For example, vacuum insulation can be difficult to maintain for long periods due to material outgassing. Further, the thermal contractions caused by cryogenic cooling must be properly designed into the mechanics of the construction, including the stresses caused by dissimilar thermal expansion coefficients of different parts. The difficulty, reliability, and cost of the cryogenic subsystem are major reasons why SC motors and generators have not yet had commercial success.

A second important design point for SC motors and generators stems from the fact that SC have zero resistance (i.e., no ohmic losses) only with dc (direct current). Alternating current (ac) causes losses in all SC. It is possible to reduce the ac losses of SC by making fine wire filaments that are twisted and transposed (similar to copper Litz wire), but this has not been well accomplished for HTS, and the cost of cryorefrigeration at 4 K makes it difficult to tolerate these losses in LTS even in their best ac wire architectures. The effect of this on motors and generators is to limit, in most cases so far, the application of SC to the field winding, which is substantially dc, while the ac armature windings remain copper. However, even the field is normally ramped during motor or generator operation to different field levels by the exciter (especially for large ac synchronous generators and motors) to accommodate load or grid changes, and this will cause some losses in the SC field that must be known and accommodated by the cryogenics. Although there have been some designs and a few prototypes with SC armatures (Sugimoto et al., 2007), the ac losses are quite high with concomitant cryogenic penalty. Since there are options for the field other than electromagnets, for example, permanent magnets or induction, SC motors or generators are immediately placed into competition with those machine types for improving the field. It is conceivable to produce a hybrid machine with a permanent magnet field and SC armature (Sugimoto et al., 2008).

A third point affecting design and construction, which is related to the second, is the effect of ac magnetic fields on SC. Just as ac currents cause heating of SC, the same is true for impinging ac magnetic fields. Such fields are generated by the armature reaction, equipment (e.g., motor drives or grid) connected to the armature, and machine motions. This requires the construction of a field damper, or EM shield around the SC field (normally a highly electrically conductive shell, usually copper or aluminum). Since the field is usually rotating, the damper mechanical properties must also be adequate. The desired dc component of the magnetic field penetrates the damper completely.

A fourth characteristic affecting SC machine design is the potential loss of superconductivity in the SC winding in a phenomenon called quench. Quench can be caused by locally exceeding any of the critical parameters of the SC wire (temperature, current, or magnetic field), which can in turn be caused by either exceeding the design values of these parameters or degrading of the SC wire properties. Superconductors, especially HTS, are usually poor electrical conductors in the normal state. Since they are carrying a large dc current without loss in the SC state, they will immediately experience large ohmic heating on quench. If this phenomenon is not accommodated adequately, it will likely damage the machine by destroying the SC wire or insulation. Within LTS designs, quench is commonly dealt with by transferring the current into conductive materials surrounding the SC (e.g., copper, often called a stabilizer) and by transferring the electrical energy into a dump resistor. The quench normal-zone propagation velocity is high in LTS due to its low heat capacity at 4 K, which helps complete the process and minimizes damaging heat localization, and this may be aided by quench-triggered heaters around the SC coil. For HTS, the much higher heat capacity at 20–40 K is a two-edged sword: it slows the normal-zone propagation velocity, making protection much harder by tending to localize the heating, but it is harder to produce a quench since larger energy transients are required to raise the temperature above the critical temperature T_c. Completely satisfactory quench protection schemes for HTS machines are a subject of active research (Markiewicz, 2008; Sivasubramaniam, Zhang, et al., 2008). Quench protection usually adds software, hardware (e.g., temperature sensors), and control electronics to the design, and these add cost and complexity compared to conventional copper designs.

A fifth design and construction difference comes from the properties of the SC itself versus its conventional competitor, copper. Copper is very ductile, malleable, and fairly strong. While the workhorse LTS material NbTi (niobium titanium) is reasonably ductile and strong, it is not as malleable and forgiving of harsh treatment as copper, and most other SC wires are far worse. This has led to the need and practice of attaching the SC material to reinforcing materials of various types. Even so, the bend radii and strain tolerances of SC wires are considerably less than copper, and therefore, care must be taken in design and construction to protect the SC wire from all forces it will see in normal operation and under allowed overspeed or fault conditions.

Many other design details are necessary for the successful manufacture of motors and generators, such as number of electrical phases and slots for the armature winding, pole count for the field winding, rotor dynamics, cooling of conventional components, insulation, bearings, torque transfer, and vibration (Boldea, 2005). Those that are not specifically needed to implement superconductivity in these machines are not discussed here. However, some of these may be influenced by SC, depending on the design. For example, pole count of an SC field influences the geometry of the field winding in design-dependent ways. Pole count is always an even number, with 2 (i.e., a dipole B field in Eqn (10.1)) being the lowest and most common and 4 (quadrupole) also being common, and it is usually largely determined by desired machine rotational speed relative to line frequency and is influenced by machine size, cost, torque, and efficiency considerations. However, an SC dipole field implemented as saddle-shaped SC coils on a cylindrical rotor surface may be difficult to wind and thereby influence the pole count decision, although this design geometry is not required (Dawson, Wang, & Laskaris, 2003). This coil-winding problem comes under the category of difficult SC wire properties, dealt with in more detail in Section 10.5.1.

10.2.2 Benefits sought for these machines

The first benefit of SC motors and generators usually considered is efficiency increase. Indeed, the normal ohmic losses in the SC field are removed, except for the ac loss when the field is ramped to a new current level. Lower ohmic losses can have a secondary benefit by simplifying the exciter, which supplies the current to the field windings to produce B in Eqn (10.1) and needs to supply much less power to an SC field, although its ceiling voltage may need to be just as high (determined by SC coil inductance). A further benefit from nearly lossless field excitation is better part-load efficiency than conventional excited-field machines, which must maintain excitation with ohmic losses even at no load, but permanent magnet machines also have this benefit. However, the field ohmic losses are a small portion of the total losses (from armature, bearings, magnetic core hysteresis, eddy currents, windage) in the overall machine, and are generally smaller than the armature losses, for example (Schiferl, Flory, Umans, & Livoti, 2006). It is also true that the efficiency of very large conventional motor and generators is quite good, often 98–99%, leaving small room for improvement.

Since SC is capable of providing much higher magnetic fields than copper, one can increase the power density of a motor or generator, which will reduce the size and weight for a given power level or enable a higher power use/production within a given constrained volume. This follows from the basic machine equation

$$P = \pi^2 n\sigma B D^2 L \tag{10.2}$$

where P is the machine power rating, n is the rotor speed, D is the machine diameter, L is the active machine length, σ is the armature current loading per unit circumferential dimension of the stator (essentially, πD), and B is the airgap magnetic induction (the airgap is the radial distance between the rotor and the stator). The form of this equation can be reasoned as follows: power is normally given by armature (current) × (voltage); the Bn factors in Eqn (10.2) arise from the $\partial B/\partial t$ term in Eqn (10.1) and thereby give the armature E, which produces its voltage; σ gives the armature current; the remaining factors characterize the machine geometry. Equation (10.2) is not a formal design equation, which may be obtained elsewhere (Kalsi, 2011), but is properly illustrative of the present point. Power density increases linearly with B. This may have little practical value for machines that are unconstrained in weight or size but may have high value in applications (e.g., planes, ships, submarines) where size and weight are quite limited or produce high cost. Size and weight can also be substantially reduced by increasing B past the point where the magnetic core materials in the machine are saturated and therefore cannot provide their normal field focusing/concentration, which leads to its substitution by lighter, nonmagnetic structural material. This is called an air-core design, which is accomplished almost exclusively by SC designs.

Further benefits, which may be application dependent, can be enumerated. Since a major cause of field wear-out in machines is temperature cycling, and since an SC field is kept at cryogenic temperatures during all phases of operation, it is likely that an SC field will have longer life than a conventional wound field. The high B in SC machine allows large airgaps and air-core machines, and this leads to high short-circuit ratios (low synchronous inductance) in these machines, which in turn leads to higher reactive power (VAR) capability with potential for greater stability in grid-connected situations (Sae-Kok & Yokoyama, 2007; Suryanarayana, Bhattacharya, Raju, & Prasad, 1997). Short-circuit ratio is defined as the ratio of the field current for rated armature voltage on open circuit to the field current for rated armature current on short circuit, both with the machine running at rated speed. Large airgaps may also allow for the large frame-bending distortions sometimes found in high-torque machines. Since the field on the rotor has its own cooling, the need for rotor cooling ventilation is removed, allowing simpler cooling for the rest of the machine, eliminating the noise associated with rotor ventilation, and removing windage losses associated with cooling the rotor. The fact that the SC rotor field has a continuous EM shield may increase the negative sequence capability of the machine. Imbalance among the three armature phases causes negative sequence (armature) currents that cause an airgap magnetic field rotating backward with respect to the rotor at twice its speed, which will cause heating of the rotor field, but it is damped by the EM shield.

10.3 Types of SC motors and generators

Researchers have explored many types of motors and generators for the applicability and benefits of SC in their construction. A primary realization from this exploration is that the motor or generator must be "large" because the aforementioned cryogenic system must be added to the construction (see also Section 10.5). These cryogenics add significant cost to the machine, which can be born economically only by machines large enough to have benefits such as efficiency gain (see Section 10.2.2) sufficient to overcome those costs. The quantification of large is impossible without knowing the details of the machine cryogenics design, machine costs versus conventional competitors, and the proposed benefits for the machine. Researchers have often used 1000 HP ($\sim$1 MW) and $\sim$10 MW as lower limits for motors and generators, respectively, but the relevant facts may vary greatly.

A very common machine design used for SC motors and generators is the ac synchronous design. Figure 10.1 shows a typical such machine. As implied, these machines rotate synchronously with the frequency (usually 50 or 60 Hz) connected to the armature. The field winding is located on the rotor and is SC, whereas the armature is on the stator and is normally copper for reasons previously explained. It is unusual but possible to rotate the armature and leave the field stationary to achieve the benefit of a stationary SC field (Laskaris & Sivasubramaniam, 2010). "Stator" and "rotor" lose their traditional meaning in this case, and brushes are required to take substantial power to or from the rotating armature. The armature can be SC if the ac losses can be beneficially tolerated (Jiang, Majoros, Hong, Campbell, & Coombs, 2006; Sugimoto et al., 2007).

The ac synchronous design allows two distinct types, which may also apply to other designs: iron core and air core. Iron-core designs (Sivasubramaniam, Laskaris, & Bray, 2007) follow conventional motor and generator design philosophy by using soft magnetic materials, normally silicon–iron, to concentrate and direct the magnetic flux in the machine to the most advantageous locations to produce optimal voltage in the armature windings and reduce eddy-current heating in other metal parts in the machine. However, SC fields can produce magnetic fields far in excess of the saturation magnetization of known soft magnetic materials (e.g., >2.5 T), and in this case the magnetic core materials can no longer contain and direct the flux but may instead mostly act as a source of eddy-current heating. In this case, air-core construction is the best approach; in this construction the magnetic iron is replaced with nonmagnetic structural materials that hold the armature and field coils in place. This usually increases the airgap of the machine and lowers its size, weight, and synchronous reactance, often beneficially. Although the field flux is less contained within the armature, it is still contained by the machine back-iron, and the SC field can supply any needed flux magnitude (B in Eqn (10.2)). However, the machine's structural materials become unconventional and perhaps untested and expensive. It is worth noting that a few machine designs have compromised between the iron-core and the air-core designs by placing some iron within the machine to direct flux away from sensitive HTS windings (Goddard, Lukasik, & Sykulski, 2009; Sugimoto et al., 2007).

Another type of SC construction can be called trapped-field machines. In these, an SC material, usually HTS, is made to trap magnetic flux, for example, by cooling it in a magnetic field. Larger fields than are available from conventional permanent magnets, which are typically around 1 T or less, can thereby be produced. The SC magnet can then be used as the field magnets in motors or generators (Matsuzaki et al., 2007; Zhou et al., 2012). Such machines can be thought of as superpermanent magnet machines. A practical issue with such machines is that their field magnetism is permanent only as long as the SC magnet is kept cold, and the SC magnets must be remagnetized if they are warmed above their transition temperature. This leads to complex subsystems of coils and power supplies to remagnetize the machine field after such a fault, and this distracts from their value.

Motors and generators are often divided into two types: synchronous (operating in lockstep with the line frequency) and asynchronous. The most popular type of asynchronous generator and motor is the induction machine. In these machines, the field current is induced into the closed-circuit field windings on the rotor (often shaped like a "squirrel cage") by the ac currents in the armature windings and the rotation of the field. The asynchronous rotation of the rotor (called the slip frequency with respect to the line frequency) causes the motor or generator torque and concomitantly causes ac currents on the field windings. This ac field current makes this design not a good choice for an SC machine because of ac losses, although there have been a few SC designs of this type (Nakamura et al., 2006).

Reluctance generators or motors use variations in the magnetic properties of the rotor, and thereby the reluctance of the magnetic flux path, to create at least part of the machine torque. Salient poles on the rotor can be used to create this situation, and SC materials in bulk form can be used to enhance this effect, since SC is diamagnetic. Diamagnetism in SC, often called the Meissner effect, is the property of an SC that causes it to reject magnetic flux and have a negative magnetic susceptibility, the opposite behavior of the soft magnetic materials normally used in machine construction to channel magnetic flux (see Chapter 2 for a more complete description). Such a machine has the disadvantage of requiring control of potential flux trapping by the SC through fault situations (like a trapped-flux machine), but there have been a few designs (Chu & Torii, 2005).

As with conventional machines, both radial and axial magnetic flux designs for SC motors and generators are possible. The more common design is radial, since it is generally easier to construct concentric cylinders for the rotor and stator, and this design allows high power more easily. The axial design favors either stationary or rotating discs on a shaft next to each other. Various SC designs have employed the axial approach (Okazaki, Sugimoto, & Takeda, 2007; Sugimoto et al., 2007).

Design approaches that allow the SC field to remain stationary have gained some popularity because it is an added complication and expense to take the cryogen on and off the rotor, where the field normally is located. A design which allows this is an ac homopolar inductor alternator (Kwon et al., 2007; Sivasubramaniam, Laskaris, Shah, Bray, & Garrigan, 2008; Sugimoto et al., 2007). These machines use a stationary SC field, attached to the stator, to induce magnetism in a high permeability rotor. The variable reluctance paths designed into this rotor then apply a time-varying magnetic flux to the armature winding to produce the armature voltage in accord with Eqn (10.1). Another advantage of this design is that it allows solid rotors without coils or other additions, which are particularly helpful for high-speed operation, although low-speed designs are also possible. A disadvantage of this design is that magnetic field flux magnitudes are limited by the saturation magnetization of the rotor material, and the design utilizes this flux less efficiently (being homopolar) than a more traditional dipolar machine. In a dipolar machine, the flux magnitude goes through zero as the field dipole rotates, changes to its opposite directional sign, and passes through the armature. In a homopolar design, the flux oscillates from a high to a low value (as dictated by the variable reluctance paths) without changing to the opposite sign. Total $\partial B/\partial t$ (Eqn (10.1)) is thereby limited.

Another design with stationary SC coils is the dc homopolar machine, usually envisioned as a motor. It is a true dc machine that obtains its torque from the Lorentz force between an axial current traversing the rotor and a radial magnetic field produced by the SC. Advantages are quietness (no ac magnetic field effects) and easy speed/power control. Disadvantages are that it is inherently a high-current, low-voltage machine, and these currents must be taken on and off the rotor with brushes, which have been a reliability problem. These machines have been designed and built (Marshall, 1983; Thorne et al., 2002).

Although less common in SC machine discussions, SC linear machines deserve mention (Gieras, Piech, & Tomczuk, 2012). One can think of them as radial rotary machines that have been cut axially and flattened. As motors, such machines are designed for specific manipulations since their length is limited. However, they can be quite long, as when they are used as tracks for maglev trains. As usual and expected, they can also function as generators, and some have suggested them for generation when the driving force is reciprocating, as wave forces tend to be. Advantages of SC versions over conventional versions have not so far been convincing.

Although not strictly a motor or a generator, SC synchronous condensers deserve mention, since their construction is almost identical to a rotary SC motor or generator (Kalsi, Madura, & Ingram, 2005). As usual, the field winding is SC and the armature is copper, and the major benefit sought is efficiency from the removal of the field ohmic losses. The purpose of the synchronous condenser is not to generate or use any real power but instead to generate or remove reactive power (VARs) on the grid to which it is connected. By this means, the voltage is regulated to specifications in a grid. The rotor is kept spinning near synchronously by a small "pony" motor or power electronics, which overcome the usual rotor losses (bearings, windage, eddy currents).

Although not usually the motivation to call a motor or generator SC, it is worth noting some other useful components that may be SC in these machines. The machine bearings may be constructed from SC bulk magnets, chiefly HTS for the higher operating temperature (Werfel et al., 2012). Magnetic bearings give lower frictional losses and longer life, and the SC versions have the advantage of being inherently stable, unlike conventional commercial magnetic bearings, which must be made stable through electronic control of the electromagnets. Of course, these are more expensive than conventional bearings. There must be a second set of backup bearings in place in case of magnet failure, and this adds to the cost, which retards their use. HTS current leads may also be usefully employed in these machines for power connections to the SC windings (Tsuzuki et al., 2010). The advantage is lower heat transmission to the SC coils due to the low thermal conductivity of HTS. Finally, it is possible that the machine exciters may become cryogenic with some SC parts (Bray & Garces, 2012) to gain efficiency. None of these aforementioned components would likely be SC in a conventional non-SC machine because the cost of providing cryorefrigeration would be excessive. In an SC machine, which already has a cryogenic subsystem, their use is more likely to be cost-effective.

10.4 Prototypes built to date

It is not the purpose of this section to try to enumerate all the SC prototype motors and generators that have been built to date. There are many, especially small, test machines, and other reviews are available (Barnes, Sumption, & Rhoads, 2005; Gieras, 2009; Kalsi, 2011; Kalsi et al., 2004). Instead I will focus on the larger machines that illustrate the current state of the art for prototypes of the various possible designs.

Before the discovery of HTS, several prototype LTS motors and generators were successfully built, chiefly using the workhorse material NbTi. All these machines must and did use liquid He cooling at 4.2 K. A number of test machines of several types were constructed in the 1960–1970s (Smith, Kirtley, & Thullen, 1975). An example of a large, successful motor was the 3000 HP SC dc homopolar motor built by GE (Marshall, 1983), which was used to drive the US Navy test ship Jupiter C. This motor used liquid metal current collectors to deal with the high currents and concomitant brush problems of these machines. General Atomics built a 3.7 MW dc homopolar LTS motor for the US Navy and is working on a 36.5 MW version (Schaubel et al., 2006). A further example from GE is the first utility-type, 3600 rpm, 20 MW, ac synchronous generator (Keim, Laskaris, Fealey, & Rios, 1985) to be fully load-tested in a pump-back mode (Fealey, Jones, Keim, & Laskaris, 1985).

Of particular note in LTS machines is the Japanese LTS generator project called SuperGM, funded by MITI (Ministry of International Trade and Industry) over a 12-year period beginning in 1988 (Oishi & Nishijima, 2002). The project members designed and created one conventional stator and three LTS rotors, each with varying design parameters, in the 70 MVA class. All three versions were load-tested between 1997 and 1999, ending in a test connected to a 77-kV conventional grid. The machines set the current power record for SC generation of 79.7 MW and demonstrated a leading reactive power of 82 MVAR, about double a conventional generator of similar real power. One machine set the record for continuous SC machine operation of 1500 h. The top demonstrated efficiency was 98.3%, including the refrigeration power, and this led to a projection that a 200 MVA LTS generator could be about 0.5% more efficient than a conventional one.

Other large LTS machines were designed and partially constructed but not carried through final construction and load testing (e.g., Intichar & Lambrecht, 1983). LTS machines have demonstrated expected performance gains compared to conventional versions. However, none have become commercial, chiefly because of the cost of the LTS 4 K cryogenic subsystem.

The discovery of HTS gave renewed life to the dream of SC motors and generators because of relieved cryogenic costs for operation up to such still-hoped-for temperatures as 77 K (liquid nitrogen). It did not generally change the design of these machines. Reliance Electric built and tested a 125 HP HTS ac synchronous motor (Schiferl, Zhang, Driscoll, Shoykhet, & Dykhuizen, 1997) and a 1000 HP, 4-pole, air-core version (Zhang, Driscoll, & Dombrovski, 2002) with the usual HTS rotors with conventional stators. AMSC designed, built, and tested a 4-pole, 1800 rpm, 5000 HP ac synchronous motor (Gamble, Kalsi, Snitchler, Madura, & Howard, 2002); a 5 MW, 230 rpm ship propulsion motor (Eckels & Snitchler, 2005); and a 36.5 MW, 120 rpm ship propulsion motor (Gamble, Snitchler, & MacDonald, 2011). The latter is currently the largest tested HTS motor. GE engineers built and tested a 1.7 MW, 28-pole, 214 rpm HTS prototype hydrogenerator (Fair, Lewis, Eugene, & Ingles, 2010; GE, 2013). Low-speed motors or generators are generally built with high pole numbers. AMSC has also built and tested a 10 MVAR HTS synchronous condenser (Kalsi et al., 2005).

Siemens built in 2001 and tested for 1.5 years a 400 kW, 4-pole, 50 Hz HTS motor (Frank et al., 2003). It had a short-term output of 590 kW at 1500 rpm and an overall efficiency of 97%. In 2006 they also finished and tested a 2-pole, 3500 rpm, 4 MVA generator (Nick, Frank, Klaus, Frauenhofer, & Neumüller, 2007). It had an efficiency of 98.4–98.7% (depending on power factor), about 2% higher than a conventional machine of this size. GE built and no-load tested a 2-pole, 3600 rpm, 1.8 MVA HTS generator in 2003 (Ackermann et al., 2003). In 2007, the University of Southampton built a 100 kW, 2-pole HTS ac synchronous generator and tested it with its HTS field mounted on an iron-core rotor, operating 61.7–77 K (Wen et al., 2011).

GE pioneered the ac homopolar inductor alternator design for HTS implementation of a high-speed 1 MW HTS generator for the US Air Force (Sivasubramaniam et al., 2009). This machine was successfully load-tested to 1.3 MW at greater than 10,000 rpm. This work was followed in Korea by development and testing of a 1 MW, 2-pole, 3600 rpm homopolar motor (Kwon et al., 2007). Japanese work in this area has produced prototype HTS homopolar motors with up to 365 kW output, axial flux design, and with HTS armatures (Okazaki et al., 2007; Sugimoto et al., 2007). They have also produced a hybrid version with a permanent magnet field, HTS armature, and 400 kW output at 250 rpm (Sugimoto et al., 2008).

Some small bulk HTS trapped-field motors have been built and tested in Japan (Zhou et al., 2012) and elsewhere. These are of both the radial and the axial type.

10.5 SC wire and cryorefrigeration requirements

The two most unique features of SC motors and generators are the SC wire itself and the cryogenic subsystem, a required enabling technology. Both have strong effects on machine performance, reliability, and cost and are discussed below.

10.5.1 SC wires for motors and generators

The most obvious benefit of replacement of copper with SC wire is the removal of ohmic losses, at least for the nearly dc field windings. Bray (2009) pointed out requirements for the SC wires for various applications and used them to understand why the commercialization of SC wire is very difficult. Here I will emphasize those requirements that are particularly important for SC motors and generators. It is of primary importance that one choose an SC wire whose critical SC parameters (critical current, magnetic field, and temperature) allow SC operation of the wire within the temperature, currents, and fields designed for the motor or generator. It is important not to design to the SC wire limits but to allow an engineering safety margin (often around half) for each critical parameter. It is also important to know the maximum temperatures, currents, and fields to which the SC wire is exposed, not just average values, and use these to select the SC wire.

Perhaps of next importance are the mechanical properties of the SC wire. These are first tested when winding the SC coils for the fields or armatures within the motor or generator. The winding involves handling and bending the SC wire. The allowable bending radius must be known to preserve the SC properties. Bending places the material on the inside of the bend into compression and on the outside into tension, which is why most wire designs try to place the SC close to the neutral wire center. Coil sealing materials (e.g., epoxy impregnation) may place SC wires under additional strains through differential contractions during curing. During machine operation, the wires in the SC fields or armatures of these machines always carry substantial current in substantial magnet fields and are therefore subject to Lorentz forces. For a solenoid or circular coils, these Lorentz forces produce hoop stresses, and other shapes produce other force configurations. For SC coils placed on a moving part, such as a rotor, additional forces are generated from accelerations, such as the centripetal force of a rotor, and these forces must be tolerated to allowed overspeeds of the machine. SC machines are warmed to ambient or fault (e.g., quench) temperatures and cooled to cryogenic temperatures over their lifetimes at least a few times. This introduces stresses in the SC wires from differential thermal expansion with parts to which they are strongly connected. All of these sources of mechanical strain must be calculated for a given design and will set a threshold strain to which the SC wire must maintain most (e.g., 95%) of its SC properties; a typical threshold number might be around 0.4% strain. Note that these strains may be compressive, tensile, or sheer, depending on the situation, and so the response of the SC wire in all these cases must be known.

Some of the electro-mechanical forces in a motor of generator are cyclical with varying frequencies (e.g., vibrations from rotor imbalance). This can lead to fatigue of the SC wire, whose low or high cycle fatigue characteristics must therefore be known and considered for in the design (Ryan et al., 2005). When the temperature excursions of the machine are cyclical (e.g., from known repeated on—off states), thermally induced cyclic fatigue must also be taken into account.

There are also requirements on the geometry of the SC wires. Compact coil winding may require particular shapes of the wire other than just cylindrical. Many HTS wires are produced in tape form, and magnetic field shape and performance may dictate the best tape widths and thicknesses for SC motors or generators. For tapes especially, the camber (straightness) of the wires is important for good coil winding. The absence of asperities or too-sharp edges is normally needed to protect insulation. Finally, the availability of good lengths of the wires (e.g., greater than around 1 km) is important to avoid excessive joints in the coils. This introduces the necessity of good SC wire joint-making. Since the coils in motors or generators are normally connected to a power supply or grid, these joints do not have to be persistent (i.e., SC with parameters equal to the base SC wire), but they do need to have maximum allowed resistances (determined in design by thermal budgets) and mechanical properties consistent with the rest of the SC wire (as described above) so as not to present a mechanical weak link.

If an SC wire is intended to be used in an ac coil, in particular the armature coil within these applications, additional geometric constraints apply. The strategy for reducing ac loss in an SC wire is similar to that used for copper Litz wire. By forming multiple small diameter filaments to make up the SC wire cross-section and further by twisting and transposing these filaments, the ac loss may be minimized. This has been accomplished for the workhorse commercial SC wire, NbTi, but has not been pushed very far in any HTS wire so far. Even with these improvements, the SC wire will sustain ac losses that may make the required cryogenic refrigeration too expensive to be cost-effective, as is generally considered to be the case for any 4 K operation, even with ac-capable NbTi.

It is necessary to mention insulation for the SC wires. Since SC wires experience a voltage during current ramping, ac use, and any fault (e.g., quench) event, they must have electrical insulation consistent with the voltage expected in any of these situations. This is similar to the expectation for insulation for copper wire in any motor or generator, but the situation is different here because of the cryogenic operating temperatures and the occasional excursions to ambient or higher temperatures. The insulation must be thoroughly tested for these conditions. Insulation failure will lead to machine malfunction through coil shorts.

10.5.2 Cryorefrigeration for motors and generators

The required cryogenic subsystem in any SC motor or generator is of critical importance and has been the major reason for the lack of commercialization of these machines. It is expensive to construct, is a potential additional source of unreliability, requires maintenance, and uses nonnegligible power that must be subtracted from final machine efficiency calculations. It introduces a fundamental design conflict between the requirement to transfer torque efficiently from the motor (prime mover) to load (generator) and the need to reduce thermal conduction as much as possible from ambient or elevated temperature components to the SC coils. The situation has been eased somewhat with the discovery of the HTS materials, which can function in these high-field machines at around 20–40 K so far, an improvement over the 4 K liquid helium (LHe) temperatures used for the LTS materials. However, the cryogenic system is still complex at HTS temperatures.

A number of good reviews of cryorefrigeration in large LHe LTS systems (Tavian, 2007) and for HTS systems (Radebaugh, 2004; Wolsky, 2002) have been written and are not repeated here. I focus on points of particular interest for motors and generators. Of course, the primary problem is cooling the SC coils, usually just the field coil but perhaps also the armature coil, to an operating temperature that is an engineering safety margin below the SC critical temperature T_c at the designed maximum magnetic fields and operating currents of the machine. This is almost always done by circulating a cryogenic fluid from a refrigerator near the machine to the SC coils and back in a closed loop. As demonstrated in MRI (magnetic resonance imaging) (Minas & Laskaris, 1995), it is possible to cool SC coils by conduction through metals to the refrigerator cold head without cryogens, but this is very difficult for moving (rotating) SC coils.

For LHe (LTS) machines, only the field winding will be cooled, since cooling the ac losses in an SC armature would be prohibitively expensive at 4 K. The cryogen of choice is LHe at 4.2 K, the only choice without resorting to much more expensive ^{3}He or other very difficult cooling techniques, such as adiabatic demagnetization. For most prototype LTS machines, the user may choose to provide a tank of prepared LHe rather than purchase an LHe refrigeration system, although these are available from vendors and were used in the SuperGM project (Oishi & Nishijima, 2002). Since the LHe vaporizes as it is circulated around the cryogenic loop through the LTS field coil, a two-phase system will result, and this may produce some balance challenges within a field coil on the rotor if not handled correctly.

Whenever the SC field coil (LTS or HTS) is placed on the rotor, its usual location, the challenge of taking the cryogenic fluid on and off the rotor arises. This is usually solved by providing a rotating cryogenic fluid coupling between the rotor and the stationary refrigerator. This coupling must work at cryogenic temperatures, have superb thermal insulation, be reliable long term, and be essentially leak free. These are difficult requirements, making this a difficult construction, but it can be done (Keim et al., 1985). It may be possible to place some portion of the cryorefrigerator, such as the cold head, onto the rotor (Ki, Kim, Jeong, Kim, & Jung, 2011), but gas cryogen or electricity for power and sensors will still have to be conducted on and off the rotor, albeit at ambient temperature. As mentioned in Section 10.3, there are some machine types with stationary field coils that eliminate the need for this coupling.

For HTS machines, more options are open. Operating temperatures extend to around 40 K with available HTS wires and are improving. Although HTS wires have transition temperatures exceeding 77 K (liquid nitrogen boiling point), these temperatures are not yet useable with present wires at the fields and currents desired in most motors and generators. These operating conditions allow hydrogen (boiling point 19 K) and neon (boiling point 27 K) as additional elemental cryogens, although the former has notable flammability and explosion hazards to consider. Neon has been successfully employed in various prototypes (Zhang, Haran, Laskaris, & Bray, 2011). Mixed refrigerants (e.g., neon and nitrogen) have been considered to cover intermediate temperatures (Laverman & Lai, 1993). In addition, gaseous helium may be circulated from the cryorefrigerator to the HTS coils as the heat transfer medium at any temperature above 4 K. This eliminates any problems from two-phase flow and rotor fluid imbalance but eliminates the heat of vaporization as a cooling agent. In LHe cooling, two-stage cooling is usually employed with Gifford—McMahon coolers to reach 4 K, but single-stage cooling is possible at HTS temperatures, although Gifford—McMahon coolers remain popular. If cooling of an HTS armature is attempted, the cooling requirements will increase greatly, and other refrigerator types such as reverse-Brayton may be desirable (Radebaugh, 2004).

Another problem that cryorefrigeration introduces for these machines is cool-down time. When cooled from ambient, the refrigeration power designed into SC motors and generators for normal operation can take a number of days to cool the machine to operating temperature. It can be unacceptable to a user to wait this long for machine availability, but larger refrigeration power to shorten this time would be very expensive. Although this remains an issue, a couple of approaches can help. First, the machine can be kept cold by running the refrigerator continuously. The machine is thereby always available at the expense of powering the refrigerator at times when the machine is not in use. Second, an auxiliary cooling tube(s) can be built into the machine to allow for relatively cheap liquid nitrogen to be poured into the machine to take it quickly to around 77 K. This requires some additional construction and careful design to accommodate rapid thermal contraction of cooled machine parts.

In order to minimize the required cryogenic refrigeration power, the cryogenic subsystem almost certainly requires a vacuum subsystem to provide thermal insulation for the cold sections and connecting cryogenic piping. The vacuum resides between double-walled constructions required for the outer walls of the cryogenic components and perhaps around the SC coil. Vacuum systems are a source of additional cost, unreliability, and maintenance. Occasionally, foams have been considered for insulation at HTS operating temperatures (never for LTS), but these are less effective than vacuum and therefore require larger refrigeration power which is usually not cost-effective. Emissivity reduction is also usually required (always for LTS) to reduce the radiative component of heat transfer from the hot components to the cold. This usually involves using commercial MLI (multilayer insulation) blankets or low emissivity coatings when machine forces make the use of MLI problematic.

10.6 Conclusion and future trends

From the listing of successful prototypes of SC motors and generators in Section 10.4, it should be apparent that most significant technical hurdles for building these SC machines have been overcome, and they can be produced at will. This immediately begs the question: why are there no commercial SC motors and generators at present? The answer is mainly economic (Schiferl, Flory, Livoti, & Umans, 2008). The conventional competition is well entrenched, having over a 100 year history to be refined and cost reduced, and they perform their required jobs rather well. Therefore any new machine type, such as SC, must compete with this entrenched competition and demonstrate a cost—benefit trade-off that is clearly superior. This is a difficult job. First, the monetary value of the extra benefits of the SC machine, as discussed in Section 10.2.2, must be computed. Then the extra costs of the SC machine must be similarly computed, and these include the cost of the SC wire versus copper; the cost, operating, and maintenance expense of the new cryorefrigeration subsystem; and any extra costs from any unreliability of these new components. The nonrecurring engineering costs for the design and development of a new machine also must generally be paid by future sales of the machine.

We should add to the commercialization difficulty of these machines the conservatism of the customer base. Large motors and generators, the SC targets, are highly reliable machines with high cost and important functions, such as powering a grid or running a factory process. The customer base is understandably shy about switching to a new technology that, although demonstrating clear net monetary benefits on paper, would cause severe disruptions (e.g., power blackouts or plant shutdowns) if it were to fail. This usually causes customers to desire not to be first to buy and instead want to see many hours of proven performance before any purchase. This in turn may cause would-be manufacturers to build and operate prototypes in realistic situations for many days, for which they may have to provide the machine, operating costs, and insurance for any reliability issues caused by the machine. This can be quite costly and time-consuming.

What can be done to ameliorate these economic issues? The cost of the cryorefrigeration system can be reduced somewhat by temperatures available with HTS. It would be quite helpful to operate in the liquid nitrogen range (77 K down to 63 K with lowered pressures), since that is an established industrial cryogen and process, but present HTS wires do not have sufficient current-carrying capacity at these temperatures and in the magnetic fields common in these machines. Therefore, improvement in HTS wire performance is important, but perhaps more important is HTS wire price reduction. Currently HTS wire prices for 20–40 K use are roughly two orders of magnitude higher than LTS prices. Higher HTS operating temperatures give some cost reduction in the cryorefrigeration system, but this is not enough to allow large (e.g., 10 times or more) increases in HTS wire prices. The refrigeration–wire cost trade-off must be calculated for any given SC machine design. LTS machines have access to desirable NbTi wire prices but must depend on small evolutionary improvements in cryorefrigeration for any price reduction and therefore would generally need a major design advance to produce an economically competitive machine. Magnesium diboride (MgB_2), which may operate up to 20–25 K, may offer an intermediate option between LTS and HTS if its price and performance become competitive with NbTi.

There can be no doubt that SC motors and generators offer various performance benefits (Section 10.2.2). These can be realized technically, but the costs of implementing SC must be less than these benefits. When wire and refrigeration prices are adequately low and when machine reliability equal to the competition has been demonstrated, we should see future commercialization of these SC machines. The first groups to commercialize are likely to be those that have a strong need for some property of SC motors or generators that can only be satisfied through superconductivity and who therefore place the highest value on those SC benefits. For example, weight and size reduction, while usually of minor value for most ground-based applications, may have high value for machines that are flown on planes or placed in cramped spaces like submarines or especially those that must operate in space. Large offshore wind generators also place value on size/weight reduction since it reduces tower and construction costs. Another example might be a special grid-connect situation where the extra stability of an SC generator is judged to have high value (Sae-Kok & Yokoyama, 2007).

References

Ackermann, R., Alexander, J., Gadre, A., Laskaris, T., Sivasubramanium, K., Urbahn, J., et al. (2003). Testing of a 1.8 MVA high temperature superconducting generator. In *IEEE 2003 power engineering society annual meeting, Toronto, Canada.*

Barnes, P. N., Sumption, M. D., & Rhoads, G. L. (2005). Review of high power density superconducting generators: present state and prospects for incorporating YBCO windings. *Cryogenics, 45*, 670–686.

Boldea, I. (2005). *Variable speed generators and synchronous generators*. Boca Raton, FL: Taylor and Francis.

Bray, J. W. (2009). Superconductors in applications; some practical aspects. *IEEE Transactions on Applied Superconductivity, 19*, 2533–2539.

Bray, J. W., Garces, L. J., General Electric. Co. (2012). *Cryogenic exciter*. US patent 8,134,345, 3/13/2012.

Chu, S. D., & Torii, S. (2005). Torque-speed characteristics of superconducting synchronous reluctance motors with DyBCO bulk in the rotor. *IEEE Transactions on Applied Superconductivity, 15*, 2178–2181.

Dawson, R., Wang, Y., Laskaris, E. T., General Electric. Co. (2003). *High power density superconducting electric machine*. US6590308, 07/08/2003.

Eckels, P. W., & Snitchler, G. (2005). 5MW high temperature superconductor ship propulsion motor design and test results. *Naval Engineers Journal, 117*, 31–36.

Emadi, A., & Andreas, J. C. (2005). *Energy-efficient electric motors*. New York: Marcel Dekker.

Fair, R., Lewis, C., Eugene, J., & Ingles, M. (2010). Development of an HTS hydroelectric power generator for the Hirschaid power station. *Journal of Physics: Conference Series, 234*, 1–12.

Fealey, J. A., Jones, W. D., Keim, T. A., & Laskaris, T. E. (1985). Comprehensive test and evaluation of a 20 MVA superconducting generator. *IEEE Transactions on Power Apparatus and Systems, PAS-104*, 1484–1491.

Frank, M., Frauenhofer, J., van Hasselt, P., Nick, W., Neumueller, H. W., & Nerowski, G. (2003). Long-term operational experience with first Siemens 400 kW HTS machine in diverse configurations. *IEEE Transactions on Applied Superconductivity, 13*, 2120–2123.

Gamble, B., Kalsi, S., Snitchler, G., Madura, D., & Howard, R. (2002). The status of HTS motors. *IEEE conf. publications, 2002 power engineering soc. meeting (Vol. 1*, pp. 270–274).

Gamble, B., Snitchler, G., & MacDonald, T. (2011). Full power test of a 36.5 MW HTS propulsion motor. *IEEE Transactions on Applied Superconductivity, 21*, 1083–1088.

GE. (2013). *GE successfully trials breakthrough high-temperature superconducting technology for next-generation power generation, Paris*. Available from http://www.genewscenter.com/Press-Releases/GE-Successfully-Trials-Breakthrough-High-Temperature-Superconducting-Technology-for-Next-Generation-3ef3.aspx Accessed 04.04.13.

Gieras, J. F. (2009). Superconducting electrical machines–state of the art. *Przeglad Elektrotechniczny, 85*, 1–19.

Gieras, J. F., Piech, Z. J., & Tomczuk, B. (2012). *Linear synchronous motors: transportation and automation systems*. Boca Raton, FL: CRC Press, Taylor and Francis.

Goddard, K. F., Lukasik, B., & Sykulski, J. K. (2009). Alternative designs of high-temperature superconducting synchronous generators. *IEEE Transactions on Applied Superconductivity, 19*, 3805–3811.

Hughes, A. (2006). *Electric motors and drives fundamentals, types, and applications*. Amsterdam: Elsevier/Newnes.

International Energy Agency. (2009). *Electricity/Heat in world in 2009*. Available from http://www.iea.org/stats/electricitydata.asp?COUNTRY_CODE=29 Accessed 09.05.13.

Intichar, L., & Lambrecht, D. (1983). Technical overview of the German program to develop superconducting AC generators. *IEEE Transactions on Magnetics, MAG-19*, 536–540.

Jiang, Q., Majoros, M., Hong, Z., Campbell, A. M., & Coombs, T. A. (2006). Design and ac loss analysis of a superconducting synchronous motor. *Superconductor Science and Technology, 19*, 1164–1168.

Kalsi, S. S. (2011). *High temperature superconductors to electric power equipment*. Hoboken: Wiley and Sons.

Kalsi, S. S., Madura, D., & Ingram, M. (2005). Superconductor synchronous condenser for reactive power support in an electric grid. *IEEE Transactions on Applied Superconductivity, 15*, 2146–2149.

Kalsi, S. S., Weeber, K., Takesue, H., Lewis, C., Neumueller, H.-W., & Blaugher, R. D. (2004). Development status of rotating machines employing superconducting field windings. *Proceedings of the IEEE, 92*, 1688–1704.

Keim, T. A., Laskaris, T. E., Fealey, J. A., & Rios, P. A. (1985). Design and manufacture of a 20 MVA superconducting generator. *IEEE Transactions on Power Apparatus and Systems, 104*, 1475–1483.

Ki, T., Kim, H., Jeong, S., Kim, J., & Jung, J. (2011). High temperature superconducting motor cooled by on-board cryocooler. *IEEE Transactions on Applied Superconductivity, 21*, 2217–2220.

Kwon, Y. K., Baik, S. K., Lee, E. Y., Lee, J. D., Kim, Y. C., Moon, T. S., et al. (2007). Status of HTS motor development in korea. In *IEEE conference publications, power engineering society general meeting 2007* (pp. 1–5).

Laskaris, E. T., Sivasubramaniam, K., (General Electric Co.). (2010). *Method and apparatus for a superconducting generator driven by a wind turbine*. US Patent 7821164, 10/26/2010.

Laverman, R. J., Lai, B.-Y., (Chicago Bridge Iron Tech Serv Company). (1993). *Method and apparatus for cooling high temperature superconductors with neon-nitrogen mixtures*. US Patent 5193349, 3/16/1993.

Markiewicz, W. D. (2008). Protection of HTS coils in the limit of zero quench propagation velocity. *IEEE Transactions on Applied Superconductivity, 18*, 1333–1336.

Marshall, R. A. (1983). 3000 horsepower superconductive field acyclic motor. *IEEE Transactions on Magnetics, MAG-19*, 876–879.

Matsuzaki, H., Kimura, Y., Morita, E., Ogata, H., Ida, T., Izumi, M., et al. (2007). HTS bulk pole-field magnets motor with a multiple rotor cooled by liquid nitrogen. *IEEE Transactions on Applied Superconductivity, 17*, 1553–1556.

Minas, C., & Laskaris, E. T. (1995). Structural design and analysis of a cryogen-free open superconducting magnet for interventional MRI applications. *IEEE Transactions on Applied Superconductivity, 5*, 173–176.

Nakamura, T., Miyake, H., Ogama, Y., Morita, G., Muta, I., & Hoshino, T. (2006). Fabrication and characteristics of HTS induction motor by the use of bi-2223/Ag squirrel-cage rotor. *IEEE Transactions on Applied Superconductivity, 16*, 1469–1472.

Nick, W., Frank, M., Klaus, G., Frauenhofer, J., & Neumüller, H. W. (2007). Operational experience with the world's first 3600 rpm 4 MVA generator at Siemens. *IEEE Transactions on Applied Superconductivity, 17*, 2030–2033.

Oishi, I., & Nishijima, K. (2002). Summary of development of 70 MW class model superconducting generator – research and development of superconducting for electric power application. *Cryogenics, 42*, 157–167.

Okazaki, T., Sugimoto, H., & Takeda, T. (2007). Liquid nitrogen cooled HTS motor for ship propulsion. In *IEEE conference publications, power engineering society general meeting 2006* (pp. 1–6).

Orlando, T. P., & Delin, K. A. (1991). *Foundations of applied superconductivity*. New York: Addison-Wesley.

Radebaugh, R. (2004). Refrigeration for superconductors. *Proceedings of the IEEE, 92*, 1719–1734.

Ryan, D. T., Li, L., Huang, X., Bray, J. W., Laskaris, E. T., Sivasubramaniam, K., et al. (2005). Critical current degradation in HTS wires due to cyclic mechanical strain. *IEEE Transactions on Applied Superconductivity, 15*, 3684–3687.

Sae-Kok, W., & Yokoyama, A. (2007). Installation schemes of superconducting generators in power systems. *IEEJ Transactions on Power and Energy, 127*, 104–112.

Schaubel, K. M., Langhorn, A. R., Creedon, W. P., Johanson, N. W., Sheynin, S., & Thome, R. J. (2006). Development of a superconducting magnet system for the ONR/General atomics homopolar motor. *AIP Conference Proceedings, 823*, 1819–1826.

Schiferl, R., Flory, A., Livoti, W. C., & Umans, S. D. (2008). High-temperature superconducting synchronous motors: economic issues for industrial applications. *IEEE Transactions on Industry Applications, 44*, 1376–1384.

Schiferl, R., Flory, A., Umans, S. D., & Livoti, W. C. (2006). High temperature superconducting synchronous motors: economic issues for industrial applications. In *IEEE conference publications, petroleum and chemical industry conference 2006* (pp. 1–9).

Schiferl, R., Zhang, B., Driscoll, D., Shoykhet, B., & Dykhuizen, R. (1997). Development status of a 125 horsepower superconducting motor. *Advances in Cryogenic Engineering, 42*, 977.

Sivasubramaniam, K., Laskaris, E. T., & Bray, J. W. (2007). High power density HTS iron core machines for Marine applications. In *IEEE power engineering Society general meeting 2007* (pp. 1–4).

Sivasubramaniam, K., Laskaris, E. T., Shah, M. R., Bray, J. W., & Garrigan, N. R. (2008). High-temperature superconducting homopolar inductor alternator for marine applications. *IEEE Transactions on Applied Superconductivity, 18*, 1–6.

Sivasubramaniam, K., Zhang, T., Caiafa, A., Huang, X., Xu, M., Li, L., et al. (2008). Transient capability of superconducting devices on electric power systems. *IEEE Transactions on Applied Superconductivity, 18*, 1693–1697.

Sivsubramaniam, K., Laskaris, E. T., Lokhandwalla, M., Zhang, T., Bray, J. W., Gerstler, B., et al. (2009). Development of a high-speed multimegawatt HTS generator for airborne applications. *IEEE Transactions on Applied Superconductivity, 19*, 1656.

Smith, J. L., Jr., Kirtley, J. L., Jr., & Thullen, P. (1975). Superconducting rotating machines. *IEEE Transactions on Magnetics, MAG-11*, 128–134.

Sugimoto, H., Morishita, T., Tsuda, T., Takeda, T., Togawa, H., Oota, T., et al. (2008). Development and test of an axial flux type PM synchronous motor with liquid nitrogen cooled HTS armature windings. *Journal of Physics: Conference Series, 97*, 1–6.

Sugimoto, H., Tsuda, T., Morishita, T., Hondou, Y., Takeda, T., Togawa, H., et al. (2007). Design of an axial flux inductor type synchronous motor with the liquid nitrogen cooled field and armature HTS windings. *IEEE Transactions on Applied Superconductivity, 17*, 1571−1574.

Suryanarayana, T., Bhattacharya, J. E., Raju, K. S. N., & Prasad, K. A. D. (1997). Development and performance testing of a 200kVA Damperless superconducting generator. *IEEE Transactions on Energy Conversion, 12*, 330−336.

Tavian, L. (2007). Review of cryogenics for large superconducting systems. *IEEE/CSC ESAS European Superconductivity News Forum, 2*, 1−16.

Thome, R. J., Creedon, W., Reed, M., Bowles, E., & Schaubel, K. (2002). Homopolar motor technology development. *IEEE conference publications, power engineering society summer meeting 2002, 1*, 260−264.

Tidball, R., Bluestein, J., Rodriguez, N., & Knoke, S. (2010). *Cost and performance assumptions for modeling electricity generation Technologies*. Golden, CO: NREL. Available at http://www.nrel.gov/docs/fy11osti/48595.pdf Accessed 09.05.13.

Tsuzuki, K., Kimura, Y., Miki, Y., Sugyo, D., Sano, T., Yamaguchi, K., et al. (2010). Study of bulk current leads for an axial type of HTS propulsion motor'. *Journal of Physics: Conference Series, 234*, 1−6.

Waide, P., & Brunner, C. U. (2011). *Energy-efficiency policy opportunities for electric motor-driven systems*. Paris: International Energy Agency. Available at http://www.iea.org/publications/freepublications/publication/EE_for_ElectricSystems.pdf Accessed 09.05.13.

Wen, H., Bailey, W., Al-Mosawi, M. K., Goddard, K., Beduz, C., & Yang, Y. (2011). Further testing of an "Iron-Cored" HTS synchronous generator cooled by liquid air. *IEEE Transactions on Applied Superconductivity, 21*, 1163−1166.

Werfel, F. N., Floegel-Delor, U., Rothfeld, R., Riedel, T., Goebel, B., Wippich, D., et al. (2012). Superconductor bearings, flywheels and transportation. *Superconductor Science and Technology, 25*, 1−16.

Wolsky, A. M. (2002). *Cooling for future power sector equipment incorporating ceramic superconductors* (work done for International Energy Agency).

Zhang, B., Driscoll, D., & Dombrovski, V. (2002). Construction and testing of a 1,000 hp high-temperature superconducting motor. *IEEE conference publications − petroleum and chemical industry conference. Industry applications society 49th annual* (pp. 223−229).

Zhang, T., Haran, K., Laskaris, E. T., & Bray, J. W. (2011). Design and test of a simplified and reliable cryogenic system for high speed superconducting generator applications. *Cryogenics, 51*, 380−383.

Zhou, D., Izumi, M., Miki, M., Felder, B., Ida, T., & Kitano, M. (2012). An overview of rotating machine systems with high-temperature bulk superconductors. *Superconductor Science and Technology, 25*, 1−12.

High-temperature superconducting magnetic energy storage (SMES) for power grid applications

11

T.A. Coombs
Cambridge University Engineering Department, Cambridge, UK

11.1 Introduction

11.1.1 What is superconducting magnetic energy storage

It is well known that there are many and various ways of storing energy. These may be kinetic such as in a flywheel; chemical, in, for example, a battery; potential, in a pumped storage scheme where water is pumped to the top of a hill; thermal; biochemical; or electrical. This chapter deals with the storage of electrical energy. There are two main ways of storing electrical energy: capacitors (or supercapacitors) and inductors.

All electrical components have inductance; how much inductance is a function of their geometry and their magnetic permeability. Inductance is given the symbol L, measured in Henry's and provides a simple method of calculating the voltage (V) developed when a changing current ($\mathrm{d}I/\mathrm{d}t$) is passed through it according to the following equation:

$$V = L\frac{\mathrm{d}I}{\mathrm{d}t} \tag{11.1}$$

Typically an inductor will be made from a coil with a ferromagnetic core, but a simple wire will also have inductance. In addition to allowing us to relate the current to the volts in an electric circuit, inductance is an indication of the object's ability to store energy. The higher the inductance the greater the stored energy.

For example, the coaxial cable (Figure 11.1) that carries the signal from a TV aerial to a TV has an inductance (L).

The inductance is easily calculated from Maxwell's equations and for a coaxial cable with a sheaf of diameter b (in m) and an inner conductor of diameter a (in m) comes out as in inductance L (in henrys):

$$L = \frac{\mu_0}{2\pi}\ln(b/a) \text{ per unit length} \tag{11.2}$$

Superconductors in the Power Grid. http://dx.doi.org/10.1016/B978-1-78242-029-3.00011-X

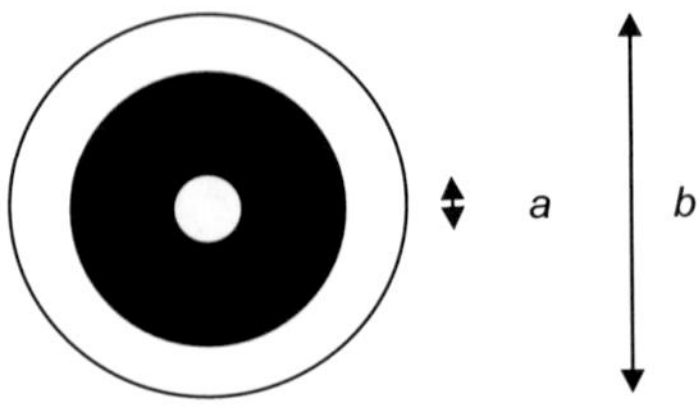

Figure 11.1 Coaxial cable.

If the cable then carries a current I (in amps) the stored energy per unit length (in J/m) is given by

$$E = \frac{1}{2}LI^2 \tag{11.3}$$

Unfortunately, μ_0, the (magnetic) permeability of free space, is rather a small number ($4\pi10^{-7}$H/m), which means that to store just 1 J of energy in a 1 m piece of coaxial cable, you would need a current of greater than 2000 A!

Much greater inductances can be achieved by winding coils or solenoids with many turns, but since it can be clearly seen from the above equation that the energy stored is a function of the square of the current it also follows that in order to store large amounts of energy it is an advantage to have high currents.

While the energy is being stored there is the potential for resistive loss (Q) in the conductor. The rate of this loss (in W) is given by the familiar equation

$$Q = RI^2 \tag{11.4}$$

where R is the electrical resistance in Ohms (Ω).

Since we must have a high I in order to store energy in the first place it follows that R must be as low as possible.

Fortunately superconductors satisfy both these requirements. Not only can they carry very high currents but, for DC currents at any rate, they have no intrinsic resistance at all.

Superconducting magnetic energy storage (SMES) devices are basically magnets in which energy is stored in the form of a magnetic field (B in Tesla), which is maintained by currents that (ideally) flow persistently (without losses) in the SMES magnets. The energy in an SMES is given by

$$E_{\text{magnetic}} = \frac{1}{2\mu_0}\int B^2 \text{dv} \tag{11.5}$$

where the volume integral is performed over all space. Thus, to increase the energy stored in an SMES device, either the magnetic field must be increased or the volume (size) of the device must be increased.

11.1.2 SMES power and energy densities

SMES devices are ideal for providing high-power densities in the range $10-10^5$ kW/kg. This is far in excess of batteries. Unfortunately, the specific energy density is very low and the only comparable source is water in the form of pumped hydroelectric. Various types of energy storage devices and their corresponding energy densities are given in Table 11.1.

The energy density in an SMES is ultimately limited by mechanical considerations. Since the energy is being held in the form of magnetic fields, the magnetic pressures, which are given by

$$P = \frac{B^2}{2\mu_0} \tag{11.6}$$

rise very rapidly as B, the magnetic flux density, increases. Thus, the magnetic pressure in a solenoid coil can be viewed in a similar manner as a pressured cylinder of compressed gas wanting to burst out at its midplane. At a magnetic field 1 T, the magnetic pressure is about 10 times that of atmospheric pressure and at 10 T it is already 1000 times atmospheric pressure.

Unfortunately, from Eqn (11.5), it is not possible to simply raise the B and therefore the stored energy indefinitely. There are two limiting constraints. The first is that the critical current (I_c) of the superconducting wire is a function of B and typically reduces as B increases. As the I_c of the superconducting wire decreases with increasing B, it requires more and more turns in the magnet, hence making the SMES heavier (i.e. lowering its energy density) and more costly since the superconducting wire itself can be quite expensive. However, I_c alone is not typically the limiting factor; the limiting factor in an SMES is often the stress experienced by the superconducting wire itself, which is caused by the magnetic pressure (Eqn (11.2)). The magnetic pressure in the magnet causes a stress in the azimuthal direction (i.e. 'hoop stress'), which in turn places a tensile stress on the superconducting wire. The maximum stored energy to mass ratio can be calculated using the principle of virial laws (see Section 2.1).

11.1.3 SMES systems

A complete SMES system comprises three primary subsystems: (1) the superconducting coil and its corresponding support structure, (2) the Power Condition Systems (PCS), which is used to transfer the AC line power source back and forth to the superconducting storage magnet, and (3) the cryogenic refrigeration system that is used to initially cool and ultimately maintain the superconducting magnet at its operating temperature. Each subsystem plays a vital role in what makes an SMES a useful energy storage medium.

11.1.4 Applications

Depending on which source you read there are a multitude of different possible applications for SMES. However it is important to recognise that despite the name SMES is

Table 11.1 Specific energy of various forms of energy storage

Storage type	Specific energy (MJ/kg)
Water at 100 m dam height (potential energy)	0.001
SMES	0.01
Compressed air at 300 bar (potential energy)	0.5
Liquid nitrogen	0.77
Battery, zinc–air	1.59
ANFO—ammonium nitrate fuel oil	3.7
Household waste	8
Battery, lithium–air rechargeable	9
Sod peat	12.8
Coal, lignite	14
Dry cow dung and camel dung	15.5
Glucose	15.55
Peat briquette	17.7
Wood	18
Coal, bituminous	24
Coal, anthracite	32.5
Biodiesel oil (vegetable oil)	42.2
Jet A aviation fuel/kerosene	42.8
Diesel fuel/residential heating oil	46.2
Crude oil (according to the definition of ton of oil equivalent)	46.3
Gasoline (petrol)	46.4
LPG butane	49.1
LPG propane	49.6
Natural gas	53.6
Methane (1.013 bar, 15°C)	55.6
Hydrogen, liquid	141.86
Uranium-235	83,140,000

Source: Haynes (2014).

really most applicable to power quality rather than energy storage. The energy stored per unit cost is relatively low, certainly lower than conventional means of generating power such as coal, nuclear, or gas and lower in most cases than the renewables, such as wind power, it might otherwise be supporting. Unless very high magnetic fields are achieved then to act as an energy storage device, the overall volume would be large.

In applications where power quality is important, however, SMES has great potential. It has a subcycle (50 or 60 Hz) response time, greater than 97% 'round-trip' conversion efficiency and, unlike batteries, it can provide power for an indefinite number of cycles.

Although the specific energy per unit mass is low (an order of magnitude less than batteries), it is higher than both capacitors and supercapacitors. In addition, the specific power (per unit mass), which due to rapid discharge time can be as high as 100 MW/kg (Tixador, 2008), is several orders of magnitude higher than that of supercapacitors, which are typically measured in tens of kW/kg (Tixador, 2008).

The chart in Figure 11.2 (Leibniz Institute for New Materials) makes it clear where SMES lies in relation to other forms of electrical energy storage and puts the application of SMES into the region between power quality and bridging power. This means that it is appropriate for preventing temporary voltage sags either on the network or in a high value application where any form of power outage would be potentially expensive, such as in an aluminum rolling mill where any interruption to the speed of the rollers could cause catastrophic results.

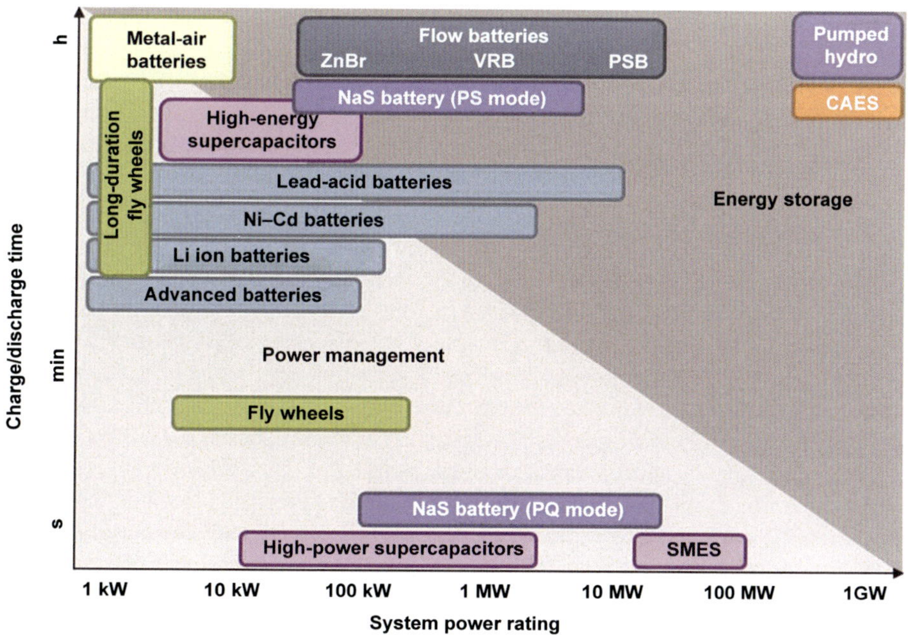

Figure 11.2 Typical power versus discharge times for various forms of energy storage (Leibniz Institute for New Materials).

Table 11.2 **Demonstrator SMES units**

Country	Year	Material	Cooling	Energy	Peak power
Japan	2005	BSCCO	Closed cycle helium	1 MJ	1 MW
Japan	2006	LTS	Helium bath	7.3 MJ	5 MW
Japan	2009	LTS	Conduction at 4 K	1 MJ	1 MW
Korea	2006	LTS	Helium bath	3 MJ	0.75 MW
Finland	2003	LTS	Conduction at 10 K	0.2 MJ	200 W
China	2008	BSCCO	Helium 4.2 K	1 MJ	0.5 MW

SMES, superconducting magnetic energy storage.

As can be seen from Figure 11.2 the power is typically of the order of tens of megawatts, making SMES ideal for load leveling in power critical applications.

SMES is ideal for nonmobile applications where a high power (megawatts are perfectly practical) is required for a relatively short space of time, such as UPS and power quality management. Examples of where SMES has been used in UPS are given in Table 11.2. The Finnish device is unusual in that it is a lab-based system being used to support a PC and the power was extracted over the relatively long time frame of 600 s as opposed to the rest, which are designed to discharge in 1 or 2 s.

11.2 Construction of superconducting magnetic energy storage (SMES): maximising energy storage and minimising cost

11.2.1 Mechanical constraints

Although, in principle at least, REBCO superconducting coils could be built in which the magnetic field is up to 100 T and, since the magnetic energy stored is a function of the square of the flux density, it is a very attractive prospect to push the magnetic field up from the 10 to 11 T demonstrated so far. In practice, mechanical constraints will limit the size of these coils. Nevertheless the world record for an REBCO coil is 34.5 T, which was demonstrated at Brookhaven National Labs in 2011 (http://www.magnet.fsu.edu/mediacenter/publications/reports/mlr/2011/MLR-18.4-web.pdf). The coil is shown in Figure 11.3; however, this was a relatively small superconducting coil measuring 64.5 mm long with an inner diameter of 14.3 mm and an outer diameter of 38 mm and it relied on a resistive coil surrounding it to provide the first 31 T.

Nevertheless following on from this success, there is a program underway in the United States to produce an SMES with a magnetic field of 25 T, which was due for completion in 2013. However, despite the high fields this will again be a relatively small coil storing as it will only be 2.5 MJ. Details of the program are given as follows.

Figure 11.3 A high-field YBCO coil (http://www.magnet.fsu.edu/mediacenter/publications/reports/mlr/2011/MLR-18.4-web.pdf).

Partners:

ABB Inc.—project lead, grid-interface power converter, system integration, and lab demonstration

Brookhaven National Laboratory (BNL)—SMES magnet design, construction, protection, and test; superconducting joints and switches

SuperPower Inc.—2G high-temperature superconducting (HTS) wire, assist BNL with coil development

University of Houston, TcSUH—2G HTS manufacturing improvements for wire cost reduction.

Specifications:

20 kW ultrahigh-field (UHF) SMES system with a capacity up to 2.5 MJ

Field of up to 25 T at 4.2 K

Capable of flexible and direct deployment in medium voltage distribution networks at 15−36 kV

Uses 2G HTS wire with high critical currents.

If we start by looking at the hoop stress in the magnet, which occurs due to the current in the solenoid interacting with the magnetic field giving rise to a Lorentz force then, if we assume that each individual turn acts independently (i.e. it is not supported by its nearest neighbor), we can calculate the magnetic pressure for a single turn of radius r (in m) and of thickness t (in m). Using the magnetic pressure and noting that the hoop stress (σ_{h}) is related to the pressure by the equation Pr/t then we have the hoop stress (in Pa), which is given by

$$\sigma_{\mathrm{h}} = \left(\frac{r}{t}\right)\frac{B^2}{2\mu_0}$$

But the stored energy from Eqn (11.1) is given by the volume of the solenoid multiplied by the magnetic pressure. Hence

$$E_{\mathrm{mag}} = \mathrm{Vol}\frac{B^2}{2\mu_0} \tag{11.7}$$

So the energy stored is

$$E_{\mathrm{mag}} = \mathrm{Vol} \cdot \sigma_{\mathrm{h}} \cdot \frac{t}{r} = \sigma_{\mathrm{h}} \pi \cdot r \cdot t \cdot h \tag{11.8}$$

where h is the height (in m) of the solenoid and σ_{h} is the maximum working stress.

However this is a theoretical limit and is rarely reached. Moon (1982) plotted the energy stored for various magnet configurations and in the paper he gives the plot that can be seen in Figure 11.4.

$$M = 0.417E^{0.895} \tag{11.9}$$

where E is the stored energy and M is the mass of the magnet. He also plots the virial limit assuming a yield stress for steel of 34.5 GN/m^3.

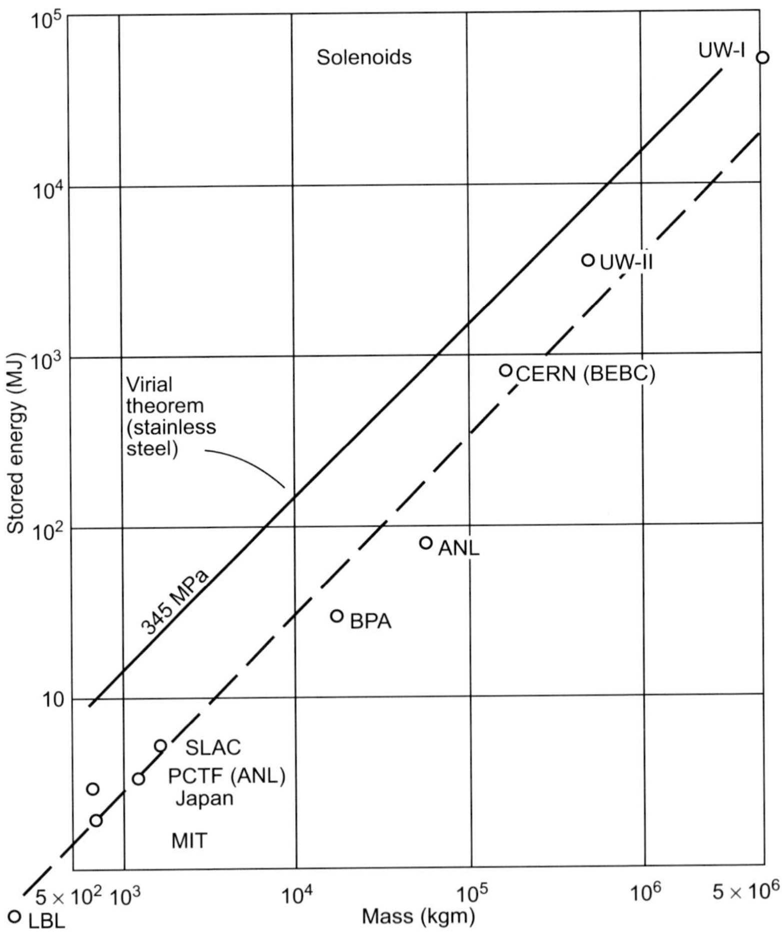

Figure 11.4 Stored energy versus mass for various sizes of solenoids (Moon, 1982).

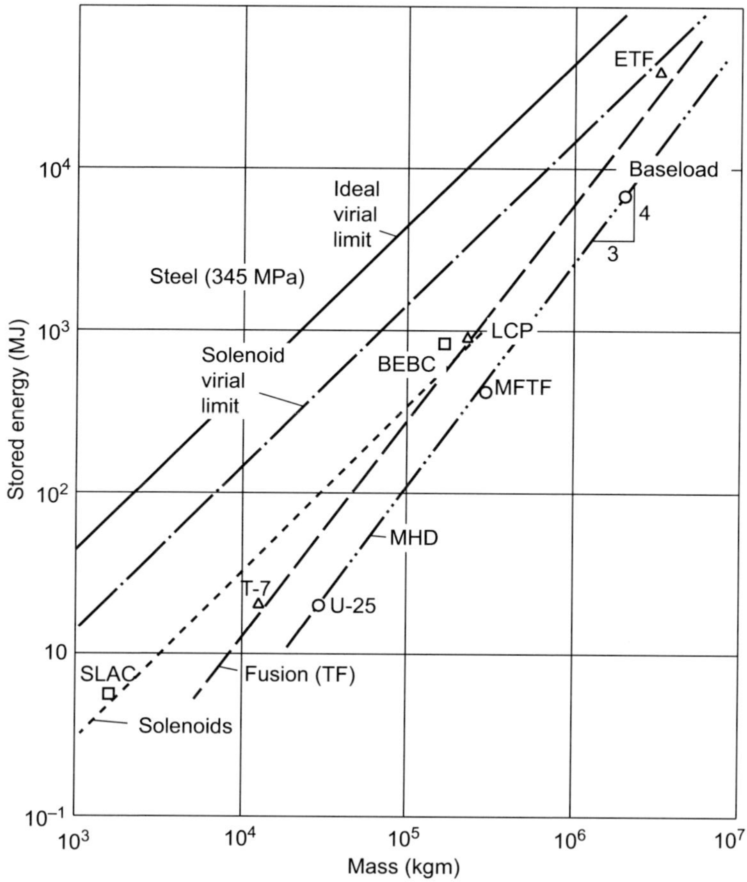

Figure 11.5 Stored energy versus mass for both solenoids and toroids (fusion magnets) (Moon, 1982).

In the same paper he has a plot (Figure 11.5) in which he shows both solenoids and toroids and also magnets for MHD. Close examination of this graph shows why toroids (on the figure they are labelled as fusion magnets) are preferred for larger scale MHD devices. As the mass gets bigger the magnet comes closer and closer to the ideal virial limit.

11.2.2 SMES cryogenic refrigeration systems

SMES are cryogenic devices, whose temperatures must be maintained low enough to ensure nondissipative operation of the superconducting wires. The cold source, which is used, is determined by a range of factors, principally the required operating temperature. The amount of cooling required is determined by the long-term losses, which are heat losses due to radiation, conduction through mechanical supports and current

Table 11.3 Most common SMES cooling methods

<table>
<tr><th></th><th>Open cycle</th><th>Closed cycle</th></tr>
<tr><td>Gas/fluid flow</td><td>This is the method currently used in large-scale magnets that use cable in conduits such as the LHC and ITER. The cryogen is typically helium and the operating range between 1.5 and 10 K</td><td rowspan="2">By adding a liquefier to the system it is possible to operate gas or liquid systems as a closed loop. MRI systems commonly use this method and it allows the system to operate autonomously over long periods of time. The downside is that the system is larger and the capital costs more expensive although the running costs should be lower.</td></tr>
<tr><td>Immersion in liquid bath</td><td>Commonly used with helium if an operating temperature of 4.2 K is required or nitrogen if 77 K is allowable, this is perhaps the most common method of all. Other candidate cryogens are hydrogen (20 K) and neon (27 K) although hydrogen is considered dangerous and neon is prohibitively expensive and poses problems with electrical isolation.</td></tr>
<tr><td>Conduction</td><td colspan="2">Commonly known as cryogen free. Systems which use a cold head coupled mechanically to the coil are becoming more common as the cost of helium rises. This method is perhaps most applicable to HTS systems operating at temperatures above 15 K as at lower temperatures there is the risk of temperature gradients, especially during discharge, although experimental devices operating at 4 K do exist.</td></tr>
</table>

SMES, superconducting magnetic energy storage.

leads, and the cyclic losses that occur due to AC loss in the superconductor during charge and discharge. The options for cooling are summarised in Table 11.3.

11.2.3 Temperature uniformity and stability

The cryogenics must also take into account the material used for the superconducting magnets and here there is a marked difference in the behaviour of LTS and HTS materials.

One factor is due to the difference in n value. LTS materials generally have a much higher n value than HTS. This leads to a higher electric field in HTS for a given current as a percentage of Jc and hence a higher idling loss.

The second factor is specific heat. This is crucial when LTS materials are being used. In general the lower the temperature the lower the specific heat (at 4.2 K typically three orders of magnitude less than at room temperature). This leads to a very low thermal inertia and consequently poor thermal stability. Minor events (e.g. vibration) may lead to hot spot formation and thermal runaway or quench. Thus LTS wires must incorporate significant additional materials (e.g. copper) for reliable operation.

11.2.4 SMES power management systems

The SMES is the storage medium and the electric utility bus is the delivery target. SMES stores energy in the form of a DC circulating current, with the discharge voltage determined by the required power delivery, whereas the target electric utility bus delivers AC (typically 50 or 60 Hz) electrical power. A PCS is required to move power back and forth between the SMES storage medium and the electric utility bus. Ideally the SMES has a fast but variable response time. Typically the power conversion system will be either line commutated or forced commutated. Line commutation is less prone to failure and often lower in cost, but its inherent weakness especially for SMES is its dependence on an adequate stiff AC voltage source. Power systems are subject to disturbances, voltage regulation difficulties, and harmonics, all of which can cause commutation problems. As a result, LC convertors are not generally suitable for feeding into weak AC systems. A more suitable alternative is forced commutation, sometime referred to as a voltage source converter, either by circuit commutation in which a voltage waveform present on a commutation capacitor is used for commutation (this voltage can be from the AC line voltage itself or if the system is driving a DC load the DC line voltage or alternatively an auxiliary voltage) or by self-commutation through, for example, a current source convertor.

A circuit diagram for a current source convertor is shown in Figure 11.6.

There are many ways of configuring the SMES, depending on the requirements of the application. Table 11.4 (Ali, Fellow, & Dougal, 2010) gives three basic configurations, Thyristor based where the SMES is being used to control the active power, current source convertor, and voltage source convertor. An advantage of the voltage source converter over the current source converter is the ability of the power conditioning system to provide both 'real' power from the SMES and 'reactive' power (volt-amp-reactive or VAR) from the capacitor, to the targeted electric utility bus.

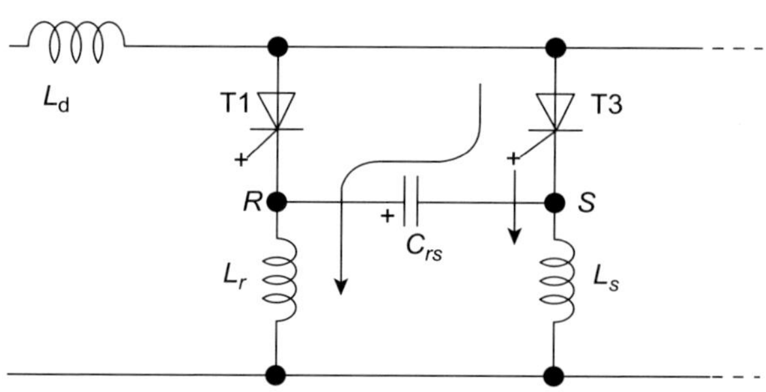

Figure 11.6 Equivalent circuit for a current source convertor (Sood).

Table 11.4 Comparison of SMES technologies

Criteria	SMES topologies: Thyristor-based SMES	SMES topologies: VSC-based SMES	SMES topologies: CSC-based SMES
Real and reactive powers control ability	The thyristor-based SMES exhibits a lagging power factor to the power system network at all times, and significant low order harmonics caused by the thyristor firing pattern. Thus, the thyristor-based SMES can control mainly the active power and has a little ability to control the reactive power, also the controls of active and reactive powers are not independent.	The VSC-based SMES allows an independent control of the real and reactive power flowing between the superconducting coil and the power system network. Also, VSC-based SMES can provide continuous rated capacity VAR support even at low or no coil current.	The CSC-based SMES allows an independent control of the real and reactive power flowing between the superconducting coil and the power system network. However, the CSC topology is able to supply a high level of capacitive reactive power. Also, CSC-based SMES is dependent on coil in providing VAR support.
Control structure	Having only one AC/DC module, the thyristor-based SMES is easier to control.	The VSC-based SMES includes not only an AC/DC circuit but also a DC/DC chopper, thus the control is complicated compared to both the thyristor-based SMISS and the CSC-based SMES.	Having only one AC/DC module, the CSC-based SMES is easier to control. Also, in the application of high power, the CSC has an additional advantage, that is, being easily paralleled of multiple bridges.
Total harmonic distortion (THD)	The total harmonic distortion is much higher than that of both the VSC and the CSC topologies.	A low total harmonic distortion can be obtained in VSC topology.	A low total harmonic distortion can be obtained in CSC topology.
Coil voltage ripple	There appears ripple in the coil voltage when using the thyristor-based SMES topology.	There appears ripple in the coil voltage when using the VSC-based SMES topology.	The superconducting coil voltage ripple is much smaller when using the C SC-based topology, especially the 12-pulse one. This implies a reduction in the superconducting coil AC losses.

SMES, superconducting magnetic energy storage.

11.3 Materials

11.3.1 Low-temperature SMES

A very early example of an SMES was the D-SMES (or distributed SMES) commercialised in the 1990s by Superconductivity Inc., which later became AMSC (American Superconductor), which used NbTi superconducting wire operating in a bath of liquid helium (LHe) and stored approximately 3 MJ. The D-SMES was designed to be a power quality device that was mounted on a standard trailer that could be driven to various industrial sites and used to supply near instantaneous power (in subpower line cycles) in the event of a temporary voltage/power sag.

On the face of it an SMES is relatively simple. It is essentially a shorted superconducting coil in which the energy is being stored in the magnetic field. Two coil geometries have been explored. In the first, the coil is wound as a toroid or doughnut shape. In the second the coil is wound as a simple solenoid. The D-SMES device (Figure 11.7) was simply a solenoid that was mounted on a lorry trailer. This was designed to support a load of between 400 and 20 kV and stored up to 3 MJ with a round trip efficiency of >97%. The primary source for loss was the nonsuperconducting power conversion system and the cryogenic refrigeration required to keep the superconducting coil cold. This SMES was for voltage sag protection and had a recharge time of approximately 90 s.

11.3.2 High-temperature SMES

There is hope for improvement using HTS materials such as bismuth–strontium–calcium–copper oxide (BSCCO), magnesium diboride (MgB_2), and yttrium–barium–calcium–copper oxide (YBCO)-coated conductors. In 2007, Higashikawa et al. (2007) proposed the SMES shown in Figure 11.8. In this analysis they optimised the SMES coil using a genetic algorithm to assess the minimum length of conductor that would be required. Two constraints were applied—electric field and flux flow loss. It was found that in order to account for the scaling of the system minimising the flux flow loss was the critical condition. The design was based on 20 MJ of storage and resulted in a much reduced volume when compared to a comparable LTS NbTi coil together with a higher operating temperature. The chief advantage was that the YBCO SMES coil was able to operate at 12 T rather than the 5 T that the NbTi coil system used.

A recent project involving DGA/CNRS and Nexans was based on BSCCO (commonly known as 2212 to describe the stoichiometry), which is a high-temperature superconductor. It was also based on a solenoid but was a demonstrator device and only stored 800 kJ. Its operating temperature was 20 K and the rated current was 315 A. The low operating temperature is explained because even though the zero-field critical temperature (T_c) of 2212 is typically 85 K, the critical current degrades rapidly in the presence of magnetic field, and lowering the temperature mitigates this deleterious effect.

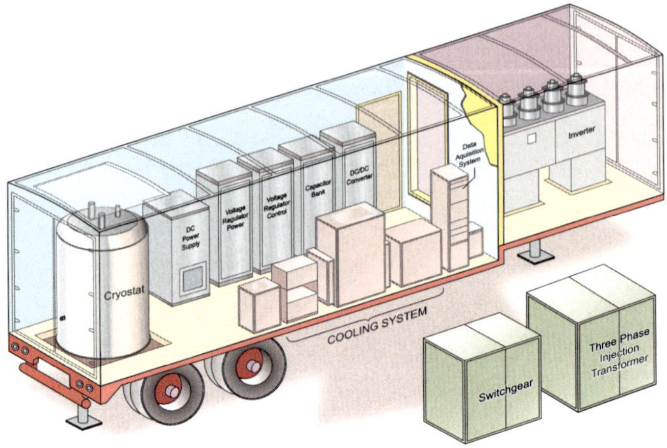

Figure 11.7 D-SMES—An early mobile SMES device. SMES, superconducting magnetic energy storage.

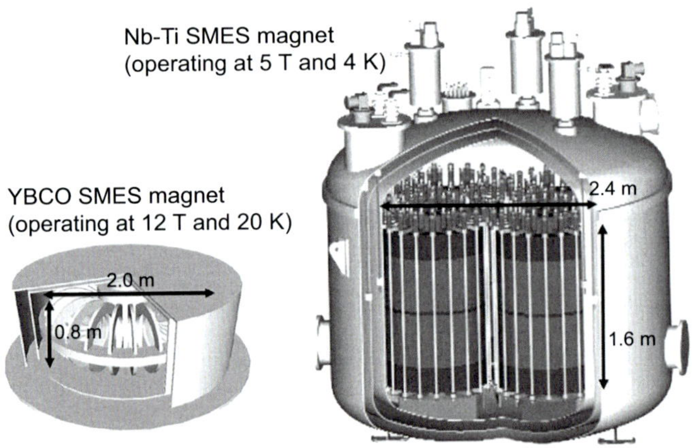

Figure 11.8 Comparison between low- and high-temperature SMES coils (Higashikawa et al., 2007). SMES, superconducting magnetic energy storage.

Larger scale SMES devices tend to be built with toroidal geometries. This reduces the stray magnetic field and makes the magnetic design simpler as there are no end effects to be considered. However, the structural design to support the hoop stress experienced by the superconducting wire due to the Lorentz force is far more complicated. Chubu electric power has a planned program of work, the target of which is to produce a coil that can store 2 GJ (Figure 11.9).

Recently though in a press release (http://www.chuden.co.jp/english/corporate/ecor_releases/erel_pressreleases/3186929_11098.html) Chubu Electric Power stated

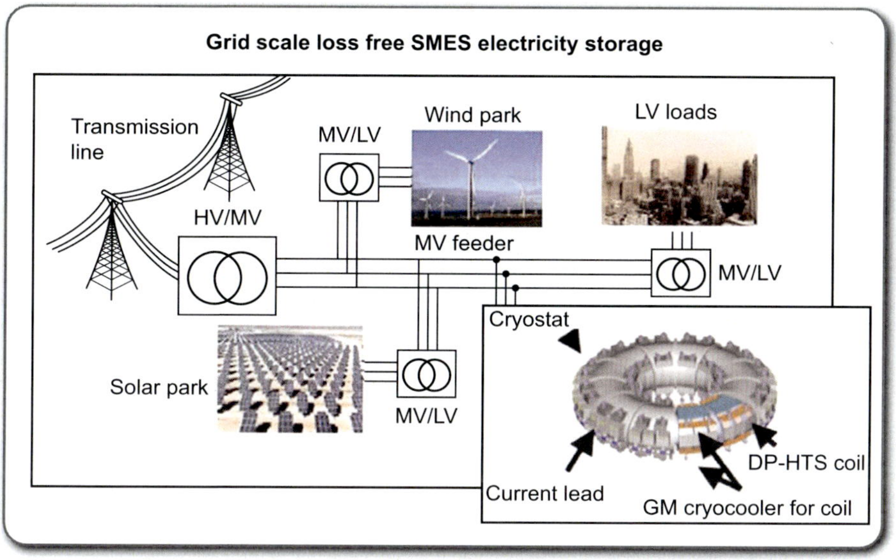

Figure 11.9 Grid scale loss free SMES electricity storage. SMES, superconducting magnetic energy storage.

their intention to build a solenoidal YBCO-based SMES coil. This will be based on a newly developed coil structure and the superconducting tape will be carrying a current of 1500 A at 4.2 K. This coil will have a magnetic field strength of 10 T and a magnetic field stress of 1700 MPa! This coil uses specially developed end plates for strengthening, which are the yellow plates in Figure 11.10.

Figure 11.10 An SMES coil wound from 2G (YBCO) HTS tape, showing required reinforcing (orange plates). SMES, superconducting magnetic energy storage.

11.4 Competing technologies

There are many different forms of energy storage. However, as seen in Figure 11.2, they fill different niche markets. Essentially there are two different criteria when considering energy storage: power, which we may split into low power and high power, and storage capacity or more pertinently whether you require short-term or long-term power transfer.

By far the most successful forms of energy storage are batteries, of which there are many types. These tend to be low-power devices delivering energy over timescales of hours rather than minutes or seconds, more pertinent to SMES. Their main limitation is the number of charging cycles that they can service before failure occurs. A major advantage of an SMES over many other storage devices is the near limitless number of charge and discharge cycles that the device can handle.

For low-power applications, short-term energy transfer devices that hold the energy as an electric field such as capacitors and supercapacitors are ideal. They have quick response times and can be cycled many times.

Devices that hold their energy magnetically such as SMES are high-cost devices and therefore are more appropriate for high-power applications, especially when the requirement is for a short-term 'boost'.

Kinetic energy storage in, for example, flywheels tends to be medium-power systems filling a range up to about 200–300 kW. Historically flywheels are limited by two things: the bursting stresses due to the centrifugal forces and the losses in the bearings. Recent developments in superconducting bearings (especially at Boeing) in which the frictional loss is minimal in comparison to the energy storage mean that flywheels can be considered for diurnal storage where power generated during off-peak times, for example, during the night, is stored for usage during peak times.

For power quality applications, the best choice is strongly dependent on the discharge time required. For a 1–2 s discharge, SMES and supercapacitors are attractive, whereas at 20–30 s, some flywheels or battery systems are less expensive. Figures 11.11 and 11.12 give sample values for the costs of different systems for both a 1 and 20 s discharge.

11.5 Markets

11.5.1 *Power quality applications*

SMES devices fill a niche market where high currents and high powers are required for relatively short amounts of time.

The cost of SMES is dependent on many things and is modest when compared to that of pumped hydro, for example. Generally speaking though the cost is reduced with scale as seen in Figure 11.13.

In addition, as the technology to manufacture high-temperature superconducting wires and tapes matures, the cost per unit of energy storage is constantly being

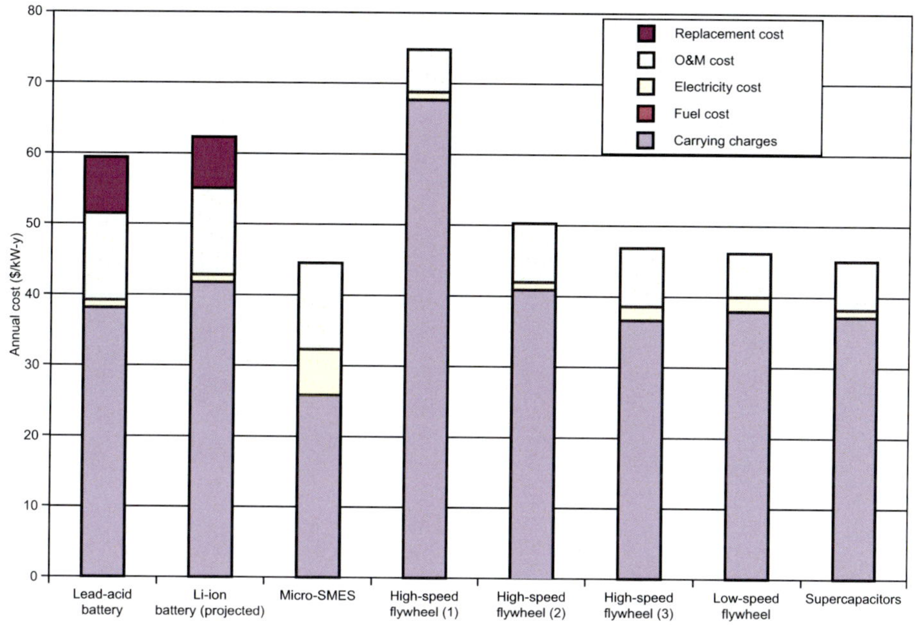

Figure 11.11 Component costs of micro-SMES for power quality applications compared to several other fast-discharge energy storage technologies. Costs are calculated based on a 1-s discharge (Schoenung & Hasselzahn, 2003). SMES, superconducting magnetic energy storage.

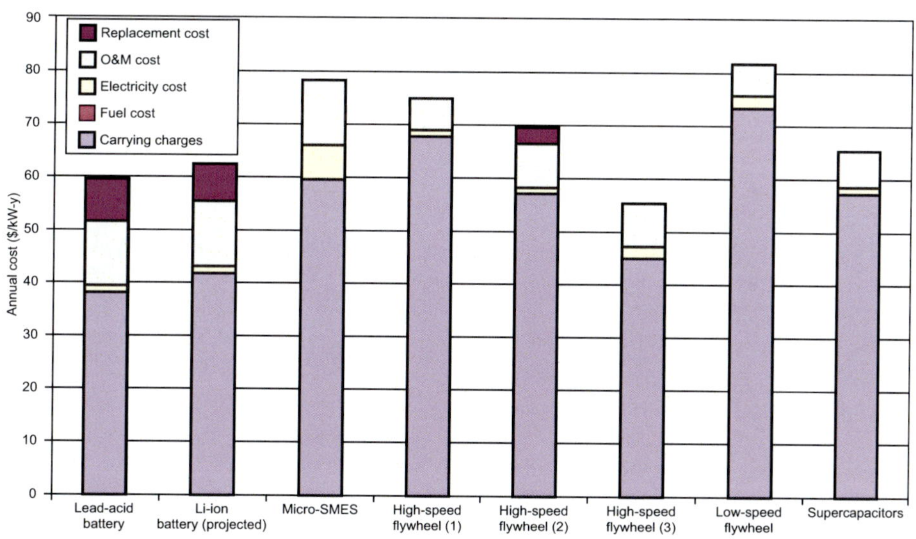

Figure 11.12 Component costs of micro-SMES for power quality applications compared to several other fast-discharge energy storage technologies. Costs are calculated based on a 20-s discharge (Schoenung & Hasselzahn, 2003). SMEs, superconducting magnetic energy storage.

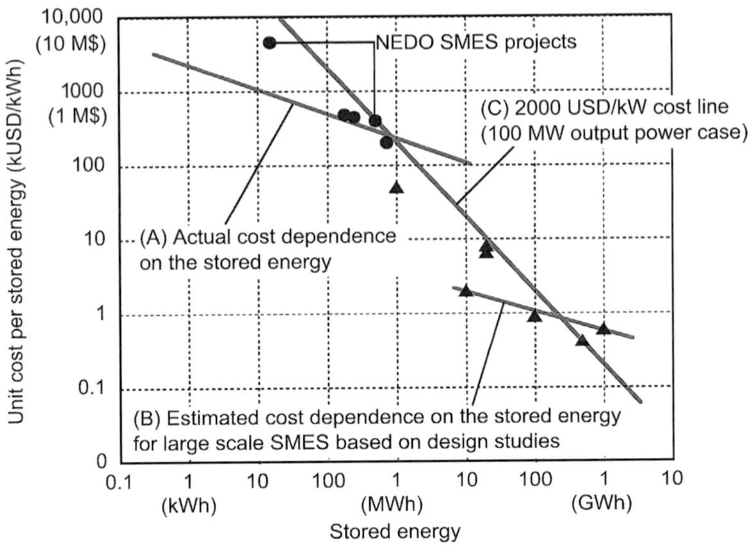

Figure 11.13 Cost per kWh of storage versus energy storage of SMES devices (Nomura et al., 2010). SMES, superconducting magnetic energy storage.

reduced. Added to that is the fact that the magnet itself can be cycled potentially an infinite number of times and that it is capable of providing very large currents in a fraction of a cycle.

Considering all of these factors it is clear that for power quality applications SMES could represent an ideal solution. In addition, there are a number of major electrical equipment manufacturers who are either actively developing SMES or who in the past have developed prototypes so as the market develops there should be manufacturers who are ready and able to exploit it.

11.5.2 Grid stability applications

A possible utilisation of SMES is in flexible AC transmission systems (FACTS). Broadly speaking a FACTS device is present in the grid in order to improve controllability and power transfer.

As far back as the 1980s, one of the first implementations of SMES was carried out in the United States by the Bonneville Power Authority who used a 30 MJ SMES (Boenig & Hauer, 1985).

The unit, which was capable of absorbing and releasing up to 10 MJ of energy at a frequency of 0.35 Hz, was designed to damp the dominant power swing mode of the Pacific AC Intertie. The reference reports that the coil was kept cold for just over 9 months during 1983 and that it was tested for approximately 120 h, 30% of which was at power outputs of 8 MW and above. The superconducting coil itself presented no problems. The main faults occurred in the convertor during initial testing but the addition of MOV protection and the repair of a faulty gate drive removed this problem.

The main problems were at the time with the helium refrigerator system, which was not fully implemented and curtailed the overall effectiveness of the system.

Although this was an interesting test and the total number of cycles 10^6 was greater than the number of cycles a diurnal storage unit would encounter over a 30 year lifetime, it was pointed out that a full-scale production unit would require at least 1000 MWh with a 200 MW convertor. This is 100,000 times the storage capacity of the coil under test.

More recently, in 2000, American Superconductor studied the efficacy of installing seven D-SMES units at key points in an area known as the North-Western Corridor, to enhance its stability (Ross, Borodulin, & Kazachkov).

This grid experienced voltage instability problems with large momentary voltage depressions, which could lead to the grid collapse. The study was carried out using PSS/E, a software package provided by Power Technology Inc. and involved simulation of the entire 7500 MW transmission system. Two critical faults were identified, both of which caused rapid voltage collapse in different parts of the network. The cause of the collapse was identified to be loss of reactive support from switched shunt capacitors. These were incapable of pushing the voltage back up once it had started to dip and the D-SMES were identified as being a candidate means for restoring the voltages. The findings of the study were that by using seven D-SMES devices with a total capacity of 21 MW and 126 MVAR, the system voltage could be recovered within 3 s of a critical fault and that this would prevent voltage collapse.

It should be noted that, in general, FACTS are required mostly for transfers of reactive power. The real power transfers, which require a superconducting magnet, are a small part. So a FACTS, which provides only reactive power such as the STACOM (D-VAR), would be sufficient and this together with pulse width modulated power converters negates the need for the real power supplied by the superconducting magnet.

11.6 Future developments

11.6.1 Dependence on material development

Is SMES a mature technology? On the face of it, it is. There are only two topologies of interest (solenoidal and toroidal) and the overall physics and methods construction are well known. However integration of new superconducting materials will make a difference to any future developments in the area. Especially materials such as coated conductors where the geometry of the material itself (a flat tape) may dictate the design and where also rapid progress is being made in development.

11.6.2 Dependence on power electronics

It would be a mistake to assume that since the cost of the associated semiconducting power electronics has been reduced in recent years this will hasten the advent of SMES solutions. It is possible that advancements may bring the cost of an SMES down sufficiently for it to be considered as a commercial proposition but at the same time competing systems will be gaining a similar advantage. The fundamental cost of the conductor is, and probably always will be, the deciding factor when adoption is being considered.

11.6.3 Commercialisation

Market progress has hitherto been hampered by the high-cost and low-energy density of SMES systems. Efforts are currently underway to bring down the cost of SMES technology. For example, a project funded by the U.S. Department of Energy (DOE) via a $4.2 million ARPA-E grant to ABB, Brookhaven, and Superpower for a 3.3 kWh proof of concept aims to enable SMES technology's use for MWh energy storage applications. The project's success would allow the technology for use in large-scale energy storage.

Asia–Pacific represents the largest potential market. Rapid economic growth in the region is necessitating the adoption of new technologies such as smart grids, micro-grids, distributed energy generation, and renewable energy integration. All of these technologies rely to a greater or lesser extent on energy storage systems of one form or another, whether it be load leveling or maintenance of power quality. In addition in developed countries such as the United States the limited level of energy storage available leads to the need to maintain a spinning reserve or in other words to overprovide on generation capabilities in order to maintain security of supply. Although solar and wind power have emerged as eco-friendly alternatives to fossil fuels, the variability of power generation continues to present challenges for the grid. All of these factors are driving the focus on SMES as a possible solution.

References

Ali, M. H., Fellow, B. W., & Dougal, R. A. (April 2010). An overview of SMES applications in power and energy systems. *IEEE Transactions on Sustainable Energy, 1*(1), pp. 38–47.

Boenig, H. J., & Hauer, J. F. (1985). Commissioning tests of the Bonneville power administration 30 MJ superconducting magnetic energy storage unit. *IEEE Transactions on Power Apparatus and Systems, 104*, 302–312.

Haynes, W. M. (Ed.). (2014). *CRC handbook of chemistry and physics* (95th ed.). Boca Raton, USA: CRC Press.

Higashikawa, K., Nakamura, T., Shikimachi, K., Hirano, N., Nagaya, S., Kiss, T., et al. (June 2007). Conceptual design of HTS coil for SMES using YBCO coated conductor. *IEEE Transactions on Applied Superconductivity, 17*(2).

Leibniz Institute for New Materials. http://www.inm-gmbh.de/en/research/interface-materials/energy-materials/.

Moon, F. C. (1982). The virial theorem and scaling laws for superconducting magnet systems. *Journal of Applied Physics, 53*, 9112. http://dx.doi.org/10.1063/1.330423.

Nomura, S., Shintomi, T., Akita, S., Nitta, T., Shimada, R., & Meguro, S. (2010). Technical and cost evaluation on SMES for electric power compensation. *IEEE Transactions on Applied Superconductivity, 20*, 1373–1378.

Ross, M., Borodulin, M., & Kazachkov, Y. (2001). Using D-SMES devices to improve the voltage stability of a transmission system.
Schoenung, S. M., & Hasselzahn, W. V. (2003). *Sandia report.*
Sood, V. K. HVDC and FACTS controllers-applications of static converters in power systems. EFIC ISBN: 978-1-4020-7890-3 (Print) 978-1-4020-7891-0 (Online).
Tixador, P. (January 2008). *Superconducting magnetic energy storage: Status and perspective.* IEEE/CSC & ESAS European Superconductivity News Forum. No. 3.

High-temperature superconducting (HTS) transformers for power grid applications

12

M.P. Staines[1], *Z. Jiang*[1], *N. Glasson*[2], *R.G. Buckley*[1], *M. Pannu*[3]
[1]Robinson Research Institute, Victoria University of Wellington, Wellington, New Zealand; [2]Callaghan Innovation, Christchurch, New Zealand; [3]Wilson Transformer Company, Melbourne, VIC, Australia

12.1 Introduction

High-temperature superconducting (HTS) transformers, defined by high-current density HTS conductors and liquid nitrogen dielectric, have many potential advantages over conventional oil-immersed transformers, such as smaller, lighter, more efficient, low fire and environmental hazard, overload capability without reduction in lifetime, low output impedance and fault current limiting (Kalsi, 2011; Pleva, Mehrotra, & Schwenterly, 2010; Wolsky, 2013), making them attractive to electrical grid operators.

In the face of this impressive list of virtues, the obvious point needs to be made: grid operators may pay a premium for HTS transformers *in capital cost*, but only when the total cost of ownership (TCO) is lower than the alternative conventional transformer, oil immersed or dry type. Smaller *is* better, but it is only worth so much, and then only in some cases. The TCO methodology is routinely used by utilities to make choices between conventional transformer options. For example, oil-immersed designs are supplanted by other options (less flammable dielectric fluids or dry-type transformers) only when the cost of mitigating the fire and environmental risks associated with mineral oil increases the TCO beyond that of the other options. TCO provides us with a quantitative tool for examining the barriers to commercialization that HTS must overcome and identifying priorities for research and development effort.

Conventional oil-filled transformers are highly optimized, reliable and efficient. The efficiency advantages of superconductors are of course greatest in DC applications. But in AC applications they generate loss, and so application to transformers, an intrinsically AC application, may seem initially less promising. Nevertheless, HTS demonstrators with significant ratings operated successfully >15 years ago (Zueger, 1998). Why have they not progressed? Understandably there has been a focus on demonstrating the technical performance of HTS conductors in transformers.

Superconductors in the Power Grid. http://dx.doi.org/10.1016/B978-1-78242-029-3.00012-1

Relatively little attention has been given to such questions as short circuit performance let alone technology readiness or commercial viability.

A basic introduction to transformers in general and HTS transformers in particular has been given in Sections 1.3.2 and 1.5.3 of the introduction to this book. In the following we first review the demands that the grid environment makes of conventional transformers. HTS transformers will need to comply with these standards, largely without exception. A review of the status and achievements of HTS transformer development worldwide follows. We then briefly review the fundamental aspects of HTS conductors and cryogenic operations that variously constrain, advantage and challenge HTS transformers. The issues associated with AC loss in HTS windings and the technology for cooling them are then reviewed. We then critically examine how HTS may measure up in terms of energy efficiency, fault, and overload and other performance standards. Grid applications of HTS transformers will always struggle to be competitive with conventional transformers at low rating where the cryogenic system presents a significant cost and energy overhead. We therefore review, with reference to ratings of 15 MVA and above, the quite limited published analysis of TCO of HTS compared with conventional transformers, considering the impact of reductions in conductor cost, and of application areas where HTS offers significant added value, such as indoor installations. We conclude with a summary and pointers to prioritize research and development areas.

12.2 Transformers and the electricity grid

In addition to their basic function of transforming voltage levels, transformers in the grid are required to give reliable, cost-effective service over a lifetime of decades, to provide voltage regulation, and to perform in transient events: handling short circuits, inrush currents, lightning strike and other voltage impulses, and overload. It is important to understand the consequences of these requirements – most are necessary conditions that HTS transformers in the grid will have to meet. The requirements are specified in various international and national standards. Some of the key requirements are summarized in Table 12.1. The table neglects specifications such as minimum efficiency performance standards relevant to smaller distribution transformers, under 2.5 MVA rating, since HTS transformers are unlikely to significantly penetrate this low-power market.

12.2.1 Short circuit current

The maximum fault current is assumed to be limited by the short circuit impedance of the transformer which is dominated by leakage reactance. The standard is written around ensuring that fault current produces a maximum copper winding temperature of no more than 250 °C at 2 s. A short circuit current duration of 2 s is required by grid operators to allow protection systems to isolate the fault while disconnecting as little healthy equipment as possible. With the fault removed the operator requires the transformer to continue in operation, implicitly at rated current. Naturally the

Table 12.1 Requirements for design and performance of medium and large oil-filled power transformers

Characteristic	Standard or recommendation	Specification
Short circuit current limitation	IEC 60076-5 power transformers – Part 5: Ability to withstand short circuit	Minimum impedance specified for different ratings, e.g. 10% for 25 to 40 MVA, fault duration 2 s
Overload capacity	IEC 60354: Loading guide for oil-immersed transformers	Maximum current (pu[a]) Cyclic loading 1.5 Emergency loading 1.8
Insulation	IEC 60076-3 power transformers: Insulation levels, dielectric tests, and external clearances in air	Basic insulation level (BIL) determined by maximum voltage U_m for winding
Voltage regulation	National regulations define allowable voltage variation at customer	Typically ±6% of nominal voltage
Efficiency	Design optimized to minimize total cost of ownership using capitalization formulae for no-load and load losses	EU regulation from mid-2015 specifies minimum peak efficiency for large transformers
Fire risk	NFPA 850 (5.1.4 Outdoor oil-insulated transformers; 5.1.5 Indoor transformers); FM3990 (Approval standards for less or nonflammable liquid-insulated transformers)	Isolation distances, barriers, fire suppression recommendations

[a]pu (per unit) specifies electrical values as a fraction of the defined value at rated load.

transformer must also withstand the dynamic stresses of the fault current in addition to the thermal stress.

12.2.2 Overload capacity

Oil immersed transformers are required to run in overload at times. Recommended load limits of 150% for cyclic loading and 180% for short-time emergency loading are given in IEC 60354 for medium-power transformers (up to 100 MVA) with specified maximum hot-spot and top-oil temperatures up to 160 and 115°C, respectively. The intent is to avoid conditions under which significant reduction in lifetime could result. The thermal time constant of the windings and oil of a large transformer may be measured in hours, but bushings and tap changers will also have maximum ratings and when these are exceeded the operator will be forced to shed load.

12.2.3 Insulation

The standards define test conditions based on the maximum voltage level specified for the transformer windings. For a 110 kV winding, for example, the insulation will be designed to withstand a standard lightning impulse test to the basic insulation level (BIL) of 550 kV peak and a short duration AC voltage of 230 kV rms.

12.2.4 Voltage regulation

On-load tap changers (OLTC) are typically incorporated in transformers to maintain voltage regulation as the load varies. The voltage tap range is typically $\pm 10\%$ in ± 8 steps. Some of the variation is to compensate for line impedance as well as the impedance of the transformer. Tap changes are made periodically throughout the daily cycle to follow the load profile. The typical number of changer operations per year for a network transformer is between 2000 and 20,000 (Dohnal, 2004).

12.2.5 Efficiency

The efficiency of large power transformers has been determined historically by the outcome of an optimization carried out by the designer using loss evaluation rates to calculate the dependence of the TCO over a term of 20, 30 or 40 years on the no-load and load losses of the transformer. In its simplest form (Galloway & Mulkey, 2004) the TCO is calculated as TCO $=$ transformer cost $+ A \times$ no load loss $+ B \times$ load loss.

The basis for determining the loss evaluation rates A and B is described in IEEE standard C.57.120-1991. The current cost of energy is a significant input, but the assumed lifetime and interest rates are also influential. Fassbinder (2013) lists average rates for European countries that show a considerable spread. On average $A = 7.16$ €/W, with the average $B = 2.20$ €/W [USD/€ ≈ 0.72], implying that the assumed average loading is 55%. According to Galloway and Mulkey typical loss evaluation factors in 2003 were $A = 2.50$ USD/W and $B = 0.80$ USD/W. The average European rates are roughly four times these values — the value of efficiency improvements depends strongly on the market, transformer type, and the customer's requirements.

In Table 12.2 typical efficiency and loss values for a range of transformers are listed. The no-load loss P_0 is almost entirely due to hysteretic and eddy current losses in the iron core, and does not vary with load. The load loss P_k, due to I^2R loss in the conductor and eddy current losses in the conductor and transformer structure, depends on the square of the current. Hence for a transformer of rating S there is a peak in efficiency given by $\eta_{\max} = 1 - \frac{2\sqrt{P_0 P_k}}{S}$, which occurs at a load $\sqrt{P_0/P_k}S$. The peak efficiency is the basis for EU regulations that will come into force in mid-2015 and will prescribe minimum efficiency requirements for large power transformers. The regulations largely reflect the performance achieved by transformers manufactured since 2004. Minimum efficiency values relevant to the potential HTS market are listed in Table 12.3.

Table 12.2 **Efficiency and loss of typical utility transformers**

Transformer type	Rated power (MVA)	Efficiency (%) at			Loss (kW) at		
		100% rated load	50% rated load	Max efficiency	100% rated load	50% rated load	No load
Generator transformer	1100	99.60	99.75	99.78	4400	1375	367
Interbus transformer	400	99.60	99.75	99.78	1600	500	133
Substation transformer	40	99.40	99.60	99.62	240	80	27
Distribution transformer	1	98.60	99.00	99.02	14	5	2

Adapted from Fassbinder (2013).

Table 12.3 Minimum peak efficiency index requirements for liquid immersed large power transformers (Part of Table 1.7 in EU regulation No. 548/2014)

Rated power (MVA)	From July 1, 2015	From July 1, 2021
	Minimum peak efficiency index (%)	
20	99.639	99.684
25	99.657	99.700
31.5	99.671	99.712
40	99.684	99.724
50	99.696	99.734
63	99.709	99.745
80	99.723	99.758
≥100	99.737	99.770

Large oil-immersed transformers are highly efficient devices, with average losses <0.3% of rated MVA. The added value to be gained by reducing this further is quantifiable. Taking the example of a 40 MVA transformer with load loss of about 200 kW, and considering the effect of a very significant change in the load loss, halving it to 100 kW would be worth between 80,000 and 300,000 USD depending on the value of the load loss factor B nominated by the customer. Taking a round figure of 1 M USD for the selling price of a 40 MVA transformer (DOE, 2014), the reduced loss would add about 10–30% to its value.

12.2.6 *Fire risk*

Fire protection and prevention of oil release are further considerations for transformer operators. A 40 MVA transformer may contain 30,000 L of oil – a significant cost and hazard. CIGRE (2013) estimate that the probability of a transformer fire is in the range of 0.04–0.25%/year, with the average probability of a transformer fire being approximately 0.1%/year. There is thus an average probability of 4% over a 40 year service life for the average utility owned transformer. Any event leading to a fire will likely destroy the transformer. The focus after prevention is then on limiting harm to other assets and to people. Indoor transformers are especially hazardous. For this reason NFPA 850 states that dry-type transformers are preferred for indoor installations, and stipulates fire suppression systems and/or fire-resistant barriers for oil-immersed transformers. Low fire risk transformers include dry type (resin impregnated windings) and less-flammable fluid type (ester-filled). Dry-type transformers are not available in higher voltage and MVA ratings. Synthetic or natural ester dielectric instead of mineral oil can be used in such applications (FM3990, 1997). Despite the higher initial cost of ester-based dielectric the elimination of the cost of installing and maintaining fire protection systems and other advantages can justify this option (Bureau of Reclamation, 2005). Schwenterly, Pleva, and Wolsky (2008) estimate the cost of oil containment and fire suppression for a 15 MVA transformer at USD 130,000.

12.3 A brief history of superconducting transformers

The development of superconducting transformers began in the 1960s using low-loss Nb–Ti wire and warm iron. However achieving acceptable efficiency at an operating temperature of 4 K proved extremely challenging (Hassenzahl et al., 2004). The development of HTS wire in the 1990s, with its higher operating temperature, encouraged renewed interest. The subsequent development of HTS transformers has been a global effort. Table 12.4 lists the key specifications of transformers of rating 500 kVA and over developed in projects spanning more than 15 years.

12.3.1 Europe

In 1997, ABB built a three-phase 18.7/0.42 kV–0.63 MVA transformer operating at 77 K using BSCCO wire. The rated low-voltage (LV) current was 866 A, and several parallel wires were used to achieve the required high current in the LV windings. The transformer was connected to the power grid and was operated for 1 year without any major problem (Zueger, 1998).

In 2001, Siemens built and demonstrated a single-phase 25/1.4 kV–1 MVA traction transformer operating at 67 K (Leghissa et al., 2002; Schlosser, Schmidt, Leghissa, & Meinert, 2003). The high-voltage (HV) winding was placed between two LV windings in a radial sequence. A significant achievement was the production of Roebel cable for the LV winding by in-plane bending of BSCCO tape, difficult to manage without affecting the critical current. The cable had 13 BSCCO wires to achieve the rated current of 360 A. The full transposition length of the cable was approximately 3 m (Hussennether, Oomen, Leghissa, & Neumüller, 2004), which is perhaps rather long to ensure that current is equally shared between the strands in the LV windings with a mean diameter of 234 and 385 mm, respectively (Schlosser et al., 2003).

12.3.2 United States

Waukesha Transformers and the Oak Ridge National Laboratory (ORNL) group built a single-phase 13.8/6.9 kV–1 MVA transformer using BSCCO wire in 1998 (Schwenterly et al., 1999). A closed-cycle cryocooler was employed to maintain the operating temperature below 25 K and two parallel BSCCO wires were used for the LV winding. A voltage breakdown occurred during open-circuit testing at 11 kV, so

Table 12.4 **High-temperature superconducting transformer projects 1996–2014 with power rating 500 kVA and above**

Group, country (year)	Rated power (MVA); phase	Rated voltage (kV)	Rated current (A)	Operating temp. (K)	AC loss × COP; power (%)	Wire technology
Kyushu Univ. Japan (1996)	0.5; 1 ϕ	6.6/3.3	76/152	77	64 W × 21; 0.27	BSCCO
ABB Switzerland (1997)	0.630; 3 ϕ	18.7/0.42	11.2/866	77	337 W × 21; 1.12	BSCCO
Waukesha USA (1998)	1; 1 ϕ	13.8/6.9	72.5/145	25	Not available	BSCCO
Kyushu Univ. Japan (1998)	1; 1 ϕ	22/6.9	45.4/145	65	159 W × 25; 0.4	BSCCO
Siemens Germany (2001)	1; 1 ϕ[a]	25/1.4	40/2 x360	67	380 W × 25; 0.95	BSCCO Roebel cable
Kyushu Univ. Japan (2003)	2; 1 ϕ	66/6.9	30.3/290	66	633 W × 25; 0.79	BSCCO
Waukesha USA (2004)	5/10; 3 ϕ	24.9/4.2	116/694	29	Not available	BSCCO
KPU, SNU Korea (2004)	1; 1 ϕ	22/6.9	44/152	65	200 W × 25, 0.5	BSCCO
Kyushu Univ. Japan (2005)	4; 3 ϕ[a]	25/1.2/0.44	160/750/909	66	7 kW × 25; 4.38	BSCCO
TBEA IEE–CAS China (2005)	0.63; 3 ϕ	10.5/0.4	35/904	77	110.7 W × 21, 0.37	BSCCO
Nagoya Univ. Japan (2010)	0.666; 1 ϕ	22/6.6	52.5/175	77	Not available	YBCO
Kyushu Univ. Japan (2014)	2; 3 ϕ	66/6.9	17.5/167	77	26.9 W × 21; 0.28	GdBCO striated conductor
RRI New Zealand (2014)	1; 3 ϕ	11/0.4	30/1390	70	336 W × 25; 0.7	YBCO Roebel cable
Planned						
CEPRI Shandong Electric Power China	25; 3 ϕ	110/10.5				YBCO

[a]Traction transformer

that the maximum power rating achieved was only 0.68 MVA. The same group built a three-phase 24.9/4.2 kV−5 MVA transformer using BSCCO wire (Weber et al., 2005). The cooling module consisted of cryocoolers and a heat exchanger employing a helium gas loop to obtain an operating temperature of 29 K, which also allowed for a large LV current of 694 A. The demonstration was not successful, with breakdown occurring during high-voltage open-circuit testing. It was concluded that a 'dry' cryocooler-based cryogenic system for transformers without a good dielectric like liquid nitrogen is not a promising approach.

A more recent project involving HTS wire manufacturer Superpower and SPX Waukesha in developing a 28 MVA fault current limiting transformer (FCLT) has suffered a setback with the withdrawal of transformer manufacturer SPX, and will now reportedly focus on confirming the characteristics of the YBCO/Cu−Ni-laminated fault current limiting (FCL) conductor rather than construction of a transformer.

12.3.3 Japan

Since the late 1990s Kyushu University in partnership with Fuji Electric and others has had a sustained effort to develop HTS transformers while acknowledging that AC loss is the biggest barrier to commercialization. To meet this challenge they have developed a coil-level transposition (CLT) method (Bohno et al., 2005; Funaki et al., 1998; Funaki & Iwakuma, 2000; Iwakuma, Funaki, Kajikawa, Kanetaka, et al., 2001; Iwakuma, Funaki, Kajikawa, Tanaka, et al., 2001; Iwakuma, Tanaka, & Funaki, 2002; Iwakuma et al., 2009; Kamijo et al., 2006, 2010). In 1996, the group built a single-phase 6.6/3.3 kV−0.5 MVA transformer operating at 77 K using BSCCO wire (Funaki et al., 1998). Two HV windings and two LV windings are alternated in the sequence of core/HV/LV/HV/LV. Both HV and LV windings were wound with parallel wires: three wires for the HV windings; six wires for the LV windings. The windings were transposed to equalize the current distribution between wires, and the transposition numbers per layer of the HV and LV windings are two and five, respectively. A power rating of 0.8 MVA was achieved at 66 K. Later they fabricated a single-phase 22/6.9 kV−1 MVA transformer operating at 65 K and using BSCCO (Iwakuma et al., 2001). The HV and LV windings comprised cylindrical coils of four layers and two layers, respectively. The windings were wound with parallel wires: two wires for HV windings; six wires for LV windings. The parallel wires in the LV windings were transposed 11 times using a different transposition method from their 0.5 MVA transformer (Funaki et al., 2001). The transformer was connected to the grid for 18 h and 6 days at the power ratings of 821 and 586 kVA, respectively. The transformer withstood an inrush current of five times the rated current. This project was followed by a third transformer, a single-phase 66/6.9 kV−2 MVA transformer operating at 66 K again using BSCCO (Bohno et al., 2005). The HV winding was wound with eight parallel wires transposed seven times to achieve 290 A and equal current distribution. The model coil was wound with 20 parallel wires and transposed 19 times achieving 480 A and good current sharing.

In 2014, the same group built a three-phase 66/6.9 kV−2 MVA transformer operating at 65−70 K and using GdBCO conductor (Iwakuma et al., 2014).

The transformer was built as a 1/10 rating model of a three-phase 66/6.9 kV−20 MVA for a distribution power grid. The HV and LV windings comprised cylindrical coils of eight layers and two layers, respectively. The LV windings were wound with eight parallel wires, and the LV windings were transposed 15 times to achieve equal current sharing between wires. A notable feature of the transformer is the use of conductor striation to reduce AC loss. The HTS layer of each wire is divided into three filaments by laser striation. The AC loss at 77 K was 26.9 W, which gives us the smallest AC loss (0.28%) as a fraction of power rating after correction for cooling penalty. A further reduction in AC loss at 65 K is expected. One of the ultimate goals of the Kyushu University group is to commercialize three-phase 20 MVA YBCO-coated conductor transformers utilizing the CLT transposition method (Iwakuma et al., 2009).

Nagoya University is developing a superconducting fault current limiting transformer (SFCLT) (Hayakawa, Kojima, Hanai, & Okubo, 2011). They have built a single-phase 22 kV/6.6 kV−0.67 MVA SFCLT using YBCO conductors as a demonstrator for a three-phase 22 kV/6.6 kV−2 MVA SFCLT. This demonstrator assumed five cycles of fault duration after which the HTS windings go normal and the transformer comes off-line.

The Kyushu University group in partnership with the Japanese Railway Technical Research Institute a built a three-phase 25 kV/1.2 kV/0.44 kV−4 MVA traction transformer using BSCCO wire and operated at 66 K (Kamijo et al., 2006). The transformer had four separate secondary windings (LV) connected in series with an expected rating of 0.9 MVA for each. The windings were wound on two transformer-core legs with each leg having two windings. Each winding was a single layer of 96 turns of eight parallel tapes vertically stacked. The tapes in the winding were transposed 15 times to achieve equal current sharing. The transformer had two tertiary windings connected in parallel with an expected rating of 400 kVA at 909 Arms (454.5 Arms for each coil). These windings were wound around two core legs with each leg having one winding. Each winding was a single layer of 35 turns with each turn composed of two parallel vertical stacks of three tapes to achieve the required current-carrying capacity. The winding was transposed five times to achieve current sharing. The AC loss values of the transformer at the power ratings of 3.5 and 4.0 MVA at 60 Hz and 66 K were exceptionally high, 6.2 and 7.0 kW, respectively (Kamijo et al., 2006, 2010). Possibly the operating point for this transformer was too high relative to the conductor critical current.

12.3.4 Korea

Korea Polytechnic University (KPU), Seoul National University (SNU), and Korean industries built and demonstrated a single-phase 22/6.9 kV−1 MVA transformer operating at 65 K (Kim et al., 2005) using BSCCO wire windings. The HV and LV windings were concentrically arranged to reduce AC loss in the HTS. The LV winding was composed of double pancake coils wound with four parallel tapes. The double pancake coils for the LV winding were transposed three times to achieve uniform current distribution in each tape.

12.3.5 China

TBEA (Tebian Electric Apparatus Stock Co. Ltd) and CAS (Chinese Academy of Sciences) built and demonstrated the three-phase 10.5/0.4 kV—630 kVA transformer shown in Figure 12.1. It operates at 77 K using BSCCO conductor (Wang et al., 2007). The LV winding was composed of 23 double pancakes connected in parallel to achieve the rated current of 909 A. No special effort was made to equalize current sharing in the parallel tapes. The efficiency of the transformer was estimated to be 98.5% based on estimates of the core and total electrical loses and a cooling penalty at rated operation. The transformer managed an inrush current of 10 times the rated current for approximately 0.2 s and was successfully grid-connected for 60 days. The transformer has been incorporated into a superconducting substation since early 2011 and will continue in operation until 2015 (Xiao, 2013).

China Electric Power Research Institute (CEPRI), a research-based subsidiary of State Grid Cooperation of China (SGCC), has started a project to build a three-phase 25 MVA, 110 kV/10.5 kV transformer using YBCO-coated conductors. They have partnered with Shandong Electric Power Equipment Company. This is the highest MVA and voltage rating HTS transformer project underway to date. The use of Roebel cable will be explored as well as current limiting and recovery performance.

12.3.6 New Zealand/Australia

The Robinson Research Institute of Victoria University of Wellington (RRI, formerly Industrial Research Limited) and partners including Australian transformer manufacturer Wilson Transformers have built a three-phase 11 kV/415 V—1 MVA transformer

Figure 12.1 A 630 kVA HTS transformer developed by a Chinese team sponsored by the Chinese Academy of Sciences.
Photo courtesy of Prof. Yinshun Wang, North China Electric Power University.

operating at 70 K (Glasson, Staines, Jiang, & Allpress, 2013; Staines et al., 2012). It is planned to grid-connect the transformer in late 2014. The LV windings were wound with a 15 strand YBCO Roebel cable to achieve a rated current of 1390 Arms with equal current sharing between the strands. The magnitude of the LV current is the largest obtained in a HTS transformer.

12.3.7 Summary

The projects have demonstrated that HTS transformers with liquid nitrogen dielectric can perform the basic functions of distribution transformers. Ratings up to 4 MVA, voltage to 66 kV, current to 1.4 kA, and length of service of over a year have been achieved. Both BSCCO and YBCO conductors have been successfully employed. Correction for cooling penalty included, most have also matched or bettered the 1% load loss of conventional 1 MVA class transformers as detailed in Section 12.7.6. However the standards for efficiency in larger transformers (see Section 12.2.5) are significantly more demanding, and relate to total loss. Loss is a strong function of the amount of conductor used, so comparisons can be difficult. Given that the selling price for a 1 MVA distribution transformer is less than 50,000 USD, HTS is hardly going to be competitive at such low-power ratings, so several ongoing transformer projects are needed in the high-voltage, medium-power class to demonstrate the commercial viability of the technology. One concern at present, underlined by the exit of Waukesha Transformers from HTS developments, could be that ongoing projects for higher MVA demonstrations are thin on the ground.

12.4 High-temperature superconducting (HTS) transformers — general principles

HTS power transformer development has mostly followed conventional designs quite closely, employing similar iron yokes and winding solenoidal LV and HV coils around them. An example of a three-phase HTS transformer is shown in Figure 12.2. Each of the three phases has a branch of the iron yoke and its own set of coils. Apart from the fact that the whole structure is not enclosed in an oil-filled tank, and that the windings for each phase are contained in cryostats, it would be immediately recognizable to a manufacturer of conventional transformers. Table 12.5 lists some consequences that flow from the combination of high-current density HTS conductor and liquid nitrogen dielectric that distinguishes HTS from conventional transformers. These range from potentially decisive advantages in efficiency, size, and weight, to possible FCL capability, to severe technical challenges in handling fault currents, to more tractable design issues. A dominating issue in the cryogenic system is the impact on efficiency of the cryogenic cooling penalty, the ratio of cryocooler input power to cooling power. These challenges and opportunities are examined in more detail in the following sections.

Figure 12.2 A 1 MVA 11/0.4 kV HTS transformer developed by a New Zealand/Australian team shown during assembly.

12.5 AC loss in transformer windings

12.5.1 General principles

AC loss in the windings of an HTS transformer is a major factor to be considered in assessing transformer efficiency because of the additional energy required to maintain the HTS windings at cryogenic temperatures. The magnitude of the losses when the cooling penalty is included requires careful management of AC loss if HTS transformers are to be more efficient than traditional devices.

There are numerous factors to be considered and mitigated during the design of the conductors, coils, and transformers. For instance, hysteric losses associated with changing magnetic fields and currents in HTS wires can be reduced by using wires of smaller lateral dimensions normal to the local field, and hence the interest in applying striated conductor to AC windings (Iwakuma et al., 2014). The AC loss can also be reduced by lowering the operating current relative to I_c. The relevant measure is the current amplitude I_a as the loss goes as the third to fourth power of the I_a/I_c ratio. This can be achieved by reducing the operating temperature from 77 to 65 K, thereby increasing I_c for YBCO conductor by more than 100% while suffering a much smaller increase in cooling penalty ($\sim$18%). Thus many demonstrations in Table 12.4 were undertaken at temperatures lower than 77 K.

There are two main issues in HTS transformer windings, configured as they invariably are as coaxial solenoids around an iron core. First is the increased AC loss associated with the radial component of the magnetic field, normal to the plane of the

Table 12.5 Potential advantages and drawbacks of high-temperature superconducting transformers in comparison to oil-immersed copper transformers

	Potential advantages	Potential disadvantages
HTS conductor	High-current density: compact windings, reduced size and weight of conductor and core, low no-load loss, low impedance, improved voltage regulation Current limiting	AC loss Thermal runaway at overload limit—rapid failure of windings Low thermal mass—rapid heating by fault currents Mechanical properties of conductor—limited formability for winding, high stress from fault current forces
LN2 dielectric	Good heat transfer Good electrical insulator No fire or environmental contamination risk Temperature-related insulation degradation absent	Cryogenic operation adds cost, complexity Cooling penalty impacts on efficiency Warm iron required for efficiency dictates nonmetallic cryostat with core penetrations Current lead and cryostat losses add to no-load and load loss Boiling reduces breakdown strength Gas evolution in short circuit Differential thermal contraction Ice formation on current leads at low loads

conductor. Significant radial fields ($\gtrsim 10$ mT) occur near the end of the windings. In the central part of the winding the field is axial and predominantly confined between the LV and the HV windings. For this reason the loss in the end turns of windings is typically many times higher than the average over the solenoid. Happily, loss in transformer windings is less than the loss in the same solenoid windings carrying current without the balanced amp-turns of the other winding (Pardo, Staines, Jiang, Glasson, & Buckley, 2014). End turn loss can perhaps be reduced using flux diverters designed to steer the local flux parallel to the plane of the conductor (Al-Mosawi, Beduz, Yang, Webb, & Power, 2001) or local variations in winding geometry (Nagai & Ishiguri, 2013). The local radial field around axial gaps in solenoid windings causes increased loss as well. Such gaps may be unavoidable because spacing between turns is required for electrical insulation, liquid nitrogen circulation, or mechanical support.

One of the available YBCO tape manufacturing processes makes use of textured Ni–W alloy substrate. The hysteretic magnetic loss of the substrate (Ijaduola, Thompson, Goyal, Thieme, & Marken, 2004) may therefore discourage its use in transformers. Recent progress in developing high tungsten content Ni–W substrate has significantly reduced this loss (Amemiya et al., 2008; Rupich et al., 2013).

The second issue is high current. The high-current (LV) windings of HTS transformers require multiple wires to be assembled into cables since HTS wire is not scalable to high current in the way that copper is by simply changing the cross-section. The wire-in-tube process for BSCCO lends itself to 4 mm × 0.2 mm wire dimensions. YBCO can be scaled by increasing the conductor width to 12 mm or greater but at the cost of increased loss. In a number of the demonstrations the increased current-carrying capacity is obtained by assembling wires in a vertical stack. In a coil configuration where each turn is composed of a vertical stack as shown in Figure 12.3, there will be an unequal current sharing between the individual wires because of the difference in inductance. This unequal current distribution will result in larger AC losses.

There are two ways to prevent unequal current distribution between conductors in a high-current winding: CLT and transposed assembled conductors or Roebel cables (Goldacker et al., 2007; Long et al., 2010), used for over a century to solve analogous problems with high currents in copper conductors.

12.5.2 Coil-level transposition

The Kyushu University group has extensively explored the use of the CLT method. They have studied the effect of varying the number of wire transpositions, coil transpositions, and the number of parallel conductors, and have demonstrated current equalization in model coils (Bohno et al., 2005; Iwakuma et al., 2009). The downside is the added complexity of winding, especially since HTS conductor is not tolerant of the in-plane bending required for conductor transposition. In addition, any gaps in the winding to facilitate transposition could be a source of additional AC loss due to local radial fields where the flux confined between HV and LV windings bulges out through the gaps.

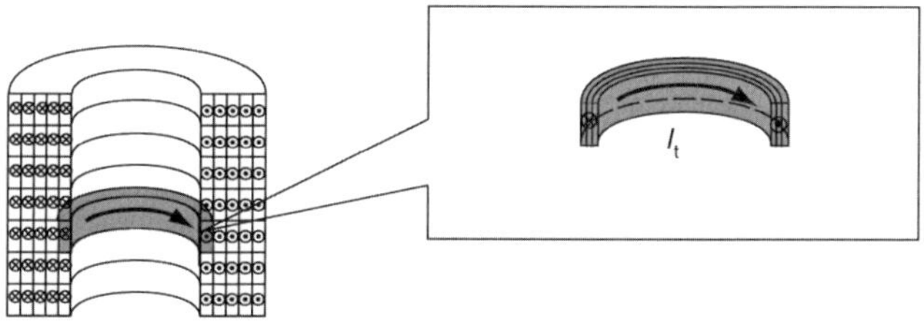

Figure 12.3 Diagram showing strands in coils wound with parallel conductors in a stack. In a coil winding where each turn is composed of a vertical stack of three wires, each wire encloses different magnetic flux, causing unequal current sharing.

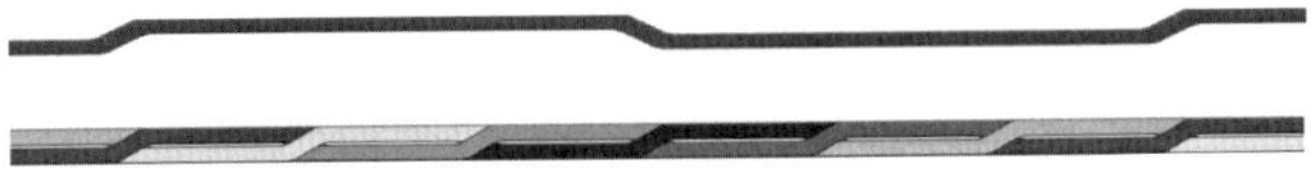

Figure 12.4 A single Roebel strand showing typical serpentine shape and six strands assembled into a Roebel cable.

12.5.3 Roebel cable

Continuously transposed Roebel cables were invented to bypass problems associated with simple vertical stacking of conductors (Long et al., 2010). In a Roebel cable, each strand is formed into a serpentine shape and these strands are continuously transposed as shown in Figure 12.4. This approach allows for each strand to be electromagnetically equivalent over a long length so the current in each strand can be equalized in a coil structure with at least one but often multiple cable pitch lengths per turn. Roebel cables have been assembled from BSCCO wires (Hussennether et al., 2004), but the limitations in the transverse bending strength of the BSCCO wires make it hard to form the serpentine shape required without loss of performance in a short enough pitch length to achieve a full transposition within a single turn of a winding. In the case of YBCO-coated conductors, the strands can be punched in the serpentine shape with a smaller period (300 mm is readily achieved with 5 mm wide strand) and assembled into a cable. YBCO Roebel cables have high-current capacity, good mechanical flexibility, and low AC loss, and can be manufactured on automated machines. The major drawback is yield, since about half the conductor is lost in cutting 5-mm-wide Roebel strand from 12 mm tape. The yield can be improved if multiple strands are cut from wider tape. Figure 12.5 shows Roebel cable winding of the 1.4 kA-rated LV winding of the New Zealand 1 MVA transformer.

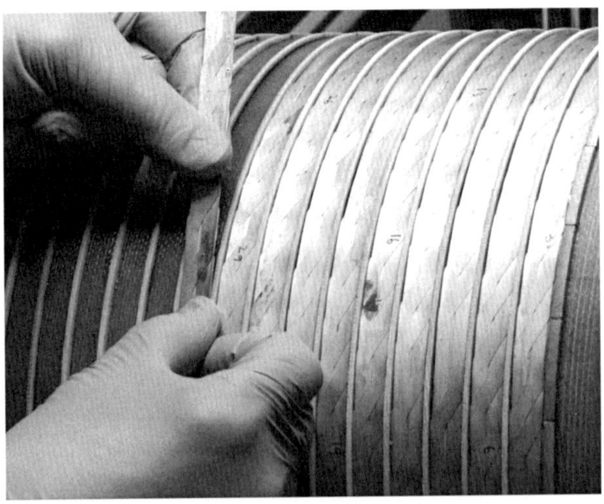

Figure 12.5 Roebel cable winding for the 1.4 kA-rated LV winding of the New Zealand/Australian 1 MVA transformer.

There are no comparative data available to compare AC losses in transformer windings wound with BSCCO and YBCO wires nor using the CLT or Roebel cable approaches. It might be expected that the better current-field performance of YBCO makes this the conductor of choice. Given the same I_c in self-field at 77 K, I_c at 65 K at 200 mT field normal to the plane of the tape, which is a better figure of merit for transformer conductor, is about 200% higher for YBCO compared to BSCCO.

Comparing windings wound using CLT with those wound using Roebel cable, if we assume that the conductor geometries are the same both, and that equal current distribution is achieved by both, then the magnetic field distribution surrounding the two windings will be the same and so the AC loss in the two windings will be identical. In practice, it is not possible to say on the basis of the AC loss results reported in Table 12.4 that Roebel cable gives superior loss performance – there are too many other variables involved, in particular the I_a/I_c ratio at rated current. CLT has not been demonstrated at current levels as high as Roebel cable; the LV current for the YBCO Roebel cable in the New Zealand transformer was 50% higher than for the highest non-Roebel winding.

12.6 Cryogenic systems for HTS transformers

12.6.1 Cryostats

A major constraint on the design of cryostats for HTS transformers is that the core must be external to the cryogenic environment. Core losses are constant and are too high to contemplate having this heat load adding to the other heat loads on the cryogenic system. Having the core outside the cryogenic environment means that there must be cryostat walls encircling each limb of the transformer core. These shells must be constructed of a nonconducting material so that they do not present as short circuits. Glass fibre reinforced composite materials are typically used.

The cryostat walls must offer excellent thermal insulation in order to minimize heat flow into the cryogenic region. One proven way to achieve good thermal insulation is by maintaining a vacuum insulation space between inner and outer shells. Practically, making a nonconducting cryostat with a vacuum insulation space means that continual vacuum pumping is necessary in order to maintain vacuum. It is very challenging to make a composite vessel that can maintain a vacuum without ongoing vacuum pumping. Standard aluminized Mylar multilayer insulation may be inappropriate in the core vacuum space because of eddy current heating from stray fields. Glass microspheres are a satisfactory alternative. There may be some cost benefit to be gained by using insulating foam instead of vacuum insulation in some areas.

An optimized cryostat for a three-phase transformer would have a single common cryogenic volume housing all three phases as depicted in Figure 12.6. This is much more challenging to design and construct compared to having a separate cryostat for each phase. A single cryostat for all three phases allows the overall transformer

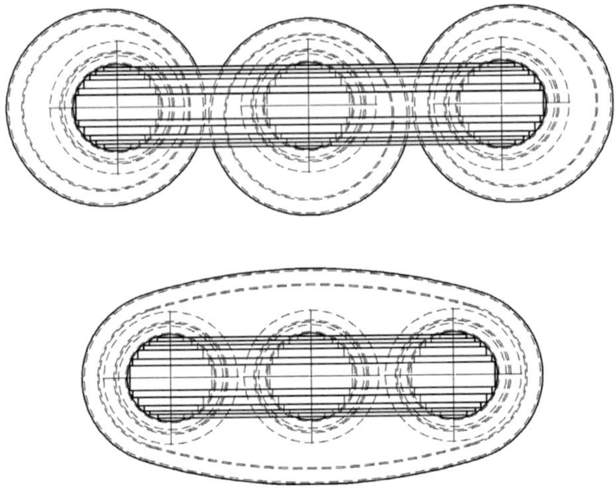

Figure 12.6 Plan view of two possible three-phase transformer designs. A single cryostat for all three phases can be more compact than having three individual cryostats.

to be more compact, minimizes vacuum and liquid nitrogen connections and pipework, and reduces the number of electrical bushings that need to penetrate the cryogenic envelope. Electrical connections between phases can be made in the cryogenic space.

12.6.2 Cooling

A practical cryogenic system for an HTS transformer must be low cost, quiet, reliable, and efficient as well as offering high availability. A good overview of the requirements for a cryogenic system for superconducting electric system applications is given by Sheahen and McConnell (2001). Cooling power required for a commercially viable HTS transformer will be on the order of several kilowatts at 70 K. For a practical grid-connected HTS transformer the cryogenic system footprint must be small and self-contained. Although conductive cooling has been attempted in the past (Schwenterly et al., 1999), recent development is focused on circulated liquid nitrogen cryogenic systems. Other cryogens (helium, hydrogen, neon, oxygen) are ruled out by cost, safety, and temperature range considerations. A circulated liquid nitrogen system provides consistent temperature distribution and scope for rapid recovery from temperature deviations that might arise from a fault as well as the benefits of a liquid dielectric. Variants involving pumped (Iwakuma et al., 2014) or thermosiphon circulation from an external subcooled vessel (Glasson, Staines, Allpress, & Badcock, 2013) as well as convective cooling from heat exchangers immersed in the transformer cryostat (Chang, Choi, Van Sciver, & Choi, 2003; Choi, 2004) have been demonstrated or proposed. At atmospheric pressure, nitrogen is liquid in a rather narrow temperature range of 63–77 K. Liquid nitrogen offers dielectric performance comparable to that provided by oil in a conventional transformer. In order to prevent

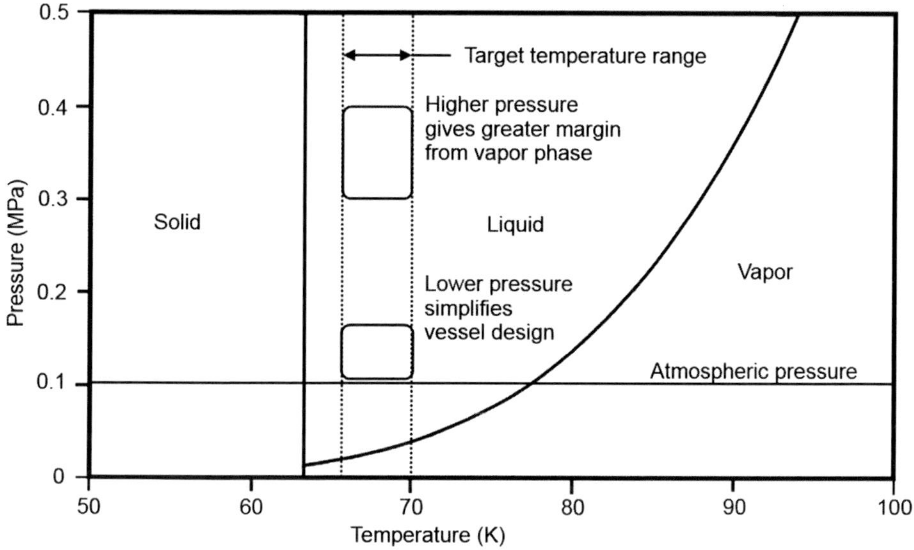

Figure 12.7 Nitrogen phase diagram depicting prospective operating regions for HTS transformers.

formation of gaseous nitrogen within a transformer and thereby preserve dielectric properties (Sauers et al., 2011), it is highly desirable to design the system so that the liquid nitrogen is always in a subcooled state in high-voltage regions. Figure 12.7 shows the pressure–temperature range favoured for transformer applications. Boiling at conductor surfaces is probably unavoidable in the event of a fault current; however, the insulation design must allow for any reduction in breakdown strength even in subcooled nitrogen. The selection of the design operating temperature has a very strong impact on the load capacity of the HTS windings – operating at 70 K gives about twice the current carrying capacity of operating at 77 K. The cryogenic system should operate above atmospheric pressure to guarantee that any leakage does not result in contamination of the cryogenic space. Elevated system pressure can also give improved margin away from the gaseous state as the boiling temperature is increased.

GM and Stirling cryocoolers are both mature and proven technologies that are capable of delivering the cooling necessary for an HTS transformer – although for larger rating transformers multiple units will be needed (Kim, 2007). Bi (2013) gives a good overview of the various cooling options for HTS power equipment ranging from 2 kW cooling for transformers up to the 22 kW cooling at 72 K for the LIPA II cable (Ravex, 2010). A key disadvantage to both GM and Stirling cryocoolers is that they both require periodic and extensive maintenance at regular intervals. Turbo Brayton systems have been developed that will deliver high efficiency and reduced maintenance requirements with cooling power up to 50.8 kW (Gondrand, 2014). Their application is more challenging in a cooling capacity small enough to suit a commercially viable HTS transformer installation (around 2–5 kW). In this cooling range, efficiency and capital cost are the main hurdles to commercial realization.

Pulse tube cryocoolers promise to deliver a low cost and low maintenance machine that may be most suited to HTS transformer application, but as with GM and Stirling machines multiple units will be needed to achieve higher cooling demands needed for larger transformers. Having multiple units can be desirable for maintaining redundancy and offering an efficient mechanism for cooling load regulation by cycling individual machines on and off.

Bulk liquid nitrogen supply is not a desirable cooling mechanism for long-term operation as it relies on large storage vessels and frequent deliveries of liquid nitrogen. Bulk nitrogen cooling does, however, provide a useful means of meeting a large cooling load with minimal capital outlay or complexity. It is therefore most suited to short-term demonstration projects or as a backup cooling mechanism to maintain temperature during a cryocooler breakdown or planned maintenance period.

It is important to note that when designing a cryogenic system, the cryocooler cost and performance are only contributing factors toward the overall system cost and performance. The balance of plant needed for heat exchange, cryogenic containment, vacuum insulation, safety, control, and circulation will mean that the overall system will be several times more expensive and significantly less efficient than the plant that produces the cooling.

Irrespective of the equipment used to extract heat from the cryogenic space, it is important to be able to maintain accurate temperature and pressure control in that space. The cooling load will fluctuate as the transformer electrical loading varies (AC loss is heavily dependent on electrical loading). It is therefore necessary to be able to vary the cooling capacity accordingly. There is a rather narrow window of allowable operating temperature: less than 64 K and there is a risk of freezing, more than 77 K and there is a risk of boiling (at atmospheric pressure). A practical target temperature range might be between 66 and 70 K. Bulk LN2 and Turbo Brayton cooling options both allow efficient and effective continuously variable cooling power over a wide range. GM, Stirling, and pulse tube cryocoolers are less efficient when operated away from their optimal frequency, so may need to be cycled off and on to achieve temperature regulation. Having multiple cryocoolers in a cooling system offers the advantage of allowing for discrete load regulation (by turning one or more units off as needed). A multicryocooler system can also provide for $N+1$ redundancy and for isolation and maintenance of individual machines, while the rest of the cooling plant remains operational. In contrast, a cooling system based on a single large Turbo Brayton cooling plant would not meet power industry standard $N+1$ redundancy requirements. A secondary cooling option of equal capacity would be needed to achieve this. This could be another Turbo Brayton plant or some other cryocooler option or bulk nitrogen supply.

Figure 12.8 shows the HTS transformer developed by the Kyushu University group (Iwakumi et al., 2014). A Brayton-cycle refrigerator with Ne gas as a working fluid was used. The refrigerator had a cooling capacity of 2.17 kW at 65 K, and its cooling penalty was 22.7.

Figure 12.8 Kyushu University 2 MVA transformer showing single cryostat containing all three-phase windings and Turbo Brayton cooler.
Photo courtesy of Prof. Masataka Iwakuma, Kyushu University.

12.7 Challenges for HTS transformers

To achieve market penetration HTS transformer will have to satisfy mandatory regulatory requirements listed in Table 12.1 based around achieving set performance levels, as well as competing with the established conventional technology on characteristics such as size, weight, efficiency, fire risk, and capital cost, which are consolidated in the TCO. Both regulations and economics provide some stern challenges for grid-connected HTS transformers.

12.7.1 Short circuit current limitation—fault impedance

The issue of fault current limitation calls into question the supposed HTS advantage of improved regulation stemming from the low impedance possible with compact high-current density windings. A first objection to low impedance HTS transformers is that they cannot be connected in parallel with higher impedance conventional transformers because the load would not be equally shared. Assuming that load sharing is achieved by connecting only low impedance HTS transformers in parallel, the problem of fault current limitation remains. Transformer impedance is the most practical and cost-effective way that a grid operator can limit fault currents, and the IEC standard recognizes this by requiring, for example, 10% fault impedance at 40 MVA rating.

HTS transformers offer the possibility of intrinsic current limiting, which might permit the reactive impedance to be reduced below the levels dictated by the standards at present, and the goal of an FCLT has been a focus of several development programs. Using commercial 2G wire with standard copper stabilizing layers will automatically provide significant resistive fault current limiting (Staines et al., 2012).

The difficulty is that the standards stipulate a fault duration time of 2 s, with the transformer remaining in operation after the fault is cleared. Fault current limiting with recovery under load has been demonstrated but the typical fault duration was only 0.1 s, and recovery under load has been demonstrated (Hayakawa et al., 2011) in MVA class windings only at a fraction of rated current, and at a rated current of 100 A in a single layer solenoid-wound 60 kVA secondary winding (Berger, Noe, & Kudymow, 2011). Although this hold time is adequate for a stand-alone fault current limiter with associated fast switch and shunt reactor it will not suffice for a grid-connected transformer.

This problem appears to have no obvious solution at present. The options are, roughly speaking, to treat the requirement as negotiable, to rate the conductor to handle the fault current, or to provide auxiliary switched copper windings.

Possibly grid operators will be prepared to compromise and relax the requirement for a 2 s fault hold time, with fault current limiting conductor used in the windings to extend the adiabatic heating time of the conductor by adding thermal mass (Schwenterly & Pleva, 2010). This would have costs for the grid operator, but they might be offset by the fault current limiting capability of the conductor.

A second approach would be to use HTS windings with a critical current 10 times, for example, the rated current so the conductor would remain in the superconducting state, assuming that the short circuit current is limited by 10% reactive impedance. Conductor cost is the major obstacle to this solution.

A third approach is to use copper auxiliary windings to carry the current after the initial few cycles until the HTS windings have cooled after the fault clearance. An ABB patent (Zueger, Bonmann, & Paul, 2002) describes such an arrangement in which an additional benefit is current limiting from the increased leakage reactance of the transformer when the room temperature auxiliary winding is connected. This solution is so far unproven, and would of course involve additional cost for the switches and auxiliary winding and dielectric. A related proposal in the patent literature (Folts, Maguire, Yuan, & Malozemoff, 2010) is to connect a conventional transformer in parallel with the HTS transformer to shoulder the fault current when the HTS transformer impedance rises as the critical current of the HTS conductor is exceeded.

Inrush current is a related issue, but a winding capable of meeting the standards for short circuit withstand should have no problem with the somewhat unpredictable transient current surges on energizing a transformer. Otherwise the problem can be handled with controlled energization of the transformer either by controlling the phase at which switches close (Zueger, 1998) or by energizing through a transiently connected impedance.

12.7.2 Overload

Overload can be accommodated by sizing the critical current of conductors to suit, at the expense of higher conductor cost. Current leads also need to be sized for the overload current, at the cost of increased heat leak into the cryostat (Schwenterly & Pleva, 2009). There is a significant difference between the consequences of overload events in oil-immersed and HTS transformers: in oil there is a cumulative and largely predictable decline in insulation properties and lifetime depending on winding temperature, in HTS there is a hard limit to overload, with no degradation of insulation until the point at which thermal runaway occurs, which leads in a matter of seconds to local destruction of the conductor. In single conductors immersed in liquid nitrogen at its boiling point the onset of thermal runaway is seen at a current amplitude only about 45% greater than the critical current (Martinez, Angurel, Pelegrin, Xie, & Selvamanickam, 2010); in a multilayer winding in subcooled liquid nitrogen with heat transfer reduced by close-packed wire and the suppression of boiling heat transfer a lower safe operating margin is to be expected. The protection of windings through early detection of incipient thermal runaway is an area where more research is needed.

12.7.3 Insulation

The insulating properties of liquid nitrogen are comparable to those of mineral oil (Gerhold, 1998). A number of studies have demonstrated reduced breakdown voltages in liquid nitrogen in the presence of bubbles. A study at ORNL (Sauers et al., 2011) shows, for example, in liquid nitrogen subcooled to 73 K at 1 bar pressure that the breakdown strength drops from about 25 kV rms/mm to about half that value above a critical flux of thermally generated bubbles. The observed increase in breakdown voltage as the pressure above liquid nitrogen at 77 K is raised from 0.1 to 0.3 MPa is attributed to the lower density of bubbles (Nishimachi, Hayakawa, Kojima, Hanai, & Okubo, 2012). Subcooling is likely to be ineffective at preventing bubbles following a fault current event however. This may require the insulation system design to eliminate the risk of breakdown even in the presence of bubbles. The reduction in breakdown voltage resulting from ice contamination of liquid nitrogen (Sumereder, Gerhod, Muhr, & Woschitz, 2003) is a potential risk that must be avoided. Commercially available composite bushings rated to 650 kV BIL have been demonstrated to perform well in liquid nitrogen and model windings successfully tested to 550 kV BIL (Schwenterly & Pleva, 2010).

12.7.4 Voltage regulation

Consideration has been given to incorporating OLTC in HTS transformers in the context of insulation design (Choi, Choi, & Baek, 2009), but there are no reports of testing of conventional tap changers in service in liquid nitrogen. Reduced transformer impedance will not remove the need for OLTC to compensate for the effect of impedance elsewhere in the network.

12.7.5 Hazards

The reduction of fire and environmental risks is a valuable advantage of HTS transformers in comparison to mineral oil-filled transformers. There are conventional solutions to these risks – dry-type transformers and less flammable fluids. HTS transformers do introduce new hazards; a large transformer might contain as much as 10 m^3 of liquid nitrogen. If the cryostat ruptured as the result of an internal fault this could pose a significant risk of asphyxiation and cryogenic burns to people in the vicinity. Mitigation is likely to take the form of confining discharged liquid and limiting the boil-off rate with suitable thermally insulating substrates to allow safe dissipation of the nitrogen gas.

12.7.6 Efficiency

AC loss has always been seen as a significant obstacle to commercialization of HTS transformers. Of the 10 examples listed in Table 12.4 of transformers of 500 kVA or higher rating that report AC loss results, the majority achieve AC loss of under 1% at rated current, converted to cryocooler input power to make a valid comparison with the conductor load loss in a conventional transformer. The cryocooler energy penalty was estimated using performance data for a large capacity GM cryocooler (Cryomech AL600). This does raise the question: is the correct cooling penalty 30, 20, or 10 depending on the cryocooler type (see Table 12.6). In practice the decision would be made on cost of ownership: do savings in energy cost and maintenance over the life of the system compensate for increased capital cost? Our aim here is to consider if HTS transformers can in principle deliver higher efficiency than competing conventional machines; the cost issues are considered later.

Can we compare the AC loss of different transformers of similar MVA rating but with different voltage and current ratings? The AC loss of the winding depends strongly on $i = I_a/I_c$, where I_a is the operating current amplitude. This is the main motivation for operating at temperatures below 77 K. A practical comparison of transformer efficiency can only be made if the quantity of wire is constrained to its optimum by cost considerations. Starting from the standard expression for the voltage induced around a transformer core it can be shown that for a winding of given IV rating, the choice of operating point i and core radius R_c determine the conductor requirements I_cL (in kA m) in each of the HV and LV windings according to

$$IV = I_cL \cdot iR_c\left(\frac{\pi f B_c}{2}\right)$$

Line frequency f is constrained at 50 or 60 Hz and the peak core flux density B_c is typically 1.7 T. The operating point i could also be regarded as constrained; i should be ~0.5 to provide the overload capability required for grid operation with the minimum of conductor.

The results reported in the table show that 1 MVA class HTS transformers already match the conductor load loss of conventional transformers and could have significantly lower losses in terms of cryocooler input power if the most efficient cryocoolers currently available were used. However this ignores the effect of cryostat and current lead heat losses, which are comparable to AC losses at 1 MVA rating (Glasson et al., 2013). As the HTS transformers are scaled up to larger MVA class the cryostat and current lead losses can be expected to be a smaller fraction of the total.

Table 12.6 Prospective cooling technologies for high-temperature superconducting transformers

Cooling technology	Available cooling capacity (at 77 K)	Cooling penalty (power input/ cooling power) at 77 K	Maintenance requirement
Gifford McMahon	600 W (www.cryomech.com)	>20 (12.5 kW rated input power for AL600)	10,000 h renew oil adsorber
Stirling	4 kW SPC-4 (www.stirlingcryogenics.com)	10, 40 kW rated input power for SPC-4	6000 h renew regenerator
Turbo Brayton	Up to 50 kW Air Liquide (Gondrand, 2014)	8–12 depending On capacity. Larger capacity can be made more efficient.	10 year maintenance interval MTBF 105,000 h
Pulse tube	200 W–1 kW Praxair	About 20 (Chen, 2010)	Theoretically very good, but only limited operational data is available
Bulk LN2 supply (subcooled by vacuum pumping)	Constrained by pumping rate	N/A	Regular LN2 delivery required

How will AC loss scale with MVA rating? Until larger MVA transformers are built, AC loss modelling is the best guide. A comparison of HTS coil modelling methods and results is beyond the scope of this chapter. However, one can have some confidence in modelling using the minimum magnetic energy variation method, which has been validated using measured losses for the New Zealand 1 MVA transformer (Pardo et al., 2014). It predicts that a 40 MVA 110/11 kV HTS transformer will better the 0.3% loss target of the EU standard (EU, 2014) when the conductor critical current exceeds about 260 A (4 mm wire at 65 K in nominal 400 mT perpendicular field, cooling energy penalty = 30). One major wire manufacturer is projecting that production wire exceeding this performance target will be available in 2015 (Iijima, 2013).

12.8 The HTS transformer value proposition – total cost of ownership (TCO)

The Waukesha/ORNL team (Pleva et al., 2010; Schwenterly et al., 2008) have carried out possibly the most complete comparison of the TCO of HTS and conventional transformers. They considered a 115 kV/13.09 kV, nominally 15 MVA, 3ϕ HTS transformer operating at 70 K with a single cryostat with cryocooler cold heads directly immersed in the liquid nitrogen. The assumed conductor cost was 50 USD/kA-m, which forced the core weight at minimum 30 year TCO to 16.4 tonnes, 80% more than the conventional transformer. Although the estimated load loss for the HTS windings was low, only 7% of the conventional, the larger core and higher value given to no-load losses meant that the capitalized energy costs of the HTS transformer were hardly reduced compared to the conventional transformer.

Energy costs were 20% of selling price for the conventional transformer, and fire protection was estimated to add almost 30% to TCO. On the positive side the HTS volume envelope was only half conventional and HTS transformer's total weight was about 60% of conventional. The HTS maximum efficiency was about 0.2% above conventional, but the high no-load loss meant that this had little impact on cost; there was hardly any real efficiency advantage despite very low load loss.

With 50 USD/kA-m HTS tape the estimated HTS selling price was 48% higher than conventional, USD 768 K compared with USD 551 K. TCO was just 28% higher than conventional, USD 874 K versus USD 681 K. They found that with 20 USD/kA-m HTS tape and a halving of cryocooler cost, the HTS selling price and cost of ownership become comparable with the figures for the conventional transformer.

Comparing the Waukesha/ORNL analysis to a study by Berger, Cherevatskiy, Noe, and Leibfried (2010), which focused only on energy savings without the constraint of wire cost, is instructive. They compared conventional and HTS designs for a 63 MVA 21/9 kV transformer where the HTS transformer had half the iron core losses of the conventional design so that the loss of the HTS transformer was a fraction of the conventional machine across the entire load range. This underlines the point: to derive the maximum cost advantage from reduced load losses, HTS conductors must also be combined with a reduction in no-load losses relative to conventional transformers. This means that the core weight as well as thermal no-load losses in the cryosystem must be reduced. At high ratings the cryosystem losses will tend to become relatively insignificant, but the requirement for low core loss will mean that high-efficiency HTS transformers will require more amp–metres of conductor than oil-immersed copper transformers.

It should be noted that copper prices are not insignificant compared to projected HTS prices. At copper conductor prices of 11 USD/kg, the price of copper is 25–50 USD/kA m for typical current densities from 4 to 2 A/mm^2. The lower current density limit corresponds to designs optimized for low load loss.

12.9 Conclusions

Barriers to commercialization of HTS transformers are of two kinds: technical/regulatory, mainly in achieving acceptable fault current performance, and economic, in achieving lower TCO than the incumbent technology based on copper conductors.

The major technical challenge appears to be achieving hold times for fault currents of 2 s and recovery under load when the fault is cleared. Network operators may be prepared to reduce the fault duration, but a transformer with an allowable fault duration of only 0.1 or 0.2 s is unlikely to find wide application in the grid, even if it offers valuable fault current limiting capability. Engineering fault current limiting conductors to prolong the fault hold time while still allowing recovery under load must be a major development priority.

The cost issue can only be addressed by reductions in the capital cost of HTS conductor and in the cryogenic system – both cryostats and coolers. Aside from capital costs, a competitive TCO can be helped by targeting:

- higher ratings where efficiency gains can be worth a larger fraction of capital cost, and applications with higher average load profiles;
- reducing core weight, because no-load loss is more expensive than load loss, and iron costs are not insignificant;
- indoor installations where elimination of fire protections costs can add maximum value;
- applications such as mobile transformers where packing more MVA in a confined space may be impossible with conventional technology.

It seems clear that large HTS transformers can better the efficiency of oil-immersed transformers, and possibly by a large margin, when coupled with high-efficiency cryocoolers. In the long term, provided acceptable fault current performance can be achieved and provided manufacturers of HTS conductor and cryocoolers achieve sufficient scale of production, the price of these essential components will come down and HTS transformers seem destined to become the most economical option for high-power transformers.

References

Al-Mosawi, M. K., Beduz, C., Yang, Y., Webb, M., & Power, A. (2001). The effect of flux diverters on AC losses of a 10 kVA high temperature superconducting demonstrator transformer. *IEEE Transactions on Applied Superconductivity, 11*, 2800–2803.

Amemiya, N., Jiang, Z., Nakahata, M., Kato, T., Ueyama, M., Kashima, N., et al. (2008). Transport AC losses in single and assembled coated conductors with textured-metal substrate with reduced magnetism. *Physica C Superconductivity, 468*, 1718–1722.

Berger, A., Cherevatskiy, S., Noe, M., & Leibfried, T. (2010). Comparison of the efficiency of superconducting and conventional transformers. *Journal of Physics: Conference Series, 234*, 032004.

Berger, A., Noe, M., & Kudymow, A. (2011). Test results of 60 kVA current limiting transformer with full recovery under load. *IEEE Transactions on Applied Superconductivity, 21*, 1384–1387.

Bi, Y. F. (2013). Cooling and cryocoolers for HTS power applications. *Applied Superconductivity and Electromagnetics, 4/1*, 97–108.

Bohno, T., Tomioka, A., Imaizumi, M., Sanuki, Y., Yamamoto, T., Yasukawa, Y., et al. (2005). Development of 66 kV/6.9 kV 2 MVA prototype HTS power transformer. *Physica C Superconductivity, 426–431*, 1402–1407.

Bureau of Reclamation. (2005). *FIST 3-32, transformer fire protection*. Available from http://www.usbr.gov/power/data/fist/fist3_32/fist3-32.pdf Accessed 26.08.14.

Chang, H.-M., Choi, Y. S., Van Sciver, S. W., & Choi, K. D. (2003). Cryogenic cooling system of HTS transformers by natural convection of sub-cooled liquid nitrogen. *Cryogenics, 43*, 589–596.

Chen, R.-L. (2010). Reliability test of a 1-kW pulse tube cryocooler for HTS cable application. *Advances in Cryogenic Engineering, 55*, 727–735.

Choi, Y. S. (2004). *Cryogenic cooling system by natural convection of sub-cooled liquid nitrogen for HTS transformers* (Electronic Theses, Treatises and Dissertations). Paper 3784. Available from http://diginole.lib.fsu.edu/cgi/viewcontent.cgi?article=2269&context=etd Accessed 26.08.14.

Choi, J.-H., Choi, J.-W., & Baek, S.-Y. (2009). The insulation design of 154 kV HTS transformer and on-load tap changers. *IEEE Transactions on Applied Superconductivity, 19*, 1972–1975.

CIGRE. (2013). *TB 537 guide for transformer fire safety practices*. Working Group A2.33.

DOE. (2014). *Large power transformers and the U.S. electric grid*. Infrastructure Security and Energy Restoration Office of Electricity Delivery and Energy Reliability U.S. Department of Energy. April 2014 Update, Table 2. (Base price given in report for 3 phase 115–13.8 kV generator step-up transformer with 75 MVA rating is US$1 000 000, with the installed price 25 to 30% higher.) Available from http://www.energy.gov/sites/prod/files/2014/04/f15/LPTStudyUpdate-040914.pdf Accessed 26.08.14.

Dohnal, D. (2004). In J. H. Harlow (Ed.), *Electric power transformer engineering*. CRC Press.

EU. (2014). *Commission Regulation (EU) No 548/2014 on implementing Directive 2009/125/EC of the European Parliament and of the Council with regard to small, medium and large power transformers*. Available from http://eur-lex.europa.eu/legal-content/EN/TXT/?uri=uriserv:OJ.L_.2014.152.01.0001.01.ENG Accessed 26.08.14.

Fassbinder, S. (2013). *Efficiency and loss evaluation of large power transformers*. European Copper Institute Publication. No. Cu0144, Available from http://leonardo-energy.stg.o-a.be/sites/leonardo-energy/files/documents-and-links/Cu0144_Efficiency%20and%20Loss%20Evaluation%20of%20Large%20Power%20Transformers_v1.pdf Accessed 26.08.14.

FM3990. (1997). *FM approvals, approval standard for less or nonflammable liquid-insulated transformers*.

Folts, D. C., Maguire, J., Yuan, J. & Malozemoff, A. P. (2010) Parallel HTS transformer device, US patent 7,724,482 B2. 25.05.10.

Funaki, K., & Iwakuma, M. (2000). Recent activities for applications to HTS transformers in Japan. *Superconductor Science and Technology, 13*, 60–67.

Funaki, K., Iwakuma, M., Kajikawa, K., Hara, M., Suehiro, J., Ito, T., et al. (2001). Development of a 22 kV/6.9 kV single-phase model for a 3 MVA HTS power transformer. *IEEE Transactions on Applied Superconductivity, 11*, 1578–1581.

Funaki, K., Iwakuma, M., Kajikawa, K., Takeo, M., Suehiro, J., Hara, M., et al. (1998). Development of a 500 kVA-class oxide-superconducting power transformer operated at liquid-nitrogen temperature. *Cryogenics, 38*, 211–220.

Galloway, D. L., & Mulkey, D. (2004). In J. H. Harlow (Ed.), *Electric power transformer engineering* (p. 82). CRC Press.

Glasson, N., Staines, M., Allpress, N., & Badcock, R. (2013). Commissioning testing of a 1 MVA superconducting transformer featuring 2G HTS Roebel cable. In *EUCAS 2013 oral presentation 2M-LS-O2*. Available from http://www.victoria.ac.nz/robinson/research/publications/Glasson_EUCAS2013.pdf Accessed 28.08.14.

Glasson, N. D., Staines, M. P., Jiang, Z., & Allpress, N. S. (2013). Verification testing for a 1 MVA 3-phase demonstration transformer using 2G-HTS Roebel cable. *IEEE Transactions on Applied Superconductivity, 11*, 5500206.

Goldacker, W., Frank, A., Heller, R., Schlachter, S. I., Bringsdorf, B., Weiss, K. P., et al. (2007). Roebel assembled coated conductors (RACC): preparation, properties and progress. *IEEE Transactions on Applied Superconductivity, 17*, 3398–3401.

Gondrand, C. (2014). Air Liquide cryogenic solutions for HTS refrigeration—focus on Turbo-Brayton. In *Int Cryo Eng Conf Int Cryo Matl Conf.* Available from http://indico.cern.ch/event/244641/session/23/contribution/59/material/slides/0.pdf Accessed 26.08.14.

Hassenzahl, W. V., Hazelton, D. W., Johnson, B. K., Komarek, P., Noe, M., & Reis, C. T. (2004). Electric power applications of superconductivity. *Proceedings of the IEEE, 92*, 1655–1674.

Hayakawa, N., Kojima, H., Hanai, M., & Okubo, H. (2011). Progress in development of superconducting fault current limiting transformer (SFCLT). *IEEE Transactions on Applied Superconductivity, 21*, 1397–1400.

Hussennether, V., Oomen, M., Leghissa, M., & Neumüller, H. W. (2004). DC and AC properties Bi-2223 cabled conductors designed for high-current applications. *Physica C Superconductivity, 401*, 135–139.

Iijima, Y. (2013). High-performance Y-based superconducting wire and their applications. *Fujikura Technical Review, 2013*, 117–121. Available from http://www.fujikura.co.jp/eng/rd/gihou/backnumber/pages/__icsFiles/afieldfile/2013/05/23/42e_32.pdf Accessed 28.08.14.

Ijaduola, A. O., Thompson, J. R., Goyal, A., Thieme, C. L. H., & Marken, K. (2004). Magnetism and ferromagnetic loss in Ni–W textured substrates for coated conductors. *Physica C Superconductivity, 403*, 163–171.

Iwakuma, M., Funaki, K., Kajikawa, K., Kanetaka, H., Hayashi, H., Tsutsumi, K., et al. (2001). Development of a 1 T cryocooler-cooled pulse coil with a Bi2223 superconducting parallel conductor for SEMS. *IEEE Transactions on Applied Superconductivity, 9*, 928–931.

Iwakuma, M., Funaki, K., Kajikawa, K., Tanaka, H., Bohno, T., Tomioka, A., et al. (2001). AC loss properties of a 1 MVA single-phase HTS power transformer. *IEEE Transactions on Applied Superconductivity, 11*, 1482–1485.

Iwakuma, M., Hayashi, H., Okamoto, H., Tomioka, A., Konno, M., Saito, T., et al. (2009). Development of REBCO superconducting power transformers in Japan. *Physica C Superconductivity, 469*, 1726–1732.

Iwakuma, M., Sasaki, K., Tomioka, A., Miyayama, T., Konno, M., Hayashi, H., et al. (2014). *Development of a 3φ–66/6.9 kV – 2 MVA REPCO superconducting transformer*, 2LPo2C-04. Charlotte, NC, USA: ASC 2014.

Iwakuma, M., Tanaka, H., & Funaki, K. (2002). Theoretical analysis on ac loss properties of two-strand parallel conductors composed of superconducting multifilamentary strands. *Superconductor Science and Technology, 15*, 1525–1536.

Kalsi, S. S. (2011). *Applications of high temperature superconductors to electric power equipment*. Hoboken: Wiley-IEEE Press.

Kamijo, H., Hata, H., Fujimoto, H., Inoue, A., Nagashima, K., Ikeda, K., et al. (2006). Fabrication of superconducting traction transformer for railway rolling stock. *Journal of Physics Conference Series, 43*, 841–844.

Kamijo, H., Hata, H., Fukumoto, Y., Tomioka, A., Bohno, T., Yamada, H., et al. (2010). Development of low AC loss windings for superconducting traction transformer. *Journal of Physics Conference Series, 234*, 032027.

Kim, S. R. (2007). Design of the cryogenic system for 100 MVA HTS transformer. *IEEE Transactions on Applied Superconductivity, 17–2*, 1935–1938.

Kim, S. K., Han, J. H., Min, W. G., Chang, T., Choi, K. D., Joo, H. G., et al. (2005). Characteristic test of a 1 MVA single phase HTS transformer with concentrically arranged windings. *IEEE Transactions on Applied Superconductivity, 15*, 2214–2217.

Leghissa, M., Gromoll, B., Rieger, J., Oomen, M., Neumuller, H. W., Schlosser, R., et al. (2002). Development and application of superconducting transformers. *Physica C Superconductivity, 372-376*, 1688–1693.

Long, N. L., Badcock, R. A., Hamilton, K., Wright, A., Jiang, Z., & Lakshmi, L. S. (2010). Development of YBCO Roebel cables for high current transport and low AC loss applications. *Journal of Physics Conference Series, 234*, 022021.

Martinez, E., Angurel, L. A., Pelegrin, J., Xie, Y. Y., & Selvamanickam, V. (2010). *Superconductor Science and Technology, 23*, 025011–025018.

Nagai, M., & Ishiguri, S. (2013). Performance improvements and optimization of new high-temperature superconducting coil with circle vacancies at edge portions. *Physica C Superconductivity, 495*, 213–216.

Nishimachi, S., Hayakawa, N., Kojima, H., Hanai, M., & Okubo, H. (2012). Pressure and temperature dependence of breakdown characteristics of sub-cooled liquid nitrogen. In *IEEE electrical insulation and dielectric phenomena (CEIDP), 2012 Annual Conference report* (pp. 440–443).

Pardo, E., Staines, M., Jiang, Z., Glasson, N., & Buckley, B. (2014). *Modelling and measurement of AC loss in a superconducting transformer*. ASC 2014. Poster 2LPo2C-01: 128. Available from http://www.victoria.ac.nz/robinson/research/publications/Pardo_ASC2014.pdf.

Pleva, E. F., Mehrotra, V., & Schwenterly, S. W. (2010). Conductor requirements for high-temperature superconducting utility power transformers. *Superconductor Science and Technology, 23*, 014025.

Ravex, A. (2010). *LIPA 2 refrigeration system design*. Available from http://www.htspeerreview.com/pdfs/presentations/day%203/applications/9_AP_LIPA_2_Cable_Project.pdf Accessed 26.08.14.

Rupich, M. W., Li, X., Sathyamurthy, S., Thieme, C. L. H., DeMoranville, K., Gannon, J., et al. (2013). Second generation wire development at AMSC. *IEEE Transactions on Applied Superconductivity, 23*, 6601205.

Sauers, I., James, R., Ellis, A., Tuncer, E., Polizos, G., & Pace, M. (2011). Effect of bubbles on liquid nitrogen breakdown in plane-plane electrode geometry from 100–250 kPa. *IEEE Transactions on Applied Superconductivity, 21*, 1892–1895.

Schlosser, R., Schmidt, H., Leghissa, M., & Meinert, M. (2003). Development of high-temperature superconducting transformers for railway applications. *IEEE Transactions on Applied Superconductivity, 13*, 2325–2330.

Schwenterly, S. W., McConnell, B. W., Demko, J. A., Fadnek, A., Hsu, J., List, F. A., et al. (1999). Performance of a 1-MVA HTS demonstration transformer. *IEEE Transactions on Applied Superconductivity, 9*, 680–684.

Schwenterly, W., & Pleva, E. (2009). *HTS transformer development*. August 5, 2009 Presentation for DOE Peer Review. Available from http://www.htspeerreview.com/2009/

pdfs/presentations/day%202/applications/7AP-HTS-Transformer-RD-Waukesha-ORNL.pdf Accessed 26.08.14.

Schwenterly, W., & Pleva, E. (2010). *HTS transformer development.* June 30, 2010 Presentation for DOE Peer Review Presentation for DOE Peer Review. Available from http://www.htspeerreview.com/pdfs/presentations/day%202/applications/6_AP_HTS_Transformer_Technology.pdf Accessed 26.08.14.

Schwenterly, W., Pleva, E., & Wolsky, A. (2008). Cost and performance comparisons between HTS and conventional utility power transformers. In *presentation to eighth EPRI superconductivity conference.*

Sheahen, T. P., & McConnell, B. (2001). *Cryogenic roadmap.* U.S. Department of Energy Superconductivity Program for Electric Systems. Available from http://web.ornl.gov/sci/htsc/documents/pdf/Cryogenic_Roadmap.pdf Accessed 26.08.14.

Staines, M., Glasson, N., Pannu, M., Thakur, K. P., Badcock, R., Allpress, N., et al. (2012). The development of a Roebel cable based 1 MVA HTS transformer. *Superconductor Science and Technology, 25*, 014002–014008.

Sumereder, C., Gerhod, J., Muhr, M., & Woschitz, R. (2003). Dielectric measurements on HTS insulation systems for electric power equipment. *Physica C Superconductivity, 386*, 411–414.

Wang, Y., Zhao, X., Han, J., Li, H., Guan, Y., Bao, Q., et al. (2007). Development of a 630 kVA three-phase HTS transformer with amorphous alloy cores. *IEEE Transactions on Applied Superconductivity, 17*, 2051–2054.

Weber, C. S., Reis, C. T., Hazelton, D. W., Schwenterly, S. W., Cole, M. J., Demko, J. A., et al. (2005). Design and operational testing of a 5/10-MVA HTS utility power transformer. *IEEE Transactions on Applied Superconductivity, 15*, 2210–2213.

Wolsky, A. M. (2013). *HTS from precommercial to commercial: a roadmap to future use of higher temperature superconductors by the power sector.* Available from http://www.superconductivityiea.org/document/HTS%20Roadmap%20Complete%20Report%20A Wolsky%2010Sept2013.pdf Accessed 26.08.14.

Xiao, L. (2013). Recent progress of the demonstration of superconducting power technology in China. In *IEEE/CSC & ESAS European superconductivity news forum (ESNF), 24, April 2013.* Available from http://snf.ieeecsc.org/sites/ieeecsc.org/files/RN29.pdf Accessed 26.08.14.

Zueger, H. (1998). 630 kVA high temperature superconducting transformer. *Cryogenics, 38*, 1169–1172.

Zueger, H., Bonmann, D. & Paul, W. (2002) Arrangement for current limiting using a superconductor transformer, US patent 6,411,479.

Implementing high-temperature superconductors for the power grid in practice: the case of China

13

L.Y. Xiao[1], L.Z. Lin[1], X.H. Zhang[2], Y. Xin[3], Y.J. Tang[4]
[1]Institute of Electrical Engineering (IEE), Chinese Academy of Sciences (CAS), Beijing, China; [2]Department of Electrical Engineering, Tsinghua University, Beijing, China; [3]Innopower Superconducting Power Cable Company, Yizhuang, Beijing, China; [4]Huazhong University of Science and Technology, Wuhan, China

13.1 Introduction

In China, most of the power sources are located at the west and northwest of the country while most of the load centers are located in the south and east. This requires a large-scale national power grid in which all the regional grids are interconnected. Power consumption requirements have increased rapidly in recent years. China's power generation capacity has increased from about 350 GW in 2003 to 1200 GW in 2012. China now has the largest power grid in the world. It has been estimated that the country's power capacity will need to increase by about 7% annually through 2025 (Zhou, 2012).

It is becoming increasingly difficult to maintain the stability and reliability of the power grid for several reasons. First, the fault current of the power grid increases rapidly as the grid's power capacity increases. This in turn causes serious instability in the power grid because of the transient instability resulting from the fault current. Inevitably this will also damage some of the electrical power equipment, leading to the potential collapse of the grid. Second, in order to reduce the magnitude of the fault current to a safe level, the architecture of the regional power grid for load centers must be changed, which results in a lower reliability for power distribution. Finally, the disconnection faults and the switching of the large loads and generators would cause a transient power imbalance, leading to further instability in the power grid.

When designing any electrical power equipment for China's grid, it is of great importance to keep land use to a minimum because China has a high population density and rigorous restrictions on land use. This means that the potential impact of large-scale power transmission over long distances on China's population centers is becoming an increasingly important issue. It is therefore important to investigate the possible applications of superconducting technology to the power grid in China. Superconducting power technologies, such as superconducting fault current limiters (SFCLs) and superconducting magnetic energy storage (SMES) can be used to improve the stability and reliability of China's power grid. Superconducting power

Superconductors in the Power Grid. http://dx.doi.org/10.1016/B978-1-78242-029-3.00013-3

cables and superconducting power transformers can also be used to reduce the space required by the equipment and enhance the capacity for power transmission over a long distance.

13.2 Research and development of superconductors in power in China

The Chinese government has supported R&D into superconducting power technology since 1996 (Xiao, Lin, & Dai, 2011). In the case of superconducting power cables, research started under China's ninth 5-year plan from 1996 until 2000. During 2001–2005, under the support from China's "863 Plan-On Superconducting Technology," the SFCL, SMES, power cables, and power transformers had been developed and operated in a live distribution system. Under the support of the same plan, during 2006–2010, development and testing operations were then carried out for a 10 kV superconducting power substation, and experiments on the use of SFCLs in the transmission grid and into large-capacity DC transmission by superconductors. In the future, the R&D for superconducting power technology will try to make use of the developments in renewable energy and further experiments will be conducted into the use of DC power cables, SFCLs, and SMES in the transmission grid. The next section presents an overview of China's R&D for superconducting power technology, followed by some more detailed case studies. The role of renewable energy in China's future power grid is also discussed.

Since 1996, the R&D on the application of superconductors in China's power grid has been supported by three separately funded government agencies:

1. China's Ministry of Science and Technology (MOST), through a project named, "863 Plan-On Superconducting Technology"
2. the Chinese Academy of Sciences (CAS)
3. the Natural Science Foundation of China (NSFC).

In addition, China's power companies, such as State Power Grid Company and South Power Grid Company, also supported superconducting power programs. Table 13.1 summarizes all of the projects related to China's power grid since 1996.

Table 13.1 shows the possible applications of superconductors in the power grid that have been successfully demonstrated in China. The following sections review the following developments:

- The 10 kV superconducting power substation, which has been in operation in Baiyin City in Gansu Province since 2011.
- The saturated-type superconducting fault limiters that have been used in the 35 kV Puji substation in Yunnan Province since 2009 and the 220 kV Shigezhuang substation in Tianjin City since 2012.
- The 360 m/10 kA superconducting cable, which has been in use in Zhongfu in Henan Province since 2012.
- The SMEs that have been in use in Hubei Province since 2008.

Table 13.1 **Progress of superconductors for power grid in China since 1996[a]**

Application	Project and the main parameters	Implementation of the project	Financial support by	Sponsor company and collaborators
Superconducting power cable	1 m/1 kA, AC	1996–1998	MOST	IEE-CAS
	6 m/2 kA,WD[b]	1998–2000	MOST	IEE-CAS
	Three-phase, WD, 75 m/ 1.5 kA/10.5 kV	2001–2005 Demonstration in Baiyin City	MOST	IEE-CAS Changtong Cable
	Three-phase, WD 30 m/ 2 kA/35 kV	2001–2005 Demonstration in Puji substation	MOST	Innopower Yunnan Power
	DC, WD, 360 m/10 kA/ 1.5 kV	2007–2012 Demonstration in Zhongfu Group	MOST CAS	IEE-CAS Zhongfu Group
Superconducting fault current limiter (SFCL)	Three-phase, 10 kV/1.5 kA	2001–2005 Demonstration in Gaoxi substation	MOST CAS	IEE-CAS Hunan power
	Three-phase, 35 kV/1.5 kA	2005–2009 Demonstration in Puji substation	MOST SGC	Innopower Yunnan power
	Three-phase, 220 kV/0.8 kA	2008–2012 Installed at Shigezhuang substation	MOST SGC	Innopower Tianjing Power
	Three-phase, 220 kV/2 kA (by YBCO tapes)	2009–2014 (To be demonstrated in Shenzhen City)	MOST Zhongtian Group	IEE-CAS Zhongtian Group

Continued

Table 13.1 Continued

Application	Project and the main parameters	Implementation of the project	Financial support by	Sponsor company and collaborators
Superconducting magnetic energy storage (SMES)	400 V/100 kJ/25 kVA (by NbTi Wires)	2001–2005 lab Test	MOST NSF	IEE-CAS
	400 V/500 kJ/150 kVA (by NbTi wires)	2001–2005 lab test	MOST	Tsinghua University
	400 V/35 kJ/7 kVA	2001–2008 used at Dynamic Simulation Lab	MOST	Huazhong University of Sci. & Tech.
	10 kV/1 MJ/500 kVA	2003–2008 Tested at Beijing Mentougou substation	CAS	IEE-CAS Beijing Power
Superconducting power transformer	Three-phase, 10 kV/400 V 630 kVA	2001–2005 Demonstration at TBEA factories	MOST	IEE-CAS TBEA Group
	Single-phase, 20 kV/10 kV/ 300 kVA	2001–2005 lab test	MOST	Zhuzhou Power
Superconducting power substation	SMES: 10 kV/1 MJ/ 500 kVA SFCL: three-phase,10 kV/1.5 kA Power cable: three-phase, 75 m/10 kV/ 1.5 kA Power transformer: three-phase 10 kV/ 400 V/630 kVA	2009–2011 Demonstration at Industrial Park in Baiyin City	CAS MOST Baiyin City Government	IEE-CAS Changtong Cable

[a]The demonstrations in the table are based on BSSCO (Bi-2223) tapes if there is no explanation.
[b]WD = Warm dielectric.

13.3 The 10 kV superconducting power substation in Baiyin City, Gansu Province

The superconducting power substation was built by integrating the following superconducting power equipment:

- a three-phase high-temperature superconducting (HTS) power cable, which is 75 m in length and carries 10 kV/1.5 kA;
- a three-phase 10 kV/1.5 kA SFCL;
- a three-phase 10 kV/0.4 kV HTS transformer with a capacity of 630 kVA; and
- a 1 MJ/500 kVA SMES.

See Figure 13.1. Each of the individual superconducting components of the system were demonstrated at substations or distribution systems from 2004 to

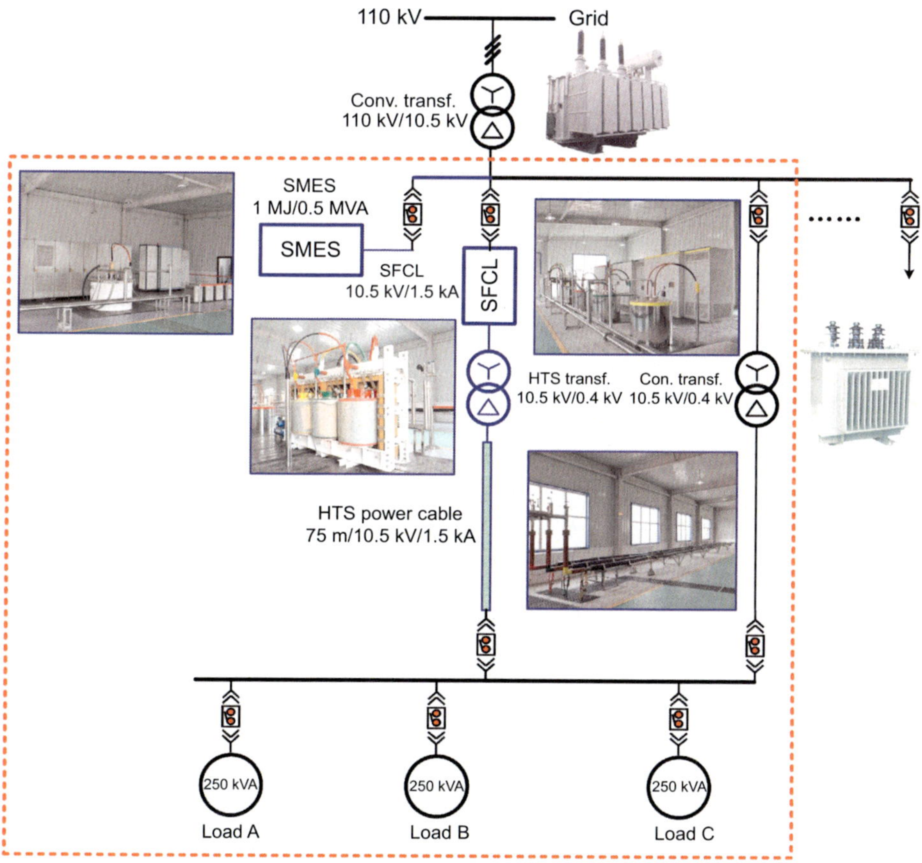

Figure 13.1 The layout of the superconducting power substation.

2008, before being combined at Baiyin Industrial Park of Gansu Province. The total budget for the substation was 120M Chinese Yuan (RMB). The following sections detail each component and the testing necessary prior to incorporation into the substation.

13.3.1 The three-phase, 75 m/10 kV/1.5 kA HTS power cable (Xiao et al., 2007)

The HTS power cable used in the substation consists of a three-phase warm-dielectric type, of which each cable is used for each phase. The conductor for the cable was made of Bi-2223 tapes manufactured by AMSC, the cross-section of the cable is shown in Figure 13.2a, and an overview of the cable after installation is shown in Figure 13.2b. In Table 13.2, some design parameters of the cable are shown. The refrigeration system of the power cable is designed into a cycle system in which the subcooled liquid nitrogen of about 70 K is used to cool the cable system. LN_2 for

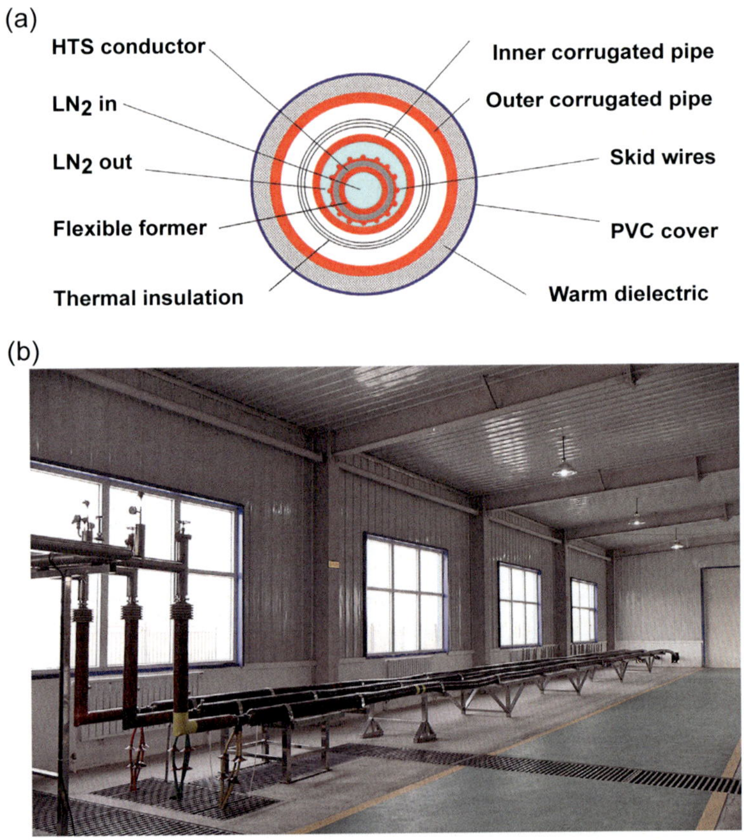

Figure 13.2 (a) The cross-section of the cable conductor. (b) A view of the three-phase, 75 m, 10.5 kV/1.5 kA high-temperature superconducting (HTS) power cable after installation.

Table 13.2 Some parameters of the three-phase, 75 m/10 kV superconducting power cable

Parameter	Tested value
Rated voltage (kV)	10.5
Rated current (kA)	1.5
Length of cable (m)	75
Operation temperature (K)	70–77
AC losses of the high-temperature superconducting (HTS) cable (W/kA m)	≤1.0
Heat losses of the cryogenic envelop (W/m)	≤2.0
Heat losses of the joint of cryogenic envelop (W)	5
Heat losses of the current lead (W/kA)	45
Joint resistance of the conductor to current lead (μΩ)	0.7
Minimum bending radius (m)	1.2
Outer diameter of the HTS cable (mm)	117

the cable is supplied by a liquid nitrogen pump. In order to keep a temperature difference within 70–77 K, the system's pressure is in the range of 1–2 bar, and then the liquid nitrogen is cooled down by a Stirling refrigerator. The cable was designed and manufactured from 2001 to 2004, and has been used in the distribution network of Gansu Changtong Cable Company in the five factories of the company's headquarters since December 2004. After operating for more than 7000 h, no damage was found in the cable. The termination and refrigeration system was then improved in order to reduce the heat losses before it was installed at the substation.

13.3.2 The three-phase, 10 kV/1.5 kA HTS fault current limiter (Xiao & Lin, 2007)

The SFCL used in the substation uses an improved bridge-type design. It employs HTS coils and a serial resistor that is shunted by IGCT (Integrated Gate Commutated Thyristors) switches as a bypass. A short-circuit fault occurs, which allows the first peak current to be limited by the HTS coil, and the IGCT switches are then switched off. The coil and resistor therefore both limit the fault current, effectively reducing the steady-state current. The coils consist of double pancakes made of Bi-2223 tapes supplied by AMSC, the four double pancakes (with inner diameter of 492 mm) at each end of the coil were wound with two parallel Bi-2223 tapes, and the other six double pancakes (with inner diameter of 502 mm) in the coil were wound with single tapes, leading to the homogeneous current distribution among the tapes.

Table 13.3 Some parameters of the three-phase 10 kV superconducting fault current limiter

Parameter	Tested value
Rated voltage (kV)	10.5
Lightning surge voltage (kV)	75
Rated fault current at 2 s (kA)	1.5
Rated current at normal state (kA)	0.25
Type of the high-temperature superconducting (HTS) coils	Double pancake
Number of double pancake	14
Inner diameter (mm)	492/502
Outer diameter (mm)	580
Height (mm)	361.6
Length of tapes in total (m)	2857
Inductance of the HTS coil (mH)	6.24
Critical current of HTS coil (A at 77 K)	600
The shunt resistor (Ω)	4
Operation temperature (K)	77
Heat losses of the dewar (W)	12.5

The inductance of each coil is about 6.24 mH; the critical current of the coils is about 600 A at 77 K. Some design parameters of the SFCL are listed in Table 13.3. An overview of the SFCL after installation is shown in Figure 13.3. The SFCL was developed from 2001 to 2005, and installed at Gaoxi substation at the Loudi City of Hunan Province. A three-phase-to-ground short-circuit test at the live grid shows

Figure 13.3 A view of the three-phase, 10 kV/1.5 kA high-temperature superconducting fault current limiter.

that the potential fault current of 3.5 kA was reduced to 635 A on the first peak. After that, the fault current limiter (FCL) operated in the live distribution grid for more than 11,000 h without breakdown. During operation, single-phase-to-ground fault occurred on three occasions, but the fault current was so low that the SFCL did not need to take any action. A new cryostat was designed for the SFCL in order to reduce heat losses.

13.3.3 The three-phase, 630 kVA/10 kV/400 V HTS power transformer (Wang et al., 2007)

The superconducting transformer used in the substation comprises HTS windings and amorphous alloys, which in comparison with conventional transformers can reduce both the copper winding losses and the iron-core losses. The primary winding is wound with eight layers with cooling channels in longitudinal direction, and each cross-section of the cooling channel is $10 \times 0.2\ mm^2$. The secondary winding is made of 23 double pancakes connected in parallel. Solenoid and double pancake coils are concentric cylindrical. The double pancake coil is located coaxially outside the solenoid one. Specifications of windings are presented in Table 13.4. Table 13.5 shows the detail parameters of the 630 kVA HTS power transformer, which were tested according to the requirements of the national standard of electric power transformer. The HTS power transformer is shown in Figure 13.4. After routine tests to comply with state standards, it was connected to the live distribution grid of the headquarters of the TBEA Group at Changji City, Xinjiang Uygur Autonomous Region.

Table 13.4 Some parameters of high-temperature superconducting windings for the three-phase 10 kV/400 V/630 kVA superconducting power transformer

Primary winding	
Winding type	Solenoid
Number of layers	8
Number of turns	262
Inner/outer diameter (mm)	488/504
Height (mm)	342.5
Secondary winding	
Winding type	Double pancake
Number of pancakes	23
Turns of each pancake	10
Inner/outer diameter (mm)	581/608
Height (mm)	355

Table 13.5 Some parameters of the three-phase 10 kV/400 V/630 kVA superconducting power transformer

Parameter	Tested value
Rated voltage (kV)	10.5
Lightning surge voltage (kV)	75
Rated fault current at 2 s (kA)	1.5
Rated current at normal state (kA)	0.25
Iron core diameter (mm)	396
Height (mm)	870
Width (mm)	780
Magnetic flux density (T)	1.275
Inner/outer diameter of cryostat (mm)	410/760
Height of cryostat (mm)	680
Operation temperature (K)	77
Heat losses of the dewar (W)	12.5
Efficiency	98.5%
Impedance	2.5%

Figure 13.4 A view of three-phase, 630 kVA/10 kV/400 V high-temperature superconducting power transformer.

During November 2005 and January 2006, it was used in the new energy factory of TBEA for more than 2 months without fault. Before it was moved in order to be integrated into the power substation, the cryogenic system was checked again and no breakdown was found.

13.3.4 The 1 MJ/500 kVA SMES (Dai et al., 2012; Xiao et al., 2008)

The SMES (as shown in Figure 13.5) of this substation was fabricated with BSSCO tape and operated at 4.2 K. The coil for the SMES consisted of 44 double pancakes. Some pancakes at either end were wound with single tape, while the remaining 38 double pancakes were wound with two sets of tape in parallel. Due to the anisotropic effect, three coils at each end were operated in parallel in order to obtain the same critical current as each of the other pancakes. The inductance of the coil is 6.28 H, the critical current of the coil is 564 A, the withstand voltage of the coil is 2 kV, and the charge or discharge time is below 2 s. The details of the design for the coil are shown in Table 13.6. The coil is cooled by liquid helium and a zero-boiling-off system in which 4 GM cryo-coolers with cooling power of 1.5 W at 4.2 K for each was used. The power conversion system of the SMES is voltage source type in which carrier phase shifting technology is used, and the equivalent switching frequency can reach up to 40 kHz, resulting in a lower lever of the harmonic waves of its output current. The SMES was tested at Beijing Mentougou substation before it was moved to the superconducting substation at Baiyin City, Gansu Province. The tested performance of the SMES is shown in Table 13.7.

13.3.5 The integration and operation of HTS power substation (Xiao et al., 2012a)

After the tests and demonstrations of the above superconducting power equipment within distribution systems or conventional power substations, they were then integrated into one superconducting power substation. The layout of this substation is shown in Figure 13.1, and a photograph of it is shown in Figure 13.6. For this particular superconducting power substation, it is suitable to use one cryogenic cooling system for all of the HTS equipment. Here a subcooled LN_2 circulation flow loop is employed to cool the power cable, FCL, and transformer. For the HTS SMES, the cryogenic system is designed to result in zero boiling off by employing four Gifford McMahon (GM) cryo-coolers of 1.5 W at 4.2 K for each.

Figure 13.5 A view of the 1 MJ/500 kVA high-temperature superconducting magnetic energy storage.

Table 13.6 Design parameters of the high-temperature superconducting (HTS) coil for the 1 MJ/500 kVA superconducting magnetic energy storage

Parameters	Value
Inner diameter (mm)	400
Outer diameter (mm)	568
Height (mm)	645
Number of double pancakes	44
Number of tapes	82
Total length of used HTS tapes (km)	16.4
Inductance (H)	6.28
Operating current (A)	564
Max. (B_r) (T)	2.39
Max. (B) (T)	5.72
Operating temperature (K)	4.2
Stored energy (MJ)	1.0
Withstand voltage (V)	2000

Table 13.7 Tested parameters of the 1 MJ/500 kVA superconducting magnetic energy storage

Parameter	Tested value
Rated voltage (kV)	10.5
Rated output power (kVA)	500
Operation frequency (Hz)	50
Operating temperature (K)	4.2
Stored energy (MJ)	1.0
Operating current of the coil (A)	564
Maximum ramping rate of the coil (T/s)	0.7
Dynamic response time (ms)	≤1 ms
Energy conversion efficiency	≥90%
Bus bar voltage fluctuation	<1%
Waveform distortion	<1.33%
Power factor	>0.95
Inductance (H)	6.28

Figure 13.6 A view of the superconducting power substation.

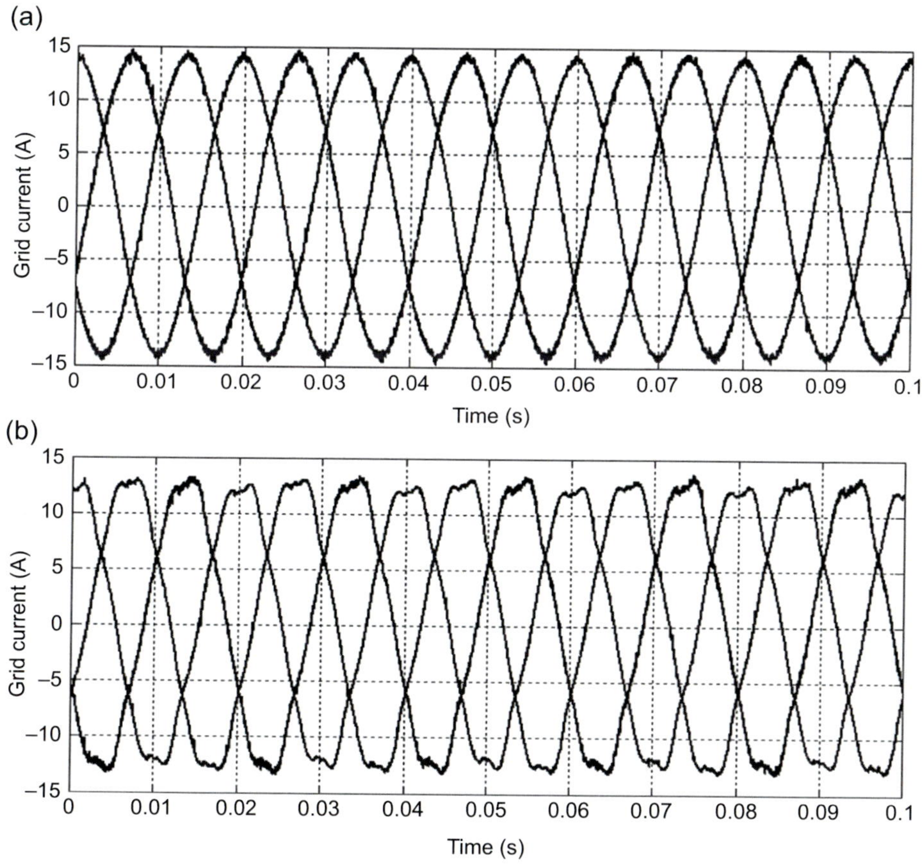

Figure 13.7 The current distortion was reduced from 5.13% (a) to 1.33% (b) with or without the superconducting power substation.

The substation has been tested since February 2011. Since then, there has not been any fault in the distribution system or electrical breakdown of the superconducting power equipment. However, if a cryogenic failure does occur in the refrigeration system, this can be recovered by replacing the cryostat of the power transformer and the LN_2 pump. The successful operation of the superconducting power substation demonstrates that the reliability and power quality of the power supply have been greatly improved by the SMES. Figure 13.7 shows an example of the improvement in the power quality produced by the superconducting power substation; the harmonic waves of the grid current have been greatly reduced from 5.13% to as low as 1.33%.

13.4 Superconducting fault current limiters (SFCLs) and the 360 m/10 kA superconducting DC power cable (Xiao et al., 2012b; Xin et al., 2010, 2013)

Recently, saturated iron-core-type and resistive-type FCLs have been developed in China. Between 2009 and 2012, saturated iron-core-type FCLs with 35 kV/1.5 kA and 220 kV/0.8 kA were installed at the Puji substation and Shigezhuang substation, respectively. From 2009, a 220 kV/2 kA resistive-type FCL made of YBCO tape has been in development for experimental operation at a transmission power grid in 2015 in Shenzen City, Guangdong Province.

13.4.1 The 35 kV/1.5 kA saturated iron-core FCL (Xin et al., 2010)

This FCL was installed at the Puji substation of South Power Grid for live-grid operation in 2009. The configuration of the iron core of the FCL was designed to effectively reduce the impedance in the normal state and enhance the impedance in the fault state. A magnetization control circuit was used to enable the iron cores to be saturated or desaturated, as required. Further, an energy release and voltage-surge protection unit can be used to promptly discharge the magnetic energy accumulated in the iron cores and to suppress the induced and surge voltages on the DC circuit when limiting the current. This design can reduce the fault current by about 50%. Figure 13.8 shows an overview of the FCL and Figure 13.9 shows an overview of the FCL after installation at the Puji substation. Once installed, a series of field tests were carried out (in July 2009). In order to fully examine its performance, current-limiting tests were conducted under an artificially imposed three-phase-to-ground short circuits. The live-grid experimental results reflected the design expectations well.

13.4.2 The 220 kV/0.8 kA saturated iron-core-type FCL (Xin et al., submitted for publication)

Although the 220 kV SFCL has the same structural configuration as the one described above, there are also significant differences between the two. The main distinction is

Figure 13.8 A view of the 35 kV/1.5 kA saturated iron-core-type fault current limiters.

Figure 13.9 An overview of 35 kV/1.5 kA superconducting fault current limiters after installation at Puji substation.

the electrical insulation. Dry electrical insulation was used for the 35 kV SFCL and the AC windings of the device were insulated with cast epoxy resin. However, oil insulation is required for the 220 kV SFCL because its AC windings should be operated at much higher voltage.

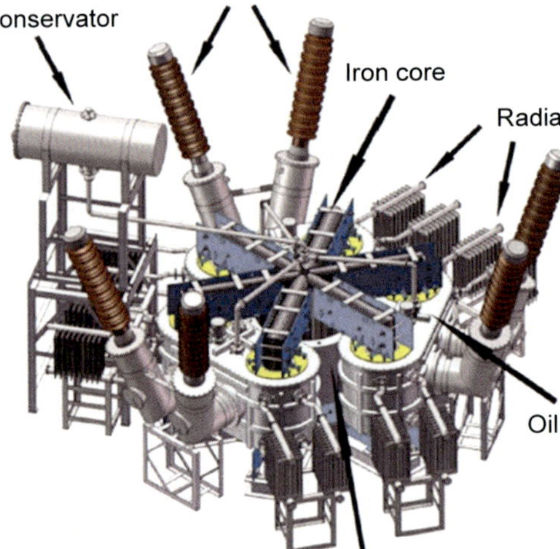

Figure 13.10 The structure of the 220 kV saturated iron-core superconducting fault current limiters.

Figure 13.10 illustrates for the structure of the main body of the 220 kV SFCL. The body comprises three main parts:

- the DC superconductor coil with its cryostat
- the iron cores
- the AC coils with their oil tanks.

Six iron frames are joined to form the hexagonal backbone of the structure. The HTS coil (the design parameters are shown in Table 13.8) is wound around the central vertical column, consisting of the six iron limbs. For each phase, the normal coil consists of two adjacent AC windings connected in series in one oil tank. The AC coils are immersed in oil in order to obtain adequate electrical insulation. The iron cores and the cryostat that host the HTS coil are open to the air. According to this design, the oil tanks sit closely around the iron limbs and the magnetic flux leakage is thus considerably enlarged when the iron cores are saturated. Fiberglass-reinforced plastic was chosen for fabrication of the oil tanks so that eddy currents on the walls of the oil tanks can be induced.

After the factory tests, the 220 kV/300 MVA saturated iron-core FCL was disassembled into five parts and shipped to Shigezhuang substation of Tianjin, China. Installation of the device was completed in the first quarter of 2012. Figure 13.11 shows the FCL after installation. Acceptance tests on this device were then carried out by Tianjing Power Company. The results showed that the SFCL was successfully restored to its functional capabilities, once reassembled and installed, and suitable for grid operation. Live-grid operation of this FCL will then proceed in order to test its performance and reliability.

Table 13.8 Specifications of the high-temperature superconducting coil for the 220 kV saturated iron-core fault current limiters

Parameters	Value
Conductor	$Bi_2Sr_2Ca_2Cu_3O_{10}$
Inner diameter (mm)	1920
Outer diameter (mm)	2080
Thickness of the ring (mm)	16
Height of coil (mm)	900
Effective turns	504
Rated current (A)	300
Total number of rings	45
Total rated magnetizing capacity (kA • turn)	176.5
Electrical insulation strength between tapes (V)	600
Electrical insulation strength of the coil (kV)	10
Total weight of the coil (kg)	800

Figure 13.11 A view of the 220 kV/0.8 kA superconducting fault current limiters after installation at Shigezhuang substation in Tianjin.

13.4.3 The 360 m/10 kA superconducting DC power cable (Xiao et al., 2012b)

In 2007, in order to demonstrate the possible application of a superconductor in a DC power transmission, we started to develop a 10 kA HTS DC power cable. The cable's capacity to carry a current of 10 kA was then tested by Zhongfu Group, an aluminum

Table 13.9 Main parameters of the cable conductor

Items	Value
Outer diameter of the former (mm)	41.0
Critical current of the high-temperature superconducting (HTS) tapes used (A)	150/90
Minimum tensile stress of the HTS tapes (MPa)	250/100
Outer diameter of the conductor (mm)	46.2
Layer of conductors	5
Pieces of HTS tapes	115
Total length of HTS tapes used (km)	46

production company. The cable is a warm-dielectric type, which uses Bi-2223 HTS tapes to form the conductor. The conductor consists of five layers of HTS tapes, in which one layer is fabricated with HTS tapes from InnoST, and four layers are fabricated with HTS tapes supplied by Sumitomo. The design parameters of the cable conductor are listed in Table 13.9.

The termination consists of a main body, a current lead, and a chamber that are designed and manufactured separately, so that they can be transported and integrated easily. For the current lead, the ratio of length to cross-section is 318.5 m^{-1} and the length is 2.9 m. This design can reduce the heat losses from the lead to 43 W/kA. In order to maintain the temperature between 70 and 77 K,

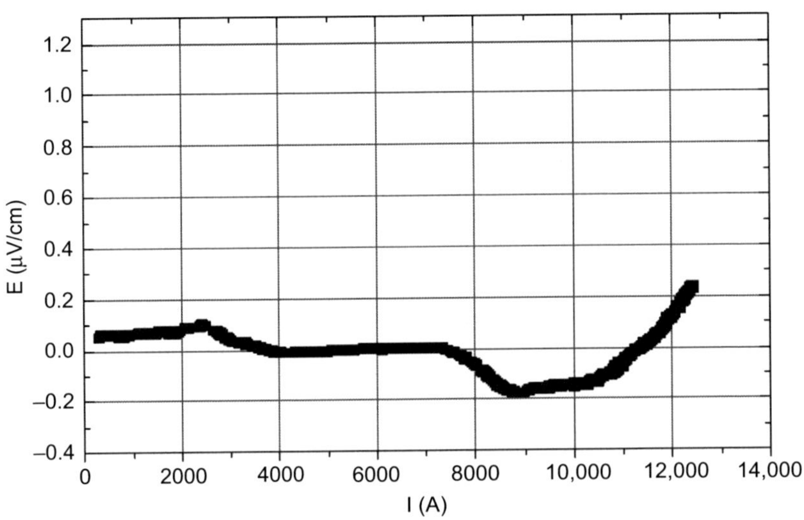

Figure 13.12 Critical current tests for the 360 m/10 kA high-temperature superconducting DC cable.

Figure 13.13 An overview of the 10 kA high-temperature superconducting DC power cable after installation.

the system's pressure is in the range of 1—5 bar, and then the liquid nitrogen is cooled by a four-cylinder Sterling refrigerator whose cooling power is 4 kW at 77 K and about 3.3 kW at 70 K.

The HTS DC cable was installed at Henan Zhongfu Industrial Company, which is an aluminum smelting plant. In order to test the bending performance of the cable, the HTS DC cable was designed to be bent at nine separate locations; three are designed to be bent along the vertical direction and six along the horizontal, and the minimum bending radius is 3 m. After installation, the cable was tested. At 77 K, the critical current of the cable by 1 μm/cm was up to 12.0 kA, as shown in Figure 13.12. The voltage drop in the cable changes as the current is increased, and repeated tests show the same result. The cable was shown to be stable, throughout 2 h of operation at 10.0 kA. The power cable, which connects the substation and the bus bar of an aluminum electrolyzing workshop, is now used as the power supply for the factory together with a conventional transmission conductor. An overview of the installed cable is shown in Figure 13.13.

13.5 Superconducting magnetic energy storage

From 2001 to 2008, HTS SMES equipment with capacities of 1 MJ/500 kVA and of 35 kJ/7 kVA and low-temperature superconducting (LTS) SMES equipment made of NbTi with capacities of 500 kJ/150 KVA and of 100 kJ/25 kVA were developed in China. The 1 MJ/500 kVA HTS SMES was discussed in Section 13.3.4. The next section focuses on the development of the 35 kJ/7 kVA HTS SMES and the 500 kJ/150 KVA LTS SMES.

13.5.1 *The 500 kJ/150 kVA LTS SMES (Jiang, Zhu, Cheng, Ren, & He, 2005; Jiang, Liu, Chu, Song, 2007)*

This SMES consists of an NbTi magnet of 500 kJ, an IGBT current source converter of 150 KVA, and three phase-shift inductors. A pair of double-spiral current leads of Bi-2223 tapes was designed for the magnet, shown in Figure 13.14.

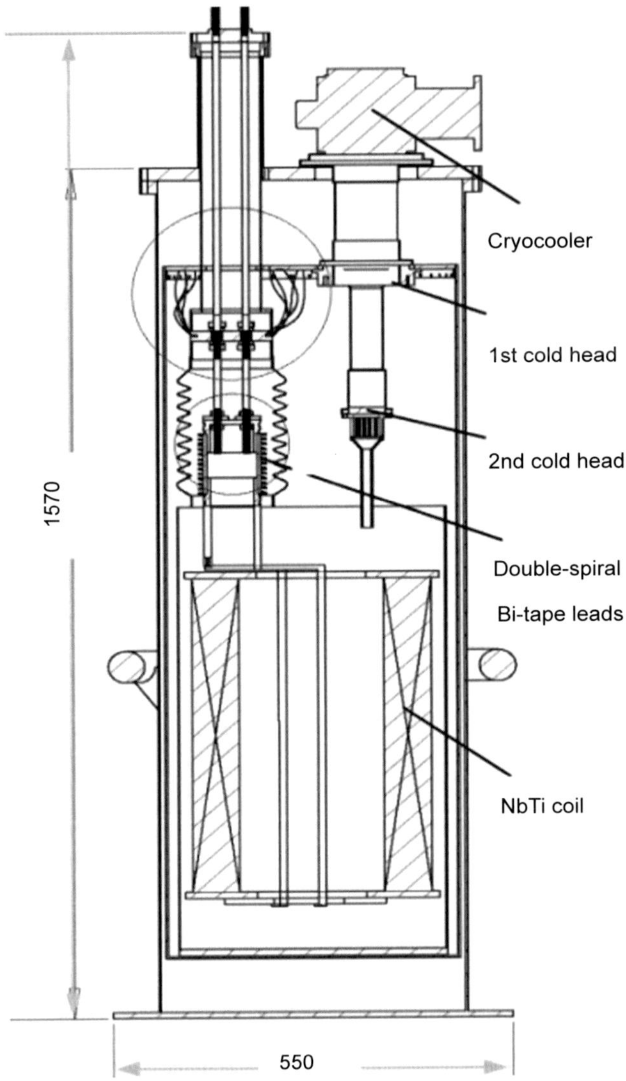

Figure 13.14 The magnet system of the 500 kJ/150 kVA superconducting magnetic energy storage with current leads of Bi-2223 tapes.

Experiments were carried out to test the compensation performances for both balanced and unbalanced voltage sags with a 110 kW load of resistance. The results, shown in Figure 13.15 and Figure 13.16, demonstrate that the load voltage recovers in less than one cycle (20 ms) whenever a three-phase or single-phase voltage sag occurs.

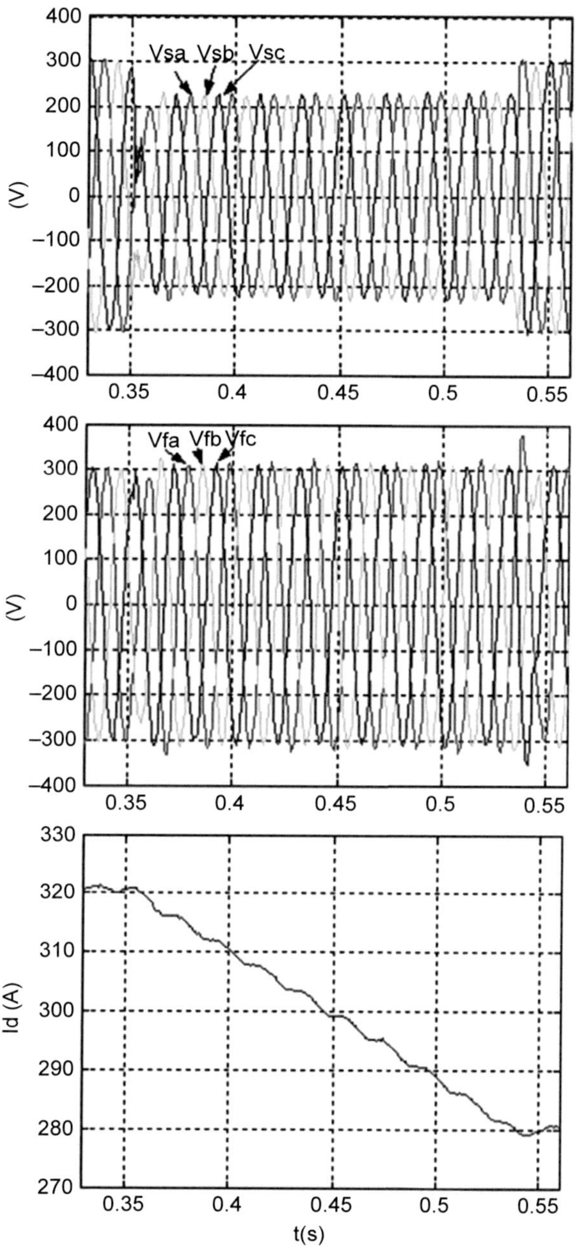

Figure 13.15 Experimental results of a balanced voltage sag.

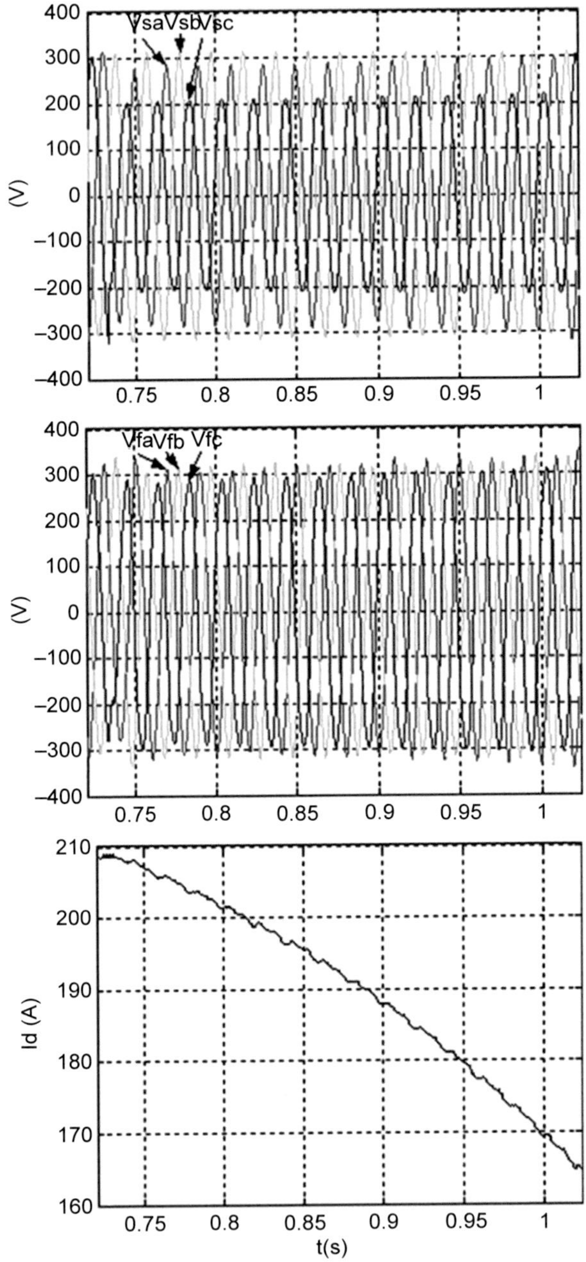

Figure 13.16 Experimental results of a single phase voltage sag.

13.5.2 The 35 kJ/7 kVA high-temperature SMES (Ren et al., 2008)

In order to investigate the possible application of SMES in a power system, a 35 kJ/7 kVA HTS SMES was developed for dynamic simulation research at Huazhong University of Science and Technology in 2005. A view of this SMES is shown in Figure 13.17 and the specifications of the HTS coil are shown in Table 13.10.

Once the SMES had been made, a dynamic experiment in a simulated power system was carried out to evaluate its performance. In this experiment, a three-phase short circuit of 350 ms was simulated on the transmission line near the generator. During the test, the SMES automatically detects the fault and compensates for the imbalance of power by reducing the oscillation in the simulated power system. Figure 13.18 illustrates the power oscillation in the simulated power system both

Table 13.10 Specifications of the high-temperature superconducting coil and the Bi-2223/Ag tape

Parameter	Value
Coil configuration	Solenoid type
Rated current (A)	100
Maximum field (T)	3
Inductance (H)	7.5
Rated energy storage (kJ)	35
Rated power (kW)	7
Conductors	Bi-2223/Ag tape

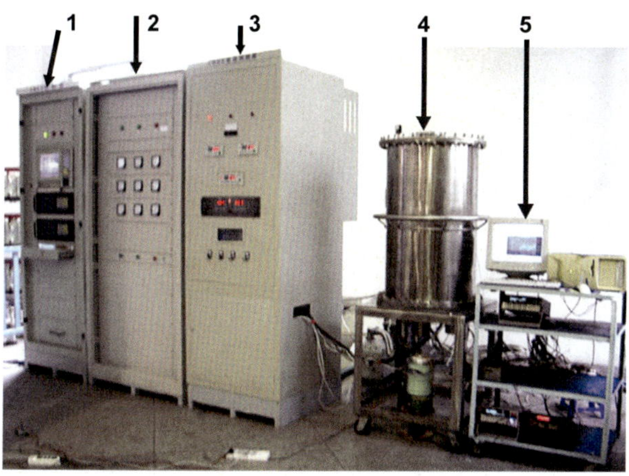

Figure 13.17 A view of 35 kJ/7 kVA high-temperature superconducting magnetic energy storage. 1, monitor and control system; 2, power conditioning system; 3, monitor of cooling system; 4, superconducting magnet and cooling system; 5, data acquisition system for experiment.

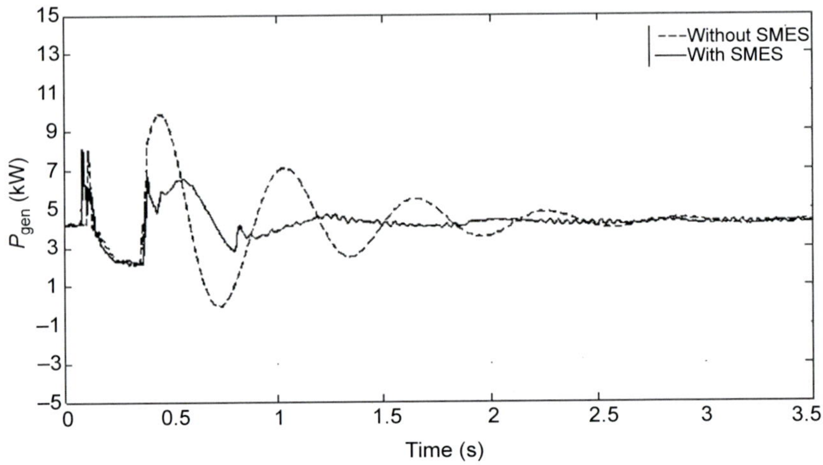

Figure 13.18 The dynamic response of superconducting magnetic energy storage (SMES) under the fault.

with and without the SMES. By comparing the two, it can be seen that the SMES dampened the power oscillation effectively. The SMES can supply a fast power support to the AC power system during the fault, which enhances the dynamic stability of the AC power system.

A magnetic field test was also conducted on the SMES system, after it was moved and reconstructed, to comply with the requirements of on-site experiments (see Figure 13.19). The field test was carried out at Laohukou Hydropower Plant in Zigui City. The results show that the mobile SMES system can operate on the power network at different locations and effectively suppress the power fluctuation of the generator terminal.

Figure 13.19 Assembled 35 kJ/7 kVA high-temperature superconducting magnetic energy storage for field test.

13.6 Future trends

In the future, given the increasing development of the power grid and of renewable energy in China, the SFCL for the transmission grid, the superconducting DC cable for long-distance power transmission, and the SMES for the transient power balance of the renewable energy appear to be the most promising near term applications in China's power grid.

China is a developing country. Its power consumption requirement is rapidly growing and expected to continue to increase at an annual rate of 7% until 2025. Therefore, the capacity of the power grid must be simultaneously increased to meet this ever-growing demand. Further, for the power grid of load centers such as Beijing, Shanghai, Guangzhou, and Shenzhen, the fault current at the key substations (220–500 kV) has reached up to about 70–100 kA using the existing architecture of the power grid. This means that the FCL will become increasingly important to the development of the power grid in China. The FCL for a 220 kV transmission system has already been developed. In the future, an FCL such as the resistive type and saturated iron-core type for a 500 kV transmission system will be developed.

From 2004 to 2014, renewable energy has been developing very rapidly worldwide. It is expected that in the near future, the primary source of energy will be renewable energy and the final energy for customs will be based on electric power. This significant change in the energy structure will have a revolutionary impact on the power grid. In order to deal with the challenges posed by renewable energy, it has been suggested that the DC power grid should be developed instead of the current AC power grid. In China, most of the renewable energy sources are located in the north and southwest areas, but most of the load centers are still located at the east and south area. Hence, a large amount of power transmission by DC will be necessary. For this reason, the development of an HTS DC power cable for long-distance power transmission is needed (Xiao, Dai, Lin, Zhang, & Zhang, 2013). For example, a superconducting power cable with ±800 kV/10 kA can transmit 16,000 MW of power (which is 2–3 times higher than the traditional HVDC), and significantly reduce the transmission losses. In recent years, R&D for HTS DC power cables has been widely conducted in Japan, Korea, Russia, and Europe. More attention should now be concentrated on HTS DC power cables than AC power cables.

Renewable energy resources do have intermittent and unstable characteristics, which lead to intermittent and unstable electric power generation output. In addition, the solar photovoltaic power generation systems do not have the mechanical inertia that the traditional hydraulic turbines or steam turbines do. It will therefore be a challenge to guarantee the real-time dynamic power balance and frequency stability of the grid when using renewable energy power. SMES could potentially be used to improve the stability of renewable energy and to level the power output from a renewable energy power plant.

13.7 Sources of further information and advice

The development of superconducting technology is a complicated undertaking involving superconducting materials, thermal physics and cryogenic engineering, superconducting magnet technology, high-voltage insulation technology, power electronics, and power systems. This chapter does not go into detail about the scientific basis and the principle of the superconducting technology for power grids. There are also other developments that have not been discussed in this chapter. For more details about the superconducting technology or related matters, one can refer to the following information:

1. Xiao L.Y., Lin L.Z., and Gong L.H. (2010), *Superconducting Power Technology at Encyclopedia of Electrical Engineering—Power Engineering,* Chapter 19, Beijing, China Electric Power Press (in Chinese).
2. Lin L.Z. (1998), *Superconductivity and Its Application*, Beijing Polytechnic University Press (in Chinese).
3. Wang Y.S. (2011), *The Basic for Superconducting Power Technology*, Beijing, Science Press (in Chinese).
4. http://iee.ac.cn/Website/index.php?ChannelID=1689.
5. http://www.innopower.com/.

References

Dai, S. T., Xiao, L. Y., Wang, Z. K., Guo, W. Y., Zhang, J. Y., Zhang, D., et al. (2012). Development and demonstration of a 1MJ high Tc SMES. *IEEE Transactions on Applied Superconductivity, 22*, 5700304.

Jiang, X. H., Liu, X. C., Chu, X., & Song, M. (2007). Double-spiral current leads of Bi-2223 tapes for a SMES magnet. *IEEE Transactions on Applied Superconductivity, 17*, 2240–2243.

Jiang, X. H., Zhu, X. G., Cheng, Z. G., Ren, X. P., & He, Y. Y. (2005). A 150kVA/0.3MJ SMES voltage sag compensation system. *IEEE Transactions on Applied Superconductivity, 15*, 1903–1906.

Ren, L., Tang, Y. J., Shi, J., Chen, N., Li, J. D., & Cheng, S. J. (2008). Cryo-cooler cooled HTS current lead for a 35kJ/7kW-class high-Tc SMES system. *Physica C, 468*, 2115–2118.

Wang, Y. S., Zhao, X., Han, J. J., Li, H. D., Guan, Y., Bao, Q., et al. (2007). Development of a 630kVA three-phase HTS transformer with amorphous alloy cores. *IEEE Transactions on Applied Superconductivity, 17*, 2051–2054.

Xiao, L. Y., Dai, S. T., Lin, L. Z., Zhang, J. Y., Guo, W. Y., Zhang, D., et al. (2012a). Development of the world's first superconducting power substation. *IEEE Transactions on Applied Superconductivity, 22*, 5000104.

Xiao, L. Y., Dai, S. T., Lin, L. Z., Teng, Y. P., Zhang, H. R., Liang, X. M., et al. (2012b). Development of a 10kA HTS DC power cable. *IEEE Transactions on Applied. Superconductivity, 22*, 5800404.

Xiao, L. Y., Dai, S. T., Lin, Y. B., Gao, Z. Y., Zhang, F. Y., Xu, X., et al. (2007). Research and development of HTS superconducting power Cable. *IEEE Transactions on Applied Superconductivity, 17*, 1652–1655.

Xiao, L. Y., Dai, S. T., Lin, L. Z., Zhang, Z. Z., & Zhang, J. Y. (2013). HTS power technology for future DC power grid. *IEEE Transactions on Applied Superconductivity, 23*, 5401506.

Xiao, L. Y., & Lin, L. Z. (2007). Recent progress of power application of superconductors in China. *IEEE Transactions on Applied Superconductivity, 17*, 2355–2360.

Xiao, L. Y., Lin, L. Z., & Dai, S. T. (2011). The prospects of superconducting power technology for future power grids. *Physics, 40*, 500–504 (in Chinese).

Xiao, L. Y., Wang, Z. K., Dai, S. T., Zhang, J. Y., Zhang, D., Gao, Z. Y., et al. (2008). Fabrication and tests of a 1MJ HTS magnet for SMES. *IEEE Transactions on Applied Superconductivity, 18*, 770–773.

Xin, Y., Gong, W. Z., Sun, Y. W., Cui, J. B., Hong, H., Niu, X. Y., et al. (2013). Factory and field tests of a 220 kV/300 MVA statured iron-core superconducting fault current limiter. *IEEE Transactions on Applied Superconductivity, 23*(3), 5602305.

Xin, Y., Wang, J. Z., Hong, H., Gong, W. Z., Zhang, J. Y., Niu, X. Y., et al. (2010). Field tests on a 35kV/90MVA superconducting fault current limiter. In *2010 international conference on power system technology, paper no.: CP1193, October 23–26, 2010, Zhejiang, China.*

Zhou, X. X. (2012). The prospects of power grid under the background of new energy revolution. In *Fragrant hill scientific meeting #436, September 27–29, 2012, Beijing, China.*

Index

Note: Page numbers followed by "f" and "t" indicate figures and tables respectively.

D

E

F

I

J

K

L

M

Printed in the United States
By Bookmasters